Microstructure and Properties of Materials

Volume 1

Microstructure and Properties of Materials

Editor

J.C.M. Li

Univ. of Rochester

Published by

World Scientific Publishing Co Pte Ltd
P O Box 128, Farrer Road, Singapore 912805
USA office: Suite 1B, 1060 Main Street, River Edge, NJ 07661
UK office: 57 Shelton Street, Covent Garden, London WC2H 9HE

British Library Cataloguing-in-Publication Data
A catalogue record for this book is available from the British Library.

MICROSTRUCTURE AND PROPERTIES OF MATERIALS

Copyright © 1996 by World Scientific Publishing Co. Pte. Ltd.

All rights reserved. This book, or parts thereof, may not be reproduced in any form or by any means, electronic or mechanical, including photocopying, recording or any information storage and retrieval system now known or to be invented, without written permission from the Publisher.

For photocopying of material in this volume, please pay a copying fee through the Copyright Clearance Center, Inc., 222 Rosewood Drive, Danvers, Massachusetts 01923, USA.

ISBN 981-02-2403-6

Printed in Singapore.

PREFACE

This is an advanced text book on the microstructure and properties of materials, the first volume of a possible 3-volume set. While there are many elementary text books in materials science, there are very few advanced text books. In our graduate school courses, we rely on our own notes or some conference volumes and journal articles. One of the reasons is that there are so many different kinds of materials and each has its own microstructure property charateristics. So it is difficult for a single person to be expert in all the materials. Thus the idea of a multi-author collection appears good. I am inviting the best authoritative expert that I can find in each material area and since they are all busy people, it has taken longer than expected to finish the task. Hence this is the first volume which should be a good supplement to your microstructure course. If you are working with a certain material area in one of the chapters, you will find a rich source of design ideas and applications as well as a good understanding of how does the microstructure affect the properties.

Chapter 1 on aluminum alloys presents microstructural optimization and critical considerations in design applications. Chapter 2 on Nickel-base superalloys reviews the compositional, microstructural and processing advances in increasing their maximum use temperature. Chapter 3 on metal matrix composites discusses the strengthening mechanisms of metals dispersed with short fibers or particles. Chapter 4 on polymer matrix composites contains the details of the microstucture property relationships of high performance fibers, polymer matrix material and the advanced composties made therewith. Chapter 5 on ceramics matrix composites describes the fibers and matrix materials used, the processing techniques involved and the mechanical properties under different loading conditions. Chapter 6 on inorgainic glasses describes the influence of second phases, both glassy and crystalline on their properties. Chapter 7 on superconducting materials shows the importance of twins, grain boundaries, dislocations and stacking faults. Chapter 8 on magnetic materials introduces the domain structure and its effects on the soft and hard magnetic properties.

Material problems are the bottle necks of most industries. New materials are created daily. But the principles and the relation between properties and microstructure remains the same. The more we know about these relations the easier will be to find new materials with desired properties.

James C. M. Li
Rochester, NY
October, 1995

CONTENTS

Preface   v

Chapter 1   Microstructure and Properties of Aluminium Alloys   1
C. P. Blakenship, Jr, E. A. Starke and E. Hornbogen

Chapter 2   Nickel-Base Superalloys   51
N. S. Stoloff

Chapter 3   Metal Matrix Composites   107
R. J. Arsenault

Chapter 4   Polymer Matrix Composites   181
Jang-Kyo Kim and Yiu-Wing Mai

Chapter 5   Ceramic Matrix Composites   247
P. G. Karandikar, T. W. Chou and A. Parvizi-Majidi

Chapter 6   Microstructure of Inorganic Glasses   351
R. H. Doremus

Chapter 7   Microstructure and Properties of Superconducting Materials   381
C. S. Pande

Chapter 8   Magnetic Materials   415
C. D. Graham, Jr

Subject Index   443

Author Index   461

Chapter 1
MICROSTRUCTURE AND PROPERTIES OF ALUMINUM ALLOYS

C. P. Blankenship, Jr. and E. A. Starke, Jr.
Department of Materials Science and Engineering,
University of Virginia, Charlottesville, VA 22903
and
E. Hornbogen
Institute für werkstoffe,
Ruhr Universität Bochum, D-44780, Bochum, Germany

I. INTRODUCTION

A little over 100 years ago Charles Martin Hall, an American, and Paul Heroult, a Frenchman, independently developed a process which allowed the economical production of aluminum by electrolysis from a fused salt bath. Although aluminum was a "late comer," its versatility has resulted in it replacing many older, more established materials, so it now is consumed, on a volumetric basis, more than all other non-ferrous metals combined, including copper, lead and zinc [1]. Aluminum is light, ductile, has good electrical and thermal conductivity and can be made strong by alloying. An advantageous chemical property of aluminum is its reactivity with oxygen which leads to the formation of a dense layer of Al_2O_3 on the surface which shields the base metal from further environmental interaction. However, problems are encountered if the layer is disturbed, for example by second phases, plastic deformation which fractures the protective layer, or by friction (tribo-chemical reaction).

Aluminum alloys are classified as heat-treatable or non-heat-treatable, depending on whether or not they precipitation harden. However, the properties of both classes of alloys depend on their structural characteristics and can be associated with different levels of structure:

(A) *The atom:* The low density is due to the low atomic mass (A_{Al} = 27). By alloying with Mg (A_{Mg} = 24) or Li (A_{Li} = 7), for example, the density can be reduced still further. Other elements, e.g., Si (A_{Si} = 28), have little effect.

(B) *The phase:* The ductility and formability of aluminum is due to the high symmetry and thermodynamic stability of the fcc lattice (high stacking fault energy), Fig. 1. Other phases of interest include aluminum solid solutions, intermetallic compounds (Al_3Ti), non-metallic compounds (AlN), quasi-crystals and metallic glasses.

(C) *The microstructure:* The high strength of precipitation hardened alloys is associated with an ultra-fine dispersion of particles (d <10 nm). Other microstructural features, e.g., grain boundaries, may have a beneficial effect on strength, but may have a detrimental effect on fracture resistance, Fig. 2.

There are a number of properties that distinguish aluminum and its alloys from other metallic materials and these include:

(1) An intermediate melting temperature (T_m = 930K) of the fcc crystal, which results in substitutional diffusion being slow but not impossible at ambient temperatures (20°C~1/3 Tm[K]).

(2) The high stacking fault energy (SFE) of the fcc crystal is not considerably lowered by solute additions: γ_{Al} = 200mJm^{-2} > γ_{Ni} > $\gamma_{\gamma\text{-Fe}}$ > γ_{Cu} > γ_{Au}. It follows that cross slip of dislocations is easy and crystal plasticity is high. Microscopic plasticity is similar to that of γ-Fe above its brittle transition temperature. However, unlike Fe, aluminum stays ductile down to at least 4K. There is a low probability for the formation of twins during plastic deformation or recrystallization. This implies crystallographic textures different from low SFE alloys (Cu, Au, γ-Fe). Dislocation rings may form by condensation of vacancies during quenching or annealing. This phenomenon has not been observed to the same extent in other fcc metals.

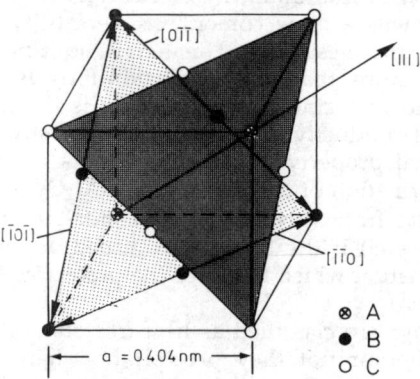

Fig. 1. The face-centered cubic (fcc) lattice with (111) slip planes and <110> slip directions; stacking sequence ABC....

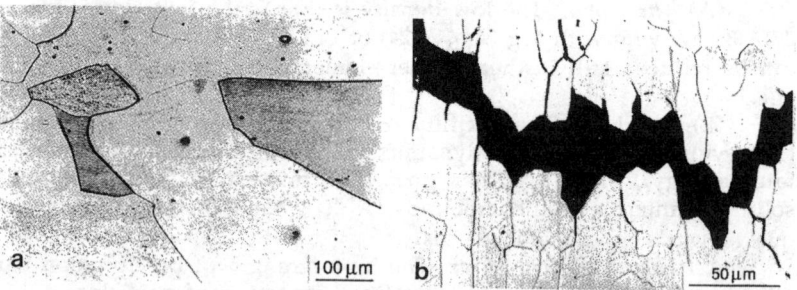

Fig. 2. Light micrographs of (a) grain structure of 99.9 Al; (b) intercrystalline stress corrosion cracking in an Al-Cu-Mg-Li alloy.

(3) Aluminum forms intermetallic compounds with most metals. Solute elements like Be, Si, Zn, Ga, Ge, etc., solidify by eutectic crystallization without forming compounds. Alkaline metals, except for Li, form no compounds and are completely insoluble in the solid state. Compounds with small atoms AlB_2, Al_4C_3, AlN, Al_2O_3 are ceramic phases with predominately covalent and ionic bonding. The solubility of small elements is very limited in liquid aluminum. The covalently bonded elements Si and Ge behave as metals in liquid and solid solution with aluminum, Fig. 3. There exists no case of complete miscibility in the solid state. Miscibility is largest with Zn which forms no compounds with aluminum.

(4) Quasicrystalline phases, i.e., non-periodic, ordered structures, were first found in certain Al-(transition element) alloys. Formation of metallic glasses seems to be more difficult than in some other metals, but glasses have been found in $Al-T_1-T_2$, where, for example, T_1 = Ni; T_2 = Y, Table 1, Fig. 4.

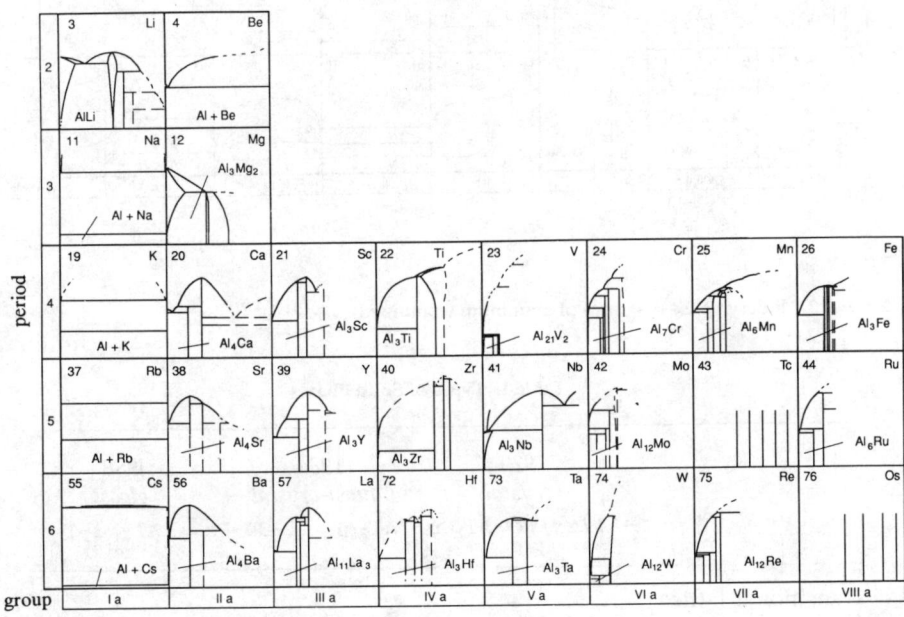

Fig. 3. Binary phase diagrams of aluminum.

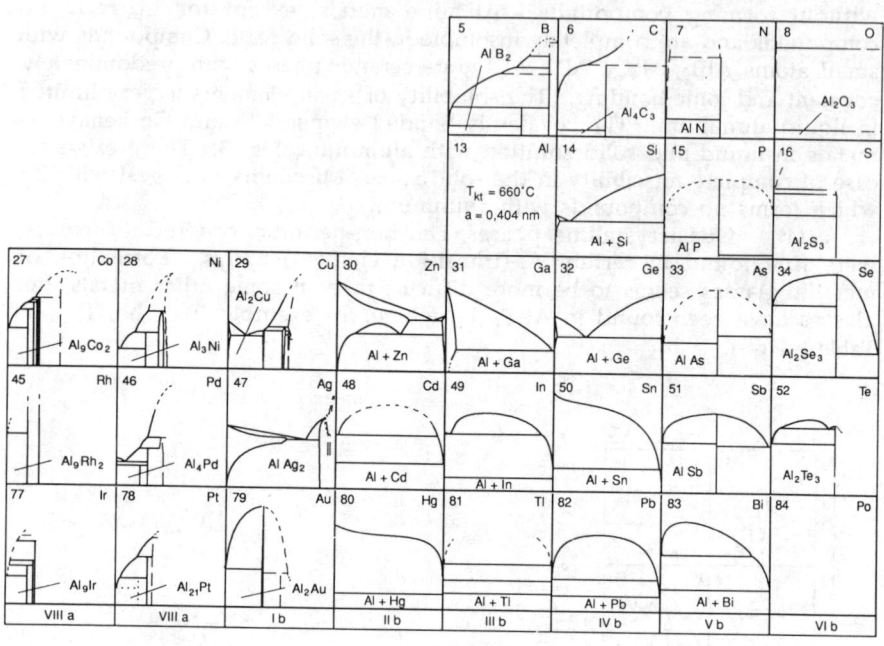

Fig. 3. Binary phase diagrams of aluminum (continued).

Table 1. Types of Solid Phases

	1920 crystal α-Al(Mg), Θ-Al$_2$Cu	1984 quasi-crystal Al-Mn, Al-Cu-Fe	1988 glass Al-Ni-Y
transitional lattice	+	–	–
long-range order	+	+	–

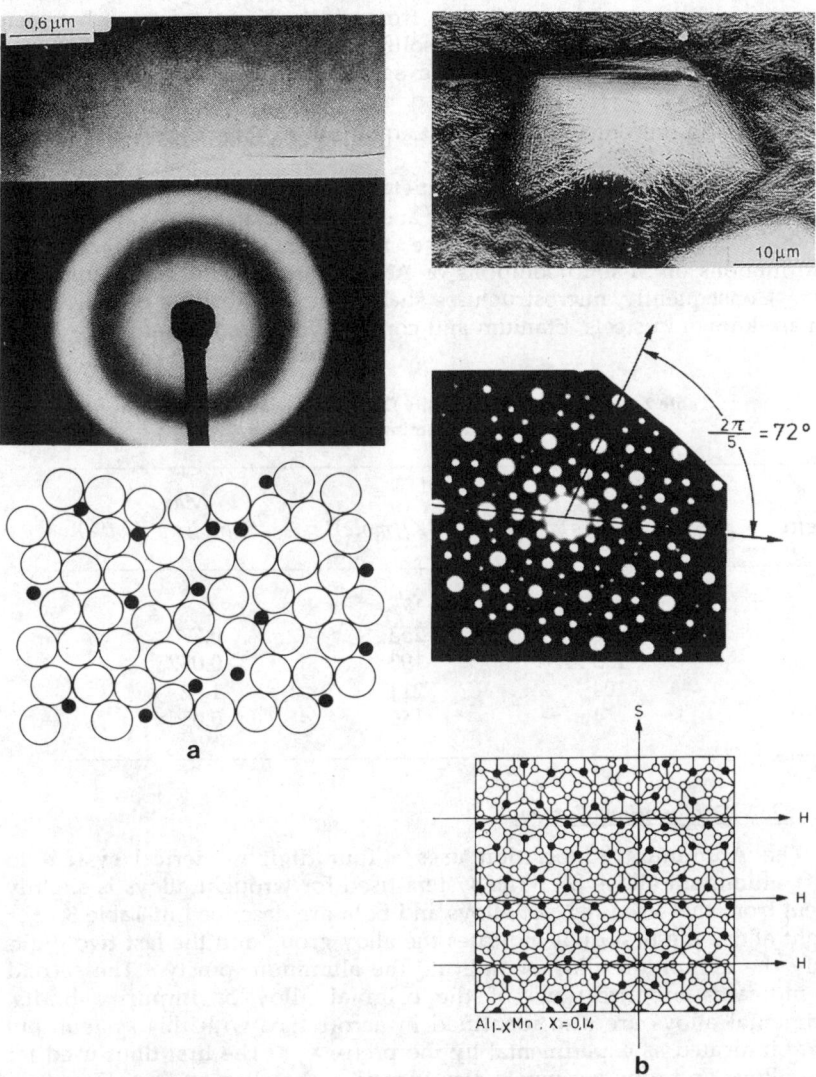

Fig. 4. Non-equilibrium structures:
(a) TEM of AlYNi-glass, electron diffraction, planar structure model; (b) SEM of icosahedral quasi-crystal of an AlFeCu alloy, electron diffraction showing five-fold symmetry, and planar structure model for Al-14at.%Mn.

(5) Diffusion of transition elements (Cr, V, Fe, Co, Ni) in aluminum is anomalously slow, which originates from electronic interactions between aluminum and these elements. Low solubilities of the transition elements (except for Mn) most likely have the same origin as the low diffusion coefficients, Table 2.

(6) The ratio of E/ρ is almost equal for Al, Mg, Fe and Ti but less than Cu, Pb and Zn.

(7) The most important non-metallic compound is Al_2O_3 which is useful as a protective surface layer, fiber and ceramic material.

(8) No polymorphous phase transformations or martensitic transformations of Al-solid solutions or Al-rich intermetallic compounds are known. Consequently, microstructures such as pearlite, bainite or martensite which are known in steels, titanium and copper alloys are absent.

Table 2. Diffusion and Solubility Data for Transition Metal Additions in Aluminum [73,74]

Element	Do (cm^2/s)	Q (KJ/mole)	Max Solubility (wt%)
Co	250	175	~0
Cr	1850	253	0.77
Fe	135	193	0.052
Mn	104	211	1.82
Ni	4	146	0.05

Alloy Designation System

The Aluminum Association uses a four digit numerical system to identify aluminum alloys [2]. The system used for wrought alloys is slightly different from that used for cast alloys and both are described in Table 3. For wrought alloys the first digit indicates the alloy group and the last two digits identify the aluminum alloy or indicate the aluminum purity. The second digit indicates modifications of the original alloy or impurity limits. Experimental alloys are also identified in accordance with this system, but they are indicated as experimental by the prefix X. The first digit used for casting alloys and foundry ingots also identifies the alloy group. However, the second two digits identify the aluminum alloy or aluminum purity. The last digit, which is preceded by a decimal point, indicates the product form (i.e., .0 for casting and .1 for ingot). A serial letter before the numerical designation indicates a modification of the original alloy or impurity limits. The serial letters are assigned alphabetically, but omitting I,O,Q and X. The X is reserved for experimental alloys.

Table 3. Aluminum Alloy Designation Systems

Alloy type*	Four-digit designation
Wrought alloys	
99.00% (min) aluminum	1XXX
Copper	2XXX
Manganese	3XXX
Silicon	4XXX
Magnesium	5XXX
Magnesium and silicon	6XXX
Zinc	7XXX
Others	8XXX
Casting alloys	
99.00% (min) aluminum	1XX.X
Copper	2XX.X
Silicon with added copper and/or magnesium	3XX.X
Silicon	4XX.X
Magnesium	5XX.X
Zinc	7XX.X
Tin	8XX.X
Others	9XX.X

*Designations are based on aluminum content or main alloying elements

II. STRENGTHENING MECHANISMS

The strength of pure aluminum limits its commercial usefulness, and a major function of alloying is to improve this property. For structural use, the strongest alloy which meets minimum requirements for other properties such as corrosion resistance, ductility, toughness, etc., is usually selected if it is cost effective. Consequently, composition is first selected for strength.

Structural alloys are rarely strengthened by one hardening mechanism alone, Table 4. There are six mechanisms which are relevant to Al-alloys: (i) solid solution hardening, (ii) work hardening and substructure hardening, (iii) grain size hardening, (iv) precipitation and particle hardening, (v) texture hardening, and (vi) fiber reinforcement. Magnesium is the only element that offers significant solid solution hardening in aluminum. Texture hardening and fiber reinforcement will not be discussed in detail in this chapter. However, sharp textures which result in slip systems of low resolved shear stress also result in high yield strengths and fiber reinforcement may be described, in the simplest case, by the law of mixtures. The four major strengthening mechanisms are listed according to the geometric dimension (i = 0, 1, 2, 3) of the obstacles which impede gliding (and/or climbing) of dislocations [3].

$$\sigma_y = \sigma_\perp + \Sigma\Delta\sigma_i = \sigma_\perp + \Delta\sigma_0 + \Delta\sigma_1 + \Delta\sigma_2 + \Delta\sigma_3 \qquad [1a]$$

There is, however, not a simple additivity of the contributions $\Delta\sigma_i$ from the individual mechanisms to the yield stress σ_y. $\sigma_\perp$, the yield stress of pure (99.99) Al, amounts to 16 MPa at ambient temperature and 60 MPa at -200°C.

Table 4. Elementary Hardening Mechanisms in Al

		Geometric dimensions	Obstacle	Density	Designation	
1	$\Delta\sigma_0=\Delta\sigma_{ss}$	0	solute atom	m^{-3}	C	solid solution hardening
2	$\Delta\sigma_1=\Delta\sigma_D$	1	dislocation	m^{-2}	ρ_D	work hardening, substructure hardening
3	$\Delta\sigma_2=\Delta\sigma_B$	2	grain boundary	m^{-1}	S_B^{-1}	grain size hardening
4	$\Delta\sigma_3=\Delta\sigma_p$	3	particle, pore	m^0	f_p	precipitation hardening
5	$\Delta\sigma_c$	–	crystal anisotropy	–	–	texture hardening
6	$\Delta\sigma_M$	–	discontinuous and continuous fibers	–	–	fiber reinforcement

A quantitative approach to the understanding of a high yield strength σ_y of Al-alloys requires a differentiation between "hard" and "soft" obstacles, Table 5. This classification depends on whether glissile dislocations are bent to semicircles or not. Dislocations are not bent in semicircles in pure aluminum, aluminum solid solutions, or alloys containing only shearable precipitates. The addition of "hard" obstacles causes dislocations to loop and bypass the nonshearable particles. This behavior can be described by the following expression:

$$\sigma_y = \sigma_\perp + \Delta\sigma_s + \Delta\sigma_H \qquad (1b)$$

which describes a solid solution matrix containing shearable particles and a dispersion of small non-shearable particles. Whether a certain type of particle is sheared or looped, depends on the size of the particle and the nature of the hardening mechanism. The critical particle diameter, d_c, for the transition from shearing to looping or bypassing can be estimated by examining F, the force a single particle is able to exert on a single dislocation [4]. For $F < Gb^2$, shearing occurs, and for $F \geq Gb^2$ bypassing occurs.

Table 5. Examples of Hard and Soft Obstacles in Al

Hard obstacles (H)	Soft obstacles (S)
particles $d > d_c$	particles $d < d_c$
pores	solute atoms, vacancies
sessile dislocations	short range order, clusters, antiphase domain boundaries (APB)

Particle-strengthened aluminum alloys may contain ordered or disordered particles that may be coherent, partially coherent, or incoherent with the matrix. For ordered, coherent particles, $F = \gamma d$ where γ is the antiphase boundary energy, and d is the particle diameter. The critical diameter may be expressed

$$d_c = G_\alpha b^2 C/\gamma \tag{2a}$$

where G_α is the shear modulus of the solid solution and b the Burgers Vector of a matix dislocation. For disordered, coherent particles which have a critical resolved shear stress, τ_β, that is different from the matrix, τ_α,

$$d_c = G_\alpha bC/|\tau_\alpha - \tau_\beta| \tag{2b}$$

For particles which have a different crystal structure from the matrix,

$$d_c = 4\pi G_\alpha bC/G_\beta \tag{2c}$$

C in the above equations is a dimensionless factor determined by the shape of the particle and is close to unity for spherical particles. These expressions do not consider that dislocations may, in some cases, travel in pairs or pile up at a particle if planar slip occurs. For the latter solution, the right side of Eqs. 2a-c should be multiplied by N, the number of dislocations in the pile up.

A mixture of two "hard" species H_1 and H_2 introduces a new smaller effective spacing S_{eff}: $1/\sqrt{S_{eff}} = 1/\sqrt{S_1} + 1/\sqrt{S_2}$ and the following addition of the individual contributions $\Delta\sigma_i$:

$$\sigma_y = \sigma_\perp + \Delta\sigma_s + (\Delta\sigma^2_{H1} + \Delta\sigma^2_{H2})^{1/2} \qquad (3a)$$

The contribution from dislocation density, ρ_D, is defined as:

$$\Delta\sigma_{H1} = \Delta\sigma_D = G_\alpha b \rho_D^{1/2} \qquad (3b)$$

The contribution from particles that are bypassed by dislocations can be described by the modified Orowan equation:

$$\Delta\sigma_p \approx Gb/S_p \approx Gb\sqrt{f_p}/d \qquad (3c)$$

for $S_p \gg d$. S_p is the spacing of the particle centers, and f_p is the volume fraction of particles.

Interaction with several dislocations (pile-up, cell structure formation) is required to derive the grain size dependence of the yield stress:

$$\Delta\sigma_B = k_y/\sqrt{S_B} \qquad (3d)$$

where k_y is the Petch factor, i.e., a parameter associated with the relative hardening contribution of the grain boundaries, and S_B is the grain diameter. The pile-up model is only applicable when the aluminum alloy contains shearable particles because the high stacking fault energy of Al prohibits localized slip.

A comprehensive equation can be established for the different contributions to the yield strength of Al-alloys for the first four elementary hardening mechanisms listed in Table 4:

$$\sigma_y = \sigma_\perp + \Delta\sigma_s + (\Delta\sigma_D^2 + \Delta\sigma_p^2)^{1/2} + k_y/\sqrt{S_B} \qquad (4)$$

where

$\sigma_\perp$ = pure aluminum
$\Delta\sigma_s$ = solid solution, shearable particles
$\Delta\sigma_D$ = dislocations, subboundaries
$\Delta\sigma_p$ = hard, nonshearable, particles
k_y/S_B = grain boundaries.

III. NON-HEAT TREATABLE ALLOYS

The 1XXX, 3XXX, 5XXX and some of the 8XXX series alloys are non-heat-treatable. These alloys are primarily strengthened by elements in solid solution and by deformation structures. Magnesium is added to non-heat-treatable alloys for its solid-solution strengthening effect. It also enhances work hardening and makes aluminum more anodic. In general, solid-solution alloys are more resistant to corrosion than two-phase alloys. Al-Mg alloys have a high resistance to corrosion, particularly in seawater and alkaline solutions. However, when Mg exceeds the solid solubility in binary alloys it precipitates at grain boundaries as Al_3Mg_2, which is anodic to the matrix and promotes intergranular attack.

As mentioned in the introduction, aluminum combines readily with transition metals, Ti, V, Cr, Mn, Fe, Co, Ni, and Zr to form intermetallic phases with little or no solubility in the aluminum matrix. The intermetallic phases increase the strength by enhancing work hardening during working operations, and by refining the grain structure. They increase the work hardening since they are usually incoherent with the matrix, are nondeformable, and must be looped or bypassed by moving dislocations. This increases the dislocation density and blocks dynamic recovery processes. During hot working the high stacking fault energy of aluminum aids in the development of a well-defined subgrain structure which increases the strength as described in the previous section. The smaller (< 0.6μm) intermetallic particles aid in the stabilization of the substructure. The intermetallic particles do not usually add a component of particle strengthening because of their low volume fraction, large size and interparticle spacing. The temper designations for aluminum alloys [2] are listed in Table 6.

Intermetallic compounds larger than 1 μm, called constituent phases, reduce both ductility and fracture toughness. In addition, dense dislocation tangles (deformation zones) develop around these coarse particles during deformation and act as nucleation sites for recrystallization when subsequent annealing operations are performed. The width of the deformation zone, λ, can be related to the particle diameter, d, and true strain, ε, by [5]:

$$\lambda = Ad\varepsilon^{(n/n+1)} \tag{5a}$$

where

$$A = \frac{K\sqrt{2}}{8\tau_{cr}} \left[\frac{4\tau_{cr}^{(n+1)}}{\sqrt{2}nk} \right]^{n/n+1} \tag{5b}$$

and τ_{cr} is the critical resolved shear stress required to move dislocation loops away from the particle. The other parameters are from the normal strain-hardening relationship, $\sigma = K\varepsilon^n$.

Table 6. Temper Nomenclature: Symbols Added as Suffix Letters or Digits to the Alloy Number

Suffix letter F, O, H, T, or W indicates basic treatment or condition	*First suffix digit indicates secondary treatment used to influence properties*	*Second suffix digit for condition H only indicates residual hardening*
F - As-Fabricated		
O - Annealed, wrought products only		
H - Cold-worked, strain-hardened	1 - Cold-worked only	2 - 1/4 hard
	2 - Cold-worked and partially annealed	4 - 1/2 hard
	3 - Cold-worked and stabilized	6 - 3/4 hard
		8 - hard
		9 - extra hard
W - Solution heat-treated		
T - Heat-treated stable	1 - Partial solution plus natural aging	
	2 - Annealed cast products only	
	3 - Solution plus cold-work	
	4 - Solution plus natural aging	
	5 - Artificially aged only	
	6 - Solution plus artificial aging	
	7 - Solution plus stabilizing	
	8 - Solution plus cold-work plus artificial aging	
	9 - Solution plus artificial aging plus cold-work	

It is customary to control both composition and cooling rates in order to prevent large primary phases from forming. For example, commercial Al-Mn alloys most often contain less than 1.25 wt% Mn, although as much as 1.82 wt% is soluble in pure aluminum. The 1.25 wt% limit is imposed because Fe, present in most aluminum alloys as an impurity, decreases the solubility of Mn in aluminum. This increases the probability of forming large primary particles of Al_6Mn, which can have a disastrous effect on ductility.

Silicon is a principal addition to most aluminum casting alloys because it increases the fluidity of the melt. In the solid state, the hard silicon particles are the major contributer to the strength of non-heat-treatable casting alloys. Additional improvement in strength can be obtained by minor additions of Mg or by trace alloy additions, e.g., Na, that refine the cast structure. The latter also minimizes porosity and increases ductility. The exact mechanism associated with these modifications is unclear, but Na may depress the eutectic temperature and thus increase the nucleation rate of Si by segregation to the Si-interface, or tie up P impurities which are associated with coarse Si particles [6]. Sometimes, however, P is added to hypereutectic Al-Si alloys to refine the size of the primary Si plates. The P reacts with the Al to form small insoluble particles of AlP that serve as nuclei for the precipitation of Si. Strontium additions also have been shown to refine the cast microstructure of Al-Si alloys and suppress the formation of primary Si in hypereutectic alloys. This improves both ductility and toughness.

IV. HEAT TREATABLE ALLOYS

The 2XXX, 6XXX, 7XXX and some 8XXX series alloys are considered heat treatable. These alloys contain elements that decrease in solubility with decreasing temperature, and in concentrations that significantly exceed their equilibrium solid solubility at room and moderately higher temperatures. One of the transition elements Cr, Mn, or Zr, is added to age hardenable wrought alloys to control the grain structure; however, the grain and deformation structure plays only a secondary role in the strengthening of this class of materials.

A normal heat-treatment cycle after deformation processing includes a soak at a high temperature (dispersoids containing the elements Cr, Mn or Zr have already precipitated at this point), followed by rapid cooling or quenching to a low temperature to obtain a solid solution supersaturated with both solute elements and vacancies. The next step involves aging at room temperature (natural aging) or at an intermediate temperature (artificial aging), Fig. 5 and Table 6.

Quenching is accompanied by a change in free energy which increases progressively as the difference between the solutionizing temperature and the quenching temperature increases. The volume free energy change is the driving force for precipitation and is associated with the transfer of solute atoms to a more stable phase. However, when precipitation occurs there are other factors which increase the free energy, i.e., formation of the interface between the matrix and the precipitate requires an increase in the surface free energy, and, if there is a volume change or interfacial strains associated with the precipitate, there is an increase in elastic strain energy. The change in free energy when a precipitate forms, ΔG, is the sum of these free energy changes and can be expressed mathematically as:

$$\Delta G = V\Delta G_V + A\gamma + V\Delta G_S \tag{6a}$$

where V is the volume of the new phase; ΔG_V is the free energy decrease due to creation of a volume, V, of the precipitates and is therefore negative; A is the area of the interface between the matrix and the precipitate; γ the energy of the new surface formed; and ΔG_S is the increase in elastic strain energy per unit volume of precipitate.

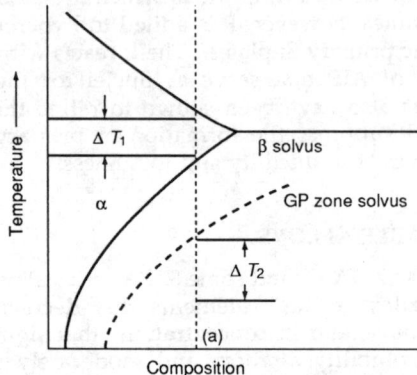

Fig. 5. Phase diagram of a hypothetical alloy system showing the β solvus and the GP zone (solute cluster) solvus. For composition (a), ΔT_1 is the temperature range for solution heat treatment, and ΔT_2 is the temperature range for precipitation heat treatment.

The critical increase in free energy, ΔG^*, required for a nucleus to become an equilibrium or metastable precipitate is known as the activation energy barrier for nucleation and can be expressed as:

$$\Delta G^* = (16\pi\gamma^3) / (3[\Delta G_V - \Delta G_S]^2) \tag{6b}$$

The nucleation rate, N, can be expressed as:

$$N = C \exp(-Q/kT) \exp(-\Delta G^*/kT) \tag{7}$$

where C is the number of nucleating sites per unit volume, Q is the activation energy for diffusion, k is the Boltzmann constant and T the absolute temperature.

There are a number of requirements for an effective age hardenable alloy. First, there has to be a sufficient volume fraction of the second phase for the desired strength. An examination of aluminum binary diagrams suggests that the most attractive alloying additions would be, in decreasing volume fraction of second phase, Ag, Mg, Li, Zn, Ga, and Cu. The required volume fraction will depend on the effectiveness of the second phase on impeding dislocation motion, i.e., whether it is a "soft" or "hard" phase, and its size and spacing. Second, the aging potential must be adequate and nucleation of the precipitates should be as close as possible to homogeneous. This is usually accomplished by the addition of a second alloying element. The addition of the second alloying element may: (i) reduce the solubility of the first element, (ii) increase the diffusion rates by trapping vacancies, (iii) increase the driving force for nucleation, or (iv) reduce the activation energy against nucleation. Although the selection of the second alloying addition has historically been by trial and error, Ryum [7] pointed out that the "± rule", which recommends the addition of atoms with positive and negative deviation from the atomic size of aluminum, may be useful in alloy design. Combinations of Mg, Li, Zn and Cu, which satisfy this rule, are present in many age-hardenable aluminum alloys.

Homogeneous nucleation of the equilibrium phase is normally very difficult because of high interfacial energy. Excluding rapid solidification processes, nature offers two means to overcome this barrier:

(1) Reducing the activation barrier via a precipitation sequence that involves several metastable phases having a lower interfacial energy, Fig. 6. This occurs in many age hardenable aluminum systems and the normal precipitation sequence is: super-saturated solid solution -> solute clusters (Guinier-Preston Zones) -> transitional structure -> final structure.

(2) Heterogeneous nucleation at 0-, 1-, or 2-dimensional defects which helps to accommodate the structure of the interface.

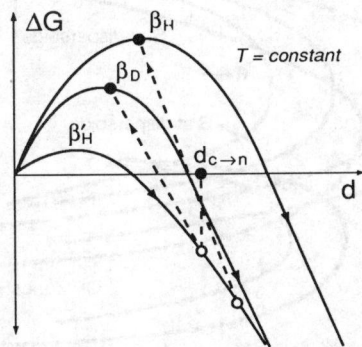

Fig. 6. Energy of nuclei ΔG vs diameter d for: β_H homogeneous nucleation of the stable phase; β_D heterogeneous nucleation of β at defect (dislocations, grain boundaries); β_H' homogeneous nucleation of the less stable phase $d_{c \to n}$ size to which β' must grow to nucleate (in-situ).

In defect-aided nucleation, sites are not determined by statistical fluctuations, but by the type, density, and distribution of effective defects. Vacancy-aided nucleation comes closest to homogeneous nucleation, i.e., it may produce very fine particles if small vacancy clusters and not dislocation rings determine the sites. In nucleation at dislocations and grain boundaries, the precipitate distribution depends on the distribution of the non-equilibrium defects. The formation of precipitate free zones, PFZ's, is usually due to solute and vacancy depletion in the environment of grain boundaries. The boundary acts as a sink for vacancies and the site for heterogeneous precipitation, thus forming a PFZ adjacent to the grain boundary. A major therapy for PFZ's in 2XXX and some 8XXX alloys has been dislocation-aided nucleation, provided by cold work prior to aging.

Precipitation kinetics of equilibrium precipitates, transition precipitates, and GP zones all show characteristic C-curve kinetics [8], Fig. 7. At high temperatures, nucleation and growth rates are low because the driving force for precipitation is small. At low temperatures nucleation and growth rates are low because diffusion is slow. As noted by Eq. 7 the nucleation rate depends on the number of nucleation sites per unit volume. For homogeneous nucleation these are essentially the number of atomic sites per unit volume and for heterogeneous nucleation these are dislocations,

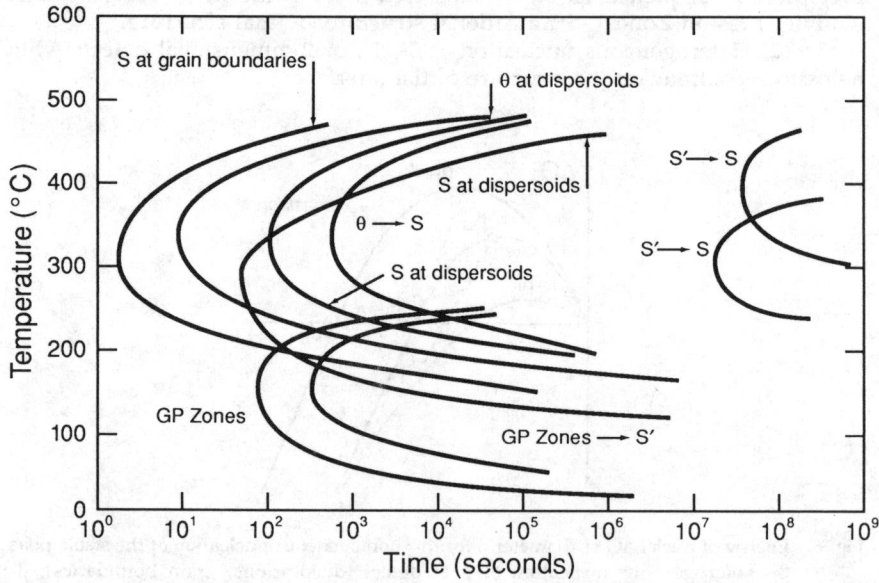

Fig. 7. Schematic representation of nucleation curves for precipitation in the Al-Cu-Mg-Mn alloy 2024 from Ref. 8.

grain boundaries, and surfaces of primary phases. When the activation energy for precipitation and the number of nucleation sites are both considered, heterogeneous nucleation is favored at low undercoolings and homogeneous nucleation is favored at high undercoolings. In some alloy systems, notably the 2XXX alloys, room-temperature deformation is sometimes applied after the solution heat treatment and prior to the aging treatment. This increases the defect density and thereby the number of nucleation sites for heterogeneous precipitation. Consequently, the number density of precipitates is greatly increased and precipitates are more homogeneously distributed throughout the matrix, Fig. 8.

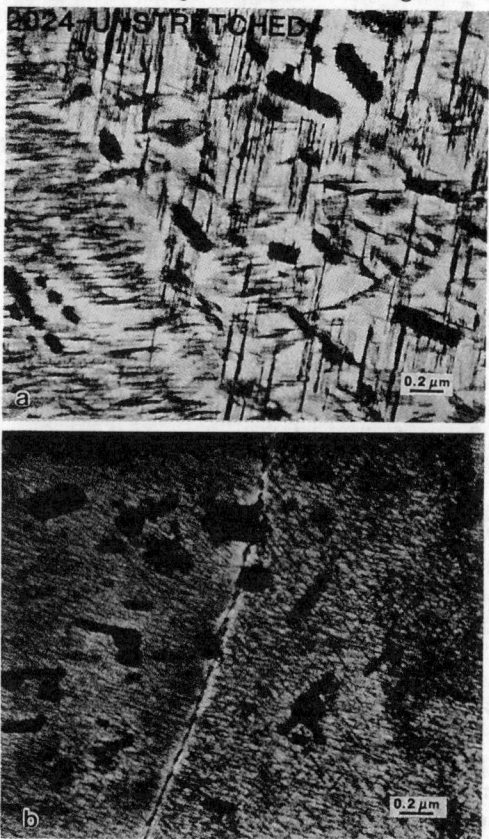

Fig. 8. Transmission electron micrographs showing the effect of plastic deformation on the precipitate distribution of 2024 (a) undeformed, aged 12 h at 190°C and (b) stretched 6% and aged 12 h at 190°C.

The first heat-treatable aluminum alloy was based on the Al-Cu system and resulted from the pioneering work of Alfred Wilm in 1906. After quenching from high temperature, and during aging at room temperature, Cu atoms cluster into GP zones and the strength is significantly improved over the as-quenched condition. The aging sequence in Al-Cu alloys is usually described as: supersaturated α -> GP-I -> θ'' -> θ' -> θ (Al_2Cu). The addition of Mg to Al-Cu alloys enhances both the rate and magnitude of natural aging. This enhancement probably results from complex interactions between the solute elements and the vacancies. For high Mg concentrations, and artificial aging, new ternary phases are formed further enhancing strength with the aging sequence described as: supersaturated α -> GP zones -> S' -> S (Al_2CuMg). Al-Cu-Mg alloys (2XXX) are one of the three major heat-treatable Al alloy systems.

Aluminum-magnesium binary alloys, containing sufficient Mg, undergo precipitation harding in the sequence: supersaturated α -> GP zones -> β' -> β (Al_3Mg_2). However, most commercial alloys contain additional alloying elements in order to improve such properties as strength and weldability. Medium-strength Al alloys, having good weldability and corrosion resistance, can be obtained by adding Si in balanced amounts to form the quasi-binary Al-Mg_2Si, or with an excess of Si above that needed to form Mg_2Si. These alloys strengthen appreciably during room-temperature aging and also comprise one of the three major heat-treatable Al alloys systems (6XXX series). Small additions of Cu are sometimes added to improve mechanical properties and Cr is added to offset the adverse effect that Cu has on the corrosion resistance.

The highest strength commercial alloys are based on the Al-Zn-Mg system, 7XXX series (although recently developed Al-Cu-Li-Mg-Ag alloys, 2094 and 2095, are stronger than any conventional aluminum alloy) [9,10]. The major strengthening precipitate in the Al-Zn-Mg system with high Zn:Mg ratio is $MgZn_2$, Fig. 9. For low Zn:Mg ratios $Al_2Zn_3Mg_3$ may form. Copper additions considerably increase the strength of Al-Zn-Mg alloys. Concentrations up to 1 wt% do not appear to alter the basic precipitation mechanism and, in this range, probably add a component of solid-solution strengthening. Low Cu-containing alloys are readily weldable but are not as resistant to stress-corrosion cracking as the higher Cu-content alloys. In excess of 1 wt%, Cu participates in the precipitation process and decreases the coherency of the precipitate when aged to peak strength. In the quaternary Al-Zn-Mg-Cu system, the phases $MgZn_2$ and AlMgCu form an isomorphous series, with Al and Cu substituting for Zn in $MgZn_2$. The improvement in resistance to stress corrosion cracking as Cu is added may be related to a reduction in the electrochemical activity of the precipitates as their Cu content increases.

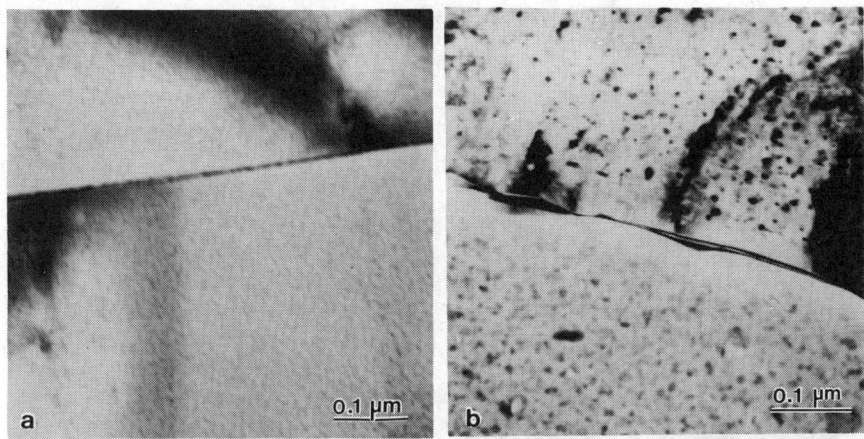

Fig. 9. Transmission electron micrographs of 7075 solution heat treated and cold water quenched and aged for (a) 4 hr at 120°C resulting in GP zones and no PFZ, and (b) 4 hr at 120°C + 12 hr at 165°C resulting in η' and η (MgZn$_2$) precipitates and a wide PFZ. Courtesy of Thomas H. Sanders, Jr.

Lithium is an attractive addition to Al since each weight percent added decreases the density by three percent and increases the modulus by six percent. In binary and most high Li (2-3 wt% addition) alloys, Li causes precipitation of the metastable phase Al$_3$Li, δ', during the age-hardening heat treatment. Most of the commercial alloys in this class, e.g., 8090 and 2090, contain additions of Cu with and without Mg. 8090 contains S' in addition to δ', while 2090 contains θ' and T$_1$ (Al$_2$CuLi) in addition to δ'. The exact nature of the phase equilibria of the quaternary Al-Li-Cu-Mg alloys depends on the relative concentrations of all three alloying elements. For example, the addition of small amounts (0.5 to 1.0 wt%) of Mg to a high Cu alloy suppresses the formation of θ' and introduces the S' phase. Since S' contains no Li, δ' precipitation is not markedly influenced by the Mg addition. In the latest generation of high strength Al-Li alloys (2095-type with 1-1.5 wt% Li) the high Cu/Li ratio produces a high volume fraction of T$_1$ without δ'. Both S' and θ' are also present in these alloys, Fig. 10.

Aluminum alloys are selected by the designer because they offer an attractive combination of properties and are cost effective. Achievable properties ultimately depend on the alloy's microstructure which is controlled by chemical composition and processing. Our understanding of microstructure/property relationships has aided in the development of improved aluminum alloys which has led to a significant increase in usage during the last fifty years.

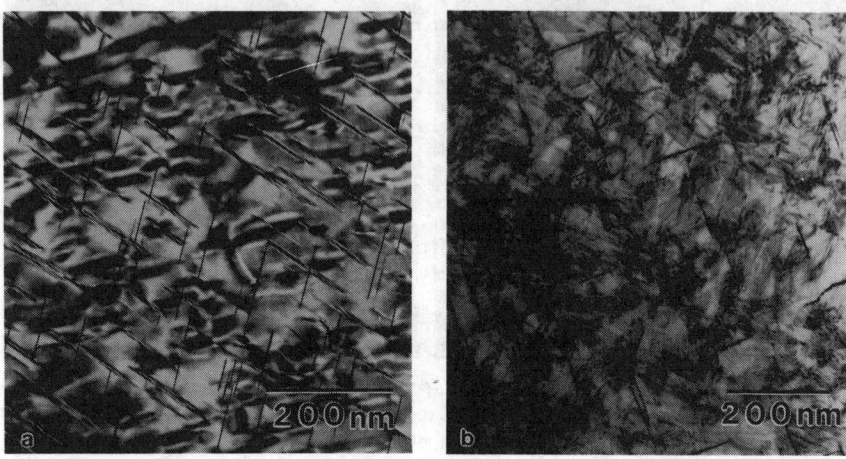

Fig. 10. Transmission electron micrographs of X2095-T8, (a) B near [110] showing 2 out of 4 T_1 variants parallel to the beam, (b) B near [001] showing θ' and S' precipitates and residual strain contrast from the T_1 plates.

V. STRUCTURE-PROPERTY RELATIONSHIPS

The appropriate microstructure for an aluminum alloy application can be identified after a designer has specified what properties will be critical in the design of the component. Primary design criteria may involve tensile and yield strength, fracture toughness, fatigue resistance, or electrical/ thermal conductivity, just to name a few. Using the concept of microstructural design, the first step is to identify the primary properties of interest. Next, the microstructural features that allow each of the goals to be accomplished are identified. The final step involves optimizing the microstructure to provide a suitable balance of the important properties.

Microstructure may have an appreciable influence over a number of design critical mechanical properties. These properties are listed in Table 7, along with desirable microstructural features and their functions. Many other physical and mechanical properties are of interest to the design/manufacturing community, but microstructural modifications only have minor effects on their properties.

Liquidus and solidus temperatures are determined by the bulk composition of the alloy, but the microstructure, e.g., particles or dendritic structures, influences the incipient melting temperature. If an alloy is heated at a rapid rate to a temperature near the solidus temperature, there may not be enough time for dissolution of second phase particles, or homogenization of composition in interdendritic regions if either is present. Under these conditions, local melting occurs at the interface of the particles or in the interdendritic regions causing embrittlement.

Stiffness, or Young's modulus, is usually not altered significantly by changes in the microstructure of an alloy with a fixed composition. Alloying elements, whether present in solid solution or as particles of varying size and distribution seem to have little or no affect on Young's modulus. One notable exception is the addition of a significant volume fraction of dispersoids of high modulus by rapid solidification technology (8009), powder metallurgy or conventional composite manufacturing (2024/SiC/15). Although this property is relatively insensitive to microstructure and composition, a few elements exert a modest influence on stiffness. The graph in Fig. 11 illustrates the effect of various alloying elements on alloy stiffness [11].

Table 7. Property-Microstructure Relationships in Aluminum Alloys

Property	Microstructural Feature	Function of Feature(s)
Yield Strength	uniform dispersion of fine, hard particles, fine grain size	inhibit dislocation motion
Toughness	no constituent particles, clean grain boundaries	encourage plasticity, inhibit void growth
Ductility	no constituent particles, clean grain boundaries, low dislocation density	encourage plasticity, inhibit void growth
Creep	thermally stable particles on grain boundaries, large grain size	inhibit grain boundary sliding
Fatigue Crack Initiation	no shearable particles, fine grain size	limit magnitude of slip steps at surface
Fatigue Crack Growth	shearable particles, no anodic phases or interconnected hydrogen traps, large grain size	encourage crack closure, branching, deflection and slip reversibility
Pitting	no anodic phases	prevent preferential dissolution of second phase particles
Stress Corrosion Cracking/HE	no anodic phases, or interconnected hydrogen traps	prevent crack propagation due to anodic dissolution or hydrogen embrittlement
Conductivity (electrical, thermal)	no particles with elastic strain fields, low solute concentration, low dislocation density	limit scattering sites for electron migration

Density is another property that is more or less insensitive to microstructure. However, compositional changes have a profound effect on alloy density. Elements heavier than Al increase the density of the alloy and elements lighter than Al decrease the density of the alloy relative to the 2.7 g/cc density of pure Al. In aerospace and automotive applications, both high stiffness and low density are important design properties. It is interesting to note that both Be and Li additions simultaneously increase stiffness and decrease density. This fact is one reason Al-Li alloys have been the subject of much research in the aerospace materials arena. The cost of these alloys is somewhat prohibitive for the automotive market at this time.

In light of the treatise on strength in Section II, microstructure-strength relationships will not be discussed here. The effect of microstructure on the following properties will be discussed: fracture toughness, fatigue resistance, and stress corrosion cracking resistance.

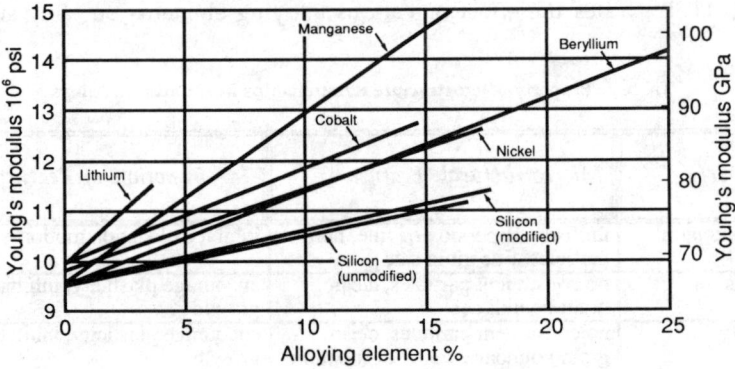

Fig. 11. Influence of selected alloying additions on Young's modulus from Ref. 11.

Fracture Toughness: For high toughness, good tensile elongation, or good formability for that matter, a microstructure must accommodate significant plastic deformation, whether it be at a crack tip, in a tensile bar or in a formed part, without encouraging failure processes to operate, i.e., void growth, slip band cracking, grain boundary separation. Figure 12 shows the typical relationship of decreasing toughness with increasing strength for 2XXX and 7XXX series alloys [12].

The "toughness tree," as presented in Fig. 13 (after Staley [13]), is an effective tool for classifying the contribution of microstructural features to toughness. The metallic and non-metallic inclusions on the right-hand side of the tree are treated as extrinsic effects because they are controlled more by defects due to processing than by composition. Metallic and non-metallic inclusion content is minimized by controlling composition, melt temperature, filtering, or direct-chill processing. Hydrogen levels are kept low by minimizing exposure to moisture, and deformation processing often closes porosity.

The features on the left of the "toughness tree" are termed intrinsic because they are characteristic of the composition and microstructure of the alloy. These features also depend on the response of the material to deformation processing and heat treatment. The left side of the tree is supported by three distinct classes of particles. They are classified by their size range and function in the microstructure as shown in Table 8. Figure 14 shows the trend of the amount of Fe+Si on the toughness of 7075 sheet. The same trend holds for 2024 [14] and Al-Li alloys [15].

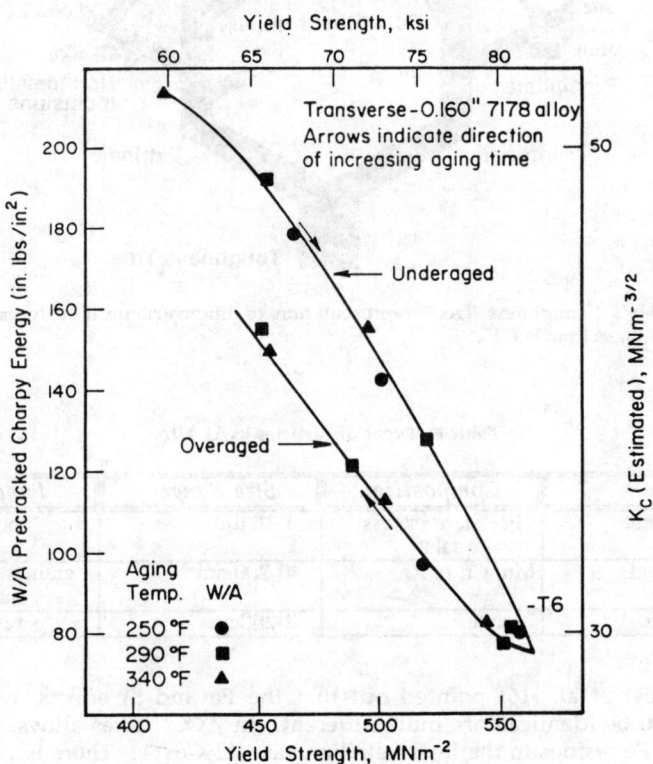

Fig. 12. Strength-toughness relationship for alloy 7178 from Ref. 12.

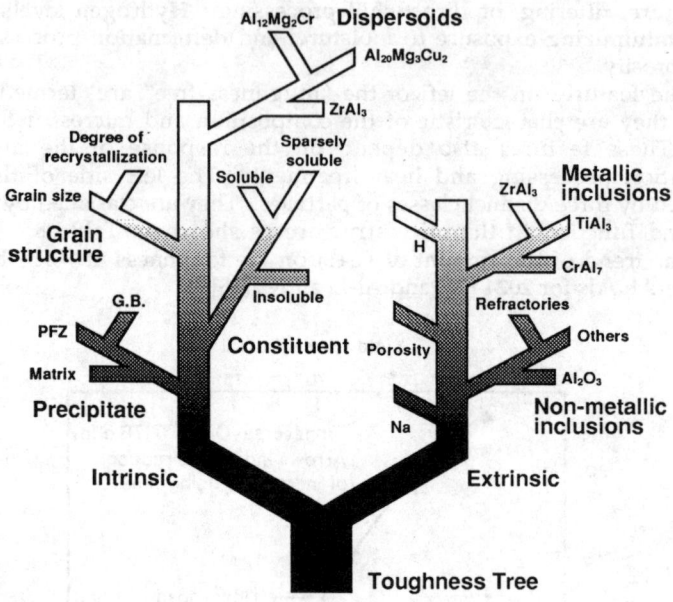

Fig. 13. Staley's "Toughness Tree" representation of microstructural features that affect toughness from Ref. 13.

Table 8. Types of Particles in Al Alloys

Particle	Composition	Size Range	Function
Constituent	Fe, Si, or excess solute	1-10 μm	none
Dispersoid	Mn, Cr, or Zr	40-200 nm	grain structure control
Precipitate	solute	2-200nm	strength

Quist et al. [16] pointed out that the Fe and Si effects, while often thought to be identical, are quite different. In 7XXX series alloys, essentially all of the Fe resides in the intermetallic phase Al_7Cu_2Fe. There is virtually no effect of Fe on strength since Cu is not the major strengthening addition in

7XXX series alloys. However, increasing Fe content causes a precipitous decrease in fracture toughness. The effect of Si is somewhat more complicated. Silicon results in the formation of AlFeSi and Mg_2Si constituent phases. Formation of Mg_2Si removes Mg from solid solution and thus lowers the amount of Mg available for precipitation of the strengthening phase: $MgZn_2$. Silicon additions lower the strength and thus increase toughness. Therefore, when reducing Si levels to increase fracture toughness, Mg levels must also be reduced to avoid a strength increase and an associated drop in toughness.

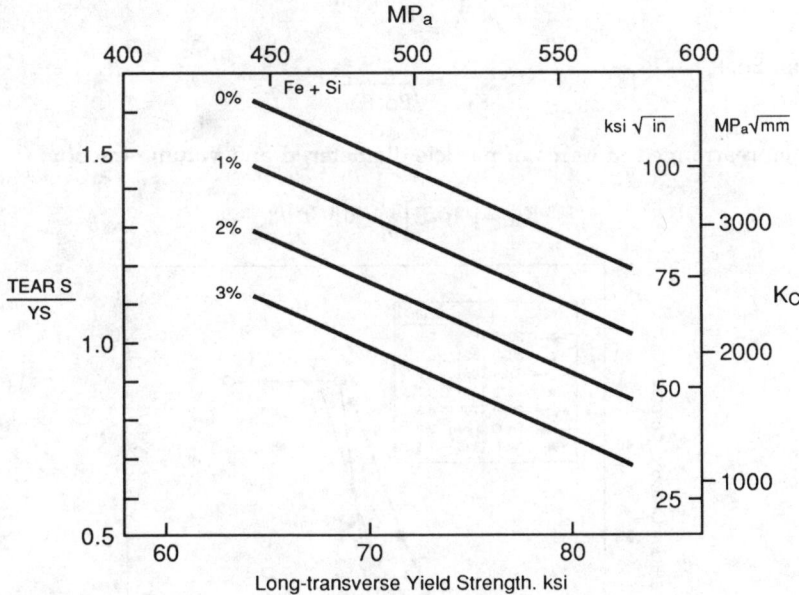

Fig. 14. Effects of total Fe + Si on toughness of alloy 7050 sheet from Ref. 13.

Constituent particles participate in the fracture process through void formation at particle/matrix interfaces or by fracturing during primary processing. Through metallographic sectioning experiments, Low et al. [17] observed that 0.07 plastic strain (uniaxial tension) in 2XXX and 7XXX series alloys caused nearly half the constituent particles to fracture (see Fig. 15, which was compiled by Hahn and Rosenfield [18] and contains data from Gurland [19]). Within the plastic zone of a crack tip (~ crack tip opening displacement, δ) plastic strains may exceed 1. Hahn and Rosenfield used the following definition of crack tip opening displacement to develop a correlation among strength, microstructural features and K_{IC} [18]:

$$\delta = \frac{0.5K^2}{E\sigma_y} \quad (8)$$

where K is the stress intensity factor, E Young's modulus, and σ_y the tensile yield strength. Crack extension is proposed to occur when the plastic zone is on the order of the distance between cracked constituent particles, λ:

$$\lambda = \delta \quad (9)$$

From Eq. 8,

$$K_{IC} = \sqrt{2\sigma_y E \lambda} \quad (10)$$

When rearranged in terms of particle diameter, d and volume fraction, f:

$$K_{IC} = \left[2\sigma_y E \left(\frac{\pi}{6}\right)^{1/3} d \right]^{1/2} f^{-1/6} \quad (11)$$

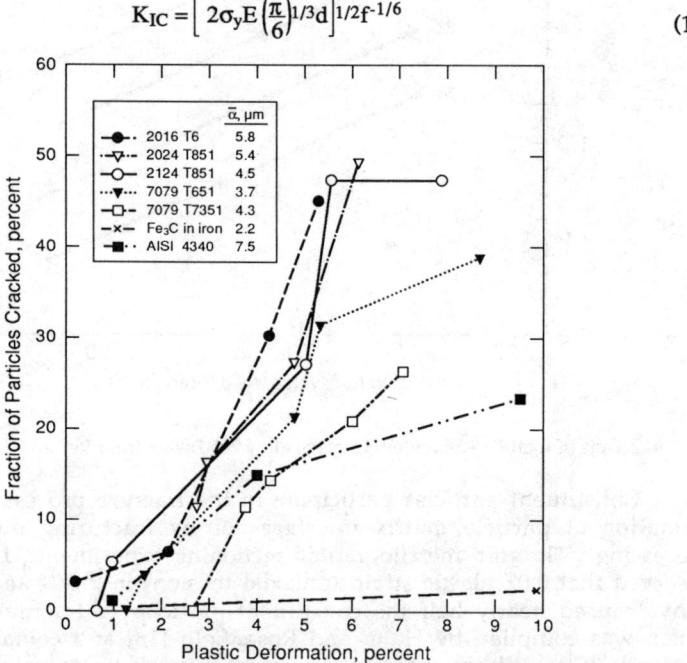

Fig. 15. Influence of plastic strain on the fracture of cracked second phase particles in tensile specimens of various aluminum alloys and 4340 steel from Ref. 18.

Figure 16 shows some data from various researchers that supports the proposed relationship between microstructural features and fracture toughness. However, this relationship suggests that toughness increases with increasing strength, which is contrary to experimental observations.

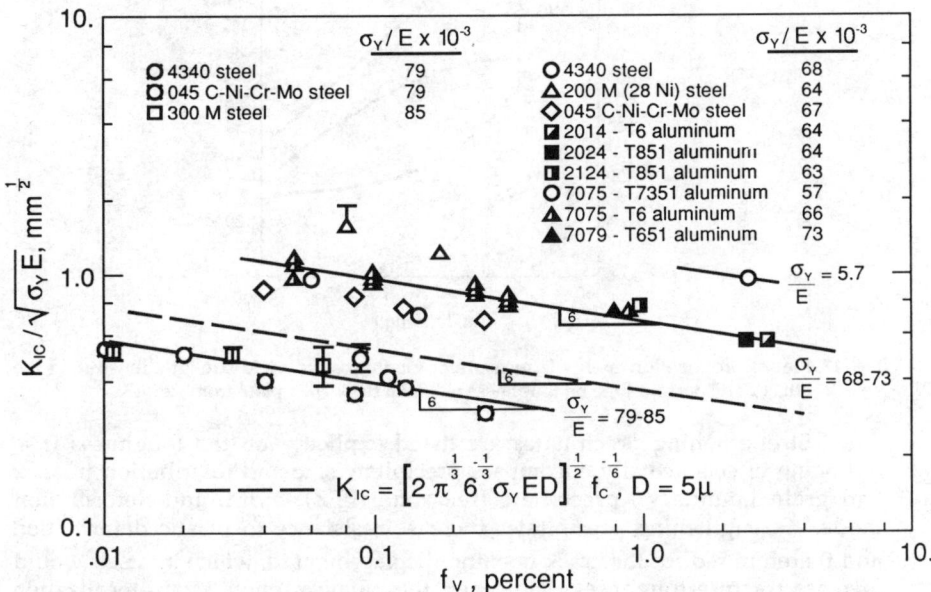

Fig. 16. Influence of the volume fraction of large ($\geq 1\mu$ size) second phase particles on the plane strain fracture toughness of commercial alloys from various researchers and compiled in Ref. 18.

Dispersoid forming elements are added to control grain size during recrystallization, or inhibit recrystallization altogether. However, depending on particle size and the details of the particle/matrix interface, dispersoids may play an unfavorable role in the fracture process. Specifically, void coalescence between constituent particles may occur by the formation of void sheets by nucleation of fine voids at dispersoids [20]. The strength-toughness relationships in Fig. 17 were determined for 7075 variants containing different dispersoid forming elements. Since Zr particles are smaller and coherent with the matrix (strong interface), they tend not to participate in the fracture process.

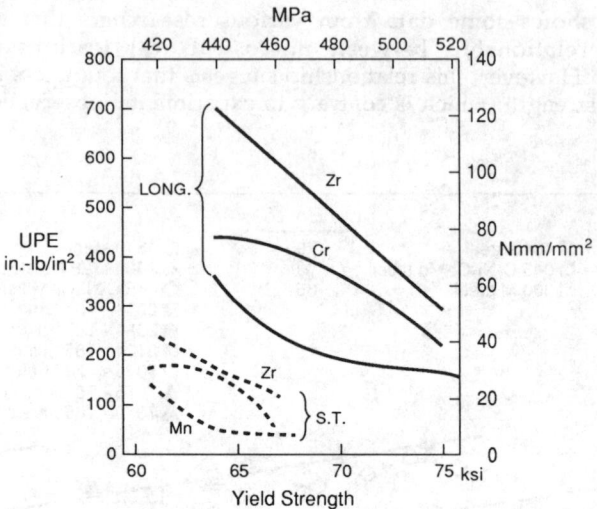

Fig. 17. Plot of propogation energy (a measure of toughness) vs. yield strength, illustrating the effect of dispersoid type on toughness of 3 inch thick 7075 plate from Ref. 13.

Strengthening precipitates are listed explicitly on the toughness tree, but aging effects actually encompass precipitate size and distribution (matrix and grain boundary), precipitate free zone (PFZ) width, and deformation mode. Strengthening precipitates increase resistance to plastic deformation and therefore reduce the crack opening displacement, δ, which by itself would increase fracture toughness. However, they may promote strain localization which leads to premature crack nucleation and low fracture resistance. During aging, heterogeneous precipitation occurs at grain and subgrain boundaries resulting in soft, solute-denuded PFZ's in the matrix adjacent to the boundaries. Therefore, strain localization occurs in Al alloys by two distinctly different mechanisms:

(1) Cutting of shearable particles
(2) Preferential deformation in soft precipitate free zones.

For the first case, strain localization may be quantified by examining the dislocation-particle interactions. Duva et al. [21] argued the best way to quantify the slip intensity in deformed microstructures is to calculate the number of dislocations, N, expected in a slip band. The authors started by defining the strength of a particular slip plane, τ, in the following manner:

$$\tau = \tau_o + C_p\sqrt{V_f}(r_p - Nb/2)^{1/2} + \frac{C_B Nb}{L} \qquad (12)$$

where τ_o is the Pierls stress, N the number of dislocations gliding on the slip plane, V_f the particle volume fraction, r_p the particle radius, and L the slip length. C_p is related to the particle hardening mechanisms, and as an example, the case of order hardening is shown here:

$$C_p = \frac{\gamma^{3/2}}{2b}\sqrt{\frac{3}{\pi G b^2}} \qquad (13)$$

where γ is the antiphase boundary energy.

C_B is related to the matrix shear modulus as follows:

$$C_B = \frac{G}{\pi(1-v)} \qquad (14)$$

The last term in Eq. 12 should be considered a back stress at the end of a pileup of length L containing N dislocations. The slip intensity relation was derived based on the assumption that if slip initially weakens a plane (by reducing cross sections of strengthening precipitates), slip will continue on that plane until it regains its initial strength (by work hardening). The following equation can be derived, appropriate for the case of slip plane softening:

$$N = \sqrt{V_f}\ \sqrt{r_p}\ L\ \frac{C_p}{C_B b} \qquad (15)$$

Salient features of the model include increasing slip intensity for increasing volume fractions and increasing particle diameters (below d_c), both of which are observed experimentally.

Recent alloy design trends have focused on reducing the volume fraction of void nucleating particles, i.e., by minimizing Fe and Si contents, optimizing solute content to prevent the formation of coarse primary phases, and use of Zr as the dispersoid forming addition to control the grain size and degree of recrystalllization. Experimental observations [12-16] have shown that for this class of alloys, strain localization in the matrix and in PFZ's and grain boundary precipitates control fracture behavior.

Jata and Starke [27] developed a model to describe transgranular fracture in microstructures that exhibit strain localization in slip bands. The model related the fracture toughness to slip band width, W_{SB}, and slip band spacing, S_{SB}. The slip band width decreases and the slip band spacing increases with increasing strain localization when particle shearing occurs (as predicted by Duva et al. [21]). The crack tip opening displacement was defined:

$$\delta = 2(Nb)\sin\alpha = \frac{0.5 K^2}{E\sigma_y} \quad (16)$$

where α is the angle between the slip planes and the crack plane, and K is the stress intensity factor. The shear strain in the slip band was then defined:

$$\gamma_{SB} = Nb/W_{SB} \quad (17)$$

Substituting $\gamma_{SB}W_{SB}$ for Nb, and assuming that initial crack extension occurred when γ_{SB} reached a critical value $\gamma_{SB}c$, led to the following equation for plane strain fracture toughness:

$$K_{IC} = (4\sigma_y \gamma_{SB}c W_{SB} \sin\alpha)^{1/2} \quad (18)$$

Taking into account the number of active slip bands which would be determined by the plastic zone width, D, and the slip band spacing, S_{SB}, Eq. 18 became:

$$K_{IC} = \{8E \sin\alpha\, \sigma_y W_{SB}(D/S_{SB})\gamma_{SB}c\}^{1/2} \quad (19)$$

Roven [28] recently simplified Eq. 19:

$$K_{IC} = \{2S_{GB}W_{SB}\sigma_y E/Mtan\alpha S_{SB}\}^{1/2} \quad (20)$$

where S_{GB} is the spacing of the grain boundaries and M the Taylor crystallographic orientation factor. This equation applies to the case where strain localization occurs in the matrix, and therefore for most underaged to peakaged conditions in heat treatable aluminum alloys.

The second type of strain localization was the subject of studies by Hornbogen [22] and Ludtka and Laughlin [29]. Overaging is normally associated with a change in deformation mode from dislocation shearing to dislocation looping or bypassing of strengthening precipitates. At this stage, strain localization in the PFZ, microvoid nucleation at GB precipitates, and intergranular fracture may occur and control the fracture toughness of the material. Hornbogen [22] used the critical strain to fracture concept of Hahn and Rosenfield [18] to develop a relationship between fracture toughness and strain localization in the PFZ similar to Eqs. 19 and 20. For this case, crack initiation and propagation takes place through the soft PFZ :

$$K_{IC} = \{CE\sigma_{ypfz}e_{fpfz}d_{pfz}/S_{GB}\}^{1/2} \qquad (21)$$

where d_{pfz} is the width of the PFZ and C is an empirical constant with the dimension of length. Ludtka and Laughlin [29] examined several 7XXX series alloys with different solute levels. They observed that fracture toughness decreased as the term ($\sigma_M - \sigma_{PFZ}$) increased, where σ_M is the precipitation hardened yield strength and σ_{PFZ} is the strength in the PFZ. They claimed that this term was a qualitative measure of the tendency toward intergranular fracture, because it represented a propensity for strain localization near the grain boundary.

Sugamata et al. [30] combined the concepts leading to Eqs. 20 and 21 to describe mixed mode fracture in 8090 as a function of aging time. A weighted average was used based on the amount of intergranular and transgranular failure observed on the surface of the fractured compact tension samples, and the results are shown in Fig. 18.

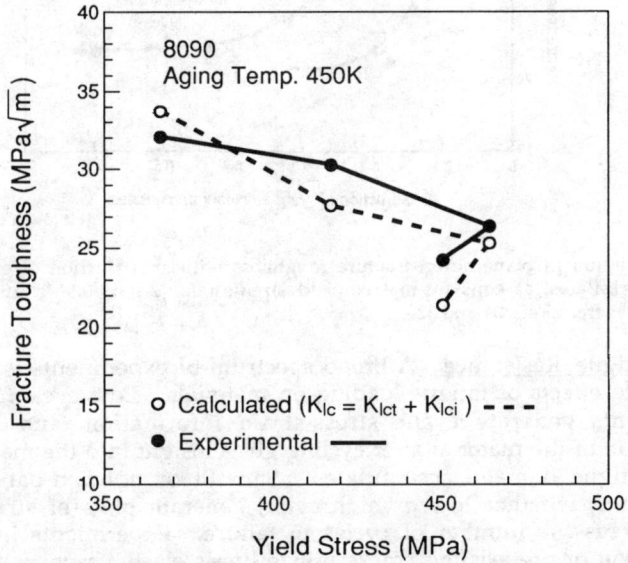

Fig. 18. Comparison of toughness values predicted by weighted average of transgranular and intergranular fracture models and experimental results.

In addition to the strain localization in the PFZ, dimpled rupture may occur at equilibrium grain boundary precipitates in severely overaged microstructures [31,32]. Figure 19 illustrates the relationship of areal density of equilibrium δ grain boundary particles on toughness.

Fracture resistance is maximized by reducing the volume fraction of constituent particles, and relying on Zr additions for grain structure control. Aging times and temperatures should be chosen to minimize grain boundary precipitation. Strain localization may be minimized by increasing the volume fraction of nonshearable particles.

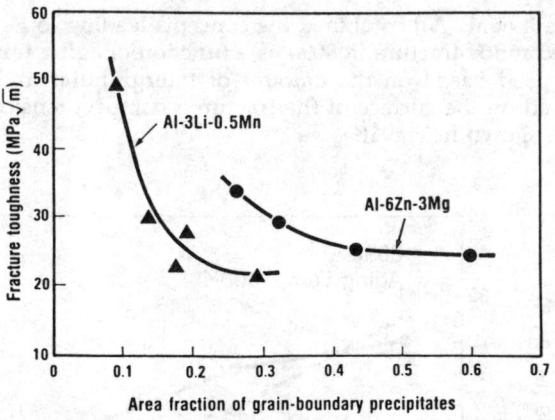

Fig. 19. Variation in plane strain fracture toughness with area fraction of grain boundary precipitates, at constant-matrix yield strength for Al-3Li-0.5Mn and Al-6Zn-3Mg alloys from Refs. 31 and 32.

<u>Fatigue Resistance:</u> A broad spectrum of experiments is available to investigate effects of fatigue loading on materials. Low cycle fatigue (LCF) experiments generate cyclic stress-strain information, and microscopic observation of the material after cycling gives insight into the mechanisms by which fatigue damage accumulates. Smooth or notched-bar fatigue life experiments, whether low or high cycle, generate plots of stress or strain range versus the number of cycles to failure. Experiments involving the propagation of pre-existing cracks utilize linear elastic fracture mechanics to accurately describe stress intensities at the crack tip and crack growth rates at various applied stress intensity ranges. Using these experimental techniques, the effects of variables such as composition, modulus, strength, environment and microstructure on fatigue resistance may be investigated.

Cyclic Stress-Strain Behavior: To achieve an understanding of fatigue, one must first realize how a material responds to an imposed cyclic strain. Cyclic plastic strain is a measurable physical quantity that is directly related to fatigue damage [33]. If a metal is cycled a given amount in tension and compression, the cyclic stress-strain response will be in the form of a hysteresis loop. Since plastic deformation is not totally reversible, modifications to the microstructure occur during cycling [34]. As a result of these modifications, the material exhibits cyclic softening or hardening. Annealed single phase alloys (e.g., 1100-O) tend to harden to a plateau, while work hardened tempers (e.g., 1100-H18) soften due to the instability of dislocation structures under fatigue conditions [35]. Age-hardened Al-Cu alloys (e.g., 2024 T8) show initial hardening followed by saturation, then softening before fracture [36,37]. The mechanism of strength loss is debatable, but has been attributed to the following phenomena:

(1) Repeated shearing reduces particle diameters to the point of thermodynamic instability [38]
(2) Repeated shearing reduces particle cross-section until the particle is no longer an effective barrier [39]
(3) Repeated shearing causes disordering of the ordered precipitates [40].

Regardless of the details, the mechanism responsible for cyclic softening depends on both microstructure and slip distribution. Fine and Santner [36] generated cyclic stress strain curves for Al-Cu alloys with various microstructures. Cyclic hardening followed by cyclic softening was observed when the microstructure contained only GP zones. With θ' particles as the sole precipitate, the softening behavior was less pronounced and less strain localization was observed. In 2024, the distribution of Mn containing dispersoids and constituent particles homogenized slip, thereby preventing any softening from occurring.

Stolz and Pelloux [41] investigated the low cycle fatigue behavior of two categories of Al alloys: those strengthened by shearable precipitates and those strengthened by nonshearable particles. They observed a large Bauschinger effect in the alloys containing nonshearable particles, and attributed this to back stresses exerted by the nonshearable particles on the matrix. The shape of the hysteresis curves was unusual in that the compression side of the stress strain loop exhibited a change in curvature.

Prasad et al. [42] studied the low cycle fatigue behavior of 8090-T851. They observed hardening for only a few cycles, followed by cyclic stability and final softening. Their fractography indicated a change in fracture behavior with applied strain amplitude. At low strain amplitudes a high energy transgranular shear process appeared to dominate, while at high applied strain amplitude a lower energy intergranular fracture process was observed.

They attributed the intergranular fracture to localized plastic deformation resulting from particle shearing.

Smooth Bar Life: Cracks may nucleate at the surface of a smooth bar fatigue specimen, or at internal sites. The initiation event is tied to the level of observation, and in many cases, cracks exist in the microstructure, but are not large enough to be detected by the usual techniques (dye penetrants, radiography, eddy currents). Surface nucleation sites include scratches and extrusions/intrusions where persistent slip bands (PSB's) intersect the surface. Alloy microstructure, or, more specifically, the deformation behavior of a particular microstructure may affect the severity of the PSB's that intersect the surface and form slip steps. As illustrated in Eq. 15, the number of dislocations in a slip band may be predicted by examining the details of the microstructural features. The greater the tendency of the alloy to produce intense slip bands (higher N), the earlier a fatigue crack is likely to form, and the larger it is likely to be. As mentioned earlier, commercial alloys contain dispersoids, which tend to homogenize slip by providing impenetrable obstacles to gliding dislocations, thereby reducing the slip length. Figure 20 shows S-N curves for 7075 and X7075, an alloy with no dispersoids [43]. The alloy with no dispersoids failed in fewer cycles at the same stress level compared to the commercial alloy. A similar trend is observed for grain size, i.e., fine grain size (shorter slip lengths) allows a longer fatigue life than large grain size (longer slip length) material of the same composition, mainly because the crack that will propagate through the sample starts at a higher ΔK level in the large grain material (equal stress range, larger initial crack length; see FCP section).

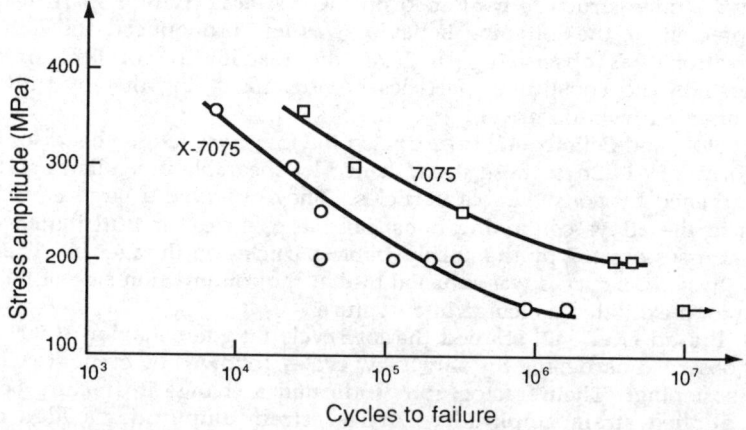

Fig. 20. S-N curves for commercial 7075 alloy and the pure equivalent (X-7075), each aged 24 hr at 100°C, from Ref. 34.

Strengthening precipitates and dispersoids influence fatigue lives of smooth bar samples by determining the slip length. However, in commercial alloys, crack nucleation often occurs at constituent particles. Nucleation may occur by debonding of the particle/matrix interface [44], or by cracking along slip bands emanating from or terminating at the large constituent particles [35]. Fatigue resistance of smooth bars may be improved by reducing the size and volume fraction of constituent particles, and reducing the slip length with fine grain size or sufficient volume fraction of dispersoids.

Fatigue Crack Growth: The resistance of an aluminum alloy to crack propagation under cyclic loading is of great interest to the design community. Assuming there are already cracks in a structure, fatigue crack propagation (FCP) data will help designers predict a safe service life for the component. This "damage tolerant design" is used by the aircraft industry to maximize service lives of structures. Fatigue crack growth data for a specific alloy can also be used to predict an applied stress intensity range below which fatigue crack propagation does not occur.

Modern study of fatigue crack growth began in 1963 when Paris and Erdogen demonstrated the functional relationship between the incremental crack extension per cycle (da/dN) and the stress intensity range (ΔK) [46]:

$$\frac{da}{dN} = C\Delta K^m \qquad (22)$$

This relationship is reasonable between ~10^{-5} and 10^{-3} mm/cycle, and constitutes the linear, or Paris region of the FCP curve. Lindley, Richards and Ritchie [47] studied a variety of alloys over a wide range of ΔK and K_{max} and constructed the summary diagram shown in Fig. 21, dividing the sigmoidal FCP curve into three regions based on the mechanism of FCP.

The discovery of the crack closure phenomenon by Elber explained the effect of load ratio ($\sigma_{min}/\sigma_{max}$, or K_{min}/K_{max}) on FCP. Subsequently, considerable effort was dedicated to separating "extrinsic" effects such as closure from the "intrinsic" material FCP behavior. Unfortunately, extrinsic effects such as closure are unavoidably associated with alloy composition and microstructure [38]. Microstructural components, e.g., precipitates, dispersoids and constituent particles, control the slip character of the alloy. Slip character, in turn, dictates deformation and fracture behavior, which tends to control the amount of closure.

Environmental effects such as corrosion fatigue and hydrogen embrittlement occur in a wide variety of environments including laboratory air [49,50]. These experiments can be performed explicitly, but often the environmental effects interfere with attempts to investigate purely mechanical FCP behavior. Carter et al. [50] showed that conditions leading to heterogeneous deformation behavior (shearable precipitates, large grain size) improved FCP resistance in vacuum, but had little effect in laboratory air. This mild environment interacted strongly with the microstructure that favored heterogeneous deformation causing the FCP curves to collapse into a narrow band [50].

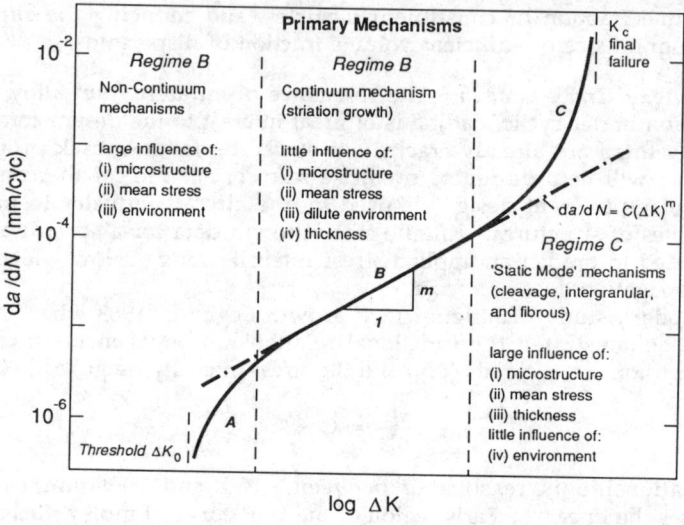

Fig. 21. Schematic representation of log da/dN vs log ΔK and suggestions of possible influence of stress, environment, and microstructure on crack growth in three different regimes, from Ref. 47.

A concerted effort has been made to study the FCP behavior of Al-Li alloys, since these alloys are targeted for aerospace applications. In general, Al-Li alloys exhibit superior FCP resistance when compared to conventional alloys [51-54]. This behavior is generally attributed to a high level of closure, crack branching and deflection. Figure 22 shows that for constant load ratio (R=0.1) FCP testing, three of the five commercial Al-Li alloys exhibit better FCP resistance in laboratory air than Al-Zn-Mg alloy 7150 [55]. 8090-T8X and 2090-T8X exhibit comparable and better FCP resistance than 2124-T351. All Al-Li alloys exhibited higher closure levels than 7150-T651 or 2124-T351.

Piascik and Gangloff [46] studied environmental effects on fatigue crack growth of 2090. The authors observed identical intrinsic (constant K_{max})[1]

---

[1]Testing procedure where K_{max} is held constant and K_{min} is elevated to perform the ΔK shed to threshold. The load ratio increases, thereby ensuring a high R in the threshold region.

growth rates for 2090 in vacuum, helium, and oxygen environments. Hydrogen embrittlement was proposed as the mechanism for environmental fatigue cracking in moist air, water vapor, and salt water. In addition, 2090 was shown to exhibit better intrinsic environmental FCP resistance than 7075-T651 and similar resistance to 2XXX series alloys.

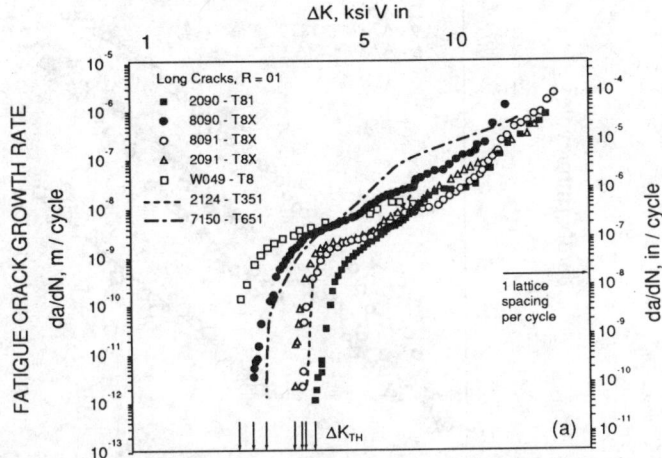

Fig. 22. Fatigue propagation curves for commercial Al-Li alloys compared with conventional 2124 and 7150, from Ref. 54.

Slavik et al. [57] investigated the influences of microstructure on inert environment (dynamic vacuum better than 5µPa) intrinsic FCP for Al-Li-Cu-Mg alloys 8090 and X2095, Al-Li-Cu alloy 2090, and Al-Cu-Mg alloy 2024. Figure 23 shows the da/dN vs. ΔK behavior for all microstructures investigated. As listed in Table 9, the fracture modes are identical for all Li-containing alloys, regardless of deformation mode. "T" plane sections[2], shown in Fig. 24, illustrate the marked difference in failure modes between the ductile tearing of 2024 and the faceted slip band cracking of the Al-Li alloys.

Studies of many alloy systems by Speidel [58] suggest that normalizing crack growth rate curves by dividing ΔK by the elastic modulus effectively collapses them into a narrow band. Figure 25 illustrates the effect of normalizing based on modulus, and the correlation is reasonable, except for alloy X2095.

---

[2]The loading direction of the compact tension specimens was parallel to the rolling direction, "L", and the crack propagated in the transverse direction, "T". Therefore, the sectioning plane perpendicular to the crack growth direction is "T".

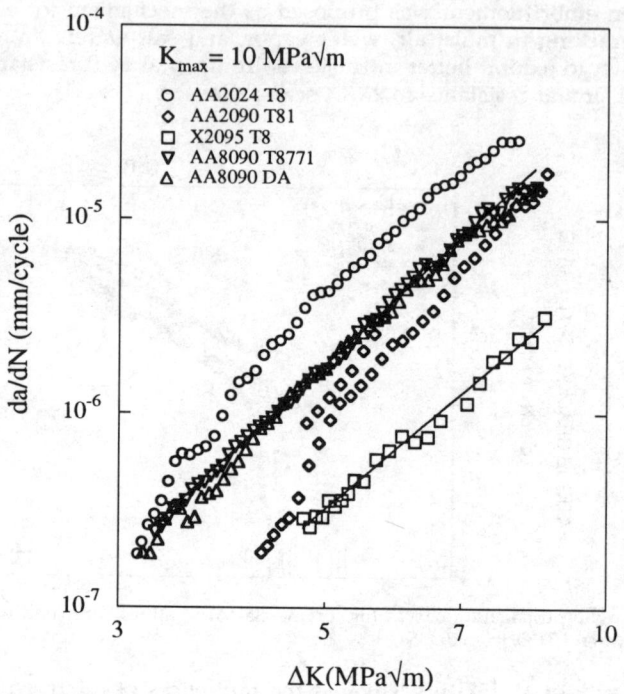

Fig. 23. Intrinsic, inert environment fatigue crack growth rates for Al-Li-Cu-Mg and 2024 alloys, from Ref. 56.

Table 9. Summary of Deformation and Fatigue Fracture Properties

Alloy	Wt Pct Li	V_f δ'	Deformation Mode	Failure Mode	da/dN* at $\Delta K = 5$ $MPa\sqrt{m}$
AA2024-T8	0	0.0	homogeneous	tearing	4.2×10^{-6}
AA8090-T8771	2.6	0.2	coarse planar	faceted	1.7×10^{-6}
AA8090-DA	2.6	0.1	homogeneous	faceted	1.7×10^{-6}
AA2090-T81	2.2	0.2	coarse planar	faceted	8.4×10^{-7}
X2095-T8	1.3	0.0	homogeneous	faceted	3.1×10^{-7}

*da/dN in millimeters/cycle

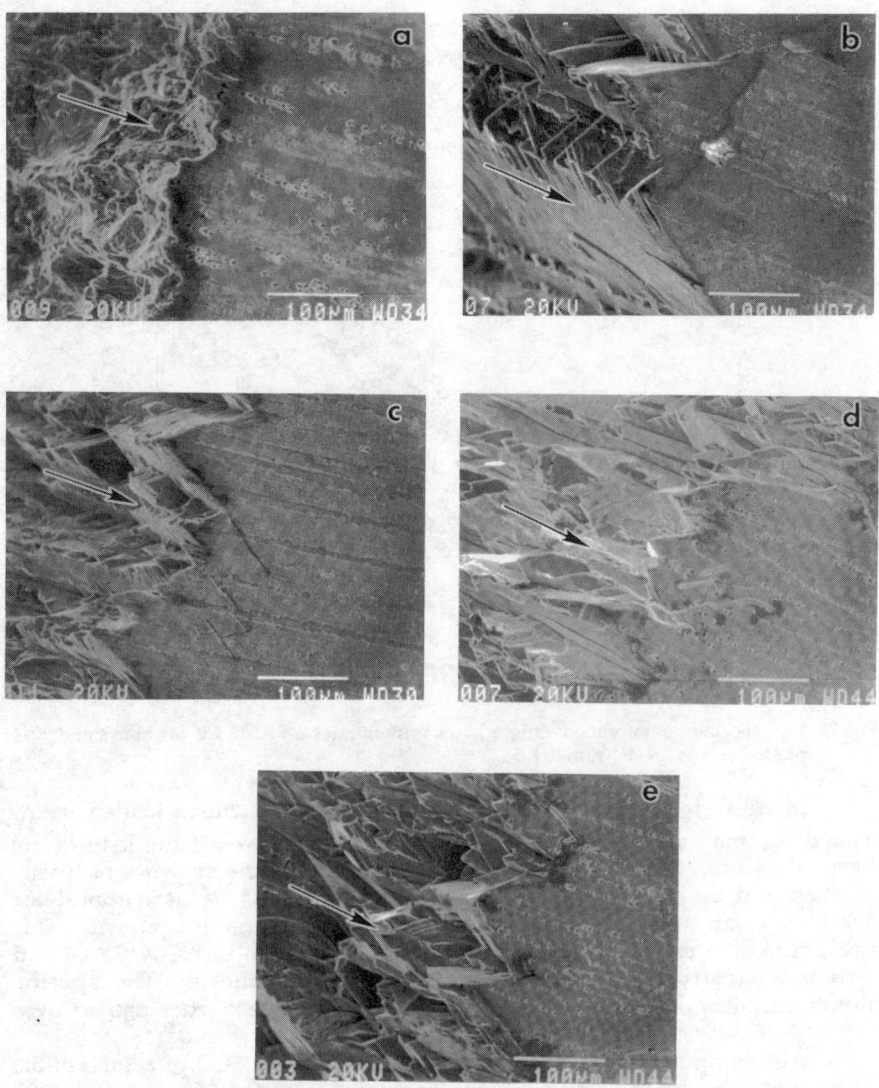

Fig. 24. Fracture surfaces and T-plane sections for (a) 2024-T8, (b) 2090-T81, (c) X2095-T8, (d) 8090-T877, and (e) 8090-DA, from Ref. 56.

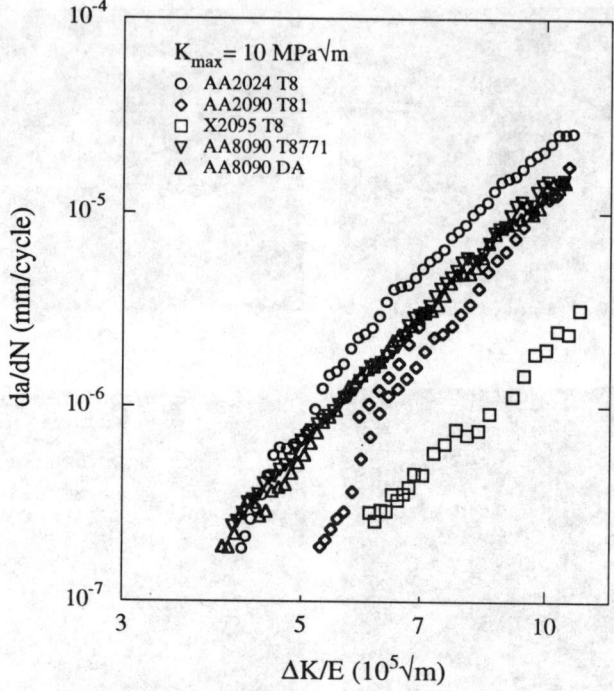

Fig. 25. Intrinsic, inert environment fatigue crack growth rates for Al-Li-Cu-Mg alloys and 2024 plotted versus $\Delta K/E$, from Ref. 56.

<u>Stress Corrosion Cracking:</u> Aluminum alloy structures loaded below σ_y and K_{IC} may still experience catastrophic failure, even if the load is not cyclic in nature. Cracks may initiate and propagate in the presence of tensile stresses and aggressive environment. As mentioned earlier, humid air qualifies as an aggressive environment for some aluminum alloys. This stress corrosion cracking (SCC) failure mode is common in 2XXX, 7XXX and certain microstructural conditions of 5XXX series alloys. The specific aluminum alloys, 7079-T6, 7075-T6, and 2024-T3, have contributed to over 90% of SCC service failures of aluminum alloys.

At this time, the exact mechanism responsible for SCC of a susceptible aluminum alloy in a particular environment remains controversial [58]. However, variations of two basic theories abound in the literature. Crack advance by anodic dissolution was first proposed by Dix [59] in 1940. In this scheme, preferential corrosion is assumed to occur along active paths, such as grain boundaries with precipitates, stimulated by stress. The second scenario, embrittlement and loss of ductility from penetration of the aluminum lattice by hydrogen atoms, was first proposed as a mechanism of SCC in aluminum alloys by Gruhl et al. [60]. There is a formidable body of evidence presented in the literature for hydrogen embrittlement (HE) as the cause of SCC [60-63]. Table 10 lists the controlling factors (microstructural and other) in the two basic theories.

Table 10. Controlling Factors in SCC Models [58]

Anodic Dissolution	Hydrogen Embrittlement
Grain boundary precipitate size, spacing, and/or volume fraction	Hydrogen absorption leading to grain boundary or transgranular decohesion
Grain boundary PFZ width, solute profile or deformation mode	Internal void formation via gas pressurization
Matrix precipitate size / distribution and deformation mode	Enhanced plasticity (adsorption and absorption arguments exist)
Oxide rupture and repassivation kinetics	

An important fact to remember is that pure aluminum does not stress corrode, and for any given system, susceptibility increases with the solute content that may be put in supersaturated solid solution [64]. This information, coupled with data and the controlling factors of the two models, suggests that microstructural alterations may influence SCC behavior for a given composition.

Wrought 7XXX series alloys are susceptible to SCC in moist air as well as aqueous environments. The alloy susceptibility is known to be sensitive to heat treatment, and therefore microstructure. However, no clear cause and effect has been established between specific microstructural features and crack velocity. Figure 26 shows v-K curves for different tempers of Al-6Zn-2Mg-XCu alloys. In the T651 tempers there is an improvement in both regions I and II with copper content. Alloys with 0% and 1% copper show only small improvements in plateau velocity, and significant strength loss with overaging, whereas alloys with 1.6% and 2.1% Cu are immune to SCC in the overaged temper. Improved SCC resistance with overaging (T7X tempers) has been attributed to a number of factors. Two of the most often cited controlling factors are matrix deformation characteristics and grain boundary precipitate size and distribution.

Sarkar et al. [65] offer the following explanation for the effects of deformation and copper content on the stress corrosion susceptibility of 7XXX alloys. Stress corrosion is a stress activated cracking process. In region I it is strongly affected by K_1; consequently, the factors that control the effective local stresses at the crack tip control the crack growth velocity in this region. Since SCC in these alloys is intergranular, the critical local stress, τ_c, necessary for a certain crack growth velocity in region I, is most likely due in part to the dislocations piled up at the grain boundary, i.e., $\tau_c = N\tau_{RSS}$ where τ_{RSS} is the resolved shear stress and N is the number of dislocations in the pile-up. In the T651 condition, the 0 pct Cu alloy exhibits an inhomogeneous planar mode of deformation and, therefore, N is very high (as predicted by Duva et al. [21]). Consequently, a high stress concentration develops at the tip of the pile-up at a high angle grain boundary and the stress corrosion cracking velocity is high. An addition of 1 pct Cu has no effect on the deformation mode; N remains similar, and stage I cracking is not affected (Fig. 26a). As the amount of copper is increased to 1.6 pct, the deformation mode changes from inhomogeneous planar slip to homogeneous looping and the effective number of dislocations in a pile-up decreases. To achieve the same local stress with fewer dislocations in the pile-up, τ_{RSS} has to be increased, i.e., the externally applied stress has to be raised. This shifts the region I line to the right in the v-K_1 graph. Increasing the copper content to 2.1 pct shifts region I further to the right because of further homogenization of deformation.

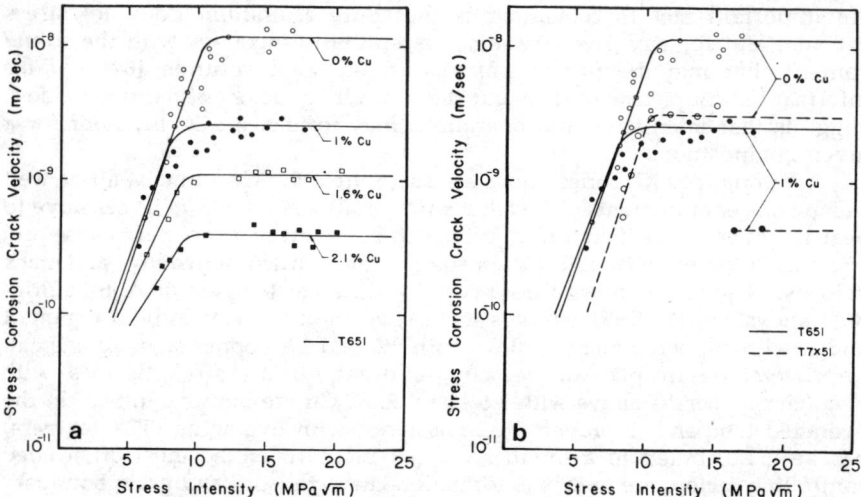

Fig. 26. Effect of stress intensity on stress corrosion crack velocity of Al-6Zn-2Mg-XCu alloys with varying copper contents in 3.5 pct. NaCl, (a) T651 temper, and (b) T7X51.

Increasing the copper content decreases the crack velocity progressively. The reduction reached two orders of magnitude for the change from 0.01 to

2.0 pct Cu. The effect can be mainly attributed to the change in the electrochemical activity of the precipitates as a function of their copper content. In the 7XXX series alloys the η phase is very active and anodic with respect to the film-covered matrix [66]. If the alloy contains copper, copper both disolves in the matrix and enters the η phase, making both more noble [67,68]. As a result, the mixed potential at the crack tip shifts to a more noble value. The decrease in the crack velocity can then be attributed to the reduced rate of dissolution of the more noble precipitates, or reduced rate of hydrogen ion reduction and hydrogen adsorption at the crack tip at the more noble potential.

The decrease in crack velocity with overaging for the 0 pct Cu alloy (Fig. 26b) may be attributed to the change of the deformation mode from inhomogeneous in the T651 condition to homogeneous in the T7X51 condition, and the corresponding reduction in the stress concentrations at the grain boundaries. For the copper-containing alloys the decrease in crack velocity may be due to the combined effects of further copper enrichment of the precipitate with overaging, and homogenization of deformation. Another possible cause of a change in crack velocity is an increase in the grain boundary area fraction covered with precipitates during overaging; this correlation was reported by Poulose, et al. [69]. In the dissolution model, assuming that the crack velocity is controlled by the dissolution of the precipitates, the larger area fraction of the anodes would result in lower anodic current density and lower crack velocity. On the other hand, in the model of cracking caused by hydrogen absorption the more negative mixed potential associated with larger anodes would be expected to result in higher rate of hydrogen ion reduction and higher crack velocity.

The results of Sarkar et al. [65] explain why copper-containing alloys such as 7075 can be made resistant to SCC by the T73 overaging treatment whereas for the low copper alloys, like 7079, a considerable amount of overaging is required with severe strength penalty to improve the stress corrosion resistance, the effect being substantial only in region I. Since 7079 contains little copper, extended overaging results in the precipitates becoming completely incoherent and the deformation mode changes from planar to homogeneous. This improves the stress corrosion behavior in the stress intensity sensitive region I but has less effect in region II because the potential difference between the precipitates and the matrix remains essentially unchanged.

The duplex aging treatment that is used to produce the T7X temper results in a 10-15% loss in strength compared to T6 [58]. A treatment referred to as retrogression and reaging (RRA) has been shown to produce T6 strength levels while imparting T7X SCC and exfoliation corrosion resistance. The proposed heat treatment schedule follows normal T6 procedures, followed by a short excursion in the 180-280°C range and reaging at a lower temperature to regain strength. Explanations for improved SCC resistance are similar to those for duplex aging [58], and at a similar level of acceptance.

Wrought Al-Mg alloys (5XXX series) are known to be quite resistant to SCC and other forms of corrosion. However, if the alloys are exposed to temperatures in the neighborhood of 40°C for a number of years (tropical climate), or 90°C-100°C for a number of days, precipitation of the beta phase (Mg_2Al_3) occurs in a semi-continuous film covering the grain boundaries [70]. This phase is anodic to the aluminum matrix, and the result is extreme susceptibility to SCC in aqueous chloride containing environments. Although anodic dissolution of the grain boundary phase was originally proposed as the mechanism for crack advance [71], Pickens et al. [72] demonstrated that HE could be the mechanism of crack advance if the cathodic half reaction was considered in the dissolution of the beta phase.

Summary: An aluminum alloy is usually selected based on its ability to provide adequate strength for an application. However, the designer must consider damage tolerance in a number of forms. Resistance to catastrophic failure due to the presence of flaws (toughness), subcritical crack growth from cyclic loading (FCP), or environmental interactions (SCC), among others, are important factors in designs that utilize high strength aluminum alloys.

Microstructural features that contribute to improved mechanical properties, as well as those that degrade alloy performance have been presented in this section. Microstructural optimization, where strength concessions are made for adequate damage tolerance, is an extremely important procedure. Some examples of this have been presented (T7X temper). In addition, examples of how adequate damage tolerance may be achieved without sacrificing strength (RRA) are also mentioned. Increased understanding of how the microstructural features influence each of the design critical properties will allow further advances in microstructural design and optimization of microstructures for applications requiring high strength and damage tolerance.

VI. PROBLEMS

1. Derive a materials selection parameter (e.g., E/ρ) for the minimum weight of a cantilever beam for a deflection limited design. Repeat the exercise if elastic deflection is not important, but yielding is to be avoided. Using the values in Table P1, how do the alloys AA7150, AA2090 and 4340 steel compare for each case? (For simplicity, assume circular cross section for the beam)

Table P1

Alloy	ρ (g/cm³)	σ_y (MPa)	K_{IC} (MPa√m)	E (GPa)	Elongation to failure (%)
AA2090	2.55	550	35	77	12
X2095	2.70	660	25	76	8
AA7150	2.82	575	30	71	12
4340 (steel)	7.84	1200	80	200	10

2. Al-Mg and Al-Li binary alloys exhibit serrated yielding at certain temperature and strain rate conditions. If the mechanism is similar to the dynamic strain aging observed in carbon steels (migration of solute atmospheres with gliding dislocations), what strain rate-temperature combinations should produce the effect in the aluminum alloys? Are the results reasonable? Explain. What effect will cold work prior to tensile testing have on the results? ($D=D_o\exp[-Q/RT]$ Li: $D_o=4.5\times10^{-4}$ m²/s Q= 139.4 kJ/mol; Mg: $D_o=4.4\times10^{-4}$ m²/s Q= 140.3 kJ/mol)

3. Calculate the spacing of hard particles required to double the yield strength of pure aluminum. What spacing would be required to produce a yield strength on the order of the theoretical strength of the aluminum crystal lattice?

4. The critical particle diameter for the transition from dislocation shearing to looping was derived for the case of order hardening ($d_c = Gb^2/\gamma$). Derive the critical diameter for the case of coherency hardening.

5. Isothermal time-temperature-transformation (TTT) and continuous cooling diagrams for precipitation processes take the form of C-curves, see Fig. 7. What two competing forces or phenomena lead to this shape? For the dissolution of $MgZn_2$ in a typical 7XXX alloy, sketch a time-temperature-dissolution (TTD) diagram.

6. Sugamata et al. predicted toughness values for AA8090 by calculating the relative contribution of failure in slip bands and grain boundaries. If the failure criteria of Hahn and Rosenfield was applied to their data, what range of toughness is predicted? [σ_y=465 MPa, E=80 GPa, K_{IC}=26 MPa√m (experimental), and 25 MPa√m (calculated by their expression)]

7. Calculate the critical flaw size for an embedded, circular crack in alloy X2095 (K_{IC} = 25 MPa√m) and AA2090 (K_{IC} = 35 MPa√m). Assume a service stress of 400 MPa in tension. If the initial crack length is 1/100 the critical flaw size, estimate the number of cycles to failure for X2095 and AA2090, assume an alternating stress of 400 MPa (Hint: use Figure 23).

Acknowledgment

The authors would like to acknowledge the significant contribution by Ms. Tana Herndon in the preparation of this chapter.

References

1. D. Altenpohl, *Aluminum, Viewed From Within* (Aluminium-Verlog, Düsseldorf, 1982) p. 1.
2. *Aluminum Standards and Data 1978 Metric SI* (Aluminum Association, Washington, D.C.).
3. E. Hornbogen and E.A. Starke, Jr., *Acta Metall. Mater.* **41** (1993) 1.
4. E.A. Starke, Jr., E. Hornbogen and C.P. Blankenship, Jr., in *Aluminum Alloys - Their Physical and Mechanical Properties* (ICAA3), ed. L. Arnberg, O. Lohne, E. Nes and N. Ryum, Vol. III (NTH-SINTEF, Trondheim, Norway, 1992) p. 279.
5. A.S. Argon, J. Im and R. Safoglu, *Metall. Trans.* **6A** (1975) 625.
6. I.J. Polmear, *Light Alloys - Metallurgy of the Light Metals* (Edward Arnold, London, 1989).
7. N. Ryum, in *Aluminum Alloys - Their Physical and Mechanical Properties* (ICAA1), ed. E.A. Starke, Jr., and T.H. Sanders, Jr. (EMAS, West Midlands, England, 1986) p. 1511.
8. L. Ives, L. Swarlyengruber, W. Boettinger, M. Rosen, S. Ridder, F. Biancaniello, R. Reno, D. Ballard and R. Mehrabian, Report NBS 83-2669 (National Bureau of Standards, Washington, D.C., 1983).
9. F. W. Gayle, W.T. Tack, F.H. Heubaum and J.R. Pickens, in *Aluminum-Lithium* (6th International Aluminum-Lithium Conference), ed. M. Peters and P-J. Winkler, Vol. 1 (DMG Informationsgesellschaft, Verlog, Germany, 1992) p. 203.
10. C.P. Blankenship, Jr., and E.A. Starke, Jr., "Structure-Property Relationships in Al-Li-Cu-Mg-Ag-Zr Alloy X2095," *Acta Metall. Mater.* (in press).
11. N. Dudzinski, et al., *J. Inst. Metals* **74** (1947-48) 291.
12. J.T. Staley, "Fracture Toughness and Microstructure of High-Strength Aluminum Alloys," AIME, Institute of Metals Division, Spring Meeting, May, 1974.

13. J.T. Staley, "Microstructure and Toughness of High Strength Alloys," ASTM Symposium on Properties Related to Toughness, Montreal, Canada, June 22-27, 1975.
14. D.S. Thompson and S.A. Levy, AFML-TR-70-171 (WPAFB, 1970).
15. R.F. Ashton, D.S. Thompson, and F.W. Gayle, *Aluminum Alloys - Their Physical and Mechanical Properties*, ed. E.A. Starke, Jr., and T.H. Sanders, Jr. (EMAS, West Midlands, UK, 1986) p. 403.
16. W.E. Quist, M.V. Hyatt and W.E. Anderson in *Properties Related to Fracture Toughness*, ASTM STP 605 (1975) 96.
17. J.R. Low, Jr., R.H. Van Stone, and R.H. Merchant, NASA Tech. Rep. No. 2, NGR 38-087-003, CMU 1972.
18. G.T. Hahn and A.R. Rosenfield, *Metall. Trans.* **6A** (1975) 653.
19. J. Gurland, *Acta Metall.*.**20** (1972) 735.
20. R.H. Van Stone and J.A. Psioda, *Metall. Trans.* **6A** (1975) 668.
21. J.M. Duva, M.A. Daeubler, E.A. Starke, Jr., and G. LÜtjering, *Acta Metall.* **36** (1988) 585.
22. E. Hornbogen, *Z. Metallk.* **66** (1975) 511.
23. S.P. Lynch, *Mat. Sci. Engr.* **136** (1991) 25.
24. S.P. Lynch, R. Byrnes and R.B. Nethercott, in *Aluminum-Lithium*, ed. M. Peters and P.J. Winkler (DGM, Oberursel, FRG, 1992) p. 391.
25. C.P. Blankenship, Jr. and E.A. Starke, Jr., *Metall. Trans.* **24A** (1993) 833.
26. S.C. Jha, T.H. Sanders, Jr., and M.A. Dayananda, *Acta Metall.* **35** (1987) 473.
27. K.V. Jata and E.A. Starke, Jr., *Metall. Trans.* **17A** (1986) 1011.
28. H.J. Roven, *Scripta Metall. et Mater.* **26** (1992) 1383.
29. G.M. Ludtka and D.E. Laughlin, *Metall. Trans.* **13A** (1982) 411.
30. M. Sugamata, C.P. Blankenship, Jr., and E.A. Starke, Jr., *Mater. Sci. Engng.* **A163** (1993) 1.
31. A.K. Vasudevan and R.D. Doherty, *Acta Metall.* **35** (1987) 1193.
32. P.N.T. Unwin and G.C. Smith, *J. Inst. Metal* **97** (1969) 299.
33. B.I. Sandor, *Fundamentals of Cyclic Stress and Strain* (University of Wisconsin Press, Madison, 1972).
34. E.A. Starke, Jr., and G. LÜtjering in *Fatigue and Microstructures*, ed. M. Meshii (American Society of Metals, 1979) p. 205.
35. C. Laird, *Mater. Sci. Eng.* **25** (1976) 187.
36. M.E. Fine and J.S. Santner, *Scripta Metall.* **9** (1975) 1239.
37. C. Calabrese and C. Laird, *Mater. Sci. Eng.* **13** (1974) 141.
38. A. Abel and R.K. Ham, *Acta Metall.* **14** (1966) 1495.
39. J.W. Martin, *Micromechanisms in Particle-Hardened Alloys* (Cambridge University Press, Cambridge, UK, 1979).
40. C. Laird and G.C. Smith, *Phil. Mag.* **7** (1962) 847.
41. R.E. Stolz and R.M. Pelloux, *Metall. Trans.* **7A** (1976) 1295.

42. N.E. Prasad, G. Malakondaiah, K.N. Raju, and P.R. Rao, in *Advances in Fracture Research*, Vol. 2 of Proc. of IFC7, ed. K. Salama, K. R. Shandar, D.M.R. Taplin, and P.R. Rao (Pergamon, London, 1989) p. 1103.
43. G. LÜtjering, in ref [29] 142.
44. J.C. Grosskreutz and G.C. Shaw in *Fracture 1969*, ed. P.L. Pratt (Pergamon, Oxford, 1969) p. 620.
45. C.Y. Kung and M.E. Fine, *Metall. Trans.* **10A** (1979) 603.
46. P.C. Paris and F. Erdogen, *J. Basic Engng.* **85** (1963) 528.
47. T.C. Lindley, C.E. Richards, R.O. Ritchie, in *Proceedings of Conference on Mechanics and Physics of Fracture* (Inst. of Physics, Metals Soc., Cambridge, 1975).
48. E.A. Starke, Jr., and J.C. Williams in *Fracture Mechanics: Perspectives and Directions*, ASTM STP 1020, ed. R.P. Wei and R.P. Gangloff (ASTM, Philadelphia, 1989) p. 184.
49. R.S. Piascik, Ph.D. Dissertation, University of Virginia (1989).
50. R.D. Carter, E.W. Lee, E.A. Starke, Jr., and C.J. Beevers, *Metall. Trans.* **15A** (1984) 555.
51. K.T.V. Rao, W. Yu, R.O. Ritchie, *Metall. Trans.* **19A** (1988) 549.
52. M. Peters, V. Backmann, K. Welpmann, in *Aluminum-Lithium Alloys IV*, ed. G. Champier, B. DuBost, D. Miannay, and L. Sabetay (Les Editions Physique, Les Ulis, France, 1987) p. C3-785.
53. A.K. Vasudevan, P.E. Bretz, A.C. Miller, and S. Suresh, *Mater. Sci. Eng.* **64** (1984) 113.
54. K.T.V. Rao and R.O. Ritchie, *Int. Mater. Rev.* **37** (1992) 153.
55. R.S. Piascik and R.P. Gangloff, *Metall. Trans.* **22A** (1991) 2415.
56. D.C. Slavik, C.P. Blankenship, Jr., E.A. Starke and R.P. Gangloff, *Metall. Trans.* **24A** (1993) 1807.
57. M.O. Speidel, in *High Temperature Materials in Gas Turbines*, ed. P.R. Sahm and M.O. Speidel (Elsevier, Amsterdam, 1974) p. 207.
58. N.J.H. Holroyd, A.K. Vasudevan, and L. Christodoulou in *Aluminum Alloys -- Contemporary Research and Applications*, eds. A.K. Vasudevan and R.D. Doherty, Treatise on Materials Science and Technology, **31** (Academic Press, New York, 1989) p. 463.
59. E.M. Dix, *Trans. AIME* **137** (1940) 11.
60. W. Gruhl, *Z. Metallkunde* **75** (1984) 819.
61. G.M. Scamans, *J. Mater. Sci.* **13** (1978) 27.
62. G.M. Scamans, *Metall. Trans.* **11A** (1980) 846.
63. R.J. Gest and A.R. Troiano, *Corrosion* **30** (1974) 274.
64. M.O. Speidel, *Metall. Trans.* **6A** (1975) 631.
65. B. Sarkar, M. Marek, and E.A. Starke, Jr., *Metall. Trans.* **12A** (1981) 1939.
66. E.H. Dix, R.H. Brown, and W.W. Binger, *Metals Handbook*, 8th ed. T. Lyman, Vol. 1 (ASM, Metals Park, OH, 1961) p. 916.
67. G.C. English and E.H. Hollingsworth, Report No. W-66-029-C, Alcoa Research Laboratories (New Kensington, PA, May 1968).

68. J. Busby, J.F. Cleave, and R.L. Cudd, *J. Inst. Metals* **99** (1971) 41.
69. P.K. Poulose, J.E. Morral, and A.J. McEvily, *Metall. Trans.* **5** (1974) 1393.
70. J.R. Pickens, J.S. Ahearn, R.O. England and D.C. Cooke, in *High Strength Powder Metallurgy Alloys II*, ed. G.J. Hildeman and M.J. Koczak (TMS-AIME, Warrendale, 1985) p. 105.
71. U.R. Evans, in *Stress Corrosion Cracking and Embrittlement*, ed. W.D. Robertson (John Wiley and Sons, NY, 1956).
72. J.R. Pickens, J.R. Gordon, and J.A.S. Green, *Metall. Trans.* **14A** (1983) 925.
73. E.A. Brandes and G.B. Brook, ed., *Smithells Metals Reference Book*, 7th edition (Butterworth, Oxford, 1992) p. 13-15.
74. J.E. Hatch, ed., *Aluminum: Properties and Physical Metallurgy* (ASM, Metals Park, OH, 1984) p. 205.

Chapter 2

NICKEL-BASE SUPERALLOYS

Norman S. Stoloff
Materials Engineering Department
Rensselaer Polytechnic Institute
Troy, New York 2180-3590

INTRODUCTION

This chapter deals with the physical and mechanical metallurgy of nickel-base superalloys. These are the most complex and effective of all nickel alloys, allowing their use at elevated temperatures under extreme environmental conditions. In fact, nickel-base superalloys are used at the highest homologous temperatures of any structural alloys. The development of nickel base superalloys has paced the construction of larger, more powerful and more fuel efficient aircraft and industrial gas turbines. When aircraft gas turbines were first developed during the 1930's in Great Britain and in Germany, available alloys were very limited in their temperature capabilities. Beginning about 1940, however, there has been a steady increase in maximum use temperature and strength levels of nickel base alloys, brought about by a combination of compositional, microstructural and processing advances. The inability to forge the strongest nickel base alloys available in the 1950's led to the introduction of investment casting (and vacuum melting), resulting in a step-wise increase in temperature capability. The pattern of superalloy improvements may clearly be seen in Fig. 1 [1]; note that modern superalloy single crystals are capable of operating at temperatures near 1100C, while early aircraft turbines had maximum use temperatures near 500C. The use of nickel alloys has expanded also to parts of reciprocating engines (turbochargers, exhaust valves) as well as to hot working tools and dies and bolting for steam generators.

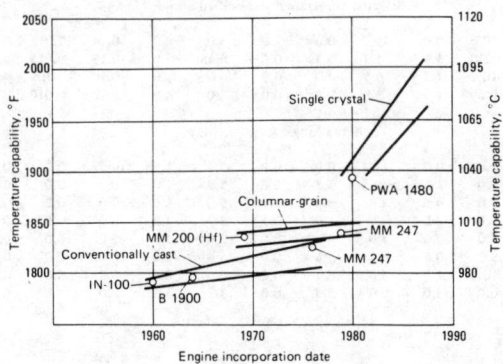

Fig. 1. Advances in temperature capability of turbine materials and processes since 1960.[1]

COMPOSITION AND MICROSTRUCTURE

Nickel is an excellent alloy base for service under extreme conditions because of the high stability of the face centered crystal structure, the high solubility of many alloying elements and the ability to strengthen by numerous means, both direct and indirect. For the most part, elevated temperature strength is obtained through a combination of solid solution hardening and precipitation of ordered intermetallic phases, with oxide particles or carbides conferring additional benefits. The most important alloying elements in nickel are aluminum, for both strength and oxidation resistance, and chromium for hot corrosion resistance. Unfortunately, the beneficial effects of chromium on environmental resistance are offset by weakening of the alloy. Therefore, while early superalloys contained up to 20%Cr, modern alloys are more likely to contain about 10%Cr. Since many superalloys contain eight or more elements, each with specific functions with respect to strength, alloy stability or environmental resistance, tradeoffs usually are necessary. The chemistry of nickel alloys will first be described, followed by a detailed review of strengthening mechanisms and surface protection schemes.

TABLE 1. SELECTED NICKEL BASE ALLOYS

Alloy	Ni	Cr	Co	Mo	W	Ta	Cb	Al	Ti	C	B	Zr	Others
Selected Equiaxed Superalloys													
Alloy713C	Bal.	12.5	0.0	4.2	0.0	0.0	2.0	6.1	0.8	0.12	0.012	0.10	
Alloy713LC	Bal.	12.0	0.0	4.5	0.0	0.0	2.0	5.9	0.6	0.05	0.010	0.10	
B-1900	Bal.	8.0	10.0	6.0	0.0	4.0	0.0	6.0	1.0	0.10	0.015	0.10	
GTD111*	Bal.	14.0	9.5	1.5	3.8	2.8	0.0	3.0	4.9	0.10	0.010	-	
IN-100	Bal.	10.0	15.0	3.0	0.0	0.0	0.0	5.5	4.7	0.18	0.014	0.06	1.0V
IN738LC*	Bal.	16.0	8.5	1.7	2.6	1.7	0.9	3.4	3.4	0.11	0.010	0.05	
IN 939*	Bal.	22.5	19.0	0.0	2.0	1.4	1.0	1.9	3.7	0.15	0.009	0.09	
IN 792	Bal.	12.4	9.0	1.9	3.8	3.9	0.0	3.1	4.5	0.12	0.020	0.10	
Rene 80	Bal.	14.0	9.5	4.0	4.0	0.0	0.0	3.0	5.0	0.17	0.015	0.03	
Rene 220	Bal.	18.0	12.0	3.0	-	3.0	-	0.5	1.0	0.02	0.01	0.01	
Selected Directionally solidified alloys													
P&W 1422	Bal.	9.0	10.0	0.0	12.0	0.0	1.0	5.0	2.0	0.14	0.015	0.10	1.5Hf
R142	Bal.	6.8	12.0	1.5	4.9	6.35	0.0	6.15	0.0	0.12	0.015	0.02	2.8Re,1.5Hf
P&W 1426	Bal.	6.5	10.0	1.7	6.5	4.0	0.0	6.0	0.0	0.10	.015	0.10	1.5Hf,3.0Re
GTD 111*	Bal.	14.0	9.5	1.5	3.8	2.8	0.0	3.0	4.9	0.10	0.010	-	
Selected Single Crystal Alloys													
P&W 444	Bal.	8.6	0.0	0.0	11.1	0.0	0.0	5.1	1.98	0.0	0.0	0.0	
NASAIR 100	Bal.	9.0	0.0	1.0	10.5	3.3	0.0	5.75	1.0	0.0	0.0	0.0	
CMSX-2	Bal.	8.0	5.0	0.6	8.0	6.0	0.0	5.6	1.0	0.0	0.0	0.0	
GE N4	Bal.	9.0	8.0	2.0	6.0	4.0	0.5	3.7	4.2	0.0	0.0	0.0	
P&W 1480	Bal.	10.0	5.0	0.0	4.0	12.0	0.0	5.0	1.5	0.0	0.0	0.0	
CMSX-4	Bal.	6.5	9.0	0.6	6.0	6.5	0.0	5.6	1.0	0.0	0.0	0.0	3.0Re,0.1Hf
P&W 1484	Bal.	5.0	10.0	2.0	6.0	8.7	0.0	5.6	0.0	0.0	0.0	0.0	3.0Re,0.1Hf
SC16*	Bal.	16.0	0.0	3.0	0.0	3.5	0.0	3.5	3.5	0.0	0.0	0.0	

*Primarily Industrial Gas Turbine Alloys

The compositions of several representative nickel-base alloys are listed in Table 1[2]. They fall into three categories: equiaxed, directionally solidified and single crystals. Typically, each solute element serves more than one function. In addition, deleterious elements such as silicon, phosphorus, sulfur, oxygen, and nitrogen must be controlled through appropriate melting practice. Other trace elements (e.g. selenium and lead) are held to very small (ppm) levels in critical parts. Most nickel-base superalloys contain 10-20% chromium for surface protection, up to about 8% aluminum and titanium combined for strength, 5-10% cobalt and small amounts of boron, zirconium and carbon. the latter three elements provide improved resistance of grain boundaries to fracture at elevated temperatures. Other common additions are molybdenum, niobium, tungsten, tantalum, and hafnium, all of which play dual roles as strengthening solutes and carbide formers. Chromium and aluminum, in addition, are necessary to improve surface stability, through the formation of Cr_2O_3 and Al_2O_3 respectively. The functions of the various elements are summarized in Table 2. The achievement of desired properties in superalloys is largely dependent upon the amount, size and shape of γ', grain size and shape and carbide distribution. In polycrystals the presence of boron or zirconium at the grain boundaries provides additional strength at high temperatures.

TABLE 2

EFFECTS OF SEVERAL ELEMENTS IN NI-BASE SUPERALLOYS

Element	Effect(a)
Chromium	Oxidation and hot corrosion resistance; solid solution strengthening
Molybdenum; tungsten	Solid solution strengthening; form M_6C Carbides
Aluminum; titanium	Form γ', $Ni_3(Al,Ti)$, hardening precipitate Ti forms Mc carbides as well; Al enhances oxidation resistance
Cobalt	Raises γ' solvus temperature
Boron, zirconium	Improve rupture life through increases in ductility; B also forms borides;
Carbon	Forms Mc, M_7C_3, $M_{23}C_6$ carbides
Niobium	Forms γ' Ni_3Nb, hardening precipitate; forms δ orthorhombic Ni_3Nb; forms Mc carbides
Tantalum	Solid solution strengthening; forms Mc carbides; enhances oxidation resistance
Rhenium	Improves creep strength, environmental resistance. May partition $+0\gamma'$
Hafnium	Forms MC carbides, improves grain boundary ductility; increases vol. fraction of γ/γ' eutectic, improves oxidation resistance, improves oxidation and hot corrosion resistance.

(a) Not all these effects necessarily occur in a given alloy

The major phases that may be present in nickel-base alloys are summarized in Table 3[3] and are described in more detail below:

TABLE 3

Some Constituents Observed In Wrought Heat-Resistant Alloys

Phase	Crystal structure	Lattice parameter, nm	Typical Formula(s)
γ'	fcc (ordered $L1_2$)	0.3561 for pure Ni_3Al to 0.3568 for $Ni_3(Al_{0.5}Ti_{0.5})$	Ni_3Al $Ni_3(Al,Ti)$
η	hcp (DO_{19})	$a_o = 0.5093$ $c_o = 0.8276$	Ni_3Ti
γ''	bct (ordered DO_{22})	$a_o = 0.3624$ $c_o = 0.7406$	Ni_3Nb
$Ni_3Nb(\delta)$	orthorhombic (ordered Cu_3Ti)	$a_o = 0.5106-0.511$ $b_o = 0.421-0.4251$ $c_o = 0.452-0.4556$	Ni_3Nb
MC	fcc	$a_o = 0.430-0.470$	TiC NbC
$M_{23}C_6$	fcc	$a_o = 1.050-1.070$ (varies with composition)	$Cr_{23}C_6$ $(Cr,Fe,W,Mo)_{23}C_6$
M_6C	cubic	$a_o = 1.085-1.175$	Fe_3Mo_3C
M_3B_2	tetragonal	$a_o = 0.560-0.620$ $c_o = 0.300-0.330$	Ta_3B_2 Nb_3B_2
Laves	hexagonal	$a_o = 0.475-0.495$ $c_o = 0.770-0.815$	Fe_2Nb
σ	tetragonal	$a_o = 0.880-0.910$ $c_o = 0.450-0.480$	FeCr FeCrMo

1. <u>Gamma matrix</u> (γ). The continuous matrix is an fcc nickel-base phase which usually typically contains a high percentage of solid-solution elements such as chromium, iron, cobalt, rhenium, molybdenum, and tungsten. All nickel-base alloys contain this phase as the matrix.

2. **Gamma prime** (γ'). Aluminum and titanium are added in amounts required to precipitate ordered fcc γ', (Ni$_3$Al, Ti) which precipitates coherently with the austenitic gamma matrix. Other elements, notably niobium, chromium, rhenium and tantalum also enter γ'. This phase is required for high temperature strength and creep resistance.

3. **Gamma double prime** (γ"). Nickel and niobium combine, in the presence of iron, to form bct Ni$_3$Nb, which is coherent with the gamma matrix while inducing large mismatch strains. This phase provides very high strength at low to intermediate temperatures, but is unstable at temperatures above about 815°C. Tantalum may replace a portion or all of the niobium for improved properties[4,5].

4. **Carbides**. Carbon, added in amounts of about 0.05 - 0.2w%, combines with reactive elements such as titanium, tantalum and hafnium to form MC carbides. During heat treatment and service these tend to decompose and generate other carbides such as $M_{23}C_6$, and/or M_6C, which ten to form at grain boundaries. Carbides are formed in all superalloys except single crystals; they provide only indirect strengthening.

5. **Grain boundary γ'**. In the stronger alloys, heat treatments and service exposure produce a film of γ' along the grain boundaries; this is believed to improve rupture properties.

6. **Borides**. Boron segregates to grain boundaries, resulting in the formation of a relatively low density of boron particles.

7. **Topologically close packed phases**. For some compositions, and under certain conditions, platelike phases such as σ, μ, and Laves may form; these cause lowered rupture strength and ductility.

In nickel-base superalloys, most of the above phases (except γ") are usually present. Several Ni-Fe superalloys such as IN 706 and IN 718, on the other hand, contain γ" Ni$_3$Nb as the principal precipitate, as well as γ'. Further, oxide dispersion strengthened alloys contain a few vol % of a dispersed phase such as Y_2O_3 in a gamma-gamma prime matrix, while composites (mechanically incorporated) will contain tungsten or tungsten-alloy fibers in a gamma-gamma prime matrix.

The Gamma Matrix (γ)

The elastic moduli and diffusivity of nickel, two factors that promote creep-rupture resistance, are not remarkable. However, the gamma matrix is readily strengthened for the most severe temperature and time conditions. Some superalloys can be utilized at 0.85 T_M (melting point), and for times up to 100,000 utilized at 0.85 T_M (melting point), and for times up to 100,000 hrs. at somewhat lower temperatures. These conditions can be met because of[6]:

1. The high tolerance of nickel for solutes without phase instability, due to its nearly filled d-shell.
2. The tendency, with chromium additions, to form Cr_2O_3 having few cation vacancies, thereby restricting the diffusion rate of metallic elements outward and oxygen, nitrogen and sulfur inward.
3. The additional tendency, at high temperatures, to form a surface film of Al_2O_3, which displays outstanding resistance to oxidation. Other elements, notably Ta, Re and Y, promote environmental resistance.

Gamma Prime

Gamma prime is an ordered intermetallic compound of nominal composition Ni_3Al, stable over a relatively narrow range of compositions, Fig. 2. It precipitated as spheroidal particles at volume fractions of less than 25% in early nickel base alloys, (see Fig. 3a)[7]. Later, cuboidal precipitates were noted in

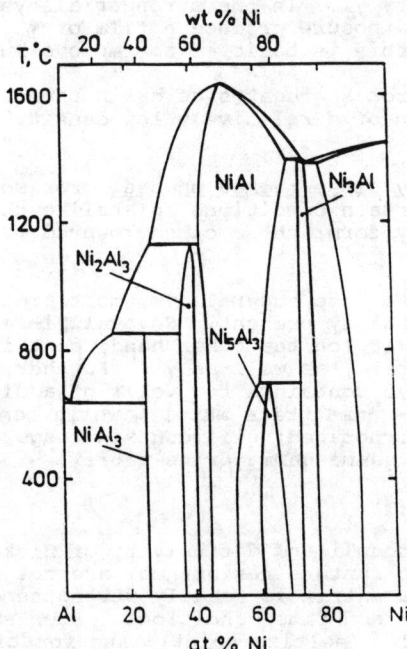

Fig. 2. Ni-Al Phase Diagram.

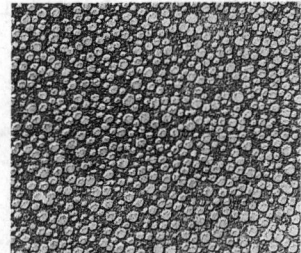

b. Typical Spherical and Cooling γ' in S-R Tested U500. 5,450X

c. Typical Cubical γ' in IN-100 S-R Tested at 1500F (815C) 13,625X
Mihalisin[14]

Fig. 3. a) spheroids in Udimet 500, typical of early, low f alloys[7]   b) Cuboids IN 100, typical of later, high f alloys[7]

alloys with higher Al+Ti contents, (Fig. 3b)[7]. The change in morphology is related to the matrix-precipitate mismatch. It is observed that γ' precipitates as spheres at 0-0.2% mismatches, becomes cuboidal for mismatches of 0.5-1%, and is plate-like at mismatches above about 1.25%. Modern cast superalloys contain 65-70% of γ', which usually is cuboidal.

We now consider the structure and properties of the γ' phase. Gamma prime is a superlattice, possessing the Cu_3Au ($L1_2$)-type structure, which exhibits long range order to its melting point of 1385°C. Although Ni_3Al exists over a fairly narrow range of composition, alloying elements may substitute liberally for either of its constituents, e.g. see Fig. 4[8]. In particular, most nickel-base alloys are strengthened by a precipitate in which up to 60% of the aluminum can be substituted for by titanium and/or niobium. Further, nickel sites in the Ni_3Al may be occupied by iron or cobalt atoms.

Unalloyed Ni_3Al deforms similarly to fcc metals, by {111} <110> slip at low temperature, and at temperatures above 400°C some slip along {100} also is observed; at 700°C {100} slip predominates[9].

Three types of stacking faults exist in the $L1_2$ structure[6]

1. Superlattice (intrinsic or extrinsic) faults
2. Antiphase boundary faults
3. Complex faults

Deformation of Ni-base alloys is affected to a great degree by the large numbers of faults induced by deformation. It will be shown in a later section that several precipitation-hardening models predict a very sensitive, dependence of critical resolved shear stress (CRSS) upon the ordering (antiphase boundary) energy of γ'. Also, creep properties of Mar-M 200 have been related to the nature of the faults in γ' left by the passage of dislocations through the particles.

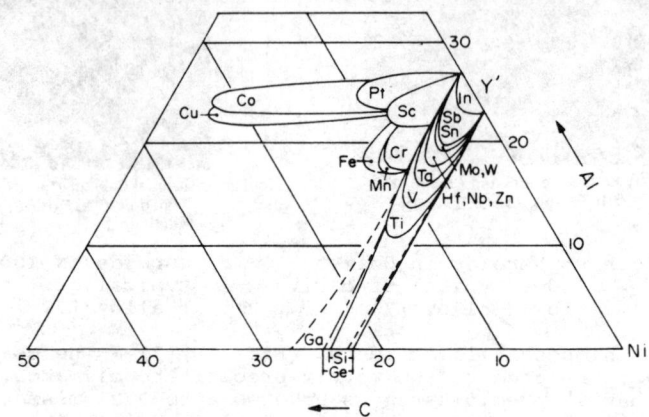

Fig. 4   Solubility of various elements in Ni_3Al at approximately 1100°C (2100°F).[8]

Both single crystals and polycrystals of unalloyed γ' exhibit a striking, reversible [10] increase in flow stress between -196°C and about 800°C (which is highly dependent on solute content as shown in Fig. 5[9]. While other superlattices exhibit a modest peak in strength over a rather narrow temperature range near T_c, the critical temperature for ordering, such peaks often are connected with a change in the degree of order with temperature [11]. However, several other superlattices: Ni_3Si, Co_3Ti, Ni_3Ge, Ni_{3Ga}, all of $L1_2$ structure, display increasing strength over a temperature range comparable to that of Ni_3Al[12]. The flow stress in all of these is fully reversible upon changing temperature.

The magnitude and temperature position of the peak in flow stress of γ' may be shifted by alloying elements such as titanium, chromium and niobium; these and other solutes dissolve readily in γ'; see Fig. 4[8]. There is no simple relation between the magnitude of flow stress increase and the change in the temperature of the peak. Tantalum, niobium, and titanium are effective solid-solution hardeners of γ' at room temperature. Tungsten, silicon and molybdenum are strengtheners at both room and elevated temperatures, while cobalt does not solid-solution strengthen γ'. All substitutional elements in Ni_3Al single crystals (Mo, Ta, Nb, Ti, W) increase the critical resolved shear stress (CRSS) for (111) {101} slip, but decrease the CRSS for (001) [110] slip, relative to binary Ni_3Al[13].

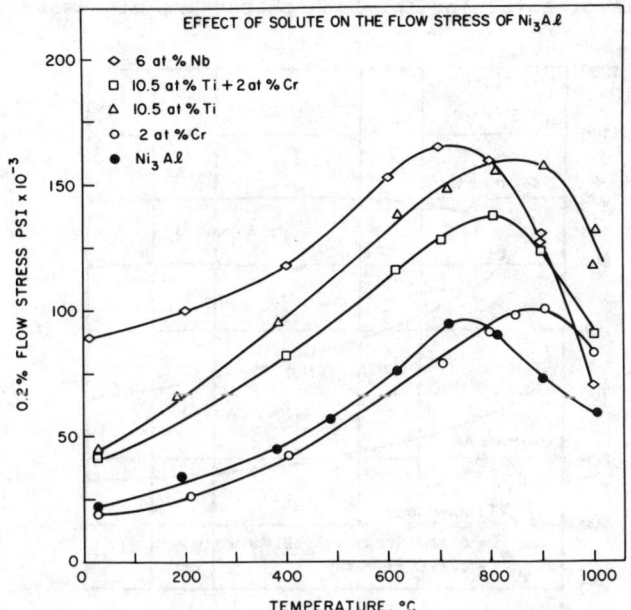

Fig. 5 Flow stress peak in gamma prime, and influence of several solutes.[9]

Gamma Double Prime

Gamma double prime, (γ''), is a bct coherent precipitate of composition Ni_3Nb that precipitates in Ni-Nb-Fe base alloys such as IN 706 and IN 718 and in Ni-Nb-Ta-Co alloy Rene´ 220[4,5]. In the absence of iron or cobalt or at temperatures and times shown in the transformation diagram of Fig. 6[14,15], an orthhombic precipitate of the same Ni_3Nb composition (delta phase) forms instead. The latter is invariably incoherent and does not confer strength. therefore, careful heat treatment is required to insure precipitation of γ'' instead of δ.

Gamma double prime often precipitates together with γ' in IN 718, but γ'' is the principal strengthening phase under such circumstances. Unlike γ', which causes strengthening through the necessity to disorder the particles as they are sheared, γ''

strengthens by virtue of high coherency strains in the lattice. Specific models of hardening for both phases are discussed in a later section.

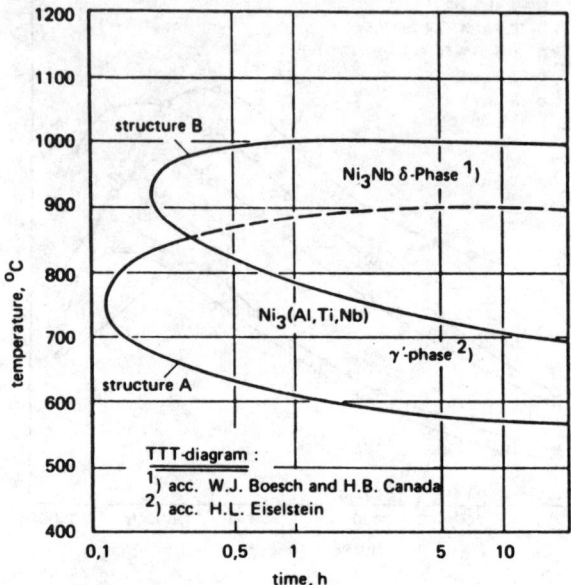

Fig. 6. Transformation diagram for INCO 718 as hot rolled bar; typical commercial heat treatment is indicated.[14,15]

The high strength and ability to process IN 718 by a variety of techniques are two of the factors that have made this the most widely used nickel based superalloy. Several successful compositional modifications of this alloy have been reported. Tantalum has been used to partially or completely replace Nb (eg alloys Rene´ 220C and Rene´ 220W which contain 1-3% Ta and 2-5% Nb)[4,5]. Iron, which acts as a catalyst to produce tetragonal γ'', may be replaced by cobalt[4,5].

Carbides

Carbides serve a number of functions in polycrystalline superalloys. First, carbides often precipitate at grain boundaries in nickel alloys, while in cobalt and iron superalloys intragranular sites are common. Early work suggested that some boundary carbides were detrimental to ductility, but most investigators now believe that carbides exert a beneficial effect on rupture strength at high temperatures. Carbon and other grain boundary strengtheners (boron, zirconium, hafnium) are not needed in single crystals and are therefore, omitted.

Carbide Types and Typical Morphologies

The common nickel-base alloy carbides are MC, M_6C and $M_{23}C_6$. MC usually exhibits a coarse random cubic or script morphology,

$M_{23}C_6$ is found primarily at grain boundaries, it usually occurs as irregular discontinuous blocky particles, although plates and regular geometric forms have been observed. M_6C also can precipitate in blocky form in grain boundaries, and less often in a Widmanstatten intragranular morphology. Although data are sparse, it appears that continuous grain boundary $M_{23}C_6$ and Widmanstatten M_6C are to be avoided for best ductility and rupture life.

MC carbides, fcc in structure, typified by TiC, TaC and HfC usually form in superalloys during solidification; they are distributed heterogeneously throughout the alloy, both in intergranular and transgranular positions, often interdendritically. Little or no orientation relationship with the alloy matrix has been noted. MC carbides are a major source of carbon for subsequent phase reactions (formation of other carbides) during subsequent heat treatment and service.

MC carbides, e.g., TiC and HfC, are very stable compounds. Their preferred order of formation in superalloys is: HfC, TaC, NbC, and TiC in order of decreasing stability. This order is not the same as that of thermodynamic stability, which is HfC, TiC, TaC and NbC, see Fig. 7[16]. Note that the MC carbides are much more stable than are M_7C_3 and $M_{23}C_6$. In these carbides M atoms can readily substitute for each other, as in (Ti,Nb)C. However, the less reactive elements, principally molybdenum and tungsten, also can substitute in these carbides. For example, (Ti,Mo)C is found in U-500, M-252, and Rene´ 77[7]. It appears that the change in stability order cited above is due to the molybdenum or tungsten substitution, which so weakens the binding forces in MC carbides that degeneration reactions, discussed below, can occur. Typically, $M_{23}c6$ and M_6C type carbides are the most stable in the alloys after heat treatment and/or service. Additions of niobium and tantalum tend to counteract this effect. Alloys with high niobium and tantalum contents (e.g. Mar M200) contain MC carbides that do not break down easily during solution treatment in the range 1200-1260°C.

$M_{23}C_6$ carbides readily form in alloys with moderate to high chromium content. They form during lower-temperature heat treatment and service, that is in the range 760-980°C, both from degeneration of MC carbides and from soluble residual carbon in the alloy matrix. Although usually seen at grain boundaries, Trans they occasionally occur along twin bands, stacking faults, and at function of temperatuure fdor severa; carbides twin ends[7]. $M_{23}C_6$ carbides have a complex cubic structure, which, if the carbon atoms were removed, would closely approximate the structure of the TCP σ phase. In fact, σ plates often nucleatestrength. Therefore, careful heat treatment is required to on $M_{23}C_6$ particles.

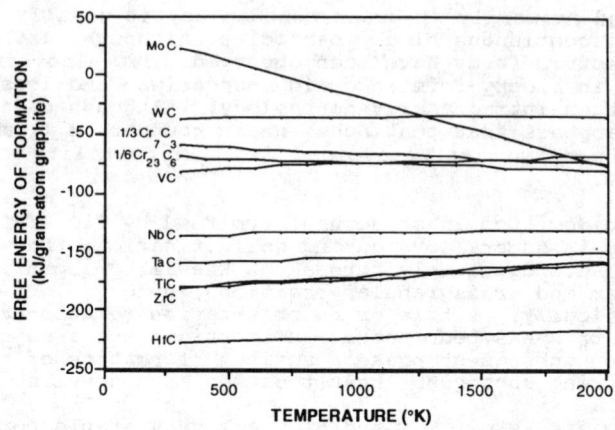

Fig. 7. Standard Gibbs Free Energy of Formation as a Function of Temperature for Several Carbides.[16]

When tungsten or molybdenum is present, the approximate composition of $M_{23}C_6$ is $Cr_{21}(Mo,W)_2C_6$ although it also has been shown that appreciable nickel can substitute in the carbide; it is possible also that small amounts of cobalt or iron could substitute for chromium[6].

$M_{23}C_6$ particles strongly influence properties of nickel alloys. Rupture strength is improved, apparently through inhibition of grain-boundary sliding. Discrete, blocky $M_{23}C_6$ particles are preferred. Eventually, however, failure can initiate either by fracture of particles or by decohesion of the carbide-matrix interface. In some alloys, cellular structures of $M_{23}C_6$ have been noted; these can cause premature failures, but can be avoided by control of chemistry or by proper heat treatment.

M_6C carbides have a complex cubic structure; they form when the molybdenum and/or tungsten content is more than 6-8 a/o, typically in the range 815-980°C. M_6C forms with $M_{23}C_6$ in Mar-M 200, B-1900 and Rene´ 80. Typical formulas for M_6C are $(Ni,Co)_3Mo_3C$ and $(Ni,Co)_2W_46$, although a wider range of compositions has been reported for Hastelloy X. Therefore, M_6C carbides are formed when molybdenum or tungsten acts to replace chromium in other carbides; unlike more rigid $M_{23}C_6$, the composition can vary widely. Since M_6C carbides are stable at higher temperature than $M_{23}C_6$ carbides, M_6C is more effective as a grain boundary precipitate to control grain size during processing of wrought alloys.

Carbide Reactions. MC carbides are a major source of carbon in most polycrystalline nickel-base superalloys below 980°C. However, MC decomposes slowly during heat treatment and service, releasing carbon for several important reactions.

The principal carbide reaction in many alloys is believed to be the formation of $M_{23}C_6$, as follows:

$$MC + \gamma \rightarrow M_{23}C_6 + \gamma'$$
or
$$(Ti,Mo)C + (Ni,Cr,Al,Ti) \rightarrow Cr_{21}Mo_2C_6 + Ni_3(Al,Ti) \qquad (1)$$

This equation cannot be balanced thermodynamically; it was assumed by metallographic observations of phase transformations at grain boundaries by Sims[17]. Reaction (1) begins to occur at about 980°C and has been observed at temperatures as low as approximately 760°C. In a few cases, it has been found to be reversible. M_6C can form in a similar manner.

Also, M_6C and $M_{23}C_6$ interact, forming one from the other[6],

$$M_6C + M' \rightarrow M_{23}C_6 + M'' \qquad (2)$$
or
$$Mo_3(Ni,Co)_3C + Cr \rightleftharpoons Cr_{21}Mo_2C_6 + (Ni,Co,Mo) \qquad (3)$$

depending on the alloy. For example, René 41 and M-252 can be heat treated to generate MC and M_6C initially; long-time exposure then causes conversion of M_6C to $M_{23}C_6$. M_6C can be formed from $M_{23}C_6$[18]. The type of refractory metal atoms present may control the reaction.

These reactions lead to carbide precipitation in various locations, but typically at grain boundaries. Perhaps the most beneficial reaction, and that controlled in many heat treatments, is reaction (3). Both the blocky carbides and the γ' produced are important, in that they may inhibit grain-boundary sliding; in any case, the γ' generated by this reaction coats the carbides and the grain boundary with a relatively ductile, creep-resistant layer.

Carbon also is in solution; at temperatures of 595-760°C its solubility limit is exceeded during cooling. Examples of very fine $M_{23}C_6$ precipitating directly on stacking faults have been noted:

$$(\gamma) \rightarrow M_{23}C_6 + \gamma \qquad (4)$$

$$(Cr,Mo,C) \rightarrow (Cr_{21}Mo_2)C_6 \qquad (5)$$

Mihalisin[19] has suggested that carbon is slowly depleted through the following reaction based upon studies of a series of experimental alloys:

$$TiC \rightarrow M_7C_3 \rightarrow Cr_{23}C_6 \rightarrow \sigma \qquad (6)$$

Borides

Boron often is present at levels of 50-500 ppm in superalloys, except single crystals. It segregates to grain boundaries, prolongs rupture life and improves rupture ductility. In one alloy, more than 120 ppm boron reacts to form two types of M_3B_2 borides depending on thermal history; one is approximately $(Mo_{0.48}Ti_{0.07}Cr_{.39}Ni_{0.03}Co_{0.03})_3B_2$ and the other is $Mo_{0.31}Ti_{0.07}Cr_{0.49}Ni_{0.06}Co_{0.07})_3B_2$[15]. Vanadium promotes the formation of M_3B_2 type borides[20]. Other borides that have been identified include Cr_2B and Cr_5B_3 in Rene′ 220C[47]. Borides are hard, refractory particles with shapes varying from blocky to half-moon in appearance; they act as a source of boron for the grain boundaries[6].

Topologically Close Packed Phases

In some alloys, if composition has not been carefully controlled, undesirable phases can form either during heat treatment or, more commonly, during service. These precipitates, known as TCP phases, are composed of close-packed layers of atoms parallel to {111} planes of the gamma matrix. Usually harmful, they may appear as long plates or needles, often nucleating on grain-boundary carbides. Nickel alloys are prone especially to formation of sigma (σ) and mu (μ). The formula for σ is $(Fe,Mo)_x$ $(Ni,Co)_y$, where x and y can vary from 1 to 7. Inhomogeneities in composition, often found in cast alloys, can cause localized TCP phase formation. Several calculational methods, based on overall composition, exist to predict the likelihood of TCP formation[22,23].

The hardness and plate-like morphology of σ causes premature cracking, leading to low-temperature brittle failure, although yield strength is unaffected. However, the effect on elevated-temperature rupture strength is particularly worrisome. Sigma formation must deplete refractory metals in the gamma matrix, causing loss of strength of the latter. Also, high-temperature fracture can occur along σ plates rather than the normal intergranular path, resulting in sharply reduced rupture life[21]. Plate-like μ also can form but little is known about its detrimental effects. The occurrence of TCP phases is controlled by adjustment of chemistry.

ALLOYING FOR YIELD OR TENSILE STRENGTH

Solid-Solution Strengthening

Commercial nickel-base alloys always contain substantial alloying additions in solid solution to provide strength, creep resistance, or resistance to surface degradation.

In the case of solid-solution strengthening, it is convenient to discuss several theories of yielding in terms of the effects of solutes on various physical or crystallographic properties, for example, lattice parameter and elastic modulus.

In general, strengthening mechanisms are considered to be independent and additive, although there is considerable controversy as to the means of superposing hardening mechanisms. For this purpose of this chapter, we will treat hardening mechanisms as essentially independent.

Lattice Parameter

Substitutional solutes produce lattice misfit, defined as the difference Δa between the lattice parameter, a_o, of the pure matrix and, a, the lattice parameter of the solute atom:

$$\epsilon = \frac{1}{c} \frac{\Delta a}{a_0} \tag{7}$$

Fleischer[24] has shown, in turn, that lattice parameter mismatch may be approximately by the relative difference in atomic volume V between solvent S and solute X.

$$\epsilon = \frac{1}{a} \frac{da}{dc} \sim \frac{V_X - V_S}{V_S} \tag{8}$$

A linear relation between flow stress and solute content was observed prior to 1960 for several solute elements in nickel[25,26]. For the same lattice strains, the larger the valency difference between solutes and solvent, the greater the hardening[27]. The strengthening influence of alloying elements persists to temperatures at least as high as 815°C[25]. Fleischer[24] suggests that valency effects may be explained by modulus differences between the various alloys, as discussed in the following section. Alternatively, the effects of valency may be felt through the decrease in stacking fault energy (SFE) of fcc alloys with increasing electron atom ratio; Beeston et al[27] have correlated the electron vacancy number, N_v, with SFE. It is generally expected that yield stress varies inversely with SFE. A more recent and comprehensive survey of hardening effects of solutes in nickel confirmed the importance of position in the periodic table, see Fig 8[28].

The solid-solution elements typically found in the gamma phase include aluminum, iron, titanium, chromium, tungsten, cobalt, and molybdenum. The difference in atomic diameter from that of nickel varies from less than 1% for cobalt to 17% for titanium.

Modulus

Fleischer's[24] suggestion that modulus differences between solute and solvent may give rise to strengthening is based on the argument that extra work is needed to force a dislocation through hard or soft regions in the matrix.

Fleischer[24,29] concluded that modulus differences and lattice misfit can be incorporated into a single equation, such that where

$$\epsilon_G^l = \left[\frac{1}{G}\frac{dG}{dC}\right]\bigg/1 + \left(\frac{1}{2}G\right)(dG/dC)$$

is a modulus misfit parameter and Z is a constant; τ_o is the critical resolved shear stress of the pure metal.

$$\tau_c = \tau_0 + \frac{G|\epsilon_G - \alpha\epsilon|^{3/2}c^{1/2}}{Z} \qquad (9)$$

Labusch[30] extended Fleischer's model to higher concentrations by use of a different type of statistical average of the interaction between solute atoms and dislocation. The resulting expression is

$$\tau_c = \tau_0 + \frac{G\epsilon^{4/3}c^{2/3}}{550} \qquad (10)$$

Although some agreement was found with Eq. 9 for Cu-base alloys, data for ductile Au, Ag and Cu-base alloy single crystals showed better agreement with $c^{2/3}$, as predicted by Eq. 10[30]. More recent efforts to fit data for Ni-base alloys to Eqs. 9 or 10 have been unsuccessful[28]. Rather, empirical relations that incorporate size and modulus misfit have been fitted to the data in Fig. 8.

Short-Range Order

Concentrated solid solutions are likely to exhibit appreciable short-range order. The energy required to shear a short-range-ordered crystal causes an increase in flow stress of the alloy.

The shear stress to move a dislocation as a consequence of short range order is[31]:

$$\tau = 16\left(\frac{2}{3}\right) \frac{\frac{1}{2}c(1-c)\nu a_s}{a^3} \tag{11}$$

where ν is the interaction energy [=$V_{AB}-\frac{1}{2}(V_{AA}+V_{BB})$] and a_s is the short range order coefficient. Since all terms in Eq. 11 are temperature independent, short range order provides an athermal increment to flow stress. However, a_s increases with decreasing annealing temperature. Consequently, the short-range-order component of flow stress is sensitive to thermal history.

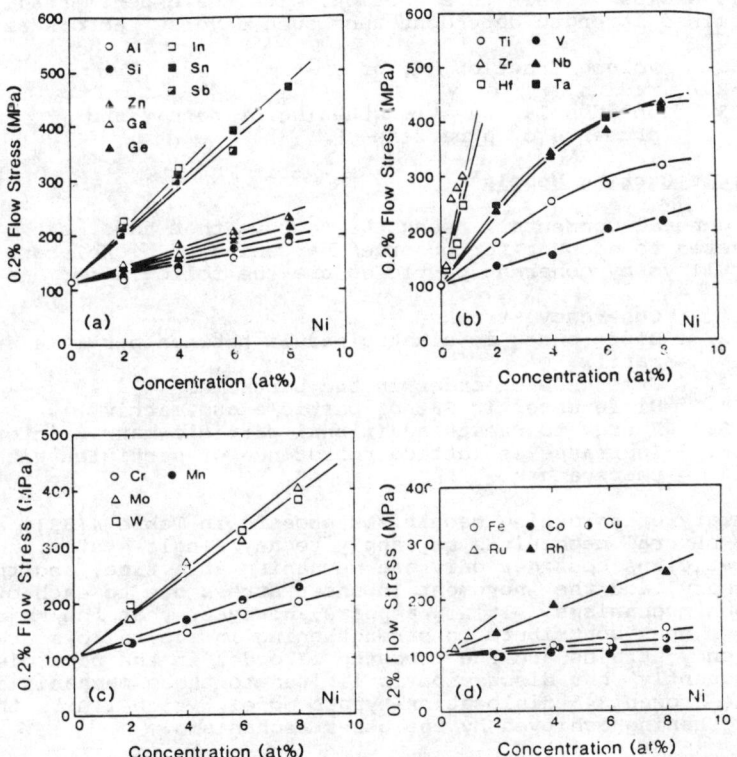

Fig. 8 Changes in 0.2% flow stress of binary nickel.[28] alloys at -196°C with additions of a) B-subgroup elements, b) transition metals of 4A and 5A groups, c) 6A and 7A groups and d) 8A groups

Nordheim and Grant[32] have suggested that short-range order exists in Ni-Cr alloys in the neighborhood of 20-25 w/o Cr. This composition range is reached in Hastelloy X and Inconel alloy 625 (22%Cr), and is approached in other γ´-lean alloys such as the early Nimonic series (19.5%Cr). Consequently, short-range-order strengthening may occur in such alloys.

PRECIPITATION HARDENING

Properties of γ - γ Alloys

The major contribution to the strength of precipitation hardened nickel-base superalloys is provided by the formation of coherent stable intermetallic compounds such as γ' [$Ni_3(Al,Ti)$] and γ"($Ni_3(Nb,Al,Ti)$]. Other phases, for example, borides and carbides, provide little additional strengthening at low temperatures due to their incoherency and the small volume fraction present. However, significant effects on creep rate, rupture life and rupture strain may be provided by these phases.

We are concerned with the properties of γ' - strengthened alloys exclusively in this section, with the experimentally determined strength dependent upon such diverse factors as:

1. volume fraction, f, of γ'
2. radius, r_o, of γ'
3. solid-solution strengthening of both γ and γ'
4. presence of hyperfine γ'

Particle Cutting Models

General Comments. Among the factors that have been suggested to account for observed hardening of nickel-base superalloys by coherent particles are the following:

1. Coherency strains
2. Differences in elastic moduli between particle and matrix
3. Existence of order in the particles
4. Differences in SFE of particle and matrix
5. Energy to create additional particle-matrix interface
6. Increases in lattice resistance of particles with temperature

A summary of hardening mechanisms appears in Table 4[33]. While several mechanisms may apply to any single system, theoreticians consider only one mechanism at a time, and then if necessary, add the increment in shear stress due to each of the various mechanisms. It now appears, however, that the major factors that contribute to strengthening in superalloys are coherency strains and the presence of order in the particles. Consequently, our discussion is limited to these mechanisms and to the Orowan[34] dislocation-bypass model, which limits the strengthening achieved by the other mechanisms.

Rather than attempt a detailed analysis of the various models proposed to account for order and misfit hardening, as has been done recently by Nembach and Neite[35] and by Nembach[36], it is our intention to discuss the principles underlying the most pertinent models. The methods used to treat hardening by solutes and by precipitates are basically similar; they depend upon calculating the force of interaction between a moving dislocation and whatever obstacles are in its path.

TABLE 4

SUMMARY OF HARDENING MECHANISMS[a] [33]

Author	Nature of Obstacles	Total Flow Stress	Conditions
		Solid Solutions	
Mott–Nabarro[1]	Misfitting atom or precipitate	$2G\epsilon c$	$L \geq \dfrac{b}{4|\epsilon|f}$
Fleischer[4,6]	Misfitting atom, modulus	$\tau_0 + \dfrac{G(\epsilon'_G - \alpha\epsilon)^{3/2} c^{1/2}}{760}$	
Flinn[10]	Short-range order	$\tau_0 + \dfrac{16(2/3)^{1/2} c(1-c)v\alpha}{a^3}$	
		Precipitates	
Copley–Kear[24]	Coherent, ordered, $\epsilon = 0$	$\dfrac{\gamma_0}{2b} - \dfrac{T}{br_0} + \dfrac{(\tau_0 + \tau p)}{2}$	$f \sim 0.6$
Gleiter–Hornbogen[21]	Coherent, ordered, $\epsilon = 0$	$\tau_0 + \dfrac{0.28\gamma_0^{3/2} r_0^{1/2} f^{1/3}}{b^2 G^{1/2}}$	
Brown–Ham[22]	Coherent, ordered, $\epsilon = 0$	$\tau_0 + \dfrac{\gamma_0}{2b}\left[\left(\dfrac{4\gamma_0 r_s f}{\pi T}\right)^{1/2} - f\right]$	$\dfrac{\pi T f}{4\gamma_0} < r_s < \dfrac{T}{\gamma_0}$
Brown–Ham[22]	Coherent, ordered, $\epsilon = 0$	$\tau_0 + \dfrac{\gamma_0}{2b}\left[\left(\dfrac{4f}{\pi}\right)^{1/2} - f\right]$	$r_s > \dfrac{T}{\gamma}$
Gerold–Haberkorn[31]	Coherent, ordered, $\epsilon \neq 0$	$\tau_0 + 3G\epsilon^{3/2}\left(\dfrac{r_0 f}{b}\right)^{1/2}$	Edge dislocation, $\dfrac{9\pi f}{16} < \dfrac{3|\epsilon|r_0}{b} < \dfrac{1}{2}$
Gerold–Haberkorn[31]	Coherent, ordered, $\epsilon \neq 0$	$\tau_0 + G\epsilon^{3/2}\left(\dfrac{r_0 f}{b}\right)^{1/2}$	Screw dislocation, $\dfrac{9\pi f}{26} < |\epsilon|r_0 < \dfrac{1}{2}$
Gleiter[32]	Coherent, ordered, $\epsilon \neq 0$	$\tau_0 + \dfrac{11.8 G\epsilon^{3/2} f^{5/6} r_0^{1/2}}{b^{1/2}}$	
Orowan[35]	Hard particles	$\tau_0 + \dfrac{Gb}{2\pi L}\phi' \ln \dfrac{L}{2b}$	$\dfrac{r_0}{b} > 30$ or incoherent ppt, $\phi' = \frac{1}{2}[1 + (1-\nu)^{-1}]$

[a] τ_0 = flow stress of matrix without obstacles, r_0 = particle radius, $r_s = (2/3)^{1/2} r_0$, T = line tension, ϵ = misfit; see text for definiton of other terms.

In order to move through a field of dispersed obstacles, a dislocation must bend to an angle $\phi \to \pi$, that is, very little bending is required for the dislocation to escape the obstacle; for strong obstacles $\phi \to 0$, as the dislocation is forced to almost double back on itself, see Fig. 9[39]. The number of obstacles per unit length of dislocation line depend upon ϕ; if $\phi \approx \pi$, the number per unit length is found by computing the number that intersects a random line. As ϕ decreases from π the dislocation sweeps out more area and therefore meets more obstacles, necessitating expressions for obstacle spacing which depend upon the applied stress τ. The most commonly used expression is:

$$L' = \left(\frac{2TL_s^2}{\tau b}\right)^{1/3} \tag{12}$$

where T is the line tension ($\sim \frac{Gb^2}{8}$ for edge and $\sim \frac{Gb^2}{2}$ for screw dislocations) and L_s is the square lattice spacing ($= n_s^{-\frac{1}{2}}$ where n_s is the number of particles per unit area of slip plane). To simplify calculations it is generally assumed that dislocations interact with a random array of obstacles of fixed strength. The limits on L' are:

$$L_s \leq L' \leq \frac{4r}{3f} \tag{13}$$

The upper limit, $\frac{4r}{3f}$, represents the spacing of random obstacles along a straight line.

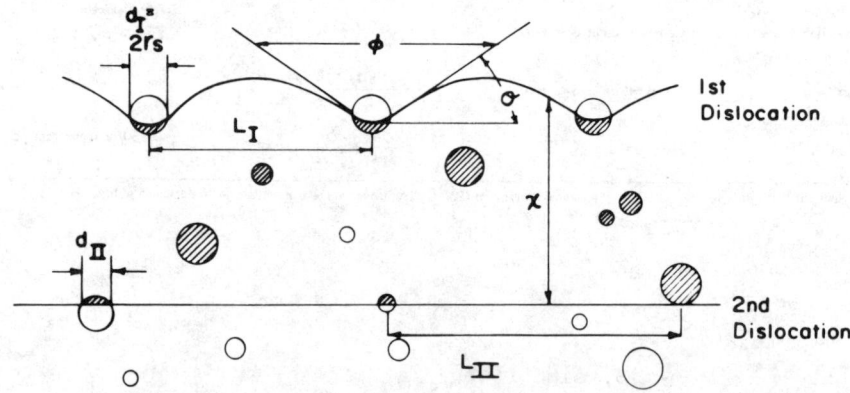

Fig. 9. Dislocation pairs interacting with ordered particles, showing effect of bend angle ϕ on obstacle spacing. Shaded areas represent APB.[39]

Order Strengthening. Consider the case of superlattice dislocation pairs interacting with particles. The calculation follows the principles first elaborated by Gleiter and Hornbogen[37] but utilizes the specific equations developed by Ham[38], and Brown and Ham[39]. As the first dislocation is just shearing the particles, see Fig. 9, the second dislocation is pulled forward by the APB remaining in all particles cut by the first dislocation. Provided that the two dislocations assume the same shape and that the separation x between the two dislocations is sufficiently small, but larger than r_s, the second dislocation may lie outside of all the particles. This situation may occur at long aging times.

In general, however, the second dislocation does come into contact with the APB in the sheared particles and is nearly straight. The more APB is cut by the second dislocation, the less effective the particles become as obstacles. Then, referring to Fig. 7, the force balances are as follows, neglecting any new particle-matrix interface formed by shear of the particles:

on dislocation 1, $\quad \tau b + \dfrac{Gb^2}{2\pi kx} - \dfrac{\gamma_0 d_I}{L_I} = 0 \quad$ (14)

on dislocation 2, $\quad \tau b + \dfrac{\gamma_0 d_{II}}{L_{II}} - \dfrac{Gb^2}{2\pi kx} = 0 \quad$ (15)

Solving Eqs. 14 and Eq. 15 simultaneously, the following relation is obtained:

$$2\tau b + \frac{\gamma_0 d_{II}}{L_{II}} = \frac{\gamma_0 d_I}{L_I} \quad (16)$$

Since the second dislocation is observed to be straight during shear by the first dislocation, f may be substituted for d_{II}/L_{II} and $(4\gamma_0 fr_s)^{1/2}/\pi T$ for d_I/L_I so that:

$$2\tau b + \gamma_0 f = \left(\frac{4\gamma_0 fr_s}{\pi T}\right)^{1/2} \gamma_0 \quad (17)$$

leading to the following relations for the applied stress τ:

$$\tau = \frac{\gamma_0}{2b}\left[\left(\frac{4\gamma_0 fr_s}{\pi T}\right)^{1/2} - f\right] \quad (18)$$

When $r_s \leq \pi f/4\Delta\gamma_0$, $\tau = 0$, since both dislocations touch the same fraction of APB. Gleiter and Hornbogen[37] also have emphasized the fact that as more and more APB is touched by the second fraction of APB. Gleiter and Hornbogen[37] also have emphasized the fact that as more APB is touched by the second dislocation, the flow stress must drop.

If the line tension T is approximated by ½Gb² (screw dislocations), then Eq. 18 reduces to:

$$\tau_c = \frac{\gamma_0}{2b}\left[\left(\frac{8\gamma_0 f r_s}{\pi G b^2}\right)^{1/2} - f\right] \qquad (19)$$

Eq. 19 cannot hold for r_s approaching zero, since τ_c cannot be negative. Nevertheless, the negative intercept can be used for an alternate computation of APB energy, as has been done by Martens and Nembach[40].

The first term of Eq. 19 is similar to an equation for order strengthening that had been proposed earlier by Gleiter and Hornbogen[37],

$$\tau_c = 0.28 \gamma_0^{3/2} f^{1/3} G^{-1/2} b^{-2} r_0^{1/2} \qquad (20)$$

except for a different constant and the dependence of volume fraction. The second term of Eq. 19 may be dropped, however, only when the second dislocation can avoid all particles[29]. The flow stress given by Eq. 19 would then reduce to one-half of the stress necessary to shear particles by single dislocations. The basic features of this model have been applied successfully to a variety of austenitic superalloys[36,37,40], both nickel and iron based.

Copley and Kear[41,42] also modified the Gleiter-Hornbogen theory; the results were applied specifically to the alloy Mar-M 200. Based on extensive electron-microscopic studies, the rate-controlling step for plastic deformation was shown to be moving dislocations from γ into γ'. Instead of setting up a force balance with the first dislocation partly through the particles, as in Fig. 9, they suggested that the first dislocation wraps around the particle, assuming its curvature, until forced in by the second dislocation. The conditions of static equilibrium for the leading and trailing dislocation of the superlattice pair about to enter a particle are:

$$\text{Dislocation 1, } (\tau_c - \tau_p)b + \frac{C}{x} + \frac{T}{r_0} - \gamma_0 = 0 \qquad (21)$$

$$\text{Dislocation 2, } (\tau_c - \tau_0)b - \frac{C}{x} + \frac{T}{r_0} = 0 \qquad (22)$$

where C/x is the force of repulsion between two dislocations, τ_p

is the friction stress of particle, and τ_o is the friction stress of matrix. T/r_o is the line tension force due to a dislocation assuming the curvature of the particle.

Solving Eqs. 21 and 22 simultaneously, the condition for static equilibrium at 22°C is obtained:

$$\tau_c = \frac{\gamma_0}{2b} - \frac{T}{br_0} + \frac{1}{2}(\tau_0 + \tau_p) \tag{23}$$

$\gamma_o/2b$ = stress to constrict the dislocation pair to the point that particle shear begins.

For dynamic conditions, the CRSS is predicted from the stress dependence of the plastic strain rate, which is related to a derived dislocation velocity-stress function. A very similar equation is obtained for τ_c:

$$\tau_c = \frac{\gamma_0}{2b} - \frac{T}{br_0} + \frac{k}{2}(\tau_0 + \tau_p) \tag{24}$$

where k is a constant dependent on the dislocation velocity of the crystal and has a value of 0.823 for Mar-M 200 at room temperature[42]. Penetration of small particles is easier than for large particles due to the line tension force. In any case, the major contribution to τ_c at room temperature is provided by the term $\gamma_o/2b$, which represents about 80% of the total flow stress computed for Mar-M 200. Leverant et al[43] concluded, however, that at high temperatures and high strain rates, where the flow stress of γ' reaches a distinct peak, both the APB energy and the flow stress of γ' are major contributors to τ_c.

In the most general case, γ should be replaced by Γ, the fault energy for shear of the particle, since faults other than APB-type faults may be produced by shear. For example, particle shearing occurs by loosely coupled intrinsic-extrinsic fault pairs in Mar-M 200 at 1400°F (760°C)[33]. In this model the influence of crystal orientation on flow stress is felt through the variation in the nature of faults produced by different glide mechanisms. Decker[44] has pointed out that, based on the critical temperature for ordering of their respective Ni_3X phases, titanium, niobium, and tantalum in γ' should not appreciably increase APB energy. However, titanium and perhaps tantalum could increase the energy of other fault types. Brown and Ham[39] have analyzed several sets of data to calculate APB energy as a function of alloy content, and found that the fault energy may be widely varied.

Misfit Strengthening. Early attempts to relate the influence of coherency strains to CRSS failed to explain the dependence of CRSS on particle size. A more complex model has been suggested by Gerold and Haberkorn[45] where the interaction between dislocations and strain fields around the particles plays a dominant role.

The calculation is similar in outline to Fleischer's[24] for solid-solution hardening and is expected to apply for matrix-particle misfit, ϵ, of approximately 0.01 in the case of spherical, coherent particles. ($\epsilon = a_{ppt} - a_{matrix}/a_{matrix}$)

The increase in flow stress due to interaction of single dislocations with strain fields is given by:

$$\Delta \tau = \frac{K}{bL''} \tag{25}$$

where K is the maximum repelling force of the strain field of a single particle on a moving dislocation and L" is the average distance between the force centers. The problem is to find appropriate expressions for K and L": K is found to be equal to or less than T, the line tension of an edge dislocation. For L', the authors use an expression in which the obstacle spacing depends upon the bend angle, $\theta = \frac{1}{2}(\pi - \phi)$ see Fig. 9.

$$L'' = \frac{r_0 \pi^{1/2}}{(\theta f)^{1/2}} , \quad \frac{9\pi f}{16} < \theta < \frac{3}{2} \tag{26}$$

The angle to which a dislocation is bent by the force K before escaping the particle is given by:

$$2 \sin \theta \simeq \frac{K}{2T} \tag{27}$$

The maximum value of K is computed to be:

$$K = 4G|\epsilon|br \tag{28}$$

For bend angles smaller than $9\pi f/16$ the dislocation must be treated as a rigid line; for bend angles near 3/2 the dislocation line is totally flexible and another expression for L" must be used. Combining Eqs. 26 and 28, the CRSS is obtained:

$$\Delta \tau = AG\epsilon^{3/2} \left(\frac{r_0 f}{b} \right)^{1/2} , \quad \frac{9\pi f}{16} < \frac{3|\epsilon|r_0}{b} < \frac{1}{2} \tag{29}$$

where A=3 for edge dislocations and =1 for screw dislocations. This equation predicts that the flow stress should increase slightly more rapidly than ϵ, because increasing misfit bends the dislocation more and makes it interact with more regions of adverse stress.

Gleiter[46] also has discussed the effect of coherency strain fields on CRSS in two-phase alloys, and by following the steps outlined above with different assumptions as to the flexibility of dislocation lines and a different averaging procedure for obstacle distribution, has obtained the following relation for flexible edge dislocations:

$$\Delta\tau = 11.8 G \epsilon^{3/2} f^{5/6} \left(\frac{r_0}{b}\right)^{1/2} \tag{30}$$

The principal difference between Eqs. 29 and 30 is the dependence of $\Delta\tau$ on the volume fraction. Data for several Ni-Al alloys have been found to be in agreement with Eq. 30.

Nembach and Neite[35] have extensively reviewed the experimental evidence bearing on lattice misfit effects on the strength of superalloys. It was concluded that there is no convincing experimental proof that misfit affects the flow stress of underaged γ'-hardened alloys, and that lattice misfits of the magnitude used in commercial alloys do not affect τ_p.

Dislocation Bypass Models

Orowan Bowing. All of the dislocation cutting models previously discussed agree that as particles grow beyond a critical size, bypass may occur by bowing, climb, or other processes. The Orowan bowing model[34] is generally considered to be most applicable for austenitic superalloys. The onset of bowing is accompanied by a loss in strength. The increment in flow stress at low temperature due to bowing is given by consideration of the radius of a curvature ρ to which a flexible dislocation can be bent by an applied stress τ; the line tension is given by:[47]

$$T = \frac{Gb^2}{4\pi} \phi' \ln \frac{L}{2b} \tag{31}$$

where $\phi' = \frac{1}{2}[1+1/1-\upsilon]$ and L is the edge to edge spacing of particles = $[(\pi/f)^{1/2} - 2]r_s$. The increment in flow stress $\Delta\tau$ is:

$$\Delta\tau = \frac{Gb}{2\pi L} \phi' \ln \frac{L}{2b} \tag{32}$$

The effect of increasing volume fraction f for a given particle size is to decrease L, leading to a prediction of increased strength. Greater hardening should also occur as particle size increases; this effect would be enhanced by coherency strains, producing a larger particle diameter in the path of a dislocation.

Critical Evaluation of Models

The results of aging studies on low volume fraction model γ-γ' alloys are not directly applicable to commercial nickel base superalloys because of the much higher volume fraction of γ' present in the latter. Also, particle sizes tend to be larger in the superalloys. Another complication is the orientation and strain rate dependence of stress-strain behavior in high volume fraction alloys[48].

Each of the models outlined suffers from shortcomings, not the least of which is the fact that the microstructures of nickel-base superalloys are too complex to expect a single mechanism to operate over all ranges of stress and service temperatures. We shall distinguish between alloys in which there is little or no mismatch between γ and γ', that is, Ni-Cr-Al type, and high mismatch alloys, that is, Ni-Al-Ti type.

Alloys with No Lattice Mismatch. The principal elements of the Brown-Ham model of pairs of dislocations interacting with ordered particles have been directly confirmed in high voltage electron microscopic experiments of Nimonic PE 116[49]. Specifically, the leading dislocation of a pair bows out strongly between γ' precipitates while the trailing dislocation remains nearly straight. The spacing of obstacles along the leading dislocaiton is in reasonable agreement with Eq. 12. When there is little or no mismatch, considerable evidence suggests that the volume fraction of γ' is the most significant variable controlling flow stress and creep resistance. The volume fraction of γ' varies from 0.2 in γ'-lean alloys, such as Nimonic 90A, to 0.6 in Mar-M 200 and 713C. Newer alloys possess up to 70%γ'. The flow stress of binary Ni-Al aged to peak hardness[10] and of Ni-Cr-Al alloys containing fractions of γ' between 0.4 and 0.6 is remarkably insensitive to temperature. The yield stress of Mar-M 200 is nearly constant from room temperature to 1380°F (750°C)[31].

It is clear that no single model of yielding may be applied over the entire range of volume fractions of γ' and service temperatures of nickel alloys. In polycrystalline alloys containing a low volume fraction of coarse particles of γ', the temperature dependence of yielding appears to be controlled by the γ matrix, hardened by nonequilibrium hyperfine γ' precipitates. The rate controlling step is the stress to move dislocation through the hardened matrix; the large γ' particles are easily sheared. The hyperfine particles redissolve at temperatures over 1290°F (700°C), γ is weakened and the flow

stress drops rapidly. Similarly, in single crystals the
temperature dependence of yielding depends upon the relation
between particle size and spacing between cross slip events[42].

Alloys containing a large volume fraction of γ' behave
similarly to pure γ' in that the flow stress increases with
increasing temperature. If the alloy contains about 50% primary
coarse γ' in γ, the strength characteristics are intermediate.
Flow strength is moderately high at low temperature; a shallow
peak in flow stress is reached near 1290°F (770°C), and strength
falls off at a temperature somewhat higher than for a leaner
alloy, see Fig. 10[50]. Note that an intermediate volume
fraction of 20% produces the highest strength at 21°C. Most of
the lower temperature yield strength is due to "hyperfine" γ'
(50-100Å diameter).

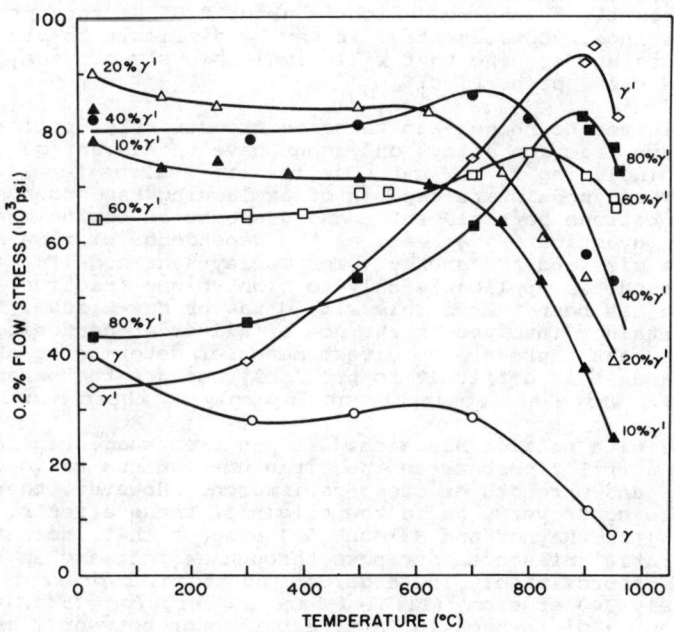

Fig. 10. Flow stress of Ni-Cr-Al alloys containing
varying proportions of gamma prime.[50]

All of the order-strengthening theories predict an increase in flow stress with increasing particle size, r_o, for constant volume fraction. This has been confirmed in Ni-12.7 a/o Al alloys[51]. However, other evidence on the effect of particle size has been conflicting; it has been shown for an 18Cr-6.5Al-3.3Nb alloy that an increase in size of γ' from 0.05 μ_m to 0.5 μ_m by varying aging time reduced the room temperature flow stress by about 13%[52]. On the other hand, hardness of Ni-Cr-Ti alloys increases initially with particle size and then decreases[53]. Dislocations initially shear the particles; when they grow larger, the particles are avoided by a bypass mechanism. Consequently, we conclude that so long as particles are being cut, the flow stress increases with increasing particle size.

For Ni-Cr-Al-Ti alloys containing 10-20% γ', the smallest γ' particle size gave optimum creep resistance at 1290°F (700°C); also, size was more important than volume fraction in determining creep life [54]. Small size is achieved in conjunction with small interparticle spacing, which is about 0.05 μm for optimum creep resistance. Consequently, it may be difficult to produce a particle size and spacing that will simultaneously provide good tensile and creep properties.

The Gleiter-Hornbogen-Ham theories are strictly applicable to low volume fraction alloys only, but have the advantage of explicitly including f, r_o, and γ_o in the expressions for flow stress. These models are capable of explaining the observed transition between particle shear and dislocation bowing observed by several investigators as well as the dependence of flow stress on particle size and APB energy. The Copley-Kear model[41,42] on the other hand, is applicable only to high volume fraction alloys, and has been tested only with data for Mar-M 200. The major uncertainty involved in the use of all order-hardening theories is that there are no direct means of determining APB energies, and it is difficult to precisely measure the parameters f, r_o and L', which are so important in applying these models.

Alloys with Lattice Mismatch. It has been suggested that there is a correlation between the titanium-aluminum ratio of superalloys and strength or creep resistance. However, there is considerable controversy as to the origin of these effects. Phillips[55] and Raynor and Silcock[56] suggest that increasing the Ti-Al ratio influences strength through an increase in APB energy from approximately 150 ergs/cm^2 (no titanium present) to approximately 240 ergs/cm^2 (Ti:Al=1) and 300 ergs/cm^2 (Ti:Al=8). In this view, a difference in lattice parameter between γ and γ' as high as 0.5%, which accompanies high titanium additions, is relatively unimportant as a strengthening mechanism. Rather, mismatch is the driving force in the growth and coalescence of γ' particles. A large mismatch, corresponding to a large interfacial strain energy, may render the γ' precipitate thermally unstable even in the absence of applied stress.

Applied stress further lowers the mismatch to stabilize the precipitate, particularly when the stress axis differs from <111>.

Conversely, Decker[44] and Decker and Mihalisin[57] argue that a high mismatch can markedly increase peak hardness by aging. Increasing mismatch from 0.2 to 0.8% doubled peak-aged hardness of several Ni-Al ternary alloys, which is in agreement with the theory of Gerold and Haberkorn[45]. Munjal and Ardell[54] found excellent agreement between the Brown-Ham[39] model and experimental results for Ni-12.19A%Al single crystals tested in compression between 77 and 373°K. Since misfit changes considerably with temperature, and no significant change in $\Delta\tau$ was observed over the same temperature range, it was concluded that the contribution of coherency hardening is negligible in this system. While the relation between coherency strain and low-temperature tensile strength is still in doubt, optimum creep resistance seems to depend upon zero mismatch. For example, creep-rupture life of Ni-Cr-Al alloys tested at 700°C and a stress level of 21,200 psi reaches a maximum at zero mismatch[58]. These confirmed[59] results are attributed to high phase stability at low mismatch. Therefore, the Gerold-Haberkorn theory must be confined to temperatures low enough so that the growth of γ' is not possible. It is clear that misfit strengthening is particularly important in γ'' strengthened alloys (e.g., IN 718 and IN 901), but these alloys are rarely utilized for their strength at temperatures above 815°C.

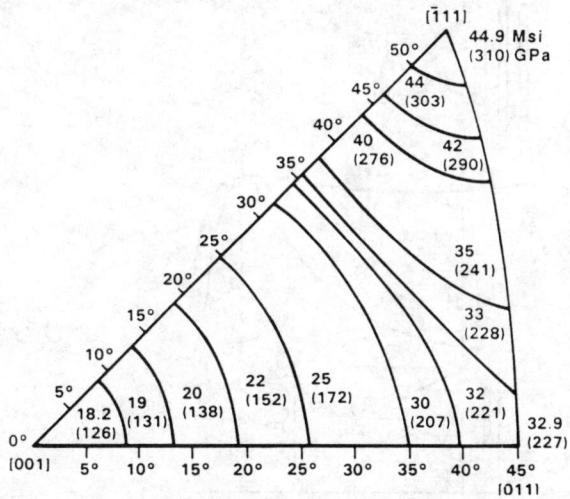

Fig. 11   Effect of orientation on elastic modulus of single crystal superalloy PWA 1480, room temperature.[60]

SINGLE CRYSTALS

The hardening mechanisms described above do not take into account interactions of dislocations with grain boundaries, and therefore are implicitly applicable only to single crystals. However, none of them take into account orientation effects, which can be substantial below 760°C. For example, the elastic modulus of single crystal superalloy PWA 1480 varies widely with orientation, see Fig. 11[60]. As a result, mechanical properties such as thermal fatigue resistance and creep rate also will be affected by orientation. Since particles are sheared by dislocations, the yield stress of the γ' (itself dependent upon orientation) may be important. Further, a tension-compression anomaly is observed in both Ni_3Al[61] and in PWA 1480[62] which may depend upon whether the applied stress acts to constrict or extend Shockley partials on {111}.

Shah and Duhl[62] have proposed a model for superalloy single crystal behavior which accounts for the above experimental observations. The model is shown schematically in Fig. 12. The principal parameters in their model are the particle radius, r, the mean free distance between cross slip events on the leading dislocation of a superlattice dislocation pair, λ, and the crystal orientation. In general,

$$\sigma \alpha \left(\frac{1}{r} + \frac{1}{\lambda} \right) \tag{33}$$

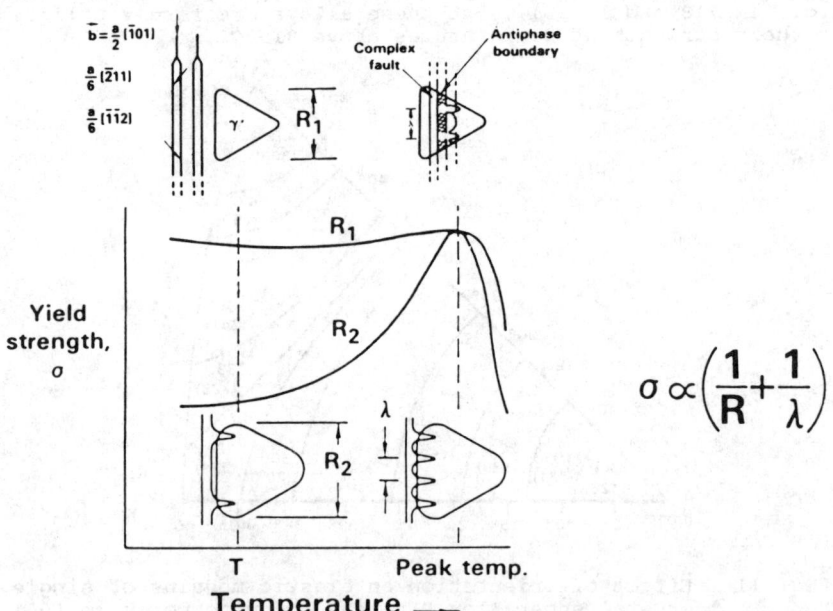

Fig. 12. Model of yield strength as a function of temperature for a single crystal superalloy with fine and coarse gamma prime, deforming by {111} slip.[62]

For <100> oriented crystals, when γ' particles are coarse, λ<r at low temperatures, but at high temperature more cross slip occurs and λ is lowered. This results in a higher stress to shear γ' and the flow stress increases as temperature increases. For fine γ', λ > r at low temperature, and r controls the strength to high temperatures; therefore, a plateau in yield stress with temperature is observed until a high enough temperature is reached for cross slip to occur readily; λ then decreases and controls behavior at high temperature. For <111> oriented crystals, massive cube slip readily occurs, and thermally activated cross slip of dislocation segments is not relevant. The particle size, r, thus controls at all temperatures, and $\sigma_{ys} \propto \frac{1}{r}$.

POLYCRYSTALS

Metals and alloys tested at temperatures below about $0.5T_m$, the absolute melting temperature, are further strengthened by the resistance of grain boundaries to dislocation motion. The Hall-Petch relation[62]:

$$\sigma_y = \sigma_0 + k_y d^{-1/2}$$

(34)

where σ_y is yield stress, σ_o is a lattice friction stress, d is grain diameter and k_y is a measure of the grain boundary resistance, demonstrates that significant strengthening can be obtained for fine grained alloys when k_y is large. Factors tending to increase k_y are solute hardening and difficult cross slip. Therefore, solutes such as cobalt which lower the stacking fault energy of nickel are expected to increase the contribution of grain boundaries to yield or flow stresses. At high temperatures, on the other hand, a coarse grain size provides for improved creep resistance, one of the factors contributing to the superior strength of cast or wrought superalloys.

ALLOYING FOR CREEP RESISTANCE

The outstanding creep resistance of conventional nickel-base superalloys is primarily a function of the precipitation of γ'. However, when the solution temperature (solvus) of γ' is approached during service, the particles may grow or even go into solid solution. When this occurs, strength is rapidly lost. Another cause of reduced strength is grain boundary sliding, which occurs readily in fine-grained alloys. However, it is possible to avoid these problems by several approaches: production of columnar grained or single crystal materials or by utilizing oxide particles or refractory wires as supplemental strengthening phases. Therefore, the most creep resistant alloys at temperatures above 1000°C are single crystal superalloys, ODS alloys and composites. In this section, the general factors influencing creep resistance are considered, and the specific alloy design principles are reviewed with respect to single crystals, ODS alloys and composites.

Steady-State Creep

Steady-state creep resistance in crystalline, single-phase solids depends upon the diffusivity D, stacking-fault energy γ_{SFE}, elastic modulus E, temperature T, and stress σ according to a formula of the form[63,64]:

$$\dot{\epsilon} = A\left(\frac{\sigma}{E}\right)^n f(\gamma_{SFE}) e^{-Q/RT} \tag{35}$$

where $f(\gamma_{SFE})$ is a function of SFE and Q is the activation energy for creep. On one model, $\dot{\epsilon}$ is dependent on $(\gamma_{SFE})^{3.5}$, while another formulation incorporates γ_{SFE} into the stress exponent, n, such that as γ_{SFE} increases, n decreases[65]. Typical solid-solution alloys reveal an exponent n with values 3-7 and with Q equal to the activation energy for self-diffusion at temperatures above half the melting point. Consequently, high creep strength is favored by solute additions which raise the modulus or lower the SFE and which lower the diffusivity. Tungsten and molybdenum serve to raise the modulus and lower the diffusivity of nickel-base alloys, while cobalt is effective in lowering SFE.

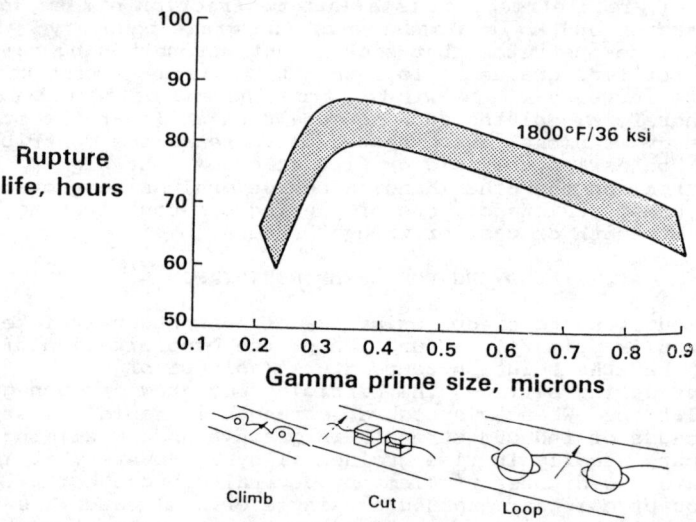

Fig. 13. Effect of γ' size on creep rupture life of a single crystal superalloy at 982°C(1800°F).[66]

The particle size affects creep rupture life, as shown in Fig. 13 for a single crystal superalloy[66]. When second-phase particles are present, the apparent activation energy for creep is much higher than the activation energy for creep (or self-diffusion) of the matrix. Thus the activation energy for steady-state creep of Mar-M 200 and other nickel-base superalloys is as high as twice that of unalloyed nickel and considerably higher than for solid-solution alloys of nickel. These apparent discrepancies can be eliminated either by considering the temperature dependence of E[59] for by replacing σ in Eq. 35 by $(\sigma-\sigma_o)$, where σ_o is a friction stress[67]. In either case, the activation energy for creep becomes very close to that for self-diffusion. Similar differences between activation energies for creep of a multiphase alloy and the activation energy for self-diffusion of the matrix have been noted for dispersion-strengthened alloys such as TD Nickel. Grain aspect ratio seems to play a role in these alloys, as Q and n both increase with increasing GAR[68].

An observation that <112> slip is responsible for particle shear suggests that crystal orientations with a low Schmid factor for <112> slip are desirable for good creep resistance. Supporting this conclusion is the observation that single crystals with a <111> tensile axis have unusually long creep lives[69]. However, later work showed that orientation effects on creep behavior of single crystals is strongly dependent upon the presence or absence of minor elements(B, Zr, C), see Fig. 14[70].

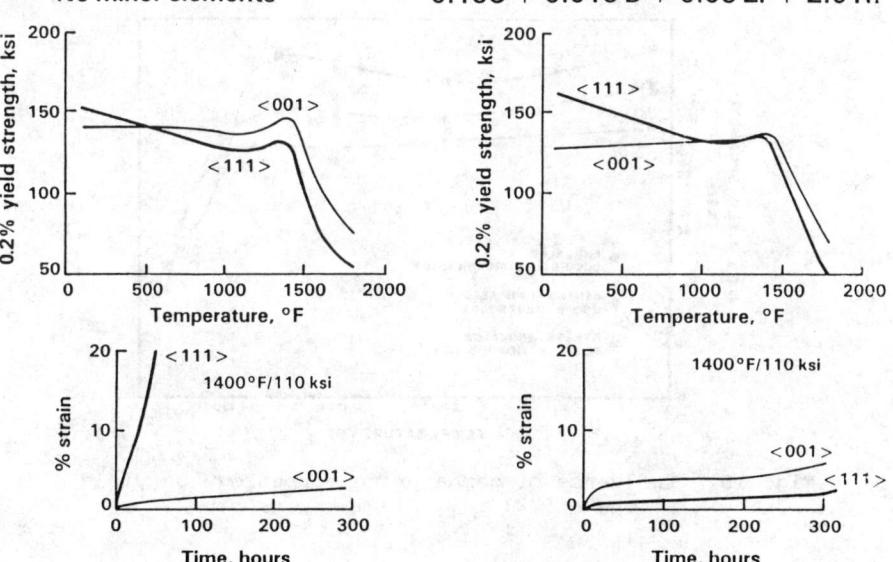

Fig. 14. Effects of orientation and temperature on creep behavior of Mar-M200 single crystals, with and without minor elements.[70]

Influence of γ' Morphology

The morphology of γ' in nickel-base alloys can be modified by annealing under stress[71]. In <100> and <110> orientation, either plates or rods of γ' may be generated, depending upon the sense of the applied stress. Tensile annealing produces γ' plates for the <100> orientation, while compressive annealing causes rods to form. In the <110> orientation, the opposite occurs, while <111> oriented crystals show no change in morphology under tension or compression. The sign of the lattice misfit also influences stress coarsening behavior; the results described above are for alloys with negative misfit. Morphological changes in γ' can affect yielding behavior of U-700 crystals[72]. The yield strength of <100> crystals is increased by rod or plate formation, with plates providing the greater effect to 1400°F (760°C), see Fig. 15[72]. At higher temperature morphology has little effect on strength. However, in creep-rupture tests a substantial improvement in properties of <100> crystals has been reported for a Ni-Al-Mo-Ta alloy[73]. Specimens in the solution treated condition (air cooled) exhibit lower steady state creep rates and longer rupture lives than material given a standard heat treatment. A pre-strain under creep conditions leads to still further improvement in properties due to the formation of γ' plates or rafts during primary creep. The molybdenum content is critical, with the creep strength maximized at the solubility limit of molybdenum in γ[74]. In summary, optimum strengthening due to γ' rafting in NI-Al-Mo-X alloys is achieved in homogeneous alloys tat are saturated with Mo and that exhibit large negative γ/γ' misfit.

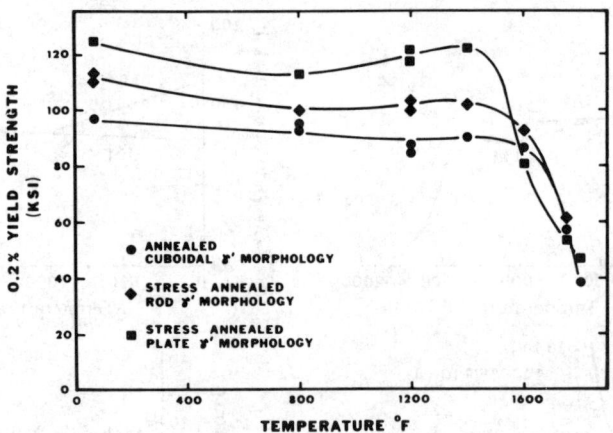

Fig. 15. Influence of gamma prime morphology on yield strength.[72]

Grain Size Effects

The interaction among various microstructural parameters affecting the creep of Ni-20%Cr-X alloys has been examined by Gibbons and Hopkins[75]. At high volume fractions of γ', increasing grain size caused a sharp decrease in secondary creep rate, see Fig. 16[75]. The same study showed that hardening by γ' containing Nb+Al was more effective than that observed with Ti+Al along. Increasing volume fraction, f, at constant grain size, produced a large decrease in creep rate up to f=0.2, with little further change to f=0.3[75]. Decker[44] reported a linear increase in rupture strength (100 hrs) with f at several test temperatures. The proportion of fine γ' (<0.5μ) interspersed with coarse γ' is particularly important in DS Mar M200 + Hf[76]. Much of the coarse γ' present in cast alloys can be replaced by fine γ' by increasing solution temperature in the range 1187-1250°C. The final distribution of γ' size in any superalloy is determined by the sequence of solution and aging treatments as well as coating cycles (if applicable). In IN 738LC the rupture life at constant stress also increases with volume fraction of γ´, Fig. 17[77].

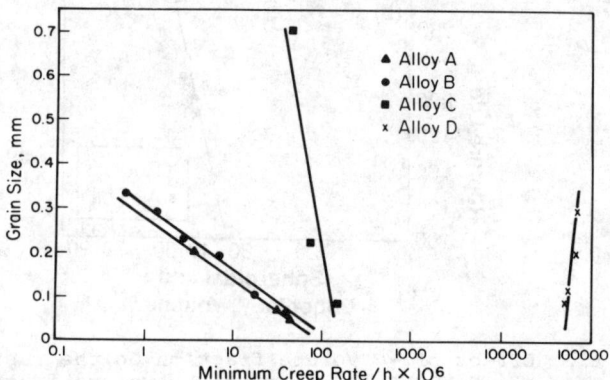

Fig. 16. Influence of varying grain size on minimum creep rate of Ni-20-Cr-Ti-Al alloys. Alloy A: 3Ti, 2.2Al. Alloy B: 2.6Ti-1.58Al. Alloy C: 1.63Ti - 0.85Al. Alloy D: 1.06Ti, 0.65Al.[75]

Grain Size and Component Thickness

The elevated temperature strength of superalloys is also very dependent upon the relation of grain size to component thickness. For example, rupture life and creep resistance increase with increasing ratio of component thickness to grain size[78]. Provided the ratio is kept constant, life and creep resistance of wrought superalloys increase with grain size[6]. Cast superalloys show the same dependence of life and creep resistance upon the ratio of thickness to grain size. As a result, large grains are to be avoided in thin sections to maintain high creep-rupture resistance.

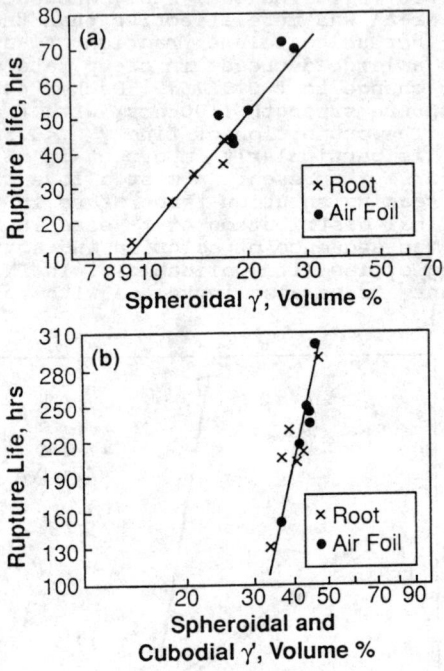

Fig. 17. Effect of γ' volume fraction on the rupture life of IN 738LC a) spherical γ' volume fraction vs. rupture life at 586 MPa/760°C and b) rupture life vs. total γ' volume fraction at 345MPa/829°C.[77]

In modern cast superalloys, a balance must be struck to avoid excessively fine grains which decrease creep and rupture strength, and excessively large grains which lower tensile strength.

Grain Boundary Chemistry

Improvement of creep properties by very small additions of boron and zirconium (see Table 1) is a notable feature of nickel-

base superalloys. Boron and zirconium can increase life of
Udimet 500 at 870°C by 13 times, elongation by 7 times, rupture
stress by 1.9 times, and n (stress dependence of creep rate) from
2.24 to 9[44].; Magnesium additions from 0.01 - 0.05% also have
resulted in improved properties and forgeability in wrought
alloys[44]. This is believed due primarily to the magnesium
tying up sulfur, a grain boundary embrittler.

The mechanism for the boron effect on properties is unclear.
It is believed that boron and zirconium segregate to grain
boundaries because of their large size misfit with nickel. Since
cracks in superalloys propagate along grain boundaries, the
importance of grain boundary chemistry is apparent. Although
early work suggested that boron and zirconium influence rupture
properties through their effects on carbide and γ' distribution,
recent work on a P/M superalloy has revealed no such effect[70].

Boron and zirconium also improve rupture life of γ'-free
alloys, cobalt alloys, and stainless steels, so that
microstructural alterations cannot, in any case, apply to all
systems. Boron also may reduce carbide precipitation at grain
boundaries by releasing carbon into the grains. Magnesium may
have a similar effect in a nickel-chromium-titanium-aluminum
alloy in which intragranular MC was noted[44]. Finally,
segregation of misfitting atoms to grain boundaries may reduce
grain-boundary diffusion rates, consistent with the findings of
Tien and Gamble[71] on the formation of denuded zones by Nabarro-
Herring type diffusion. Direct evidence for a lowering of grain
boundary diffusivity by 0.11%Zr in Ni-20%Cr alloy over the range
800-1200°K has been reported by Schneibel et al[80]. This effect
was accompanied by the precipitation of Ni_3Zr at grain
boundaries.

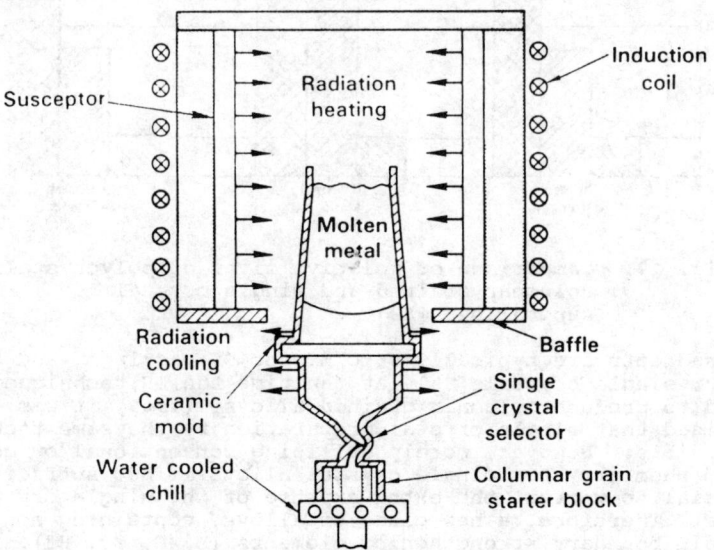

Fig.18. Schematic of Directional Solidification (DS) process.[81]

DIRECTIONALLY SOLIDIFIED ALLOYS AND SINGLE CRYSTALS

Directional solidification (DS) to produce columnar grains served to eliminate a major source of weakness at high temperatures - grain boundary sliding. A schematic diagram of the apparatus used to produce other columnar grains or single crystals is shown in Fig. 18[81]. The DS process requires control of testing parameter such as thermal gradients, G and growth rate $, to avoid defects. In general DS must be carried out at high gradients and relatively low growth rates. Compositions of several columnar grained alloys are listed in Table 1. Note that hafnium is present in most of the DS alloys listed. This is a consequence of two beneficial effects; the suppression of γ/γ' eutectic formation at grain boundaries and a marked improvement in transverse properties of thin-walled casting (i.e., hollow turbine blades).

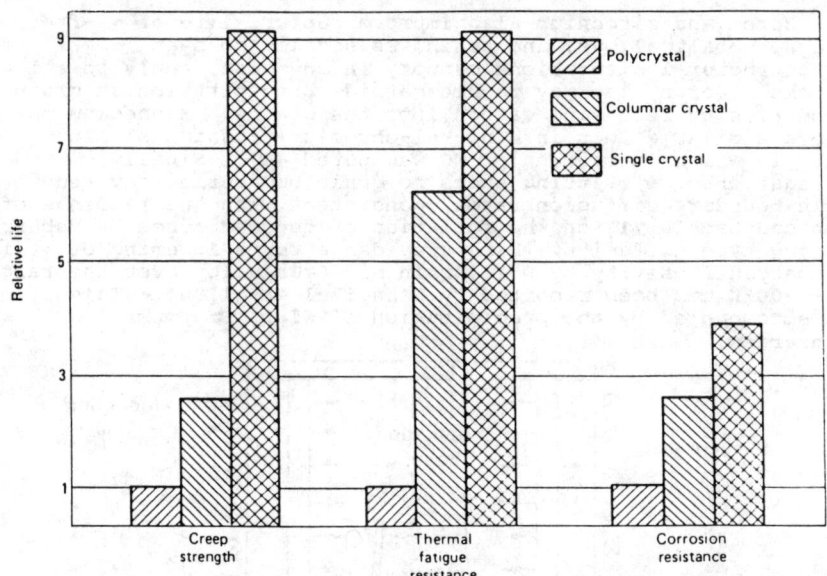

Fig. 19. Comparison of relative lives of polycrystals, DS columnar grained and single crystal superalloys.[83]

Gradients are typically 35°C/mm for DS castings, and twice that for single crystals[82]. At the time the DS technique was applied to produce columnar grained alloys, (1969) it was recognized that single crystal preparation by the same technique was feasible. However, merely utilizing conventional or columnar grained chemistry for single crystal alloys is not sufficiently beneficial to warrant the extra expense of the single crystal process. Therefore, a new class of alloys, containing none of the grain boundary strengthening elements (B, C, Zr, Hf), was developed, see Table 1. The elimination of these elements raised the melting points of the alloys, thereby permitting higher solution treatment temperatures. This in turn led to more complete solutioning of γ', and increased ability to control the size of the particles produced by heat treatment. A comparison of mechanical properties of and resistance to corrosion of single crystals vs columnar grained alloys and polycrystals is shown in Fig. 19.[83]

Another feature of the single crystal alloys is the use of rhenium as a strengthener (see Table 1). Rhenium hardens the gamma matrix and may enter the γ' precipitates. However, Re may also promote σ phase formation [84], so its use in superalloys needs to be offset by lowering the quantity of chromium and/or tungsten, which also promote σ.

Other factors to consider with respect to rhenium is its very high cost and high density. Nevertheless, the improvements in creep properties and temperature capability of Re-containing alloys has justified its use in spite of these disadvantages. A particularly instructive comparison between first generation and second generation simple crystal alloys is provided by CMSX2 and CMSX4, (see Table 1). The latter has lower Cr, 3% Re, higher overall refractory metal content (Mo+W+Ta+Re) and increased Co (9%). As a result, CMSX 4 exhibits a 35°C increase in temperature capability over CMSX 2 based on creep-rupture resistance. It also exhibits extremely good hot corrosion resistance. Further, the elimination of all grain boundaries improves thermal fatigue resistance relative to columnar grained alloys, see Fig. 20[60].

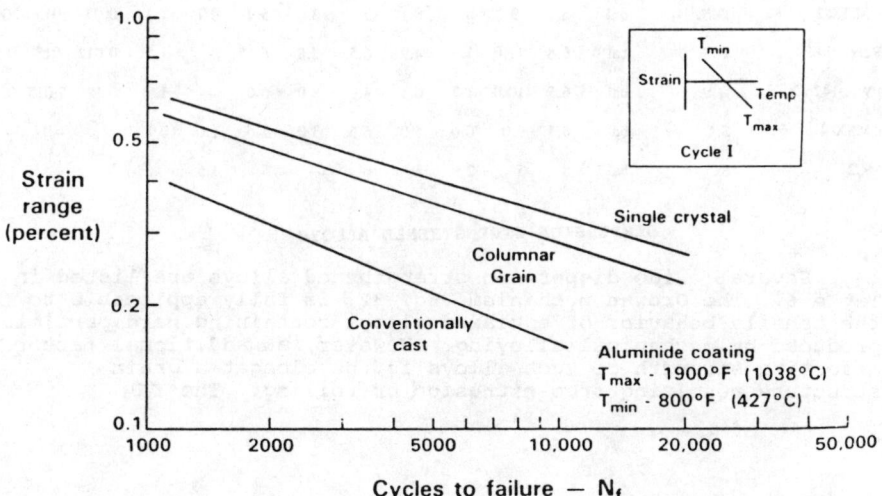

Fig. 20. Thermal fatigue of columnar grain vs conventionally cast superalloys.[60]

NEW SUPERALLOYS

Development of nickel base superalloys continues at a rapid pace. A summary of compositions of several new alloys appears in Table 5. Among the objectives of alloy development is the production of more stable alloys, alloys with higher creep and/or fatigue strengths, the replacement of single crystals by less expensive columnar grained alloys, and the development of new single crystals with enhanced temperature capability. New polycrystalline alloys include Rene′ 220W and 220C, N18 and Haynes 242. New columnar grained alloys include Rene′ 142 and PWA 1426. Finally, new single alloys include CMSX 4 and the French alloy MC2. Note that there is a strong tendency to use tantalum in the new alloys, and single crystal alloys do not always contain rhenium. Hafnium continues to be necessary in columnar grained alloys and is even finding its way into equiaxed alloys, although in small quantities.

TABLE 5
COMPOSITIONS OF RECENTLY DEVELOPED ALLOYS
wt%

	Type	Ni	Cr	Co	Mo	W	Ta	Re	Al	Ti	Hf	C	B	Zr
Rene′ 220W	P.C.	bal	18.0	12.0	3.0	--	3.0	--	0.5	1.0	--	.02	0.01	0.01
N18	P.C.	bal	11.2	15.6	6.5	--	--	--	4.4	4.4	0.5	.02	.015	.03
Haynes 242	P.C.	bal	8.0	--	25.0	--	--	--	--	--	--	--	--	--
CM 1862	D.S.	bal	6	9	0.5	8	3	3	5.7	0.7	1.4	0.07	.015	.005
Rene′ 142	D.S.	bal	6.8	12.0	1.5	4.9	6.5	2.8	6.15	--	1.5	0.012	.015	0.02
PWA 1426	D.S.	bal	6.5	10.0	1.7	6.5	4.0	3.0	6.0	--	1.5	0.1	0.015	0.1
CMSX-4	S.C.	bal	6.5	9.0	0.6	6.0	6.5	3.0	5.6	1.0	0.1	--	--	--
MC2	SC	bal	8	5	2	8	6	--	5	1.5	--	--	--	--

DISPERSION-STRENGTHENED ALLOYS

Several oxide dispersion strengthened alloys are listed in Table 6. The Orowan mechanism, Eq. 32, is fully applicable to the tensile behavior of equiaxed alloys containing hard particles produced by mechanical alloying. However, an additional factor affecting strength of such alloys is the elongated grain structure resulting from extrusion or rolling. The Y_2O_3

particles in mechanically alloyed materials are fine (approximately 30 nm diameter) and quite uniformly dispersed. The hardening due to these must be added to the strengthening effects of grain boundaries and subboundaries, as well as solid solution additions. An additional factor is the grain aspect ratio, GAR, the ratio of grain length D to width d. At high temperatures, tensile strength varies approximately linearly with GAR[68]. It should be noted that transverse creep properties are relatively poor in alloys with elongated grains.

TABLE 6

NOMINAL COMPOSITION OF SELECTED MECHANICALLY ALLOYED MATERIALS

Alloy designation	Ni	Fe	Cr	Al	Ti	W	Mo	Ta	Y_2O_3	C	B	Zr
MA754	Bal	...	20	0.3	0.5	...	...	...	0.6	0.05	...	...
MA758	Bal	...	30	0.3	0.5	...	...	...	0.6	0.05	...	...
MA760	Bal	...	20	6.0	...	3.5	2.0	...	0.95	0.05	0.01	0.15
MA6000	BAL	...	15	4.5	2.5	4.0	2.0	2.0	1.1	0.05	0.01	0.15

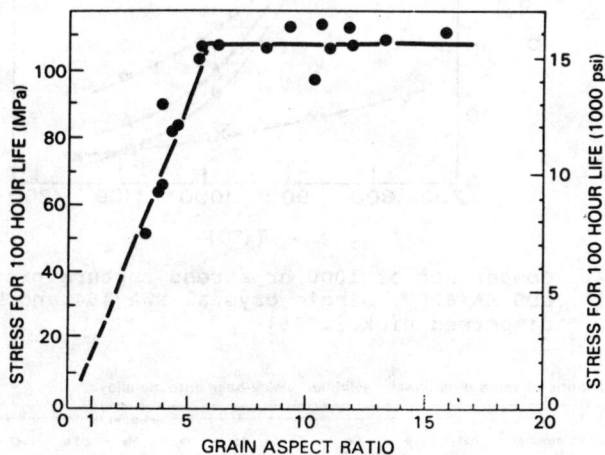

Fig. 21. Influence of grain aspect ratio on stress for 100 hr life for MA 753 at 1040°C.[85]

Creep and stress-rupture behavior also correlate well with GAR, as shown in Fig. 21 for MA 753 at 1040°C[84] Wilcox and Clauer[68] concluded that when grains are elongated the GAR effect swamps any contribution of grain size.

The strength of ODS alloy MA 6000 is greater than that of DS Mar M200 + Hf and single crystalline PWA 454 at temperatures above 1000°C, Fig. 22[86]. A further advantage of mechanical alloying is a much higher ratio of fatigue resistance to tensile strength relative to precipitation hardened alloys.

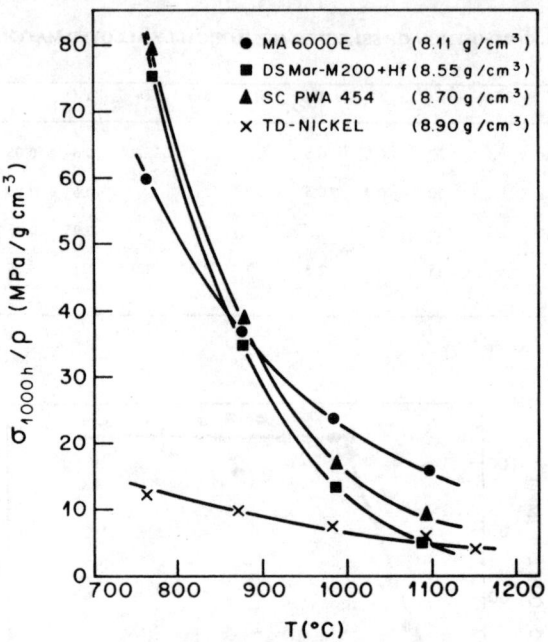

Fig. 22. Comparison of 1000 hr stress rupture properties of ODS MA 6000, single crystal PWA 454 and ThO_2-dispersed nickel.[86]

Table 7 Compositions of some directionally solidified nickel-base eutectic alloys

Alloy	Morphology	f(a)	Ni	Co	Cr	Al	Nb	Mo	Ta	C	Other
Nitac	F	0.05	69	...	10	5	...	...	14.9	1.1	...
Nitac 13	F	...	63	3.3	4.4	5.4	...	...	8.1	0.54	3.1 W, 6.2 Re, 5.6 V
Cotac-744	F	...	64	10	4	6	3.8	2	...	0.47	10 W
γ/γ'-δ (6% Cr)	L	0.3	71.5	...	6	2.5	20	...	...	...	...
γ/γ'-δ (0% Cr)	L	0.3	76.5	...	...	2.5	21	...	...	...	...
γ'/γ-Mo (AG-34)	F	0.26	62.5	...	...	6.3	...	31.2	...	...	...
γ-δ	L	0.26	66.7	...	...	...	23.3	...	...	...	...
γ'-Ni_3Ta	L	0.35	64.1	...	...	4.9	...	31	...	...	...

(a) Volume fraction.

COMPOSITE STRENGTHENING

Two classes of composite materials with a nickel or nickel alloy matrix have been developed for improved high temperature creep resistance: directionally solidified (DS) eutectics and wire-reinforced compositions. Compositions of several advanced DS eutectics are listed in Table 7. Note that several different fibers have been incorporated into a superalloy matrix: for example, TaC, NbC and molybdenum, and each of these alloys can be heat treated to increase strength by controlled precipitation of γ' in the γ matrix. However, the Mo-reinforced alloys are the most responsive of these to heat treatment.

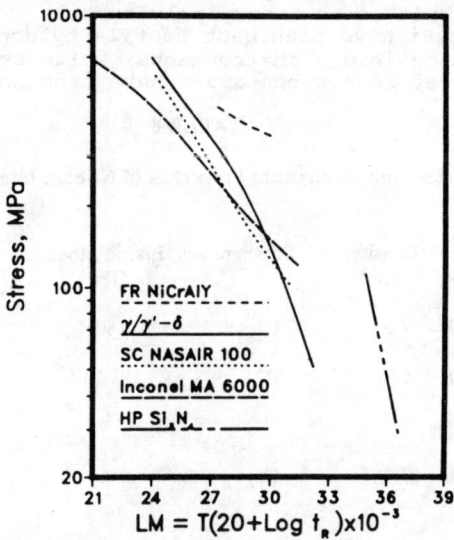

Fig. 23. Larson-Miller plot comparing fiber reinforced NiCrAlY with other high temperature materials.[89]

Concurrent with the development of aligned eutectics was a program to develop wire reinforced nickel base alloys. Ceramic fibers such as B and SiC proved to be too reactive with superalloys to produce adequately stable microstructures. However, tungsten and various tungsten alloy fibers (W-Re-HfC, W-ThO$_2$) have been effective in improving strength of nickel-base alloys for at least short term (100 hr) exposures, see Fig. 23[89]. A temperature advantage of up to 175°C is indicated at stress levels to 350MPa. Further, the high cycle fatigue resistance relative to tensile strength of tungsten fiber-reinforced alloys is substantially higher than for either precipitation hardened alloys (Nimonic 75, 95, 105) or solid-solution strengthened Hastelloy X.

INTERMETALLIC COMPOUNDS

A more recent attempt to achieve superior high temperature properties in nickel-base alloys is through the development of alloys based on single phase intermetallic compounds. Three such nickel-rich systems are of sufficient interest to mention here: Ni_3Al, NiAl and Ni_3Si. A compilation of physical properties appears in Table 8. Note that NiAl has the highest melting point (1640°C) and the lowest density of the three. Unfortunately, all three alloys are brittle as binary polycrystals, when tested in air. Ni_3Al and Ni_3Si have been made ductile by doping with boron, or by testing in dry environments, while repeated attempts to ductilize NiAl at room temperature have been unsuccessful.

TABLE 8

Physical and Mechanical Properties of Ni-Base Intermetallics

Compound	Tm °C	Density g/cm³	Structure	Elastic Modulus* GPa	Slip Systems 25°C
Ni_3Al	1390	7.5	$L1_2$	180	{111}<110>
NiAl	1640	5.86	B2	185-240	{001}<001> {110}<001> {112}<111>
Ni_3Si	1284	7.25	$L1_2$	--	{111}<110>

*25°C, polycrystals

The tensile and creep behavior of NiAl and Ni_3Al have been widely studied; the most remarkable aspect of strength behavior of the latter is the unusual rise in flow stress with increasing temperature shown in Fig. 5[9]. A series of new alloys, containing Cr, Zr, or Hf, have been produced with strength comparable to those of some early superalloys such as Waspaloy. The new alloys are nominally single phase, although small percentages of γ or β phase also may be present. NiAl seems to have the best chance for development as a high temperature structural material in spite of its open CsCl structure. Appreciable strengthening of NiAl single crystals has been achieved by a combination of solid solution and precipitation hardening see Fig. 24[88]. Other advantages of NiAl include low density, high melting point and high thermal conductivity. However, low ductility and fracture toughness, arising in part from too few independent slip systems, continue to hamper development efforts.

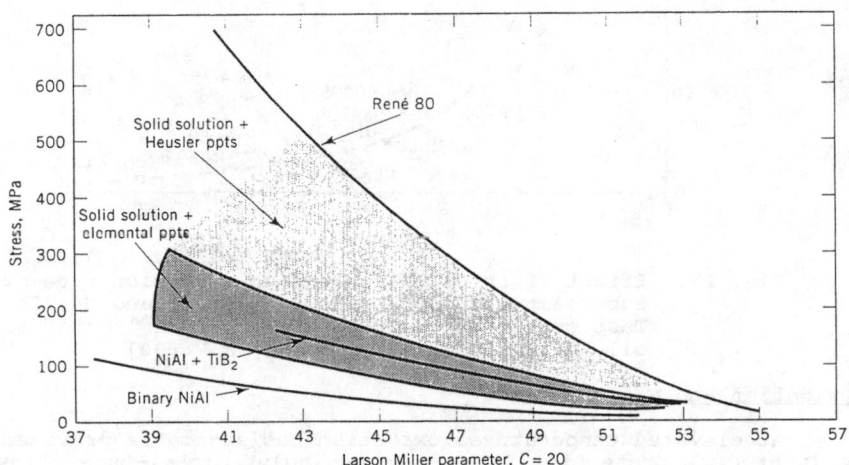

Fig. 24. Larson-Miller plot of stress rupture data for various NiAl alloys compared to superalloy René 80.[88]

A potentially more promising route to high temperature structural intermetallics is through reinforcement with filaments. Al_2O_3, especially as continuous filaments, has been shown to strengthen and provide limited increases in toughness.[92]

ALLOYING FOR SURFACE STABILITY

Low Temperature Corrosion

Several elements contribute to the surface stability of nickel base alloys, depending upon the media to which any alloy are exposed. High chromium contents are required for low temperature resistance to corrosive media such as aqueous solutions and acids, and a series of Hastelloy and Inconel alloys has been developed for such applications. Hastelloy C and IN 600 are typical: the former contains 16.5%Cr and 17%Mo as principal components for corrosion resistance while IN 600 contains 15.5% Cr and 8% Fe. Other Hastelloy alloys contain up to 28%Mo, sometimes with small additions of tungsten.

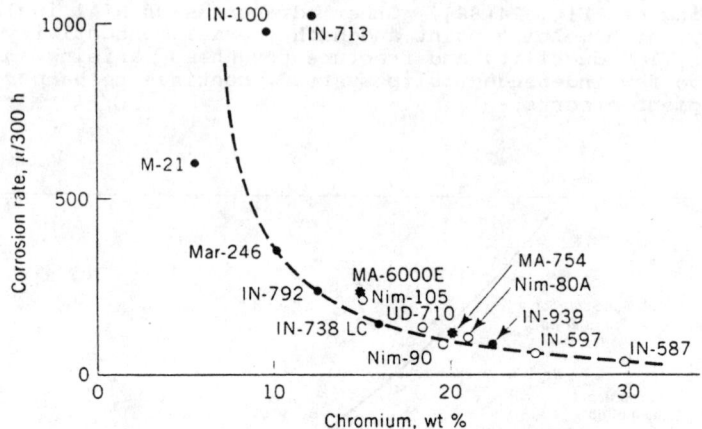

Fig. 25. Effect of chromium content on corrosion rate of superalloys and ODS alloys MA 6000E and MA 754. Test conductors: 300h, 850°C, extra light fuel oil (0.3-0.4%S) 15 ppm Na, 5ppm V.[90]

Oxidation and Hot Corrosion

At elevated temperatures oxidation resistance is provided by Al_2O_3 or Cr_2O_3 protective films. Accordingly, nickel-base alloys must contain one or both of these elements even where strength is not a principal factor. For example, Hastelloy X, one of the most oxidation and (hot) corrosion resistant of all nickel-base alloys, contains 22%Cr, 9%Mo and 15.8% Fe as principal solutes. Since Hastelloy X is essentially a solid-solution alloy when placed into service (carbides precipitate after long-term exposure), the alloy is much weaker than superalloys containing γ' or γ as strengthening precipitates. Chromium is known to degrade the high temperature strength and lower the antiphase boundary energy of γ' (see Fig. 5), so that there has been a strong incentive to lower chromium content in modern superalloys. Thus the level of chromium decreased from 20% in earlier alloys to as little as 9% in Mar-M 200. Unfortunately, this compositional change degraded hot corrosion resistance, see Fig. 25[90], to the point that superalloys utilized in gas turbines had to be coated. Further,. as turbine blade temperatures exceed 1000°C, Cr_2O_3 tends to decompose to CrO_3, which is volatile and, therefore, not protective. To some extent the loss of oxidation resistance has been compensated for by raising Al contents although the latter resides primarily in γ'. (Aluminum in small quantities promotes the formation of Cr_2O_3). However, Al_2O_3 is less protective than Cr_2O_3 under sulfidizing conditions, so that coatings have become indispensable in both aircraft turbines and more recently in industrial turbines. Other elements that contribute to oxidation and hot corrosion resistance are tantalum, yttrium and lanthanum. The rare earths appear to improve oxidation resistance by preventing spalling of oxide, while the mechanism for improvement with tantalum is not known.

Yttrium is now widely utilized in overlay coatings of the NiCrAlY type of superalloys.

Finally, it must be pointed out that molybdenum and tungsten are considered to be the most deleterious solutes from the point of view of hot corrosion resistance. Nevertheless, one or both of these elements is required for strength (e.g. see most of the γ'-strengthened alloys in Table 1), so that alloying for improved surface stability is often in conflict with alloying for strength. The two most prominent solutes which provide both strength and surface stability are aluminum and tantalum.

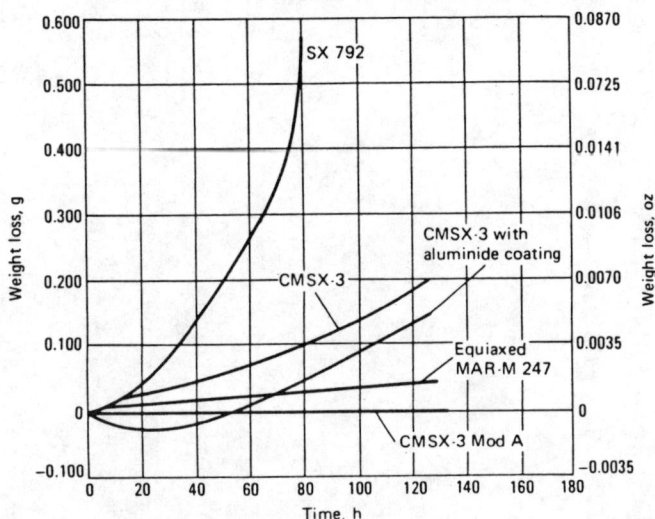

Fig. 26. Dynamic oxidation data for equiaxed and single crystal superalloys at 1177°C(2150°F).[91]

A comparison of cyclic oxidation data for several Ni-base, in Fig. 26[91]. Note that the best resistance to both cyclic oxidation is provided by a modified single crystal alloy CMSX-3 mod. A that is uncoated. Equally good was equiaxed MAR-M247.

SUMMARY

Nickel alloys are very diverse in type, composition and methods of processing. Nickel is an extremely stable alloy base, so that many elements can be dissolved without change in crystal structure. For high strength at low and intermediate temperatures (to 815°C) γ'' or γ' particles are precipitated from solution. For temperatures between 815°C and 1100°C γ'-strengthened alloys are the strongest, especially in single crystal form. However, maximum creep and fatigue resistance is achieved in composites and ODS alloys. Intermetallic compounds based upon nickel are under development as intermediate and high temperature structural materials, but are not, as yet, comparable to the stronger superalloys in strength or creep resistance. Alloying for improved mechanical properties at low temperature is often the cause of reduced creep strength or increased susceptibility to cracking caused by an external environment. Therefore, alloy design is often based upon compromises involving the temperature range of service, applied stress levels and ambient environment.

REFERENCES

1. G. L. Erickson, *Metals Handbook,* **1**, 10th Ed., (1990), 981.

2. M. Cybulsky and P.E.C. Bryant, *Advanced Materials and Coatings for Combustion Turbines,* eds. V.P. Swaminathan and N.S. Cheruva, (ASM, Materials Park, OH, 1994), p. 23.

3. W. L. Hawkins and S. Lamb, *Metals Handbook,* **2**, 10th Ed., (ASM, Materials Park, OH, 1990), p. 428.

4. S. T. Wlodek and R. D. Field, *Superalloys 92,* eds. S. D. Antolovich, et al, (TMS, Warrendale, PA, 1992), p. 477.

5. B. H. Lawless and J. F. Baker, *Superalloys 92,* eds. S. D. Antolovich, et al, (TMS, Warrendale, PA, 1992), p. 53.

6. E.W. Ross and C.T. Sims, *Superalloys II,* eds. C. T. Sims, N. S. Stoloff and W. C. Hagel, (John Wiley & Sons, New York, 1987), p. 97.

7. R. F. Decker and C. T. Sims, *Superalloys,* eds. C. T. Sims and W. C. Hagel, (John Wiley & Sons, New York, NY, 1972), p. 33.

8. S. Ochiai, Y. Oya and T. Suzuki, *Acta Metall.,* **32**, (1984), 289.

9. P. H. Thornton, R.G. Davies, and T.L. Johnston, *Met. Trans.,* **1**, (1970), 207.

10. R. G. Davies and N. S. Stoloff, *Trans. Met. Soc. AIME,* **233**, (1965), 714.

11. N. S. Stoloff and R. G. Davies, *Prog. Mat. Sci.* **13** (1966), 1.

12. D. M. Wee, O. Noguchi, Y. Oya and T. Suzuki, *Trans. Japan Inst. Met.,* **21**, (1980), 237.

13. L. R. Curwick, PhD Thesis, University of Minnesota, (1972).

14. H. L. Eiselstein, *ASTM STP 369,* (ASTM, Philadelphia, PA, 1965), p. 62.

15. W. J. Boesch and H. B. Canada. *J. of Metals,* **2(10)**, (1969), 34.

16. J. M. Nell and N. J. Grant, *Superalloys, 92,* eds. S. D. Antolovich et al, (TMS, Warrendale, PA, 1992), p. 113.

17. C. T. Sims, *J. of Metals,* **18(10)**, (1966), 1119.

18. B. J. Piearcey and R.W. Smashey, *Trans AIME,* **239**, (1967), 451.

19. J. R. Mihalisin, *Trans. Met. Soc. AIME,* **239**, (1967), 180.

20. G. Erickson, *Critical Issues in the Development of High Temperature Structures Materials,* eds. N. S. Stoloff, D. J. Duquette and A. F. Giamei, (TMS, Warrendale, PA, 1993), p.87.

21. E. W. Ross, "Recent Research on IN-100", AIME Annual Meeting, Dallas, TX, February 1963.

22. C. T. Sims, *The Superalloys,* eds. C. T. Sims and W. C. Hagel, (John Wiley & Sons, New York, 1972), p. 259.

23. N Yukawa, M. Morinaga, Y. Murata, H. Ezaki and S. Inoue, *Superalloys, 1988,* eds. D. N. Duhl et al, (TMS, Warrendale, PA, 1992), p. 225.

24. R.L. Fleischer, *Acta Metall.,* **11**, (1963), 203.

25. R. M. N. Pelloux and N. J. Grant, *Trans. Met. Soc. AIME,* **218**, (1960), 232.

26. E. R. Parker and T. H. Hazlett, *Relation of Properties to Microstructure,* (ASM, Metals Park, Ohio, 1954), p. 30.

27. B. E. P. Beeston, I. L. Dillamore and R. E. Smallman, *Met. Sci. J.,* **2**, (1968), 12.

28. H. Mishima, et al, *Trans Japan Inst. Met.,* **27**, (1986), 656.

29. R. L. Fleischer, *The Strengthening of Metals,* (Reinhold, New York, 1965), p. 93.

30. R. Labusch, *Acta Metall.,* **20**, (1972), 917.

31. J. C. Fisher, *Acta Metall.,* **2**, (1954), 9.

32. R. Nordheim and N. J. Grant, *J. Inst. Met.,* **82**, (1954) 440.

33. N. S. Stoloff, *The Superalloys,* eds. C. T. Sims and W.C. Hagel, (John Wiley & Sons, New York, 1972), p. 79.

34. E. Orowan, *Symposium on Internal Stresses in Metals,* (Institute of Metals, London, 1948), p. 451.

35. E. Nembach and G. Neite, *Prog. in Mat. Sci.* **82**, (1985), 177.

36. E. Nembach, *Particle Strengthening of Metals and Alloys,* (John Wiley & Sons, New York, 1995), in press.

37. H. Gleiter and E. Hornbogen, *Mat. Sci. Eng.*, **2**, (1968), 285.

38. R. K. Ham, Ordered Alloys: Structural Applications and Physical Metallurgy, eds. J. H. Westbrook, et al, (Claitors, Baton Rouge, LA, 1970), p. 365.

39. L. M. Brown and R. K. Ham, *Strengthening Methods in Crystals*, eds. M. Kelly and R.B. Nicholson, (Elsevier, Amsterdam, 1971), p. 9.

40. V. Martens and E. Nembach, *Acta Metall.*, **23**, (1975), 149.

41. S. M. Copley and B. H. Kear, *Trans. Met. Soc. AIME*, **239**, (AIME, 1967), 977.

42. S. M. Copley and B. H. Kear, *Trans. Met. Soc. AIME*, **239**, (AIME, 1967), 984.

43. G. R. Leverant, M. Gell and S. W. Hopkins, *Proc. Sec. Int. Conf. Strength Met. and Alloys*, **3**, (1970), 1141.

44. R.F. Decker, Strengthening Mechanisms in Nickel-Base Superalloys, Climax Molybdenum Company Symposium, Zurich, May 5-6, 1969.

45. V. Gerold and H. Haberkorn, *Phys. Stat. Sol.*, **16**, (1966), 675.

46. H. Gleiter, *Z. Angew Physics.*, **23(2)** (1967) 108.

47. A. Kelly and R. B. Nicholson, *Prog. Mat. Sci.*, **10(3)**, (1963), 151.

48. R.R. Jensen and J.K. Tien, in *Metallurgical Treatises*, J.K. Tien and J.F. Elliott (eds) (TMS-AIME, Warrendale, PA, 1981), p. 529.

49. E. Nembach, K. Suzuki, M. Ichihara and S. Takeuchi, *Philos Mag. A.* **51**, (1985), 507.

50. P. Beardmore, R.G. Davies and T.L. Johnston, *Trans. Met. Soc. AIME*, **245**, (1969), 1537.

51. V.A. Phillips, *Philos Mag.*, **16**, (1967), 117.

52. R.G. Davies and T.L. Johnston, in *Ordered Alloys: Structural Applications and Physical Metallurgy*, (Claitors, Baton Rouge, LA, 1970), p. 447.

53. W.J. Mitchell, *Z. Metallkd*, **57**, (1966), 586.

54. V. Munjal and A.J. Ardell, *Acta Met.*, **23**, (1975), 513.

55. V.A. Phillips, *Scripta Metall.*, **2**, (1968), 147.
56. D. Raynor and J.M. Silcock, *Met. Sci. J.*, **4**, (1970), 121.
57. R.F. Decker and J.R. Mihalisin, *Trans. ASM*, **62**, (1969), 481.
58. I.L. Mirkin and O.D. Kancheev, *Met. Sci. Heat Treat.*, **10**, (1967), 1.
59. G.N. Maniar and J.E. Bridge, *Met. Trans.*, **2**, (1971), 95.
60. D.N. Duhl in *Superalloys II*, eds. C.T. Sims, N.S. Stoloff and W. C. Hagel, (John Wiley & Sons, New York, 1987), p. 189.
61. S.S. Ezz, D.P. Pope and V. Paidar, *Acta Metall.*, **30**, (1982), 921.
62. D. Shah and D. Duhl, *Superalloys 1984*, (TMS-AIME, Warrendale, PA, 1984), 105.
63. O.D. Sherby and P.M. Burke, *Prog. Mat. Sci.*, **13**, (1967), 325.
64. A.K. Mukherjee, J.E. Bird and J.E. Dorn, *Trans. ASM*, **62**, (1969), 155.
65. A.K. Mukherjee, *Treatise on Materials Science and Technology*, **6**, Plastic Deformation of Metals, R.J. Arsenault (ed), (Academic, New York, 1975), p. 163.
66. M. Gill and D.N. Duhl, *Processing and Properties of Advanced High Temperature Alloys*, eds. S. Allen et al, (ASM, Metals Park, OH, 1986), p. 41.
67. K.R. Williams and B. Wilshire, *Met. Sci. J.*, **7**, (1973), 176.
68. B.A. Wilcox and A.H. Clauer, *Oxide Dispersion Strengthening*, (Gordon & Breach, New York, 1968), p. 323.
69. G.R. Leverant and B.H. Kear, *Met Trans.*, **1**, (1970).
70. D.N. Duhl in *Superalloys 88*, eds. D.N. Duhl et al, (TMS, Warrendale, PA, 1988), p. 693.
71. J.K. Tien and S.M. Copley, *Met. Trans.*, **2**, (1971), 543.
72. J.K. Tien and R.P. Gamble, *Met. Trans.*, **3**, (1972), 2157.
73. D.D. Pearson, B.H. Kear and F.D. Lemkey, in *Creep Fracture of Engineering Materials and Structures*, (Pineridge Press, UK, 1981), p. 213.

74. D.D. Pearson, private communication.

75. T.B. Gibbons and B.E. Hopkins, Met. Sci. J., **5**, (1971), 233.

76. J.J. Jackson, M.J. Donachie, R.J. Herricks and M. Gell, Met. Trans. A, **8A**, (1971), 1615.

77. A.K. Koul and R. Castillo, in Advanced Materials and Coatings for Combustion Turbines, eds. V.P. Swaminathan and N.S. Cheruva, (ASM, Materials Park, OH 1994) p. 75.

78. E.G. Richards, J. Inst. Met., **96**, (1968), 365.

79. T.J. Gerosshen, T.D. Tillman and G.P. McCarthy, Met Trans A, **18A**, (1987), 69.

80. J.H. Schneibel, C.L. White and M.H. Yoo, Met Trans A, **16A**, (1985), 561.

81. M. Gell, D.N. Duhl and A.F. Giamei, in Superalloys 1980, J.T. Tien, et al, eds., (ASM, Metals Park, OH, 1980), p. 205.

82. J. Stringer and R. Viswanathan, Advanced Materials and Processes, (ASM Intl., Materials Park, OH, 1993), p. 1.

83. M. Gell and D.N. Duhl, J. of Metals, **39(7)**, (1987), 11.

84. R. Darolia, D.F. Lahrman and R.D. Field, Superalloys 1988, eds. D.N. Duhl et al., (TMS, Warrendale, PA, 1988), p. 255.

85. J.S. Benjamin and M.J. Bomford, Met. Trans., **5**, (1974), 416.

86. T.E. Howson, D.A. Mervyn and J.K. Tien, Met. Trans. A, **11A**, (1980), 1609.

87. J.E. Grossman and N.S. Stoloff, Met. Trans. A, **9A**, (1978), 117.

88. R. Darolia, D.F. Lahrman, R.D. Field, J.R. Dobbs, K.M. Chang, E.H. Goldman and D.G. Konitzer, Ordered Intermetallics, Physical Metallurgy and Mechanical Behavior, (Kluwer Acad. Publ, Dordrecht, NL, 1992), p. 679.

89. J.K. Tien and V.C. Nardone Fracture: Interactions of Microstructure, Mechanisms, Mechanics, (TMS-AIME, Warrendale, PA, 1984), p. 321.

90. G.H. Gessinger, Powder Metallurgy of Superalloys, (Butterworths, London, 1984), p. 282.

91. Allison Gas Turbine Division, General Motors Corp., see also Harris, et al, *Metals Handbook*, 10th Ed., 1 (ASM, Materials Park, OH, 1990), p. 995.

92. D.M. Shah and D.L. Anton, *Structural Intermetallics,* eds. R. Darolia et al, (TMS, Warrendale, PA 1993), p. 755.

Problems

1. Consider these nickel-base alloys, all with the same grain size,

    Ni-20Cr-10Co-5Al-1Ti-2Y_2O_3
    Ni-20Cr-10Co-5Al-1Ti-5Mo
    Ni-20Cr-3Fe-3Mo.

    Based upon their compositions and likely microstructures after aging,
    a) Which would have the highest creep resistance at 1000°C? Explain.
    b) Which would be the most resistant to low cycle fatigue at 20°C? Explain.
    c) Which would be strongest at 25°C in tension? Explain.

2. Use Figure 23 to determine the rupture times for SC Nasair 100 for a stress of 500 MPa and T = 1100°C. LM parameter is expressed in °K.

3. A gas turbine runs at constant speed, with centrifugal stress along the blades = 40MPa. There is 2mm clearance between the blade tips and the inner wall of the turbine casing. The blade length is 100 mm. Lab tests have shown the following steady state creep rates at σ = 40 MPa; 7.3×10^{-8}/s at 800°C and 1.29×10^{-5}/s at 950°C. Estimate the safe life of the blades at 825°C.

4. The creep rate of a nickel-base alloy is very sensitive to grain size at 900°C, σ = 50MPa, but is insensitive to grain size at 600°C, σ = 150MPa. Give a possible explanation.

5. Of the strengthening mechanisms below discussed in this chapter, classify according to whether
    a) both low and high temperature strengths of nickel are increased.
    b) primarily low temperature (25°C) strength is increased.
    c) primarily high temperature (1000°C) strength is increased.
        solid solution hardening
        precipitation hardening
        cold work
        dispersion strengthening
        composite strengthening

6. A tungsten atom diffuses to near an edge dislocation in nickel: Where will the tungsten lie with respect to the dislocation? Explain. Atomic radii are:
    r_{Ni} = 0.125 nm
    r_W = 0.137 nm

7. Nickel obeys the Hall-Petch relation. Find the yield stress for a grain diameter of 90μm, if yield stress is 250MPa for a grain diameter of 350μm and is 480MPa for a grain diameter of 30μm.

Chapter 3
Metal Matrix Composites

R. J. Arsenault

Department of Chemical and Nuclear Engineering
University of Maryland, College Park, MD 20742 USA

1. INTRODUCTION

Traditionally, metals and their alloys have been the principal engineering materials, because of their combination of strength, toughness, workability, reliability, modest cost, etc. Ceramics, although light, abrasion resistant, and with elevated temperature capability, have not been used for primary structures because of their lack of toughness. Common structural materials are primarily isotropic, and in terms of attainable density compensated strength and stiffness, their properties are reasonably equivalent (E/ρ 27 $\frac{GPa}{g}$·cm^3 and UTS/ρ 260 $\frac{MPa}{g}$·cm^3) [1]. The requirements of the aerospace industry for materials which have combinations of properties (strength, stiffness, density, durability, high temperature capability, etc.) not achievable by monolithic structural materials has been responsible for the considerable investment by the materials community in research and development of advanced composites. These materials can be tailored to take advantage of the desirable properties and to minimize the undesirable properties of their constituents.

The current interest in advanced metal matrix composites is based upon the development of high strength fibers (carbon, metals and ceramics) followed by the discovery that short fibers or whiskers (these two phrases will be used interchangeably), which are very strong, can be used to markedly increase the strength and modulus of metal composites [2]. Later, it was realized that even particulate reinforcements such as SiC, or Al_2O_3 can be used effectively to provide enhancement of properties such as strength, stiffness and wear-resistance, as well as modifying thermal expansion and electrical conductivity [3,4]. As a result, those metals reinforced with fibers or whiskers, high reinforcement price and fabrication costs, have to-date ruled out all, but for aerospace applications. In the latter case, the potential for inexpensive composites has been realized, and consequently, non-aerospace applications, such as automotive brake rotors, become attractive [5]. An additional advantage of discontinuously reinforced metal composites is that they can be shaped by standard metallurgical processes, such as forging, rolling, extrusion, etc., which also contributes to their potential modest cost [6].

In this chapter the treatment of the strengthening of composites will be exclusively that of discontinuous metal matrix composites (DMMC). Therefore, only a brief discussion of the fabrication of continuous filament metal matrix composites (CMMC) will be given, along with a discussion of processing DMMC.

2. PROCESSING

Many metal matrices have been investigated in various composite systems. Of these, the primary interest has been in Al, Mg, Cu and Ti alloys. The most popular reinforcements have been carbon, SiC, B, and Al_2O_3 fibers, W and stainless steel wires, and SiC and alumina particulate. The CMMC fabrication technique depends primarily on the matrix and reinforcement selected. Boron and SiC filament-reinforced Al and Ti composites are typically processed by plasma spraying the metal onto fiber lay-ups and using a hot pressing/diffusion bonding technique, as are metal wire reinforced superalloys, to form simple components [7-9]. Carbon and Al_2O_3 filament-reinforced Al and Mg composites are usually fabricated by liquid metal infiltration. Since fibers such as carbon and Al_2O_3 are not easily wet by molten metals, coatings or surface activation techniques are required. For example, in Al_2O_3 fiber-reinforced Al, 2-3% Li is added to the alloy in order to promote fiber wetting [10]. In another example, carbon filament-reinforced Al, TiB_2 coatings formed by chemical vapor deposition, has been used prior to pulling coated fiber tows through a molten metal bath [11]. Physical vapor deposition techniques, such as evaporation, sputtering and direct ion beam deposition have also been used to deposit a variety of metals either directly onto continuous filaments or as coatings to protect the fiber from interacting with the matrix. Such coated filaments or tapes are then laid-up and processed by heat and pressure to form plates, sheets, stringers, etc. [7]. In an innovative powder method called the powder cloth technique, matrix powders are bonded by a plastic binder and hot-rolled into semi-rigid sheets which can be handled and cut into monotapes [12]. Individual powder cloth sheets can be alternated with laid up fibers and pressed into composite shapes. The powder cloth process has been successfully employed to fabricate fully dense Ti-25Al-13Nb intermetallic alloys, unidirectionally reinforced with SiC filaments.

Metal matrix composites reinforced by discontinuous fibers, whiskers, or particulates, have been produced by a number of methods. A great deal of work has been done using standard powder metallurgy procedures to blend Al or Mg matrices and fibers, whiskers, or particulates, primarily of SiC or B_4C, followed by hot compaction and secondary processing [6,13-15]. Such methods have the advantage of using conventional procedures and equipment. Fig. 1 is a schematic of a powder metallurgical production process for SiC/Al composite. Particulate composites have also been fabricated by injecting ceramic particles into a spray of metal droplets and impinging them onto a substrate [16-18]. The projection of partly solidified droplets at high velocity leads to very fine microstructures, free of macro-segregation. Steel and Al matrices have been reinforced with Al_2O_3 and SiC in this manner.

Liquid state fabrication processes have been very popular, because of the potential for high volume, low cost production. A recent innovation involves SiC or Al_2O_3 particulate-reinforced Al produced totally by a near-conventional ingot metallurgy process [19]. The process consists of an arrangement of a reactor and a special stirrer by means of which the mixing of the ceramic particulates and the molten Al is achieved in-vacuum. Total mixing takes about an hour, after which, the mixture is cast in a conventional manner. These cast composites can be forged, rolled, extruded, etc. in a manner similar to monolithic Al alloys. Other such "foundry techniques" include the vortex method, a method consisting of vigorously stirring a liquid alloy while adding reinforcement particles in the liquid vortex, and COMPOCASTING™, a technique in which a semi-solid slurry of the metal while particulate-reinforcements are added to the surface [20,21]. An advantage of stirring the melt between the liquidus and the solidus is that the partially solidified structure drags the reinforcements along into the melt even if they are not chemically wet by the molten alloy. This mechanical entrapment is favored by a vigorous agitation, which is also supposed to promote wetting by abrasive cleaning of the

DMMC PRODUCTION

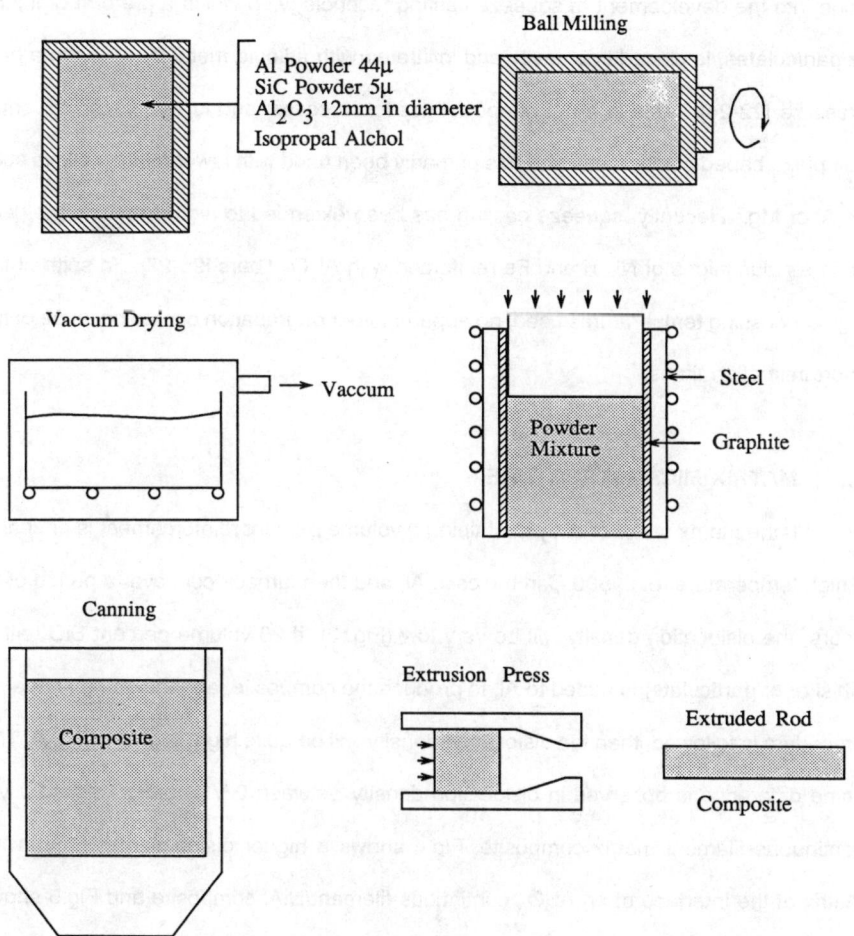

Fig.1 A schematic outline of the production of discontinuous metal matrix composites by the powder metallurgical procedure.

particulate surface. An additional benefit is the "grinding" off of corners of solidified metal dendrites to promote a more desirable cast microstructure. Considerable effort has also gone into the development of squeeze casting technology, in which a preform of fibers, or particulates, is placed in a cavity and infiltrated with a liquid metal by applying a high pressure [22-24]. This is a net shape fabrication process, and is well-suited for small complex-shaped components and has primarily been used with lower melting alloys such as Al or Mg. Recently, squeeze casting has been extended to higher melting matrices such as aluminides of Ni, Ti and Fe reinforced with Al_2O_3 fibers [25-27]. In spite of the high processing temperatures used, no apparent fiber degradation occurs because of the short infiltration times.

3. MATRIX MICROSTRUCTURE

If the matrix metal or alloy containing 0 volume percent reinforcement is anneal at a high temperature, e.g., 530°C in the case Al, and then furnace cool over a period of 12 hours, the dislocation density will be very low (Fig. 2). If 20 volume percent SiC (either whisker or particulate) is added to Al, to produce the composite, and the same annealing procedure is followed, then the dislocation density will be quite high $10^{14} m^{-2}$ (Fig.3). This same difference is observed in dislocation density between 0 V% matrix and a 20 V% continuous filament matrix composite. Fig.4 shows a higher dislocation density in the matrix at the interface of an Al_2O_3 continuous filament/NiAl composite and Fig.5 shows the slight decrease in dislocation density at distance of 200 µm from the filament interface.

This high dislocation density which is almost universally observed in metal matrix

Metal Matrix Composites 113

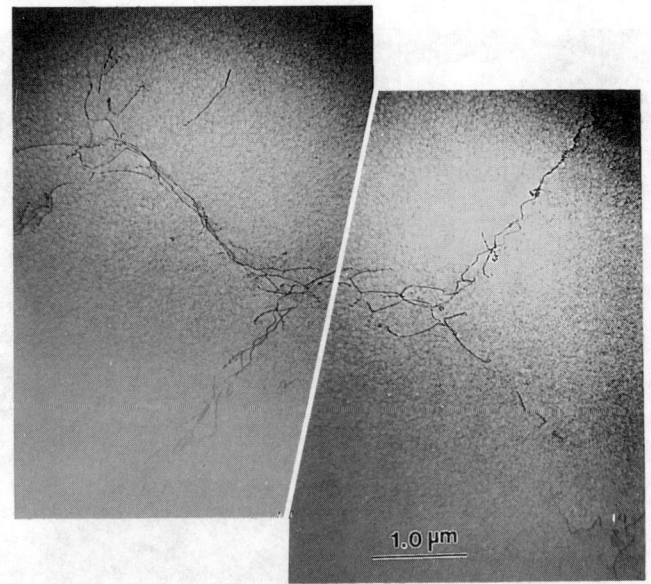

Fig.2 A transmission electron micrograph of high purity Al in the annealed condition thinned by dimpling and ion milling.

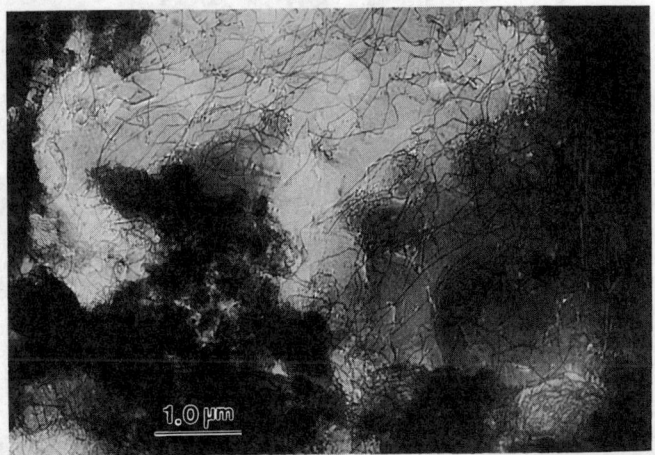

Fig.3 A transmission electron micrograph taken at an operating voltage of 1 MV.

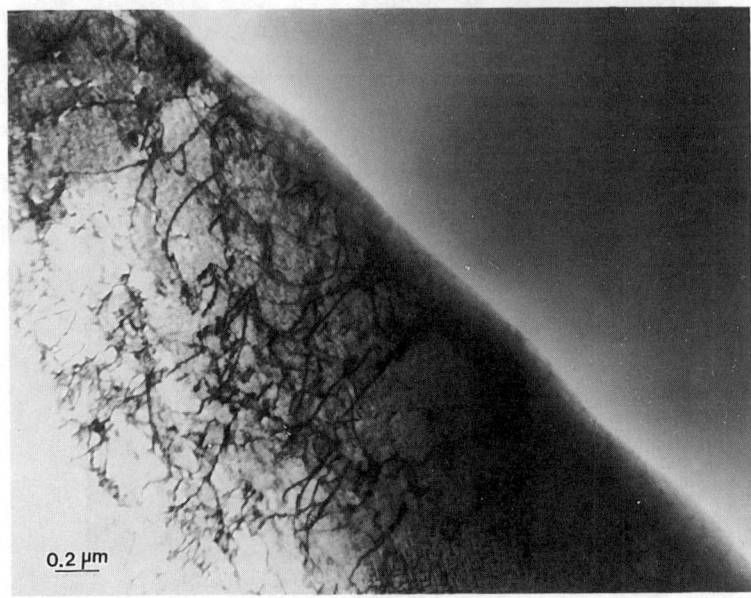

Fig.4   A transmission electron micrograph of a continuous filament Al_2O_3/NiAl composite. This shows a segment of the interface area.

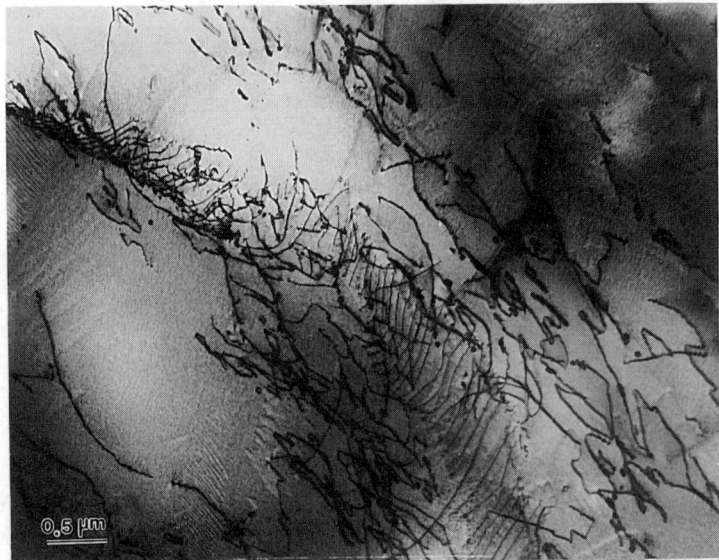

Fig.5   A transmission electron micrograph of the matrix in a continuous filament Al_2O_3/NiAl composite at a distance of 200 μm from the interface.

composites is one of the more unique and interesting aspects of metal matrix composites. Along with an increase in the dislocation density there is corresponding decrease in the subgrain size, and Fig.6 shows the change of dislocational subgrain size as a function of reinforcement particle size at a constant-reinforcement-volume fraction.

The changes of the microstructure are caused by the relaxation of the thermal residual stresses (TRS), which result from the difference in thermal coefficient of expansion of reinforcement (whiskers or particles) and the matrix (ΔCTE), and the change in temperature from the anneal or processing temperature and room temperature (ΔT). The dislocation generation and motion result in an increase in the dislocation density within the matrix (i.e., very similar to cold working) and a reduced subgrain size.

Arsenault and Shi [28] developed the first model relating prismatic dislocation punching due to TRS. This model is quite simple and it only predicts the lower limit of the dislocation density increase due to relaxation of the TRS. The dislocations generated are only those required to relax the thermal strain produced. These dislocations have been defined as the geometrically necessary dislocations. Attempts by Shi and Arsenault [29] to determine the upper limit of dislocation density proved unsuccessful. The predicted or calculated density was found to approach infinity, which is obviously incorrect.

Several other models put forward to account for the increase in dislocation density and these will be considered after the simple model by Arsenault and Shi [28] (for most of the other models are based on the premise that the number of dislocations generated are those geometrically necessary). The model is based on a DMMC, i.e., SiC whiskers in an Al matrix. Assumptions were made that both SiC and Al were elastically isotropic

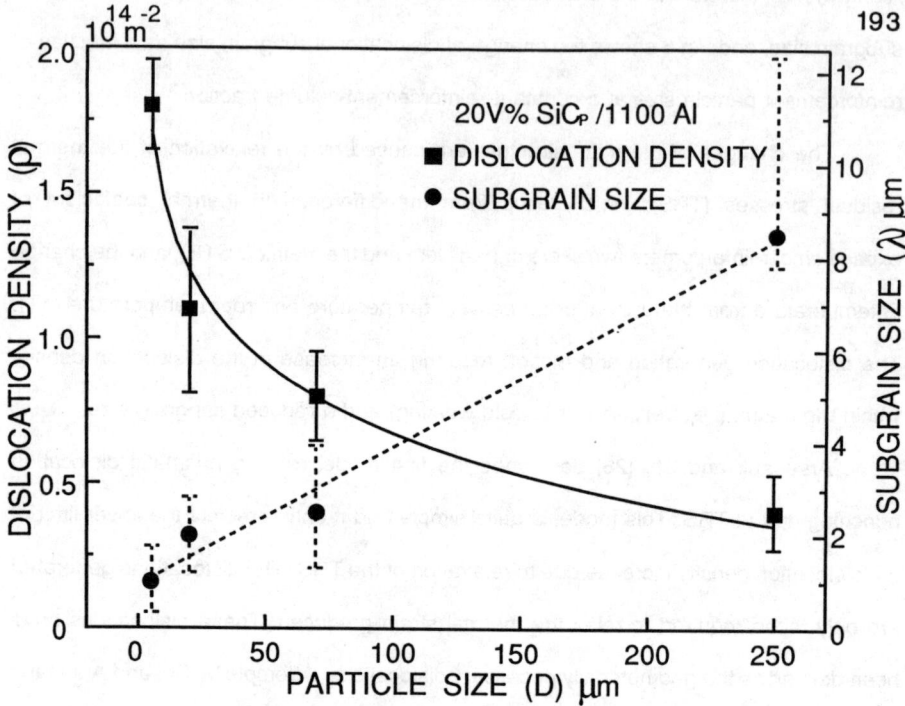

Fig. 6 The Change of dislocation density as a function of particulate size

and that the SiC reinforcement was in the form of parallelepiped particles. Prismatic punching was assumed to occur equally on all faces of the particles, as shown in Fig. 7.

Arsenault and Shi [28] then obtained the following eqn. for the increase in dislocation density

$$\Delta\rho = \frac{2V_p\varepsilon}{b(1-V_p)}\frac{S}{v'} \qquad (1a)$$

where V_p is the volume fraction of the particles, ε is the misfit strain ($\varepsilon = \Delta\alpha\Delta T$), where $\Delta\alpha$ is difference in coefficient of thermal expansion between matrix and reinforcement and ΔT is magnitude of the temperature change, S is the total surface area of a particle, v' is the volume of the particle. This means that from for the same volume fraction, in general, the smaller the particle size or the larger the surface area per unit volume of reinforcement material, the higher the dislocation density.

In general, the increase dislocation density $\Delta\rho$ due to punching can be written as

$$\Delta\rho = \frac{BV_p\varepsilon}{b(1-V_p)}\frac{1}{D} \qquad (1b)$$

where D is the diameter of the reinforcement, B is a geometric constant which is theoretically between 4 (R=∞) and 12 (R=1), and R is the aspect ratio (R=D/ℓ) where ℓ is the length or thickness of the reinforcement.

The above determination of the increase in dislocation density is based on the premise that relaxation of the TRS occurs by motion of just enough dislocations to relax the thermal residual strain. These dislocations have been defined as "geometrical dislocations", an alternative mechanism which has been put forward by Shibata et al. [30]

is that the geometrical dislocations from one particle annihilate the geometrical dislocations from the neighboring particles and the only dislocations left are those which are "statistically stored". The increase in dislocation density was obtained by relating the energy of a dislocation line to some fraction of the plastic work done during the relaxation of TRS.

For mathematical simplicity, Kim et al. [31] assume that all of the reinforcement particles are of spherical shape with radius a, and further, that a given particle is surrounded by uniformly distributed neighboring particles. This assumption allows us to use spherical coordinates whose origin is at the center of a ceramic particle with radial symmetry, as shown schematically in Fig. 8. Perfect plastic behavior is assumed for the matrix phase, and the crystallographic nature of plastic flow is neglected. Thus, the matrix is considered to yield under the condition of a constant yield stress, and the yielding is considered independent of the stress axis. First it is necessary to obtain the stress and strain associated with a misfitting spherical hard particle in the absence of plastic relaxation, i.e., under purely elastic conditions. Then, employing the foregoing assumptions and continuum plasticity theory, the solution for the elasto-plastic deformation is determined when plastic relaxation takes place in the matrix.

Consider a thick hollow sphere whose internal surface of radius a is subjected to a pressure, P_1, by the thermal misfitting ceramic particle and whose external surface of radius b is subjected to a pressure, P_2, by the surrounding ceramic particles, as well as the shear stresses and shear strains, are all zero, and the radial displacement, u, is a function of the radial distance, r. Further, the equilibrium equations in the absence of body

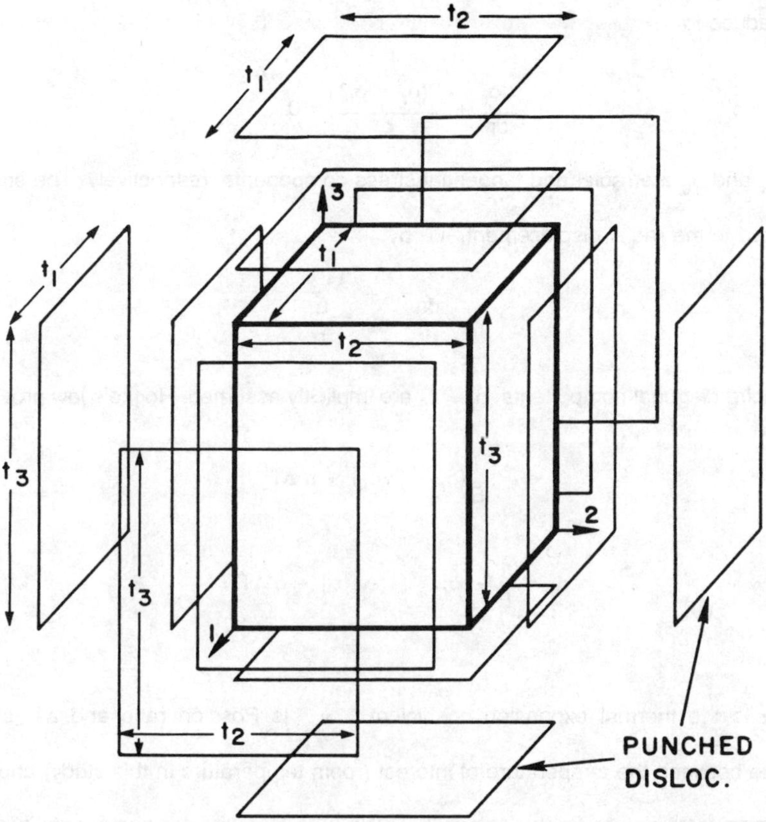

Metal Matrix Composites 119

Fig. 7 A schematic diagram of the particle and several punched dislocations

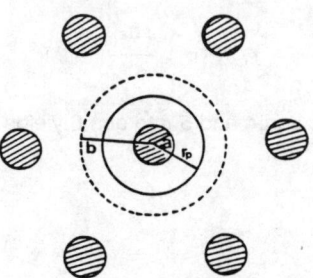

Fig. 8 A schematic diagram for an idealized ceramic configuration and plastic zone

forces reduce to

$$\frac{d\sigma_r}{dr} + \frac{2(\sigma_r - \sigma_\theta)}{r} = 0 \tag{2}$$

where σ_r and σ_θ are radial and tangential stress components, respectively. The strains are related to the radial displacement, u, by

$$\varepsilon_r = \frac{du}{dr}, \quad \varepsilon_\theta = \frac{u}{r} \tag{3}$$

For the third diagonal components, $\sigma_\phi = \varepsilon_\theta$ are implicitly assumed. Hooke's law provides

$$\varepsilon_r = \frac{1}{E}(\sigma_r - 2\nu\sigma_\theta) + \alpha \cdot \Delta T \tag{4}$$

$$\varepsilon_\theta = \frac{1}{E}[-\nu\sigma_r + (1-\nu)\sigma_\theta] + \alpha \cdot \Delta T \tag{5}$$

where α is the thermal expansion coefficient, ν is Possion ratio and ΔT is the difference between the temperature of interest (room temperature in this study) and the solutionizing temperature for the composites. We note that with the above definition for the strain, the reference state is the one at the solutionizing temperature. The strain compatibility relationship is

$$\varepsilon_r = \varepsilon_\theta + \frac{r d\varepsilon_\theta}{dr} \tag{6}$$

Substitution of the stress-strain eqn.4 and 5 into eqn.6 yields

$$\frac{1}{2}\frac{d^2}{dr^2}(r^2\sigma_r) - \sigma_r = 0 \tag{7}$$

After integration, the radial stress is given by

$$\sigma_r = A - \frac{B}{r^3} \tag{8}$$

and from eqn.1 the tangential stress is

$$\sigma_\theta = A + \frac{B}{2}r^3, \tag{9}$$

where A and B are constants.

There are several intervening steps, which given in detail by Kim et al.[31], then the elastic strain energy inside the spherical ceramic, ω_c, per unit volume of the reinforcement is given as

$$\omega_c = \frac{1}{2}[3\sigma_r(\varepsilon_r - \alpha_c \cdot \Delta T)] \tag{10}$$

where σ_r is the radial stress, e_r is the radial strain, α_c is the coefficient of expansion of reinforcement particle and ΔT is the change in temperature.

The total work consumed in the plastic zone is the sum of the plastic strain energy and the elastic strain energy. The plastic work per unit volume of an element located at distance, r, is, by definition [32]

$$\omega^P(r) = \int_0^{\varepsilon_r^P} d\omega^P(r) = \int \sigma_{ij} d\varepsilon_{ij}^P \qquad (11)$$

$$= \int_0^{\varepsilon_r^P} \sigma_r d\varepsilon_r^P + \int_0^{\varepsilon_\theta^P} 2\sigma_\theta d\varepsilon_\theta^P$$

where the usual tensor suffix notation is used in the first equation. From the incompressibility condition, $\varepsilon_r^P + 2\varepsilon_\theta^P = 0$ for the plastic strains, we have $d\varepsilon_r^P = -2d\varepsilon_\theta^P$. Substituting this relationship into eqn.11 and noting that $\sigma_r - \sigma_\theta = \sigma_y$ = a constant, we obtain

$$\omega^P(r) = \int_0^{\varepsilon_r^P} (\sigma_r - \sigma_\theta) d\varepsilon_r^P = -\sigma_y \varepsilon_r^P \qquad (12)$$

$$= \frac{2(1-2v_m)\sigma_y^2}{E_m}[(r_p/r)^3 - 1]$$

where e_r^P is the plastic strain, v_m is Poissons ratio of the matrix, E_m is Young's modulus of matrix. Finally the increase in dislocation density is some fraction of the plastic work.

$$\Delta \rho = \frac{f\omega^P(r)}{0.5\mu b^2} \qquad (13)$$

where f is the fraction of plastic work dissipated in the formation of dislocations, $\omega^P(r)$ is given in eqn. 12, $0.5\,\mu b^2$ is the energy per unit length of dislocation, and μ is the shear modulus.

From a consideration of the two models of dislocation production considered, it is obvious that a high dislocation density can be generated. However in both cases the models predict a lower limit of the dislocation density.

In the case of a reduction of the subgrain size, there is no simple model which has been developed. However, as a result of the large increase in dislocation a form of dynamic recovery takes place and some fraction of the dislocations rearrange themselves into subgrain boundaries.

4. STRENGTHENING MECHANISMS

A. Classical Theories of Continuum Mechanics

There are two basic classical theories which can be used to predict the strengthening of DMMC and these are: the shear lag model and the Eshelby model.

The shear lag model was originally developed by Cox [33], and its detailed derivation is summarized by Kelly [34]. It is best suited for an aligned short fiber composite where short fibers of uniform length and diameter (hence, constant aspect ratio) are all aligned in the loading direction and distributed uniformly throughout the material (Fig.9a). Also, it should be made clear that this is an <u>elastic</u> solution. Then a "unit cell", which is a representative short fiber surrounded by the matrix, is focused on, as shown in Fig. 9(b) where the other boundary of the surrounding matrix is taken as the mid-surface between two short fibers. Let this aligned short fiber composite be subjected to the applied uniaxial strain e along the z-direction (or axial direction) and let the axial displacements in the fiber and the matrix on the boundary of the unit cell (r=D/2) be denoted by μ and ν respectively. Then, by assuming that the difference in the axial displacements, $\mu - \nu$, is proportional to the shear stress at the matrix-fiber interface τ_0, or $d\sigma_f/dz$. One can obtain at z

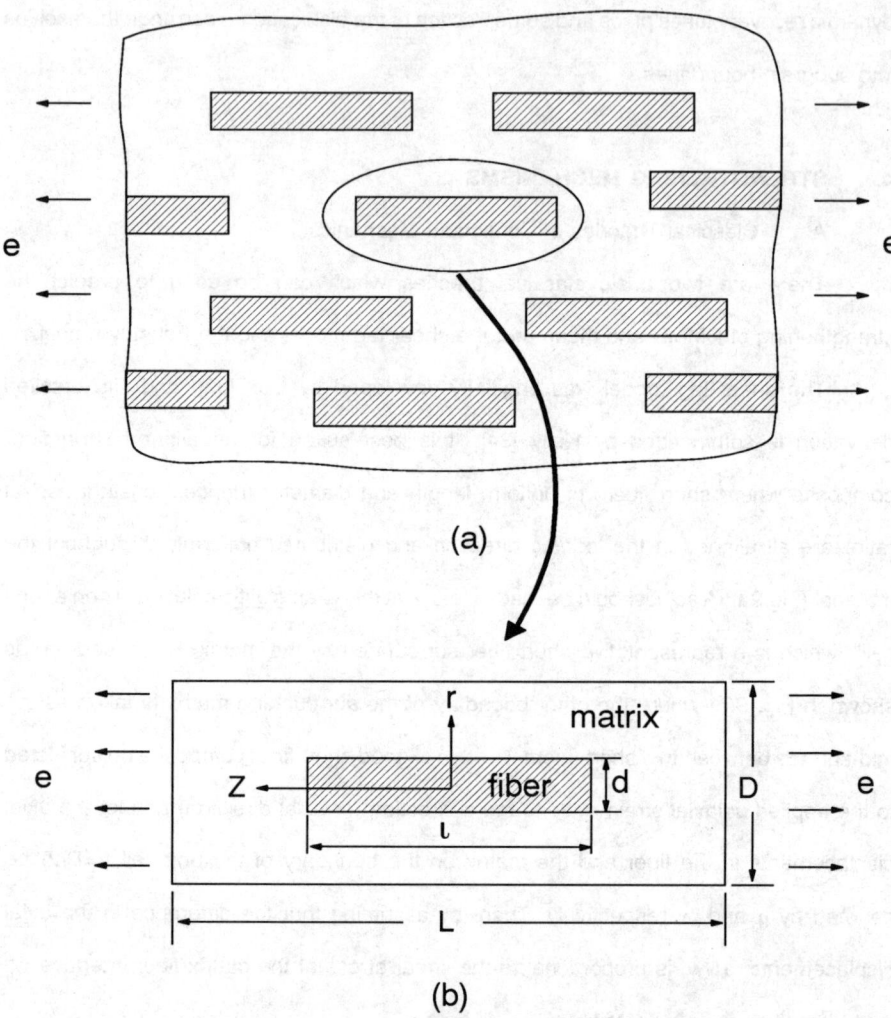

Fig.9 Shear lag model aligned short fiber composite; (a) representative short fiber, (b) unit cell for shear lag analysis.

$$\frac{d\sigma_f}{dz} = \frac{-4\tau_0}{d} = h(\mu - v) \tag{14}$$

where σ_f is the axial stress in the fiber and h is a constant which will be determined later. The first equation was derived by considering the equilibrium of force along the z-direction. It is noted in eq. that the positive direction of shear stress τ_0 is taken along the positive z-axis. In the fiber, a one-dimensional Hooke's law is valid

$$\sigma_f = E_f \frac{du}{dz} \tag{15}$$

where E_f is the axial Young's modulus of the fiber. The applied composite strain ε is equal to dv/dz. Hence, from eqs. (14) and (15) we have the ordinary differential equation

$$\frac{d^2\sigma_f}{dz^2} = h\left(\frac{\sigma_f}{E_f} - \varepsilon\right) \tag{16}$$

The general solution to eqn. (35) is given by

$$\sigma_f = E_f \varepsilon + C_1 \cosh\beta z + C_2 \sinh\beta z \tag{17}$$

where

$$\beta = \sqrt{\frac{h}{E_f}} \tag{18}$$

and C_1 and C_2 are unknown constants. Applying boundary conditions, σ_f = constant (σ_0) at z = ℓ/2 and $d\sigma_f/dz = 0$ at z = 0, we obtain the stress field in the fiber given by

$$\sigma_f = E_f\varepsilon\left\{1 + \frac{\left(\frac{\sigma_0}{E_f\varepsilon}-1\right)\cosh\beta z}{\cosh\left(\frac{\beta\ell}{2}\right)}\right\} \tag{19}$$

It is noted in eq.(19) that $\sigma_0 = 0$ was used in the original derivation of Cox [33,34], but σ_0 may not be zero for the case of strongly bonded fiber ends [35]. The average fiber stress $\overline{\sigma}_f$ is computed as

$$\overline{\sigma}_f = \frac{2}{\ell}\int_0^{\ell/2}\sigma_f dz$$

$$= E_f\varepsilon\left\{1 + \frac{\left(\frac{\sigma_0}{E_f\varepsilon}-1\right)\tanh\left(\frac{\beta\ell}{2}\right)}{\left(\frac{\beta\ell}{2}\right)}\right\} \tag{20}$$

Consider next the displacement along the z-direction at an arbitrary point (r=r) in the matrix, w, where w(r=d/2)= u, and w(r=D/2) = v. Force equilibrium at r=d/2 and arbitrary point (r=r) provides

$$2\pi r \tau = 2\pi\left(\frac{d}{2}\right)\tau_0 \tag{21}$$

The shear strain at r=r, γ is related to τ_0 as

$$\gamma = \frac{dw}{dr} = \frac{\tau}{G_m} = \frac{\tau_0}{2G_m}\frac{d}{r} \tag{22}$$

where τ is the shear stress in the matrix at r=r, and G_m is the shear modulus of the matrix. Integrating eq. (22) from r=d/2 to r=D/2, we obtain

$$v - u = \frac{\tau_0 d}{2G_m} \ln\left(\frac{D}{d}\right) \tag{23}$$

From eqs. (14) and (23), constant h is solved as

$$h = \frac{8G_m}{d^2 \ln(D/d)} \tag{24}$$

From eqs. (18) and (24), β is found as

$$\beta = \frac{2\sqrt{2}}{d}\sqrt{\frac{G_m/E_f}{\ln(D/d)}} \tag{25}$$

With β given by eq. (25) the average stress $(\bar{\sigma}_f)$ in the fiber can be calculated from eq. (20). To obtain the average stress in the composite along the loading (z) direction, σ_c can be estimated by using the rule of mixtures, where $\bar{\sigma}_m$ and $\bar{\sigma}_f$ are interpreted as the average quantities in the relevant domain

$$\sigma_c = (1 - V_f)\bar{\sigma}_m + V_f\bar{\sigma}_f \tag{26}$$

where V_f is the volume fraction of fibers. For a given applied strain ε, one can assume that

$$\bar{\sigma}_m = E_m \varepsilon \tag{27}$$

$$\sigma_c = E_c \varepsilon \tag{28}$$

A substitution of eqs. (20), (27) and (28) into (26) yields the Young's modulus of the composite E_c

$$E_c = (1-V_f)E_m + V_f E_f \left\{ 1 - \frac{\tanh\left(\frac{\beta\ell}{2}\right)}{\frac{\beta\ell}{2}} \right\} \tag{29}$$

Where $\sigma_0 = 0$ (i.e., no load transfer at fiber ends) was assumed. If $\sigma_0 = \sigma_m$ is assumed, then

$$E_c = (1-V_f)E_m + V_f E_f \left\{ 1 + \frac{\left(\frac{E_m}{E_f} - 1\right)\tanh\left(\frac{\beta\ell}{2}\right)}{\frac{\beta\ell}{2}} \right\} \tag{30}$$

It should be noted in the above derivation that the definition of applied strain e is not consistent with the law of mixtures where the average strain in the fiber should be equal to ε, which is not the case here, as seen from (eqn. 20). Despite this inconsistency, the shear lag analysis in conjunction with the rule of mixtures has been used extensively, mainly because of its simplicity. This shear lag model can be modified and made applicable to prediction of flow stress during the plastic deformation of stress-strain curve of a metal matrix composite [35]. It is known that the shear lag model tends to give a poorer approximation than the other rigorous models such as Eshelby's model [36].

Eshelby [36] proposed a simple method to calculate the stress field in and around an ellipsoidal inclusion Ω with its properties being the same as the surrounding infinite matrix, where the inclusion possesses non-vanishing eigenstrain ε^* within Ω which vanishes outside Ω. The stress inside Ω becomes constant if ε^* is constant and is expressed as

$$\bar{\sigma} = \tilde{C}_m \cdot (\bar{\varepsilon} - \bar{\varepsilon}^*) \tag{31}$$

where $\tilde{C}_m$ is the stiffness tensor of the matrix and $\bar{\varepsilon}$ is the strain field induced by $\bar{\varepsilon}^*$, and $\bar{\varepsilon}$ is constant in Ω and related to $\bar{\varepsilon}^*$ by

$$\bar{\varepsilon} = \tilde{S} \cdot \bar{\varepsilon}^* \tag{32}$$

where $\tilde{S}$ is called the Eshelby's tensor and is a function of $\tilde{C}_m$ (or the matrix Poisson's ratio if the matrix is isotropic) and fiber aspect ratio. Dot in the above equations denotes inner product between tensors. Eshelby also demonstrated that the above inclusion problem can be extended to inhomogeneity problems such as composites. Mura [37] summarized the examples that can be solved by Eshelby's model to metal matrix composites. Taya and Arsenault [38] detailed the application of Eshelby's model to DMMC.

Let us refer to Fig.10 where the mechanical properties of constituents are now expressed by $\tilde{C}_i$ (stiffness tensor) and α_i (CTE mismatch strain) with i = m (matrix), f (fiber) or c (composite). Eshelby's model applied to Fig.10 gives rise to

$$\begin{aligned}\bar{\sigma}_0 + \bar{\sigma} &= \tilde{C}_f \cdot (\bar{\varepsilon}_0 + \underline{\bar{\varepsilon}} + \bar{\varepsilon} - \bar{\alpha}^*) \\ &= \tilde{C}_m \cdot (\bar{\varepsilon}^0 + \underline{\bar{\varepsilon}} + \bar{\varepsilon} - \bar{\varepsilon}^*)\end{aligned} \tag{33}$$

where $\bar{\sigma}_0$, $\bar{\varepsilon}_0$, $\bar{\sigma}$, and $\bar{\varepsilon}$ are the stress and strain tensors of the applied (uniform) field,

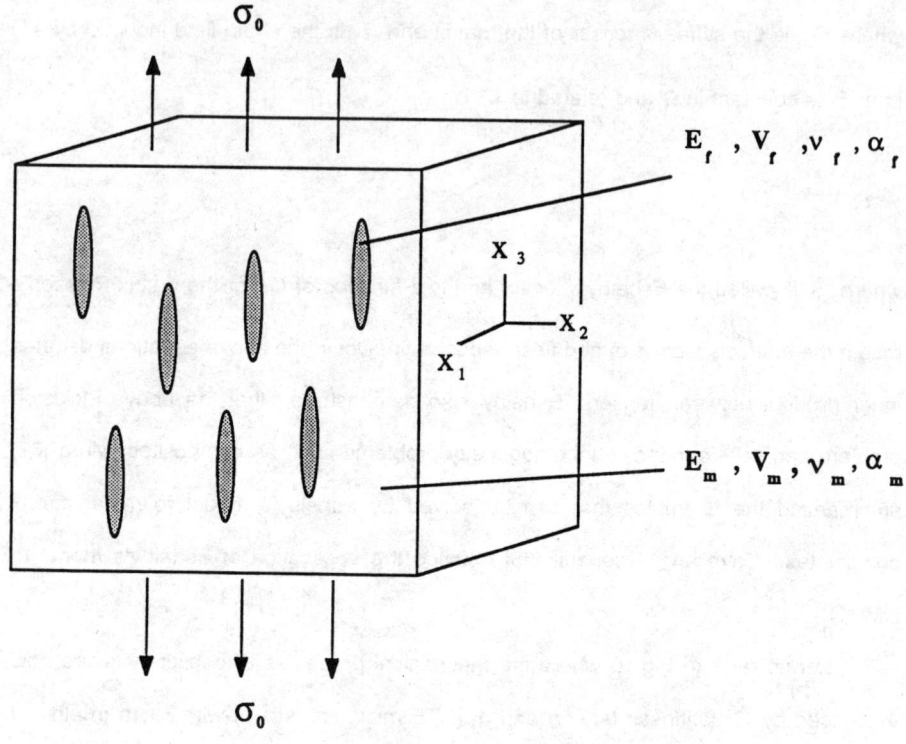

Fig. 10 An analytic model for strengthening of a composite

disturbed stress and strain field due to introduction of additional fiber, respectively, $\underline{\tilde{\varepsilon}}$ is the averaged strain disturbance in the matrix and related to [38]

$$\underline{\tilde{\varepsilon}} = -V_f(\tilde{\varepsilon} - \tilde{\varepsilon}^*) \tag{34}$$

$\tilde{\alpha}^*$ in eq.(33) is CTE mismatch strain defined by

$$\tilde{\alpha}^* = (\alpha_f - \alpha_m)\Delta T \tilde{\delta} \tag{35}$$

where ΔT is uniform temperature change in the composite, $\tilde{\delta}$ is Kronecher's delta, and $\tilde{\varepsilon}^*$ is the eigenstrain introduced fictitiously in the fiber domain Ω, yet to be determined. The uniform stress and strain satisfy Hooke's law:

$$\tilde{\sigma}_0 = \tilde{C}_m \cdot \tilde{\varepsilon}_0 \tag{36}$$

The strain disturbance $\tilde{\varepsilon}$ is also related to $\tilde{\varepsilon}^*$ by eq.(32). Thus, one can solve $\tilde{\varepsilon}^*$ from eqs. (32)-(34), and (36)

$$\tilde{\varepsilon}^* = \tilde{A}^{-1}\{(\tilde{C}_f \cdot \tilde{C}_m^{-1} - \tilde{I})\tilde{\sigma}_0 + \tilde{C}_f \cdot \tilde{\alpha}^*\} \tag{37}$$

where

$$\tilde{A} = \tilde{C}_f\{\tilde{S} - V_f(\tilde{S} - \tilde{I})\} - (1 - V_f)\tilde{C}_m \cdot (\tilde{S} - \tilde{I}) \tag{38}$$

where superscript -1 denotes inverse of 6 by 6 matrix, $\tilde{I}$ is identity matrix, and stress, strain and $\tilde{\alpha}^*$ are expressed as column vector of 6 by 1. Once the fictitious eigenstrain $\tilde{\varepsilon}^*$ is solved, one can calculate the elastic constants of a composite $\tilde{C}_c$ by the following equation of energy equivalence [36-41]

$$\frac{1}{2}\bar{C}_c^{-1} \cdot \bar{\sigma}_0 \cdot \bar{\sigma}_0 = \frac{1}{2}\bar{C}_m^{-1} \bar{\sigma}_0 \cdot \bar{\sigma}_0 + \frac{1}{2}V_f\bar{\sigma}_0\bar{\varepsilon}^* \qquad (39)$$

For prediction of composite stiffness, CTE mismatch strain $\bar{\alpha}^*$ does not come into play.

In the case of the axial applied (i.e. σ_0 along the x_3-axis, Fig.10), eq.(37) is reduced to

$$\frac{1}{2}\frac{1}{E_{cL}}\sigma_0^2 = \frac{1}{2}\frac{1}{E_m}\sigma_0^2 + \frac{1}{2}V_f\sigma_0\varepsilon_{33}^* \qquad (40)$$

Noting that ε_{33}^* is a linear function of (σ_0/E_m), for example $\varepsilon_{33}^* = B(\sigma_0/E_m)$, we can obtain the formula to calculate the longitudinal Young's modulus of an aligned fiber composite, E_{cL} normalized by matrix Young's modulus E_m given by

$$\frac{E_{cL}}{E_m} = \frac{1}{1+V_fB} \qquad (41)$$

where B is a function of the matrix Poisson's ratio υ_m, fiber aspect ratio ℓ/d and V_f [40].

Next we shall discuss the formulation based on Eshelby's model to predict the yield stress (σ_{yc}) and work-hardening rate (h) of a composite. The analytical model for prediction of σ_{yc} and h is shown in Fig. 11(a) where under high applied stress $\bar{\sigma}^0$ the matrix deforms plastically with uniform plastic strain, $\bar{\varepsilon}_p$, and fiber with CTE mismatch strain $\bar{\alpha}^*$ deforms elastically. The problem in Fig.11(a) can be converted to Fig. 11(b) where the uniform plastic strain in the matrix $\bar{\varepsilon}^p$ is removed and put into the fiber domain as eigenstrain of $-\bar{\varepsilon}^p$, for the internal stress field of Fig.11(b) is the same as Fig.11(a) [40]. Then, the change in the total potential energy of the composite δU due to small change in plastic strain $\delta\bar{\varepsilon}^p$ is calculated as

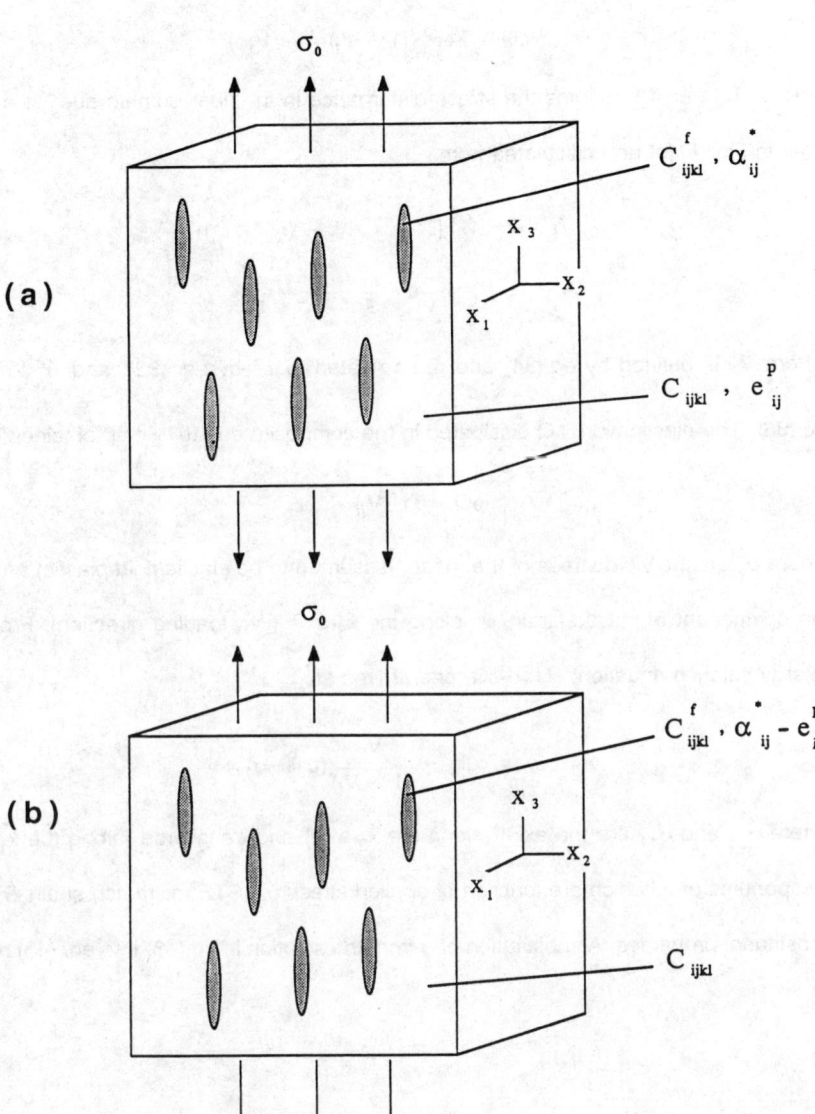

Fig. 11 Analytical Model for prediction of the yield stress and work hardening rate of a composite: (a) a composite is subjected to applied stress σ_0 and uniform plastic strain e_{ij}^p in the matrix (b) the Eshelby's equivalent inclusion problem equivalent to (a), i.e., e_{ij}^p is removed from the matrix and put into the fiber domain with $-e_{ij}^p$

$$\delta U = -\delta \bar{\varepsilon}^p \{(1 - V_f)\bar{\sigma}^0 - V_f\tilde{\sigma}\} \qquad (42)$$

where $\tilde{\sigma}$ in eq.42 denotes the stress disturbance in the fiber domain due to a single fiber in Fig. 11(b) and calculated from

$$\bar{\sigma}^0 + \tilde{\sigma} = \tilde{C}^f \{\bar{\varepsilon}^p + \underline{\bar{\varepsilon}} + \tilde{\varepsilon} - (\bar{\alpha}^* - \bar{\varepsilon}^p)\} \qquad (43)$$

$$= \tilde{C}^m \{\bar{\varepsilon}^0 + \underline{\bar{\varepsilon}} + \tilde{\varepsilon} - \bar{\varepsilon}^*\}$$

where $\underline{\bar{\varepsilon}}$ is defined by eq.(34) and $\tilde{\varepsilon}$ is related to $\bar{\varepsilon}^*$ by eq. (32), and $\bar{\sigma}^0$ to $\bar{\varepsilon}_0$ by eq.(36). The plastic work δQ dissipated in the composite due to $\delta \bar{\varepsilon}^p$ is obtained as

$$\delta Q = (1 - V_f)\sigma_{ym}\delta\varepsilon_p \qquad (44)$$

where σ_{ym} is the yield stress of the matrix (assumed to be elastic/rigid plastic) and ε_p is the component of plastic strain $\bar{\varepsilon}^p$ along the x_3-axis (i.e., loading direction). From the energy balance equation: $\delta U = -\delta Q$, one arrives at

$$\sigma_0 = \sigma_{ym} + \frac{V_f}{(1-V_f)}(\sigma_{33} - \sigma_{11}) \qquad (45)$$

where σ_{33} and σ_{11} are the axial (along the x_3-axis) and transverse (along the x_1-axis) components of σ, which are function of applied stress σ_0, CTE mismatch strain $\bar{\sigma}^*$, and constituent properties. A substitution of σ from the solution in eq.(43) into eq. (45) results in

$$\sigma_0 = \sigma_{yc} + h\varepsilon_p \qquad (46)$$

The stress-strain curves of the unreinforced matrix and a composite are schematically shown in Fig.12. The slope of σ_0 - ε curve in a linear work-hardening region is denoted by E_T, which is related to h (slope of σ_0 - ε_p curve) by

$$h = \frac{E_T}{1 - E_T/E_c} \qquad (47)$$

The work-hardening rate h predicted by the above model is reasonably accurate for initial portion of plastic stage of stress-strain curve. It is clear from Fig.12 that both yield stress and work-hardening rate increases over those of the unreinforced matrix. When the residual stress due to CTE mismatch strain is considered, σ_{yc} increases or decreases depending on the tensile or compressive yield stress for $(\alpha_f - \alpha_m)\Delta T > 0$ [39].

B. Micromechanics (FEM)

The micromechanics models discussed so far are analytical type, i.e., the formulations giving rise to closed form solutions which are of convenient form from the view point of parametric study, for one can immediately identify which parameters would greatly affect the overall properties of a composite, and which would be insensitive. However, in order to obtain detailed information of the stress and strain at a specific location within the matrix or reinforcement it is necessary to perform numerical i.e. FEM analysis. Below is brief outline of some of the areas where FEM can be applied.

Reinforcement distribution is another important parameter which can affect composite properties. In-situ experiments reveal that large strain localization exists during far-field loading of SiC/Al composites [42].

Shi et al. [43, 44] have conducted extensive FEM analysis. The evolution of plastic

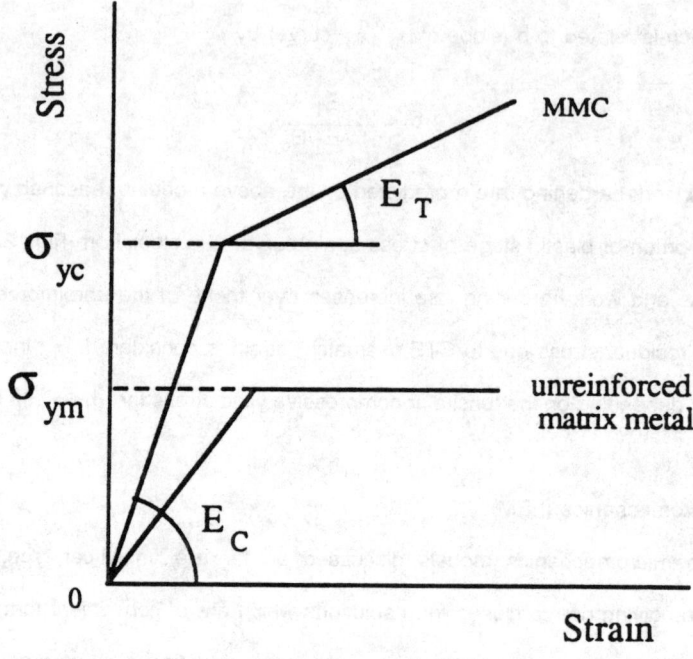

Fig. 12 Idealized stress-strain curves of an unreinforced matrix metal and dislocations metal matrix composite (DMMC).

flow in the ductile matrix of a SiC whisker (SiC_w) reinforced annealed 6061 Al was analyzed by monitoring the load-induced changes of the plastic zone generated by ΔCTE, and then a correlation was determined between the matrix microplastic flow and the global composite tensile stress-strain curve. Among the composite tensile properties, emphasis is placed on the influence of the matrix microplasticity on the composite strengthening.

The significant effect of thermal residual stresses (TRS), their generation, relation to dislocation generation and motion and the remaining (TRS), on the properties of composites can not be understated. As discussed in section A, the TRS can be determined analytically by making a few assumptions, but in order to obtain the specifics as to how the dislocation generation and motion occur about the reinforcement, i.e., the plastic zone formation upon cooling, and how the matrix residual stress changes due subsequent external deformation requires the use of FEM [45].

Before discussing specific examples, the following is a brief outline of the FEM model. Currently, 2- and 3-D FEM models were employed to model the deformation-induced changes of matrix thermal residual stresses in the whisker reinforced SiC-Al composites. The composite is assumed to consist of an infinite-periodic-staggered-array of whiskers shaped as parallelepipeds with an aspect ratio of four embedded in a matrix. Fig.13a displays the geometry of such an arrangement. Because of the periodicity, a "unit cell" model containing one eighth of the whisker can be employed by considering proper symmetry conditions (Fig.13b).

In order to satisfy the symmetry condition and maintain continuity on the cell

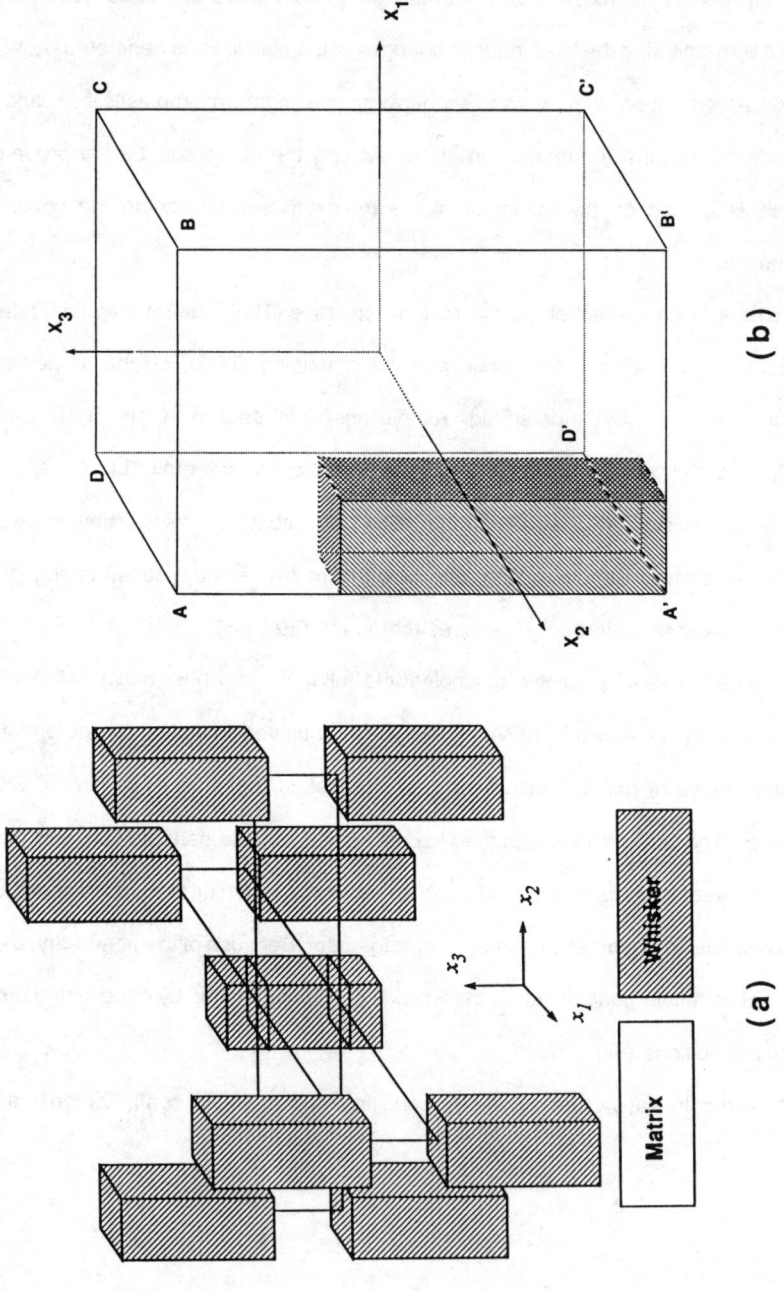

Fig. 13 Schematic of (a) composite with matrix reinforced by an infinite staggered array of aligned SiC whiskers; (b) unit cell taken from this periodic composite

boundary planes (Fig.13b), the following conditions are imposed:

On boundary plane ABCD:

$$u_3(x_1, x_2) = U_3 \tag{48}$$

where $u_i(i=3)$ is the displacement component along the ith direction, and U_3 varies with the applied stress, σ_{ij}^A, and is spatially invariant on ABCD. On boundary plane BCB'C', antisymmetric displacement conditions about $x_3 = 0$ are imposed:

On DCC'B':

$$u_i(x_1, x_3) = 2u_i(x_1, 0) - u_i(x_1, -x_3) \quad (i=2,3)$$
$$u_1(x_1, x_3) = u_2(x_1, -x_3) \tag{49}$$

On BCC'B':

$$u_i(x_2, x_3) = 2u_i(x_2, 0) - u_i(x_1, -x_3) \quad (i=1,3)$$
$$u_2(x_2, x_3) = u_2(x_2, -x_3) \tag{50}$$

On boundary planes ADD'A', ABB'A', AD'C'B', roller boundary conditions which restrict only the normal displacements are imposed. To uniquely define the boundary conditions eqs.48, 49 and 50, the following condition must be satisfied:

$$\int_D \sigma_{ij} dV = \sigma_{ij}^A \tag{51}$$

where D represents the total volume occupied by the unit cell shown in Fig.13b, and σ_{ij} is the stress tensor in the volume V. When a 2-D plane strain analysis is employed, similar boundary conditions are imposed.

The Al matrix is assumed to follow (1) the bi-linear, i.e., elastic/linear plastic, stress-strain relationship, and (2) the kinematic hardening rule. The reinforcement is taken to be perfectly elastic. The composite is assumed to be stress-free at the annealing temperature, then a temperature drop of ΔT was divided into 18 steps and was determined from linear extrapolations from those at the two neighboring temperatures, and the element nodal temperatures are uniform across the cell. This is a good approximation of furnace cooled samples. Following the cooling, the FEM mesh is subjected to an incremental external load either in tension or compression, following by unloading. At small applied strains (~ 1%), the total strain assimilated in the SiC whiskers is small due to their high modulus. Therefore, the effect of anisotropy is of second order, and is ignored. For the changes of the matrix residual stresses, as measured in this investigation, any effect of matrix texture is approximately canceled out, and is not considered.

To obtain an easier insight as to how the plastic zone changes with external loading, which forms due to relaxation of the thermal stresses, a temperature change form 50°C to 20°C was employed.

Following the temperature change of $\Delta T = 30°C$, longitudinal loading is applied to $\overline{AB}$ (in Fig.13) incrementally. Fig.14 shows the behavior of the thermally induced matrix plastic zone in a 20 V% composite under external tensile loading. In this diagram, the locations of plastic zone boundary are plotted at different load levels, the label on each line represents the level of applied traction in MPa, the hatched side of the line represents matrix that is currently plastically deforming, and we refer this enclosure as the plastic zone. Prior to external loading, the matrix has partially plastically deformed around the

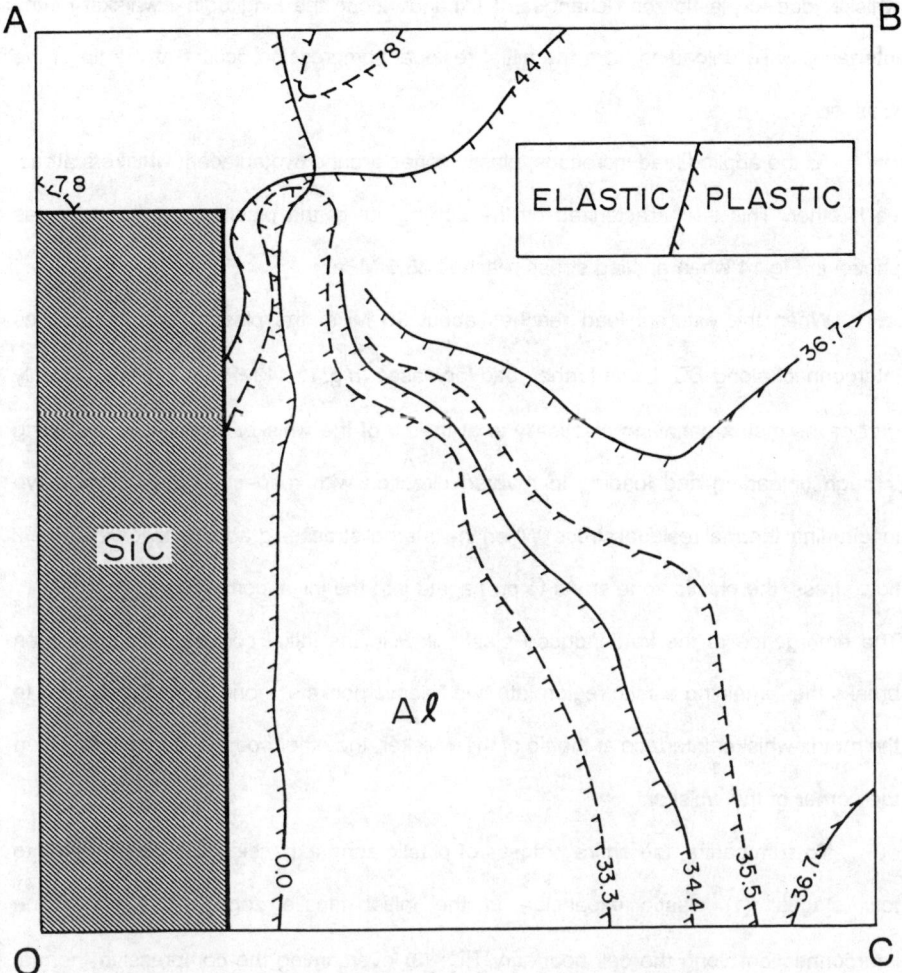

Fig.14 Evolution of the matrix plasticity of 20 V% SiC$_w$-6061 Al under external applied load in MPa. The thermal history is $\Delta T=30°C$. Each line represents the position of the plastic zone boundary at an applied load level as labeled along the boundary.

whisker due to plastic zone changes. It expands along the longitudinal whisker-matrix interface, while unloading from the initial residual compression occurs at the tip of the whisker.

As the applied load increases, plastic zones around two adjacent whiskers attract each other. This is characterized by the bowing-out of the plastic zone boundary as shown in Fig.14 when applied stress reaches 35.5 MPa.

When the external load reaches about 36 MPa, the plastic zone boundaries interconnect along $\overline{BC}$. Upon further load increases (e.g. to 44 MPa in Fig.14), the only part of the matrix behaving elastically is at the tip of the whisker where it is still going through unloading and loading in reverse direction with respect to the compressive longitudinal thermal residual stress. When the internal stresses gradually reach the matrix flow stress, the plastic zone starts to propagate into the initial compressive zone.

The emergence of the load-induced plastic flow in the initial compressive zone then breaks the remaining elastic region into two "elastic pockets" - one immediately next to the matrix-whisker interface at the tip of the whisker, the other some distance away form the corner of the whisker.

To summarize, the entire process of plastic zone expansion can be divided into four stages: (1) plastic expansion in the initial tensile zone; (2) plastic zone interconnection along the cell boundary $\overline{BC}$; (3) overcoming the compressive thermal residual stresses at the tip of the whisker; (4) fragmentation of the elastic matrix to form remnant elastic pockets at SiC_w tip surrounded by the matrix plastic flow.

The matrix microplasticity is also correlated to the composite mechanical properties. Fig.15 shows the initial portion of the stress-strain curve produced by FEM, and how it is related to the plastic zone expansion process. At the start, the stress-strain relationship is approximately linear. However, because of the thermal induced plasticity in the matrix, the thermally induced plastic zone expands even within this approximately linear region as shown in Fig.14. As the plastic zones attract to each other and interconnect, the stress-strain curve goes quickly out of the approximate linearity. Once the entire initial tensile zone is plastically deforming, the slope of the stress-strain curve approaches a constant again (Fig. 15). This is due to the fact that the rate of propagation of the plastic zone boundary is greatly reduced because of the ongoing process of overcoming the compressive residual stress in the initial compressive zone.

If the more realistic case of $\Delta T = 480°C$ is considered the results are essentially the same, the apparent Young's modulus is ~2/3 of modulus which would be measured by ultrasonic techniques or as calculated by the Eshelby method (section A).

Also from this FEM investigation [43], the difference in apparent Young's modulus between compression and tension was demonstrated as shown in Fig.16.

The composites are represented by a periodic array of transversely aligned whiskers as shown in Fig.17, i.e., an infinite two-dimensional array of SiC whiskers embedded in the aluminum matrix in the plane strain condition. In order to consider the effect of clustering, the array is transformed in such a manner that an array of periodic geometric clusters centers was selected and the neighboring whiskers (in this case, four whiskers (Fig.18)) were attracted toward their cluster centers so that a periodic array of clusters were formed as shown in Fig.19.

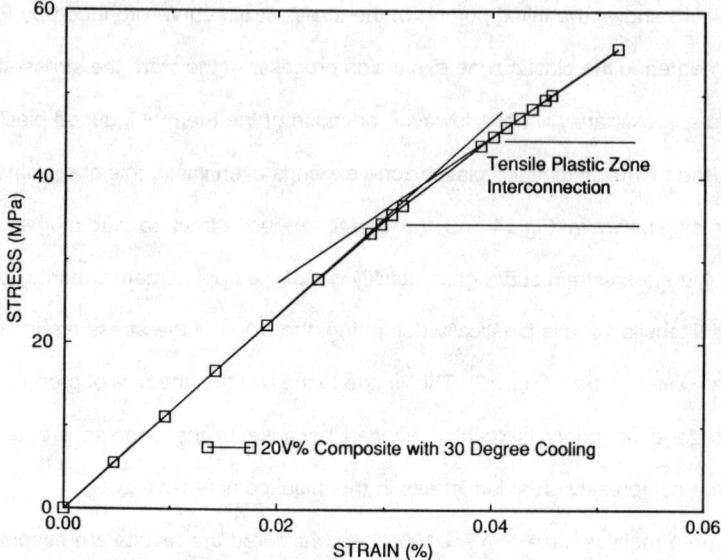

Fig. 15 The initial portion of the FEM-predicted composite stress-strain curve showing the apparent proportional limit for 20 v% composite with a thermal history of $\Delta T = 30°C$

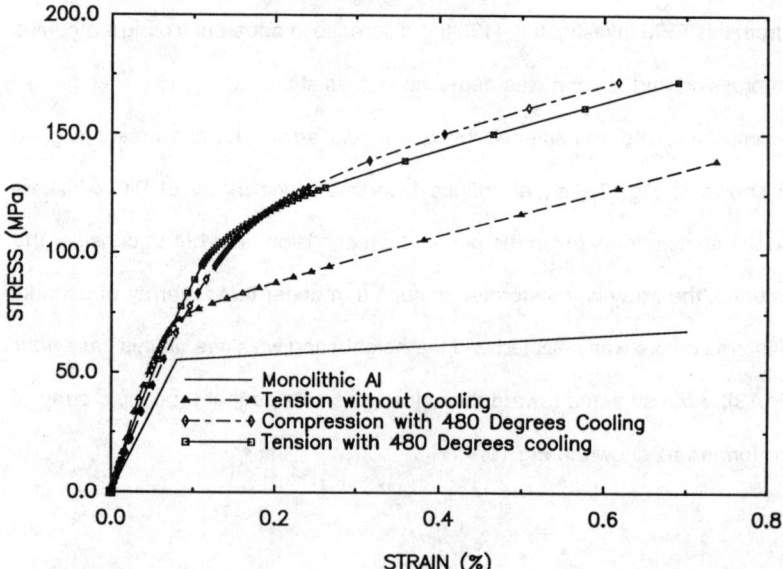

Fig. 16 Stress-strain curves of annealed 6061 Al with bilinear approximation and 20 v% SiC$_W$/Al composite with $\Delta T=0$ and 480°C, respectively

Fig.17 Periodic hexagonal arrangement of the SiC reinforcement of the whisker. The shaded areas are SiC whiskers.

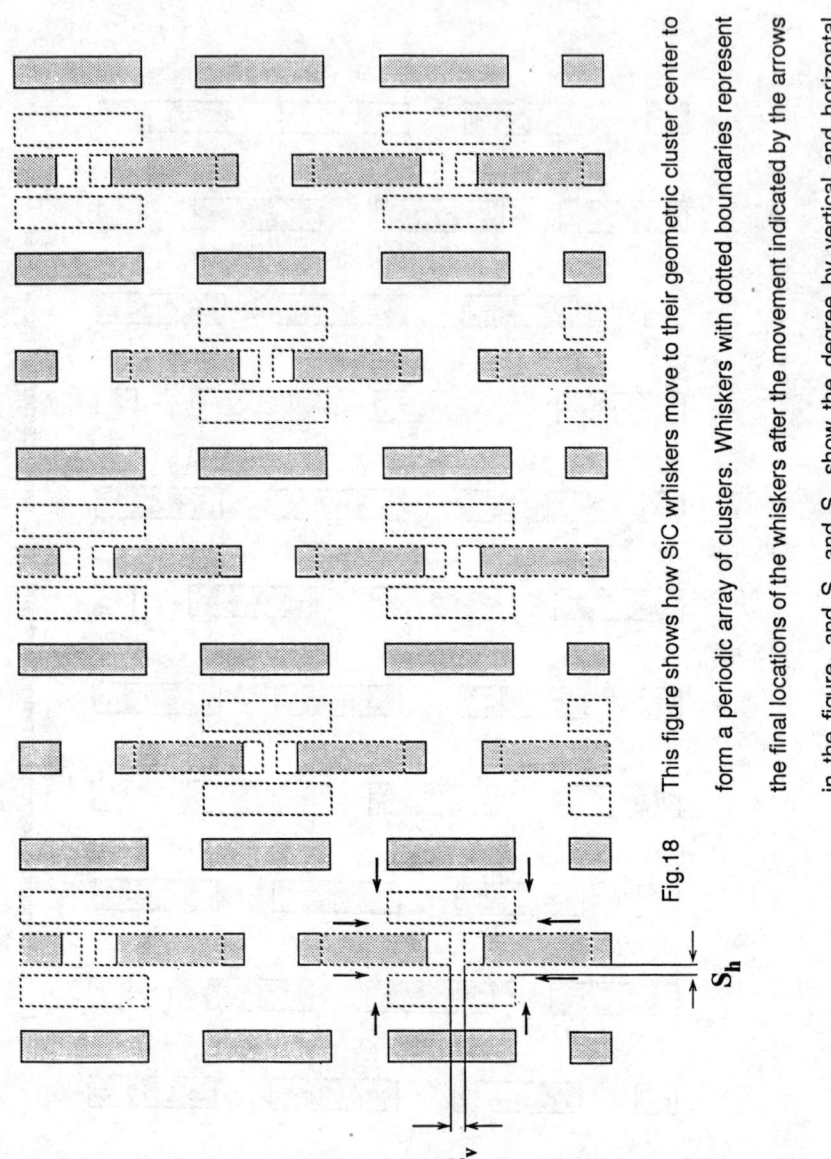

Fig.18 This figure shows how SiC whiskers move to their geometric cluster center to form a periodic array of clusters. Whiskers with dotted boundaries represent the final locations of the whiskers after the movement indicated by the arrows in the figure, and S_v and S_h show the degree by vertical and horizontal clustering.

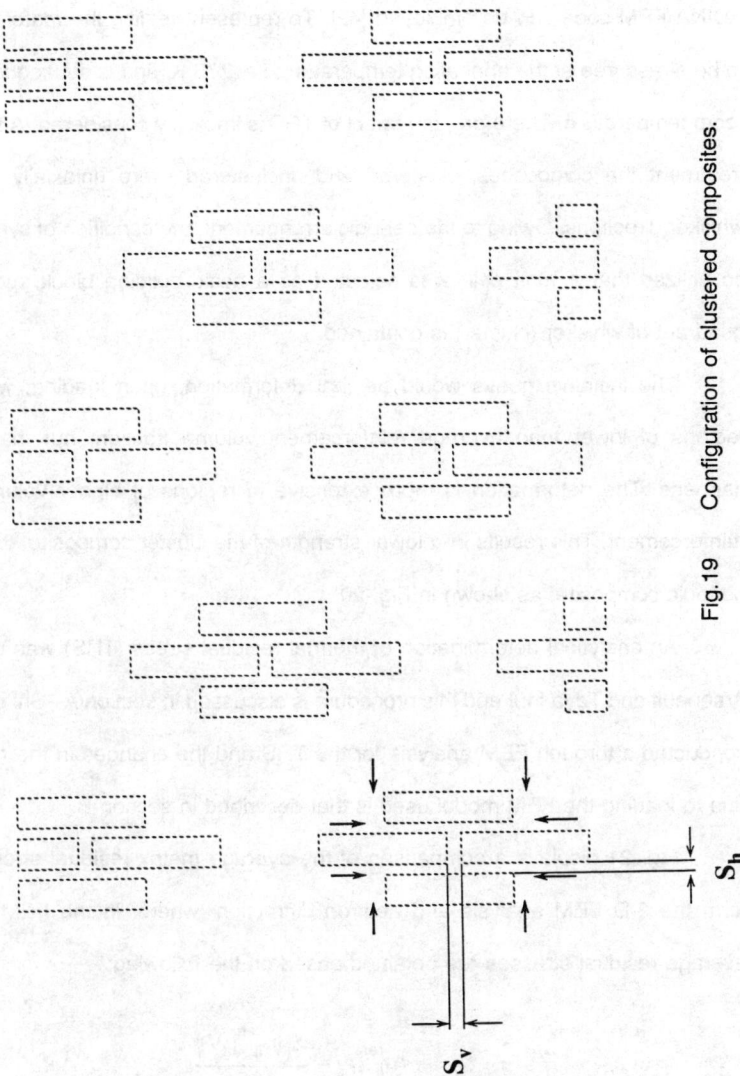

Fig.19 Configuration of clustered composites.

In applying the above geometric transformation scheme, a commercially available ADINA FEM code can be employed [42]. To represent reality, the material is assumed to be stress free at the annealing temperature, i.e. 773 K, and is subsequently cooled to room temperature. Therefore, the effect of TRS is implicitly considered. After the thermal treatment the composites, clustered and unclustered, were uniaxially loaded in the whisker directions. Owing to the periodic arrangement, the condition of symmetry can be so utilized that a "unit cell" was selected as a basic building block such that only a quadrant of whisker (cluster) is contained.

The initiative guess would be that deformation, upon loading, would begin in regions of lower than average reinforcement volume fraction, but that is not what happens. The deformation is more extensive in regions of higher volume fraction of reinforcement. This results in a lower strength of the cluster composite, than that of the periodic composite, as shown in Fig. 20.

An analytical determination of thermal residual stress (TRS) was undertaken by Arsenault and Taya [40] and this procedure is discussed in section A. Shi et al. [45] have conducted a through FEM analysis for the TRS and the changes in the residual stress due to loading the FEM model used is that described in section B.

Fig. 21 displays a comparison of the average matrix residual stresses obtained form the 3-D FEM analysis and neutron diffraction, where, in the FEM analysis, the average residual stresses are obtained based on the following:

$$<\sigma_{ij}^{res}> = \frac{\Sigma_k [\sigma_{ij}^{res}]_k V_k}{V} \tag{52}$$

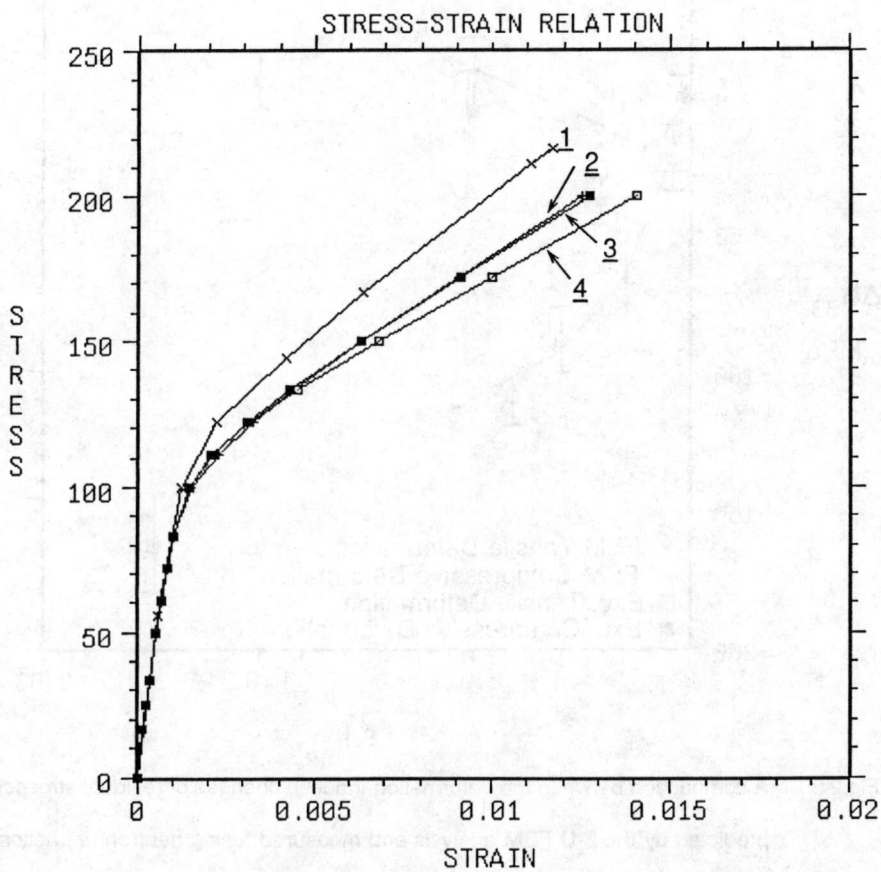

Fig.20 Stress-strain curves produced as a result of clustering: curve 1, $S_v/2 = 1$, $S_H/2H = 1$; curve 2, $S_v/2V = 0.75$, $S_H/2H = 0.429$; curve 3, $S_v/2V = 0.5$, $S_H/2H = 0.143$; curve 4, $S_v/2V = 0.25$, $S_H/2H = 0.071$.

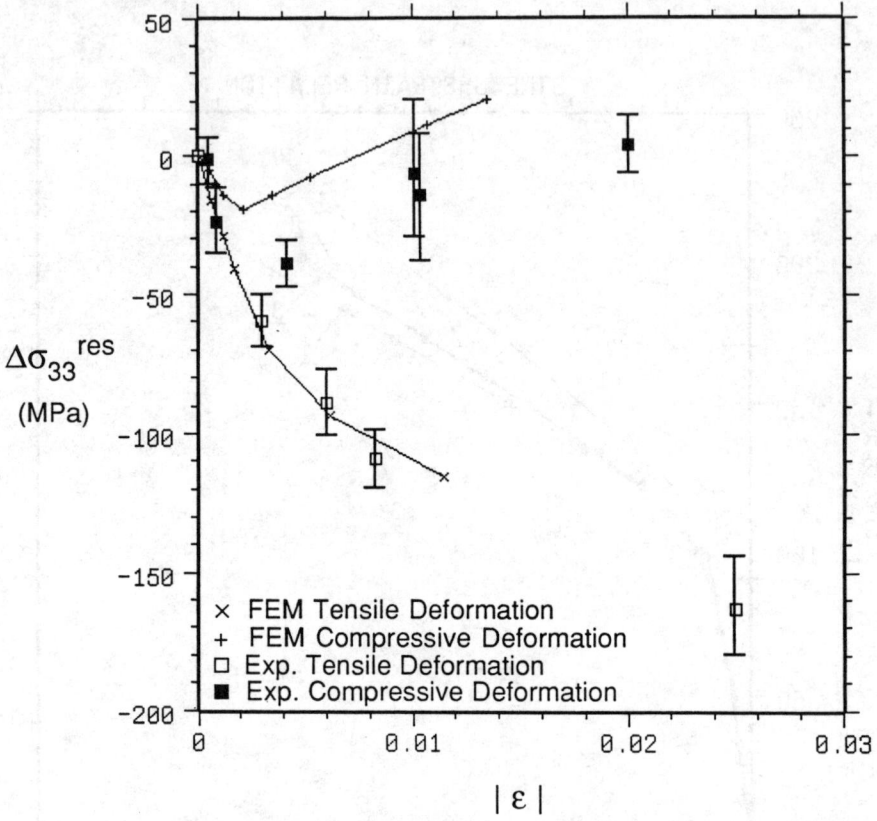

Fig.21 A comparison between the deformation-induced changes of residual stresses predicted by the 3-D FEM analysis and measured using neutron diffraction. Where $|\varepsilon|$ represents the total strain due to external loading either in tension or compression.

In Eq. 52, the $<\sigma_{ij}^{res}>$ are the average matrix residual stresses, V_k and V are the volume of the element k and the matrix in the unit cell, respectively, and the $[\sigma_{ij}^{res}]_k$ are the average residual stresses in element k. It can be seen that there is a good correlation between the experimental data and the numerical predictions. This is represented by the fact that there is an increase of residual stress with increasing external compressive deformation; while, upon tensile loading, the matrix residual stress decreases initially at a rapid rate and the rate slows down on further deformation, i.e., the changes are asymmetric. It is noted that relaxation of the residual stresses in the matrix starts immediately upon external deformation. The asymmetric response is most pronounced in the longitudinal direction (x_3).

The FEM predictions of changes in deformation-induced matrix residual stress due to deformation are in good agreement with experimental data. Regardless of the sign of the external loading, the average matrix residual stresses go through an initial reduction. After further external deformation, the residual stresses increase or decrease depending on whether the loading is uniaxial tension or compression. There are two major sources for the changes in the deformation-induced residual stresses: the matrix plastic flow and the matrix-reinforcement deformation gradients. The matrix plastic flow gradient induces monotonic changes of residual stresses. The observed deformation-induced changes in residual stresses are a combination of the two.

The analytical and numerical methods are complementary. The analytical methods are excellent defining general trends., and the steps in their derivation can be examined in detail. The numerical methods, i.e., FEM can be used to obtain detailed information concerning the stress and strain state at specific locations within the composite. However, the FEM can only be used to give general trends, not specific results.

C. Asymmetrical Apparent Young's Modulus

If a DMMC is tested in tension and compression, the characteristics of the stress-strain curves are different (Fig.22). It has been found that the compressive yield stress is higher than the tensile yield stress, but the apparent Young's modulus for compression is smaller than that for tension, i.e., the composite apparent Young's modulus is asymmetric in comparison to the field strength. The increased compressive yield stress of the composite can be account for the tensile TRS in the matrix.

To understand the asymmetric-apparent-Young's modulus of the composite, it is necessary to consider the evolution of the matrix plastic flow illustrated in the previous section. Comparing the morphology of the plastic zones under tensile and compressive loads shown in Figs.23 and 24, the composite cell can be approximately mapped as shown in Fig.25, in which the unit cell is divided into three distinct regions: the elastic matrix, plastic deforming matrix and the whisker. If we combine the two stiffer regions (the whisker and the elastic matrix) as a single imaginary phase, then two distinct domains are formed - the combined "new phase" and the plastic deforming matrix. From the geometry of these two domains, their stress state may be best obtained by the isostress and isostrain approximations when under compression and tension, respectively. Therefore, the average stiffness of the composite (E_c) can be approximated by the Reuss- or Voigt-average [46] of the stiffness responses from the two individual domains depending

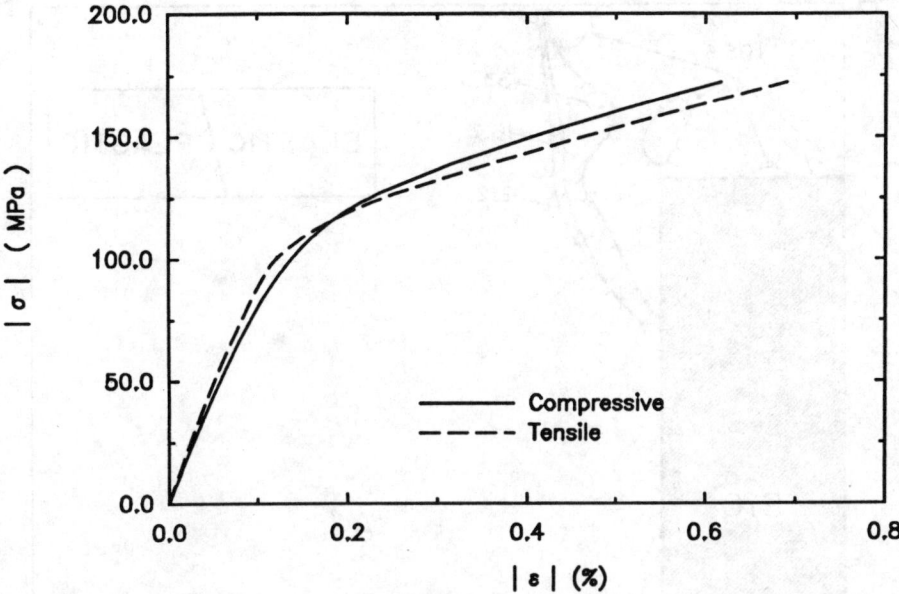

Fig.22   Stress-Strain curve under tension and compression predicted by the FEM with a consideration of ΔT=480°C, where, while comparing with those from loading along the opposite direction, the apparent young's modulus of the composite is higher when it is under tension, whereas the yield strength is greater when it is under compression.

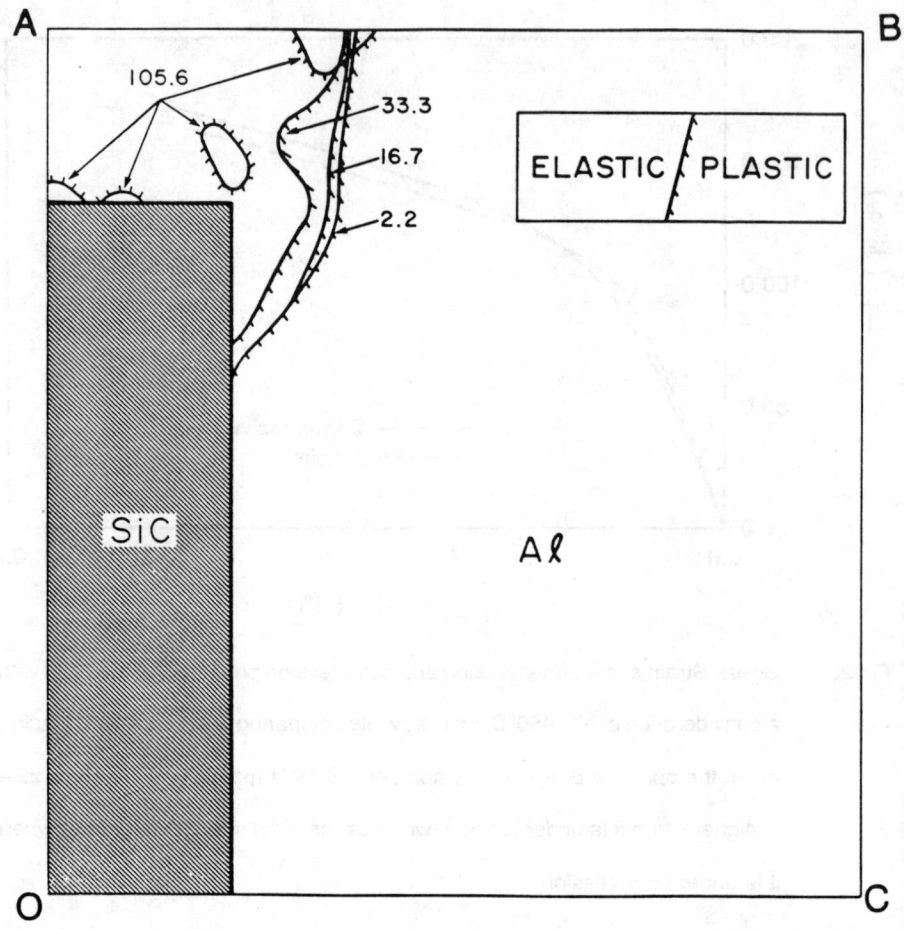

Fig.23 Process of the plastic zone expansion and interconnection in the matrix of a 20 V% SiC$_w$/Al composite with a thermal history of $\Delta T=480°C$. The applied stress (MPa) is applied incrementally.

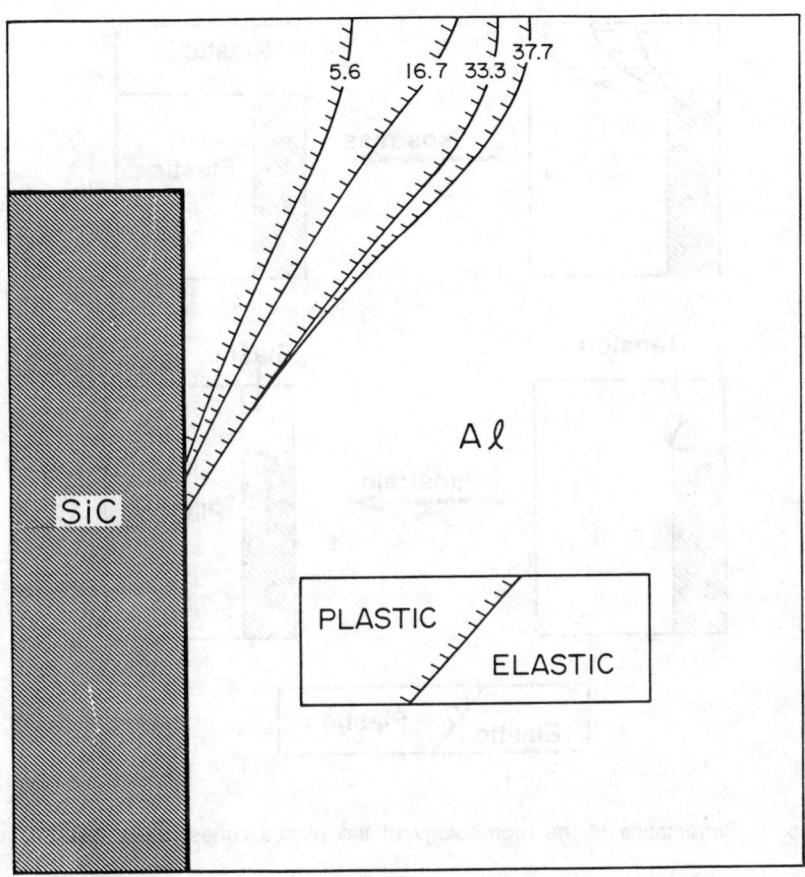

Fig.24 Matrix plastic zone expansion process of a 20 V% SiC$_w$/Al composite under compressive loading condition with a thermal history of ΔT=480°C.

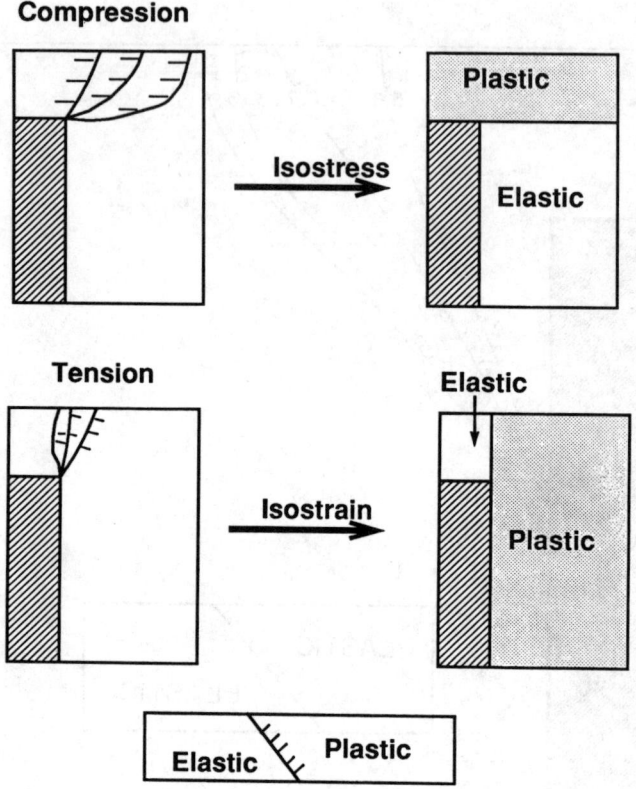

Fig.25 Schematics of the morphology of the plastic zones under different initial loading conditions, where tensile loading produces a geometry that is suitable for isostrain condition and compressive loading produces a geometry that is suitable for isostress condition.

on the loading direction, i.e.,

$$E_c = E_{ave}(1-V_p) + E_{mp}V_p \quad \text{(For tensile loading);} \tag{53}$$

$$E_c = \frac{E_{ave}E_{mp}}{(1-V_p)E_{mp} + V_p E_{ave}} \quad \text{(For compressive loading),} \tag{54}$$

where E_{ave} and E_{mp} are the average stiffness of the "new phase" and the tangential modulus of the plastically deforming matrix, respectively. V_p is the volume fraction of the plastic zone.

As shown in Fig.25, the geometric arrangements of the elastic matrix and the whisker in the "new phase" with respect of the loading direction suggest that the stress distribution between the two regions under tension and compression may again be best estimated by the isostress and isostrain approximations, respectively. Then, the average stiffness of the new phase can be obtained in the same manner, that is,

$$E_{ave} = \frac{E_w E_m}{V_{el}E_w + (1-V_{el})E_m} \quad \text{(For tensile loading);} \tag{55}$$

$$E_{ave} = (1-V_{el})E_w + V_{el}E_m \quad \text{(For compressive loading),} \tag{56}$$

where V_{el} corresponds to the proportion of the elastic matrix in the "new phase". E_m and E_w are the Young's moduli of the matrix and whisker, respectively.

Substitute E_{ave} in Eqs.55 and 56 into Eqs.53 and 54, respectively, the composite average stiffness response, E_c, under tensile and compressive load can be obtained. If we denote E_c^T and E_c^C as the composite tensile and compressive stiffness, respectively, and consider material properties: E_m = 68.3 GPa; E_{mp} = 2.08 GPa; = 483 GPa, we obtain E_c^C = 9.887 GPa; E_c^T = 50.09 GPa. This result indicates that although, from a comparison of Figs.21 and 22, a large proportion of the matrix deforms plastically under tension than that under the same amount of compression, the system still "allows" $E_c^T > E_c^C$. The magnitude of the asymmetry of the Young's modulus is exaggerated because, as compared with that of the actual plastic zone shown in Fig.24, the morphology of the plastic zone shown in Fig.25 during compression leads to a lower bound for the composite stiffness.

From this result, it can be seen that the asymmetric-apparent-Young's modulus of the composite is due to an asymmetric plastic zone expansion which induces isostress- or isostrain-type loading. This process ends when the morphology of the plastic zone can no longer be represented by the idealizations shown in Fig.25, which is when the applied stress is approaching the composite yield strength. At this stress level, the matrix plastic flow has almost spread out to entire matrix so that the morphology of the plastic zones is no longer important. Therefore, overcoming the effect of the initial TRS becomes more dominant in determining the composite tensile and compressive yield strengths. From the above analysis, it is expected that, without ΔCTE, all the asymmetry disappears. Due to the thermally induced plasticity, the composite apparent Young's modulus is lower than the modulus of elasticity of composite [44].

D.   Strengthening Due to Work Hardening of the Matrix

The classical continuum theories of composite strengthening when applied to DMMC can not account for the strength increases observed, as shown in Fig. 26. Also,

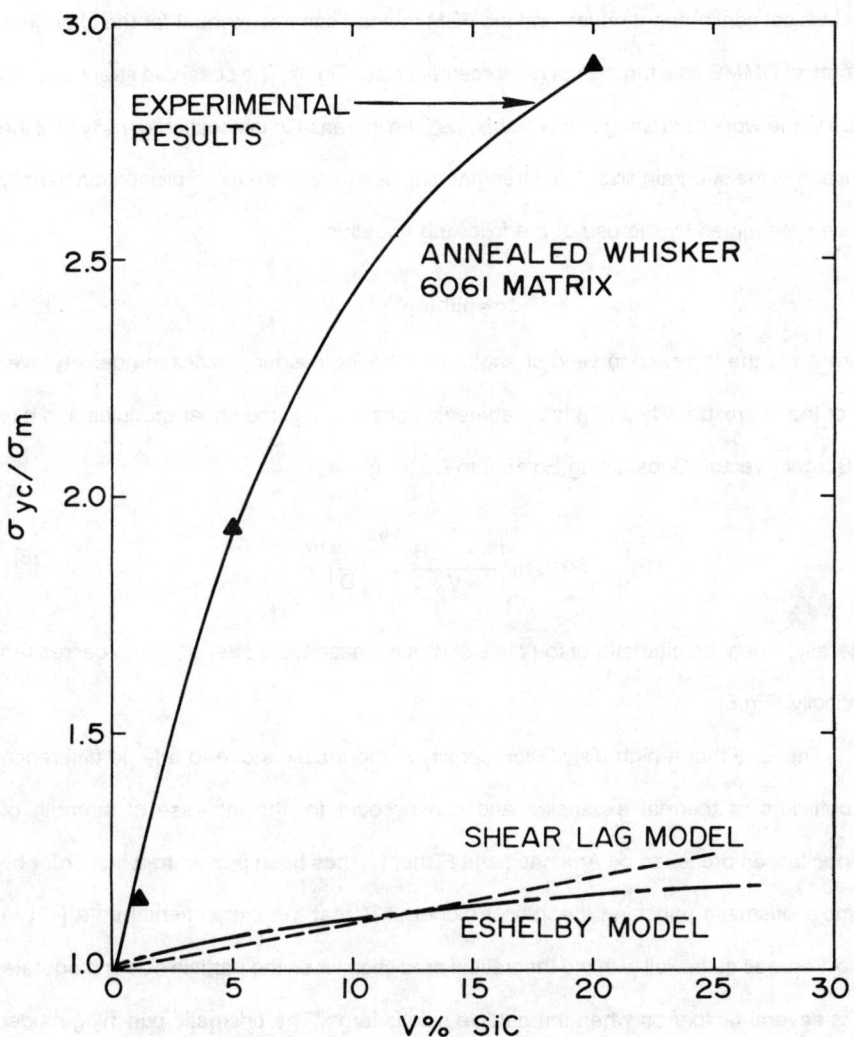

Fig.26 The ratio of the yield stress of the composite to the yield stress of the matrix as a function of volume fraction of SiC.

the classical continuum theories and the FEM method can not account for the change in strength of DMMC as a function of reinforcement size (Fig.6). The observed strengthening is due to the work hardening of the matrix, i.e., the increase in dislocation density and the reduction of the subgrain size. The strengthening due to an increase in dislocation density can be determined by the use of the following equation:

$$\Delta\sigma = \beta\mu b (\Delta\rho)^{1/2} \tag{57}$$

where $\Delta\sigma$ is the increase in yield strength, ρ is the increase in dislocation density over that of the matrix density and β is a geometric constant, μ is the shear modulus and b is the Burgers vector. Substituting Eq.1b into Eq.57 gives

$$\Delta\sigma = \beta\mu b \left(\frac{V_p}{1-V_p}\frac{B\varepsilon}{b}\right)^{1/2} \left(\frac{1}{D}\right)^{1/2} \tag{58}$$

Generally, when the diameter of the platelet was increased, the strengthening decreased drastically (Fig.6).

The idea that a high dislocation density in the matrix is due to a large difference in coefficient of thermal expansion and can account for the increase of strength of composites, as proposed by Arsenault and Fisher [3], has been proven to a first order by a simple prismatic punching theoretical model. It is that the experimental data [47] in Table 1 agrees quite well with the theoretical prediction when the particle size is moderate and is several factors off when the particle size is large. The prismatic punching model is the most efficient model. Therefore, for large particle sizes, the prismatic punching model underestimates the density of dislocations generated.

Table 1

Particle Parameter	Increase in Strength Δσ(MPa)	
	Theoretical Prediction MPa	Experimental Data MPa
R = 1 V' = 0.125 μm³	130.4	166
R = 0.25 V' = 0.5 μm³	124.2	152

At present, a detailed mechanism of how a change in subgrain size results in a change in the strength is not known. However, an empirical relationship has been developed by McQueen et al. [48], from which the predicted increase in strength due to the reduced subgrain size is ~1/2 the increase in strength due to the increase dislocation density.

An overall correlation between the increase in dislocation density and the decrease in subgrain size is shown in Fig.6 and 27. If the increase in dislocation density is inserted into eqn. 57 and the empirical data of McQueen et al. [48] is employed then there is a very good correlation between the experimentally observed strengthening and that predicted as a result of the microstructure change in the matrix.

5. THE BAUSCHINGER EFFECT IN A SIC/AL COMPOSITE

It is generally believed that the Bauschinger effect (BE) in discontinuously reinforced metal matrix composites (DMMC) results from the development of the internal matrix-reinforcement interaction stresses in the matrix [49-54]. Numerical models [50-51] have been developed to show that even without considering the intrinsic BE of the matrix

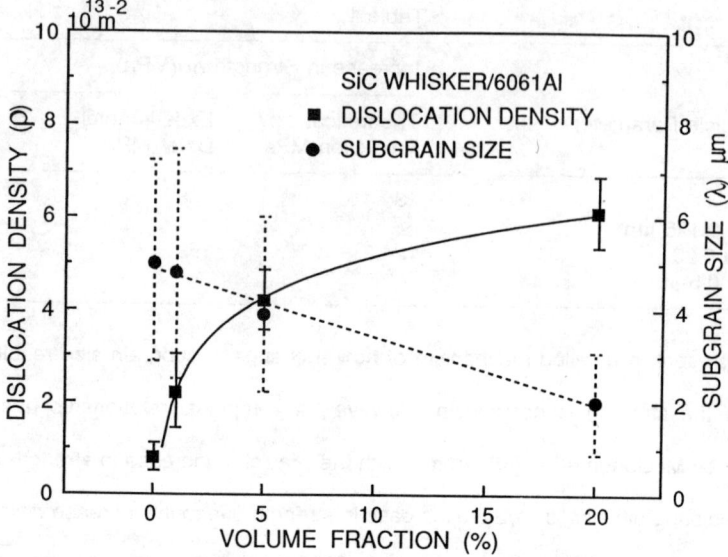

Fig. 27 The change in dislocation density as a function of volume fraction of SiC whiskers (diameter ~ 0.5 μm, length ~ 1μm)

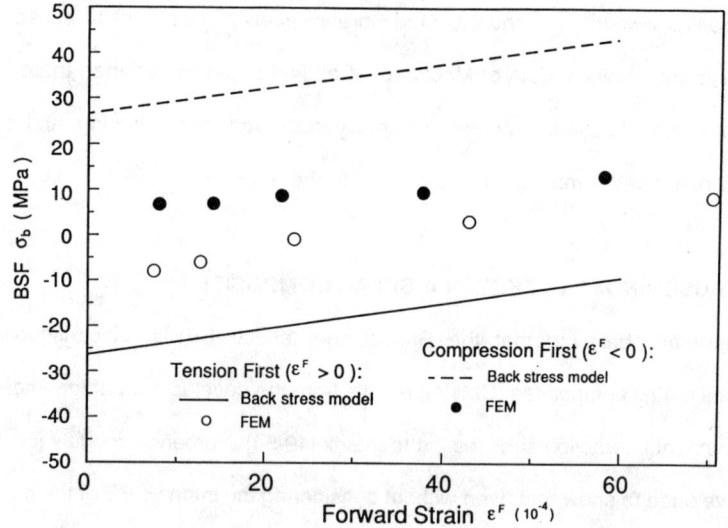

Fig. 28 A comparison of the predictions from the Back Stress Model and the FEM. The two models give a good agreement on the slopes and the general trends on the variation of σ_b while the Back Stress Model predicts a larger σ_b.

material, a significant BE exists in the corresponding composites reinforced with ceramic particles, such as the SiC. Arsenault and Wu suggested that the thermal residual stress (TRS) is responsible for the asymmetry of the composite BE [49]. The TRS is due to a difference in the coefficient of thermal expansion between the SiC and Al (ΔCTE) and the fact that samples are cooled from the annealing temperature (810 K) to room temperature (ΔT). Further evidence from finite element method (FEM) modeling by Shi and Arsenault indicates that by incorporating, the composite thermal history, the composite BE is asymmetric [51]. Without such consideration, the initial loading direction has no influence on the composite BE [52]. Withers et al. performed a parametric study of the composite BE using the mean field theory [53]. They rationalized a relationship between the influence of the TRS on the monotonic loading and that on the composite BE.

Shi and Arsenault [54] investigated the asymmetric behavior of the BE in SiC/Al composites by FEM modeling, and an analytical model based on the "back stress" from the changes in the thermal residual stress (TRS) due to the differences coefficient of expansion between the matrix and the reinforcement and the cooling from the annealing temperature. The results of that investigation are shown in Fig. 28.

An experimental investigation by Arsenault and Pillai [55] of the BE in SiC_W/Al 6061 alloy composite can be summaries as follow:

- At a small forward strain $|\varepsilon_{T_F}|$ the BSF is negative for test begun first in tension whereas it is positive if the test is begun first in compression for only the composite samples. (The BSF is $\sigma_b = \sigma_f^F - \sigma_f^R$, where σ_f^F is the flow stress in the forward (first loading) direction and σ_f^R is the flow stress in the reverse direction.

- The magnitude of BSF of the composite at a given value of ε_{T_F} is much greater (in absolute terms) than that of the alloys, i.e., approximately an order of magnitude larger.

- The BSF is greater for a test begun first in compression than if it was begun first in tension for the composite. The extent of this difference is much less in the case of the alloys, and it could be argued that the difference is within the error limits.

The magnitude of the BSF which is measured is greater than predicted by Shi and Arsenault [54], and further the BSF is predicted to be negative even at larger ε_{T_F} for the case of a test initial begun in tension [54]. If we consider another "back stress" factor arising from the work hardening of the matrix with the reinforcement present, we can use the formulation of Taya et al. [52] and Arsenault and Taya [40] and combine this with "back stress" factor arising from the residual stress [54] and obtain the following equation:

$$\sigma_b = \Delta\sigma_b^F + \Delta\sigma_b^R + 2\bar{E}_T^c \varepsilon_p \tag{59}$$

where $\Delta\sigma_b^F$ and $\Delta\sigma_b^R$ are contributions from the back stress during forward and reverse loading cycles [54] and $\bar{E}_T^c$ is tangent moduli of the composite and ε_p is the forward plastic strain. (Therefore, the elastic strain as defined by Taya et al. [52] has to be added to ε_p to obtain total strains.) Employing the results of Shi and Arsenault [54] and Taya et al. [52] we obtain Fig. 29, the predicted BSF is now much larger than that predicted by Shi and Arsenault [54], but the general trends remain the same. The BSF is always larger when the test is first begun in compression. Incorporating the additional back stress factor predicts a much larger increase in the BSF with an increase in $|\varepsilon_{T_F}|$, and is observed experimentally as shown in Fig. 29.

6. COMPOSITE DUCTILITY - THE ROLE OF REINFORCEMENT AND MATRIX

Relatively low ductility is one of the major limitation of DMMCs. The total ductility, i.e., the energy required to rupture a ductile solid is controlled by two factors: (1) the ability to withstand strain hardening before initiation of critical nuclei for final rupture (microcracks or voids); and (2) the resistance to crack extension. The experimental and theoretical efforts that have been undertaken to understand the low ductility of DMMCs can be divided into two main categories. The first group is related to the reinforcement; particle or whisker fracture, debonding and void nucleation and growth at the reinforcement-matrix interface. The second group is related to the matrix; triaxial stress state in matrix, the work hardened state of the matrix due to the ΔCTE effect, and localized matrix plastic flow. In this section, the emphasis will be on the effect of the reinforcement on the matrix, its subsequent effect on the ductility of DMMCs.

The failure of SiC/Al DMMCs as revealed by fractography may be categorized into three modes; fracture of reinforcement and /or large inter-metallics in the matrix, particle-matrix interfacial debonding, and ductile failure of matrix.

The change of ductility due to under and over-aging provides some insights into the role of the matrix in affecting the ductility of DMMCs. While the changes from matrix - to reinforcement-fracture dominated failure doubles the composite fracture-resistance (J_{IC}), only a fractional increase in strain-to-failure is obtained (see table 2) [56]. This points to the potential importance of the matrix strain hardening capacity in determining

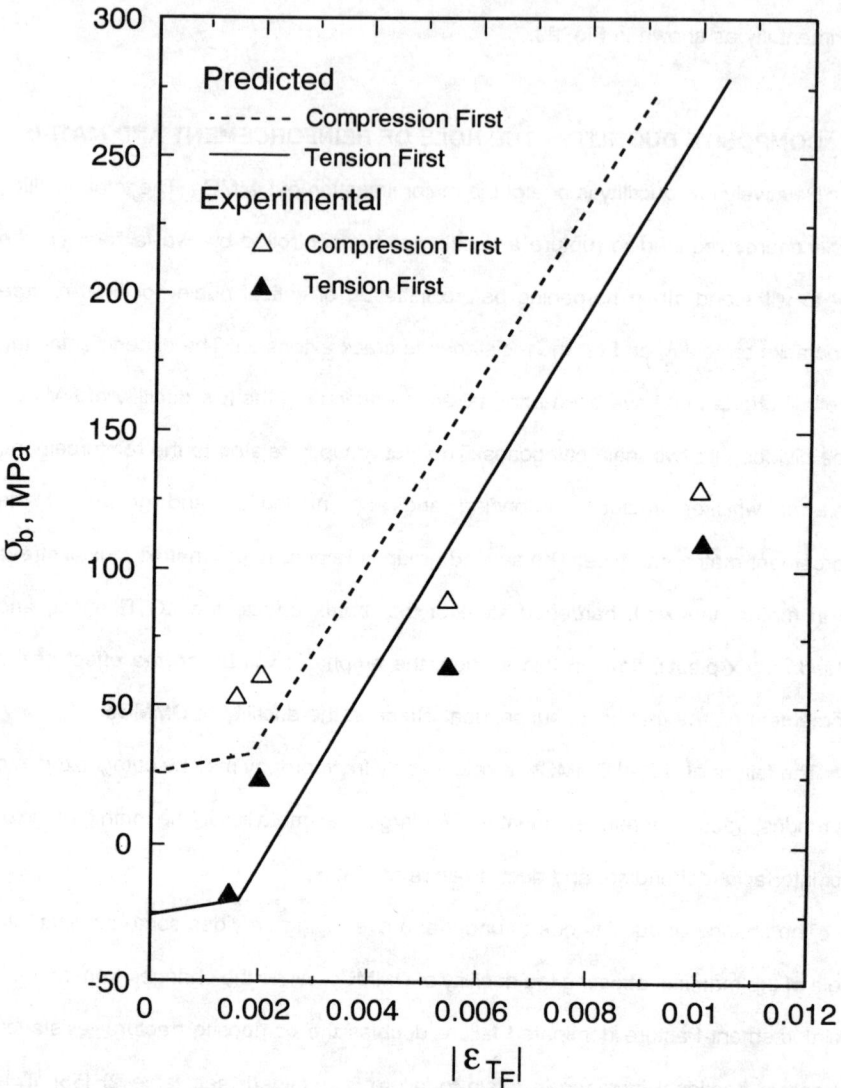

Fig. 29 A Comparison of the predicted and experimental data of the Bauschinger stress factor vs absolute total forward strain

the ductility of DMMCs. The key role of the matrix in composite failure-resistance is further demonstrated by the ability to recover most of the ductility after prestraining by re-solutionizing the composite [57]. That is, particle fracture induced by prestraining has a minimal role as compared with the matrix conditions in affecting the composite ductility. The matrix can affect the ductility of DMMCs in two ways: (a) the matrix in situ ductility is degraded by the addition of reinforcements, and (b) the matrix plastic flow preceding to the final failure is highly localized near the fracture path.

Table 2 Tensile Ductility of Underaged (UA) and Overaged (OA) Composites

Material (vol%)	Elongation (%)		J_{1c} (KJ/m^2)
	SiC/7xxxA [56]	SiC/2124 Al[a] [58]	SiC/7xxx Al [56]
UA-0 unreinforced	20	~ 20	31.0
UA-13.5 SiC$_w$	-	~ 5	-
UA-15 SiC$_p$	4.9	-	16.3
UA-20 SiC$_p$	4.3	-	11.7
OA-0 unreinforced	19	~ 18	31.5
OA-13.5 SiC$_w$	-	~ 3	-
OA-15 SiC$_p$	3.5	-	7.4
OA-20 SiC$_p$	3.4	-	5.5

[a] With 0.1 MPa hydrostatic pressure.

Arsenault [59] made an attempt to correlate the in situ matrix fracture toughness and ductility in a SiC/Al DMMCs with those of a cold-worked unreinforced alloy. He argued that the matrix is in a highly cold-worked state due to ΔCTE Therefore, the matrix in situ ductility and toughness should correspond to those in a cold-worked state. It is a general observation that ductility of a metal or alloy decreases with an increase in cold work. Since the microstructure of the matrix in 20 vol% SiC/Al composite is similar to that of an unreinforced alloy with 90% cold work [60], he suggested that the in situ matrix ductility of the composite should be comparable to the cold-worked unreinforced alloy. In a related experiment, it was shown (table 3) that cold-rolling 6061 Al alloy to 69% resulted in nearly a factor of two reduction in K_{IC}, and in the case of 139% cold-rolling there is further reduction in K_{IC}. At this level, the K_{IC} value of the cold-rolled matrix alloy is comparable to the 20 vol% SiC_p/6061 Al alloy composite [59].

Table 3 Fracture Toughness vs Cold Work

Material	Unreinforced Al [59]		20V% SiC/Al [61]
	Cold work (%)	K_{IC} (MPa·m$^{1/2}$)	K_{IC} (MPa·m$^{1/2}$)
6061 Al T6	0	43	14.5
6061 Al T6	15	31	-
6061 Al T6	69	27	-
6061 Al T6	139	~ 20	-

Localization of plastic deformation can further degrade the macroscopic ductility of DMMCs. The inherent origin for deformation localization in DMMCs attracts attention because of its importance in improving composite ductility. To assess the degree of matrix deformation localization, Arsenault et al [42] examined the dislocation density near the failure surface of a tensile sample. They found that the dislocation density was high near the fracture surface and decreased at a greater rate as a function of distance from the fracture surface in a composite with a higher SiC content. This suggested that plastic flow in a particle-reinforced metal was localized by the addition of SiC. Using stereoimaging, Davidson [62] noted that the maximum local strain near the fracture path is significantly in excess of the average strain-to-failure obtained by tensile tests. Wang et al [63] performed another in situ slip line observation followed by FEM modeling. They found that an idealized periodic clustering model [42,64] could qualitatively describe the evolution of field quantities in an actual geometry in which the FEM mesh was mapped directly from the optical images. It should be noted that the result of lower matrix triaxial stresses induced by particle clustering [64] are different from the perceptions that the matrix hydrostatic stresses are more intense in a particle cluster [57,65,66].

The periodic clustering approach [64] was also incorporated into ductility predictions [67]. By assuming the same criteria that void nucleation and its subsequent growth were governed by the effective plastic strain and the triaxial stresses [68], respectively, it was shown that the void growth stage was extended in a clustered composite resulting from a lower matrix triaxiality, and therefore the composite ductility was enhanced. This result is contrary to the suggestions by many others that homogeneous reinforcement distributions would lead to improved composite ductility (e.g. [57, 62, 69]).

The periodic clustering model [67] only accounted for the short-range interactions between particles in the cluster, long-range interactions between the clusters and the relatively homogeneous region could not be considered [43]. To study this long-range interaction (i.e., a long-range fluctuation of the internal stresses), Shi et al [43] constructed an imaginary composite as shown in Fig. 30, in which C_1 and C_2 represented the cluster and the uniform region, respectively. Because of a long-range interaction, they approximated this inhomogeneous composite by considering the smeared properties for C_1 and C_2, as if they were monolithic materials with equivalent properties from composites with predefined reinforcement volume fractions. From this approach, the short-range interactions between particles in C_is were approximated phenomenologically when the framework of mean field interactions, they obtained the following criterion [43]:

$$\left(\frac{\sigma_y^{c2}}{\sigma_y^{c1}} - K_{12}^{-1} \frac{\mu^{c2}}{\mu^{c1}} \right) \begin{cases} > 0 & C_1 \text{ yields first;} \\ = 0 & C_1 \text{ and } C_2 \text{ yield simultaneously;} \\ < 0 & C_2 \text{ yields first,} \end{cases} \quad (60)$$

where σ_y is the yield stress for the constituents, K_{12} is the ratio between the average elastic strains in C_1 and C_2, and μ is the stiffness that resists shear.

The predictions from Equation 60 may be compared with experimental results. Experimentally, strain localization in a reinforcement cluster generally initiates within an

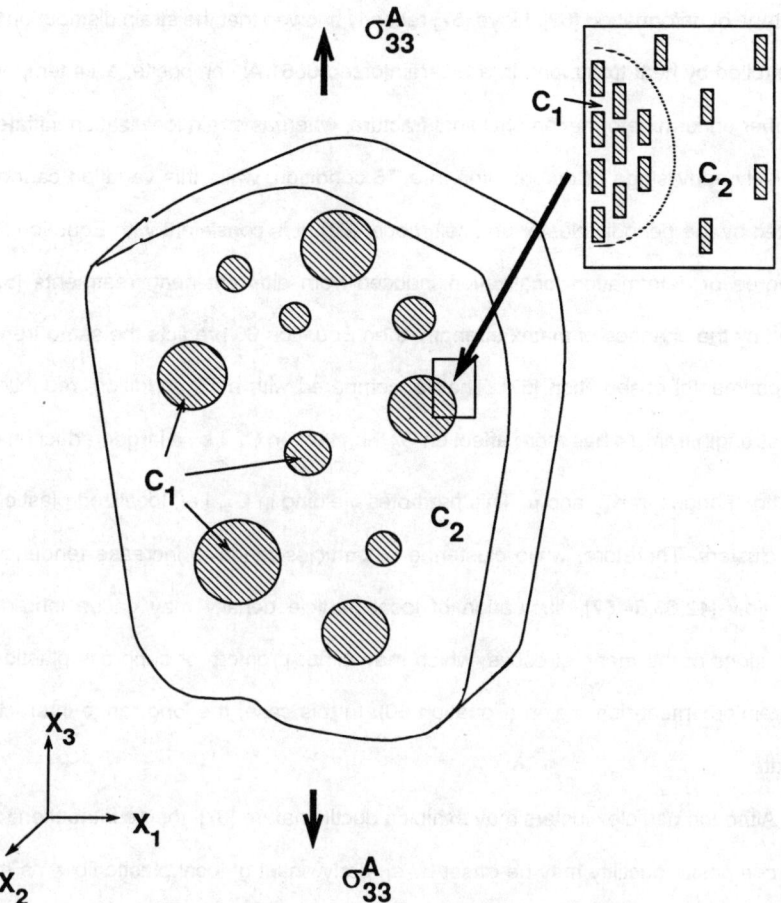

Fig.30 Schematics of mean field approximation of a composite with inhomogeneous particle distribution, where shaded spheres stand for reinforcement clusters, C_1, embedded in a uniform composite, C_2. An imaginary boundary between clustered and unclustered regions is shown on the right upper corner.

early stage of deformation [62]. Lloyd [57] recently showed that the strain distribution may be controlled by heat treatment. In a SiC-reinforced 6061 Al composite, a T4 temper led to a rather uniform deformation until final fracture, whereas strain localization initiated at a relatively early stage of deformation in a T6 condition. while this variation cannot be predicted by the periodic cluster unit cell model [67], it is consistent with Equation 60. If the degree of deformation localization induced from different heat treatments [57] is dictated by the changes of matrix strength, then Equation 60 predicts the same trend as the experimental observation [57]. That is, compared with a T6 condition, reduction in matrix strength from T4 has more effect on C_2 than that on C_1, i.e., a larger reduction in σ_y^{c2} while no changes in K_{12} and μ. This promotes yielding in C_2, i.e., localized plastic flow in the clusters. Therefore, while clustering of particles tends to increase tendency for plastic flow [42,63,64,67], fluctuation of local particle density may cause long-range perturbations of the mean stresses, which may either promote or suppress plastic flow in the reinforcement-rich region (Equation 60). In this case, the long-range interactions dominate.

Although particle clusters may exhibit a ductile nature [67], their contribution to the global composite ductility may be offset by an early onset of local plastic flow, as noted by Davidson [62], who estimated 50% local strain within clusters and yet the composite still suffered low ductility (1.6 to 2.4% strain-to-failure) because of excessive local plastic flow in the clusters. In addition, increase in the loss of strain-hardening capacity from the ΔCTE effect [59] may also offset the predicted increase in local ductility in the particle cluster [67].

From a consideration of the analysis presented above, the following summary can be drawn: the lack of strain-hardening capacity of the matrix is the major contributing factor to the low ductility of DMMCs. This may be attributed intrinsically to the exhaustion of the strain-hardening capacity from ΔCTE-induced work hardening and/or extrinsically to the high matrix triaxial stresses, which induce rapid void growth. Further study is needed to evaluate the magnitude of the contributions of these two sources. The other contributing factors are localized plastic flow and reinforcement clustering.

REFERENCES

1. D.Hull, An introduction to Composite Materials, Cambridge Solid State Science, 5 (1981).
2. S.S.Brenner, in Fiber Composite Materials, ASM, 11 (1965).
3. R.J.Arsenault and R.M. Fisher, Scripta Metall. 17, 67 (1983).
4. D.Kuhlmann-Wilsdorf, D.D. Makel, N.A.Sondergaard, and D.W. Maribo, in Cast Reinforced Metal Composites (edited by S.G.Fisherman and A.K.Dhingra), ASM, 347 (1988).
5. W.Hoover, DURAL Aluminum Composites Corp., private communication.
6. A.P. Divecha, S.G.Fishman, and S.D.Karmarker, J. Metals, 9, 12 (1981).
7. S.G.Fishman, ASTM Stand. News, 46 (Oct.1986).
8. D.W.Petrasek, D.L.McDanels, L.J.Westfall, and J.R.Stephens, Metall., 27 (Aug.1986).
9. D.W.Petrasek and R.A.Signorelli, Ceram.Engng. Sci. Proc., 2, 739 (1982).
10. A.K.Dhignra, Phil. Trans. Roy. Soc. London 294, 559 (1980).
11. R.J.Petter and R.A.Penty, J. Comp. Mater. 8, 29 (1974).
12. A.P.Divecha and S.G.Fishman, in Mechanical Behavior of Materials, ICM 3 (edited by K.J.Miller and R.F.Smith), Vol.3, p.351. Pergamon (1980).
13. J.J.Lewendowski, C.Liu, and W.H.Hunt Jr., in Processing and Properties of Powder Metallurgy Composites (edited by P.M.Kumar, K.Vedula, and A.M.Ritter), p.117. TMS (1088).

14. H.J.Rack, T.R.Baruch, and J.L.Cook, in Progress in Science and Engineering of Composites: ICCM IV (edited by T.Hagashi, Kawata, and S.Umekawa), p.1465, Jap. Soc for Comp. Mat.(1982).

15. P.K.Brindley, in High Temperature Ordered Intermetallic Alloys II (edited by N.Stoloff et al.), p.419. Mat. Res. Soc. (1987).

16. A. Leatham, A.Ogilvy, P.Chesney, and J.V.Wood, Metals and Materials, p.140 (March 1989).

17. J.Zhang, M.N.Gungor and E.J.Lavernia, in COMPOSITES Design Manufacture and Applications, ICCM/VII, (edited by S.W.Tsai and G.S.Springer), p.17-H-2, SAMPE (1991).

18. S.Caron and J.Masounave, in Fabrication of Particulate Reinforced Metal Composites (edited by J.Masounave and F.G.Hamel), p.79, ASM (1990).

19. D.M.Schuster, M.D.Skibo, and W.R.Hoover, Light Metal Age., 15 (Feb. 1989).

20. M.K.Surappa and P.K.Rohatgi, J.Mater. Sci., 16, 983 (1981).

21. R.Mehrebian, R.G.Riek, and M.C.Flemmings, Metall. Trans. 5, 1899 (1974).

22. A.Mortensen, V.J.Michaud, J.A.Cornie, M.C.Flemmings, and L.Masur, in Cast Reinforced Metal Composites (edited by S.G.Fishman and A.K.Dhingra), p.7, ASM (1988).

23. F.Delannay, L.Froyen, and A.Deruyttere, in Cast Reinforced Metal Composites (edited by S.G.Fishman and A.K.Dhingra), p.81, ASM (1988).

24. A.Mortensen, J.A.Cornie, and M.C.Flemmings, J. Metals 40, 12 (1988).

25. S.Nourbakhsh, F.L.Liang, and H.Margolin, Adv. Mater. Manufacture Proc.3, 57 (1988).
26. S.Nourbakhsh, F.L.Liang, and H.Margolin, Metall. Trans. 21 A, 213 (1990).
27. S.Nourbakhsh, F.L.Liang, and H.Margolin, Mater. Res. Soc. Proc. (edited by C.T.Liu, A.I.Taub, and N.S.Stoloff), Vol.133, p.459. Mat.Res.Soc. (1989).
28. R.J.Arsenault and N.Shi, Mater. Sci.Engng. 81, 175 (1986).
29. N.Shi and R.J. Arsenault, unpublished results.
30. S.Shibata, M.Taya, T.Mori and T. Mura, Acta Metall., 40, 3141 (1992).
31. C.T.Kim, J.K.Lee and M.R.Plichta, Metall. Trans. 21A, 673 (1990).
32. A. Mendelson, Plasticity - Theory and Applications, p.135. MacMillan Publishing Company, New York, NY (1968).
33. H.L.Cox, Br. J. Appl. Phys., 3, 72 (1952).
34. A. Kelly, Strong Solids, 2nd ed. Chapter 5, Oxford University Press (1973).
35. V.C.Nordone and K.M. Prewo, Scripta Met. 20, (43) 1986.
36. J.D.Eshelby, Proc. Roy. Soc. A241, 376 (1957).
37. T.Mura, Micromechanics of Defects in Solids, 2nd edn., Martinis Nijhoff (1987).
38. M.Taya and R.J.Arsenault, Metal Matrix Composites: Thermomechanical Behavior, Pergamon Press (1989).
39. T. Morimoto, T.Yamaoka, H.Liholt, and M.Taya, J.Engng, Mater. Tech. 110, 71 (1988).
40. R.J.Arsenault and M.Taya, Acta Metall. 35, 651 (1987).

41. Y.Takao, T.W.Chou, and M.Taya, J.Appl. Mech, 49, 536 (1992).
42. R.J.Arsenault, N.Shi, C.R.Feng and L.Wang, Mater. Sci. Engng. A131, 55 (1991).
43. N.Shi, B. Wilner and R.J.Arsenault, Acta Metall. Mater., 40, 2841 (1992).
44. N.Shi and R.J. Arsenault, Ann. Rev. Mater. Sci., 24, 321 (1994).
45. N.Shi, R.J.Arsenault. A.D.Krawitz and L.F.Smith, Met. Trans., 24A, 187 (1993).
46. W.Voigt, "Lehrbuch der Kristallphysik," p.739, Teubner, Leipzig (1928), and A.Reuss, Z.Angew. Math. Mech., 9, 49 (1929).
47. R.J.Arsenault, in Mechanical Behavior of Metallic and Ceramic Composites, (edited by S.I.Arderson et al.), 9th Risφ International Conference, 279 (1988).
48. H.J.McQueen and J.J.Jonas, Treatise on Materials Science and Technology (edited by R.J.Arsenault), Vol.6, p.394, Academic Press, New York (1975). E.227, 1146 (1963).
49. R.J. Arsenault and S.B. Wu, Mater. Sci. Eng., 96, 77 (1987).
50. J. Llorca, A. Needleman, and S. Suresh, Scripta Metall. Mater., 24, 1203 (1990).
51. N. Shi and R.J. Arsenault, J. Comp. Technol. Res. , 13, 211 (1991).
52. M. Taya, K.E. Lulay, K. Wakashima, and D. Lloyd, Mater. Sci. Eng., A124, 103 (1990).
53. P.J. Withers, W.M. Stobbs, and O.B. Pedersen, Acta Metall., 37, 3061 (1989).
54. N. Shi and R.J. Arsenault, Met. Trans., 24A, 1879 (1993).

55. R.J. Arsenault and U.T.S. Pillai, submitted for publication.

56. M. Manoharan and J.J. Lewandowski, Acta Metall. Mater., 38, 489 (1990).

57. D.J. Lloyd, Acta Metall. Mater., 39, 59 (1991).

58. G.J. Mahon, J.M. Howe, and A.K. Vasudevan, Acta Metall. Mater., 38, 1503 (1990).

59. R.J. Arsenault, "Fracture Toughness of Discontinuous Metal Matrix Composite," G.Z. Voyiadjis, ed., Damage in Composite Materials Studies in Applied Mechanics, vol.34, (Amsterdam: ASCE Elsevier), 219 (1993).

60. R.J. Arsenault, L. Wang, and C.R. Feng, Acta Metall. Mater., 39, 47 (1991).

61. C.R. Crowe, R.A. Gray, and D.F. Hasson, "Microsturcture Controlled Fracture Toughness," Proceedings of 5th International Conference of Composite Materials (ICCMV), ed. W.C. Harrigan, J. Strife, and A.K. Dhingra (PA:TMS-AIME), 843 (1985).

62. D.L. Davidson, Metall. Trans., 22A, 113 (1991).

63. Z. Wang, T.-K. Chen, and D.J. Lloyd, Metall. Trans., 24A, 197 (1993).

64. T. Christman, A. Needleman, and S. Suresh, Acta Metall., 37, 3029 (1989).

65. P.M. Singh, and J.J. Lewandowski, Metall. Trans., 24A, 2531 (1993).

66. A.S. Argon, and J. Im, Metall. Trans., 6A, 825 (1975).

67. J. Llorca, A. Needleman, and S. Suresh, Acta Metall. Mater., 39, 2317 (1991).

68. A.L. Gurson, J. Eng. Mater. Tech., 99, 2 (1977).

69. D.L. McDanels, Metall. Trans., 16A, 1105 (1985).

Problems

1. Discuss the approximations that one would make in order to apply the shear lag model for the case where the whiskers are not aligned to the tensile axis.

2. Can the Eshelby method be used to determine the effect of size of the reinforcement on the strength of a DMMC. Explain your answer in detail.

3. The σ_{yc}/σ_{ym} of an annealed SiC$_w$/Al6061 composite is much greater than the σ_{yc}/σ_{ym} of the composite and matrix if they are in the heat treated, i.e. T-6 condition. Explain why this result occurs.

4. Discuss possible methods of measuring the residual stresses in DMMC.

5. Explain why the simple prismatic model of prismatic punching due to ΔCTE effect predicts greater strengthening for a small size reinforcement as compared to a larger size reinforcement. The volume fraction of the reinforcement is held constant. A quantitative answer is also required.

6. Explain in detail and derive an expression for ductility of DMMC as function of reinforcement size.

7. In the case of continuous filament metal matrix composites, what would be strengthening contribution due to the ΔCTE effect? (Hint, this requires you to look up the rule of mixtures strengthening theory.)

8. Explain why, in detail, a model based on fracture of SiC particles can account for the larger compressive yield stress but not a smaller apparent modulus, in comparison to that of the tensile yield stress and apparent modulus.

9. Discuss the apparent contradiction in the fact that plastic deformation begins in regions of clustering of the reinforcement and not in regions of lower than average reinforcement volume fractions.

10. Explain why the apparent modulus of DMMC, as determined from the slope of the initial portion of the stress-strain curve, is much smaller than modulus of the same composite determined by ultrasonic methods.

# Chapter 4
POLYMER MATRIX COMPOSITES

JANG−KYO KIM[1] & YIU−WING MAI[2]

[1] *Department of Engineering, Australian National University, Canberra &*
[2] *Center for Advanced Materials Technology, Department of Mechanical and Mechatronic Engineering, University of Sydney, Sydney, Australia*

1. Introduction

A composite is defined as the material created when two or more distinct components are combined in various ways to produce useful properties. The fiber composite technology in particular is based on taking advantage of the high specific strength and stiffness of fibers by incorporating them in a resin matrix, where both fibers and matrix retain their physical and chemical identities, yet they produce a combination of mechanical properties that cannot be achieved with either of the constituents acting alone.

The first generation of polymer matrix composite materials was developed before their full property and structural potentials were appreciated. Typical applications have been the use of glass fiber reinforced plastics for the construction of small boat hulls and architectural panels. In these applications, the composites are used mainly as replacements for traditional materials like wood. With the introduction of high performance fibers such as carbon, aramid, boron fibers, a new high technology has been developed, aimed especially at the aerospace and military industries. They possess the desirable properties of low density (1.44–2.7 g/cm^3), and high strengths (3–4.5 GPa) and moduli (80–550 GPa). When combined with a resin binder to support the applied load, these fibers provide mechanical capability equal to or exceeding those of most metals. Bulk polymers typically have room temperature strengths 30–130 MPa, which are low compared with traditional metals like steel and aluminum alloys, and their moduli, approximately 2–4 GPa would result in unacceptably large deformation in structures of any size in their unreinforced forms.

In addition to the strengths and moduli of the fiber and matrix, the structural and mechanical performance of a composite is determined by a number of factors including aspect ratio, length distribution, volume fraction, uniformity and orientation of the fiber; stacking sequence and number of angle plies in laminates; the integrity of and adhesion at the fiber–matrix interface; the manufacturing processes of the composite components. Fabrication technology of composites in turn is dominated by the chemistry and rheology of the matrix resins and by the types and physical form of the fibers. It follows therefore that of paramount importance in the science and technology of composites is a fundamental understanding of the chemical/physical/mechanical nature of the composite constituents and the ways they are combined together.

This chapter contains the details of the microstructure–property relationship of high performance fibers, polymer matrix materials and the advanced composites made therewith. A special focus is placed on discussion of the state–of–the–art manufacturing technology, the mechanics of stress transfer in composites containing diverse forms and orientation of fibers and laminate, and the failure of composites in various loading conditions.

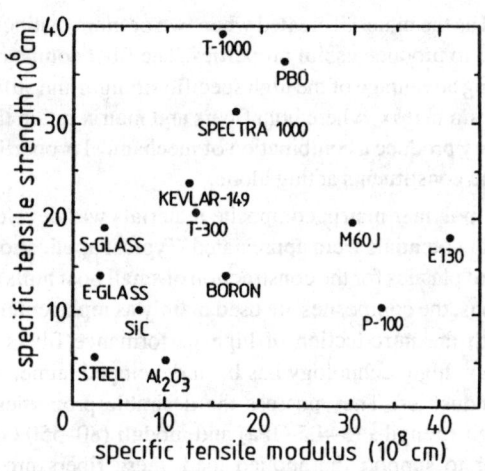

Figure 1. Specific strength versus specific modulus of a variety of reinforcing fibers: T–300, T–1000 and M60J (PAN–based carbon); E–130 and P–100 (pitch–based carbon); Spectra 1000 (polyethylene); PBO (polyphenylene benzobisoxazole).

2. Composite Constituents

2.1. Reinforcements

A reinforcement is the strong, stiff integral component of a composite which is incorporated into the matrix to improve its mechanical properties. Typical reinforcement materials include fibers, whiskers, fillers and particulates of various shapes and sizes in organic and inorganic nature. Among these, glass, carbon and aramid fibers are the most common and technologically most important for polymer matrix composites which possess both high strength and stiffness coupled with low density. Figure 1 shows the specific tensile modulus and specific strength of various high performance fibers. The use of fibers as high performance engineering materials is based on the following important characteristics [1]:

- A small diameter with respect to its grain size or other microstructural unit. This allows a higher fraction of the theoretical strength to be attained than that possible in bulk form. In general, the smaller the size, the lower the probability of having flaws or imperfections in the material.
- A high aspect ratio that allows a very large fraction of the applied load to be transferred from the matrix to the fiber across the interface.
- A high degree of flexibility which allows a variety of manufacturing techniques and textile structures to be used.

2.1.1 Glass Fibers

A variety of chemical compositions of mineral glasses have been used to produce fibers. The most common are based on silica (SiO_2) with additions of oxides of calcium, boron, sodium, iron, magnesium and aluminium. Typical compositions of three most popular glass fibers are given in Table 1 [2], and their representative properties are in Table 2, along with properties of other fibers. The designation E stands for electrical since E–glass is a good electrical insulator besides having good strength and a moderate Young's modulus; C stands for corrosion since C–glass has a better resistance to chemical corrosion; S–glass fiber is a high strength type initially developed for military applications. Its modulus is about 20% greater than that of E–glass, it is also stronger and tougher. Its creep rupture resistance is significantly better and it is able to withstand higher temperature than other glass fibers.

Table 1. Composition (weight %) of glass used for fiber manufacture. After Hull [2].

Elements	E–glass	C–glass	S–glass
SiO_2	52.4	64.4	64.4
Al_2O_3, Fe_2O_3	14.4	4.1	25.0
CaO	17.2	13.4	–
MgO	4.6	3.3	10.3
Na_2O, K_2O	0.8	9.6	0.3
Ba_2O_3	10.6	4.7	–
BaO	–	0.9	–

Table 2. Properties of fibers

Property	E–glass	S–glass	Type I carbon	Type III carbon	Kevlar 49	Polyethylene Spectra 900
Diameter (μm)	5–25	5–15	6–8	7–9	12	38
Density (g/cm^3)	2.54	2.49	1.7–1.8	1.85–1.96	1.45	0.97
Tensile strength (GPa)	2.4	4.5	3–5.6	2.4	3.6	3.0
Elongation at break (%)	3–4	5.4	1.0–1.8	0.38–0.5	2.4	3.5
Young's modulus (GPa)	72.4	85	235–295	345–520	135	117
Coefficient of thermal expansion (10^{-6}/K) axial	5.0	5.6	–0.5	–1.2	–2	–
radial	–	–	7	12	59	–

Continuous glass fibers are produced by melting the raw materials and the molten glasses are drawn through small orifices in a bushing to form a strand as shown in Figure 2. The number of filaments in the strand corresponds to the number of orifices in the bushing, and is usually between 200 and 1000. The common range of diameter for E–glass fibers manufactured for reinforcement purposes is 12 to 14 μm. The fiber diameter is controlled by adjusting the orifice size, the winding speed and the viscosity of molten glass which depends on the composition and temperature.

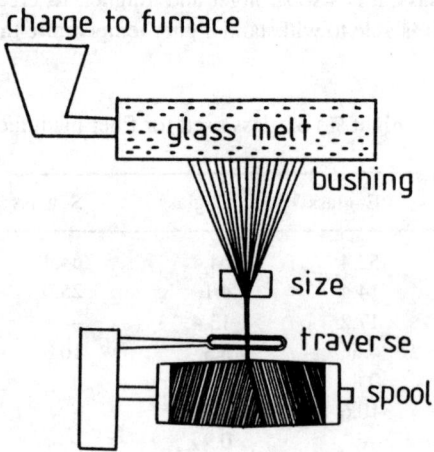

Figure 2. Schematic diagram of glass fiber manufacturing.

To protect glass fibers from abrasion and mechanical damage, sizing materials are normally added to their surface immediately after forming. The size is applied in the form of a water–based emulsion, and this emulsion may be sprayed onto the drawn strand of fibers, or the fibers are passed through a bath or over a wetted roller. For glass fibers intended for subsequent textile operation such as spooling and weaving, the size usually consists of a mixture of starch and a lubricant, which may be removed after the process. The size usually contains a coupling agent to aid bonding with the resin matrices (see Section 2.3.2).

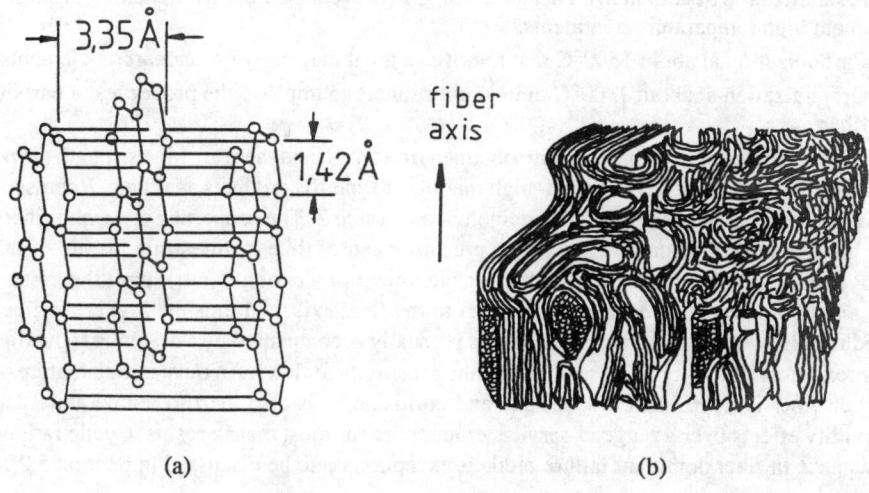

Figure 3. Schematic drawings for (a) graphite lattice structure and (b) a three–dimensional model of carbon fiber. After Singer et al. [3] and Bennett and Johnson [4].

2.1.2 Carbon Fibers

Carbon fibers are currently the predominant high strength–high modulus fibers used in the manufacture of advanced polymer matrix composites. Its properties are a direct reflection of the structures of carbon in the nanoscopic scale. Carbon in graphite form is highly anisotropic. The high bond strength between carbon atoms in the basal plane results in an extremely high modulus (about 1000 GPa) while the weak van der Waals type bond between the neighboring layers results in a low modulus (about 35 GPa) along the edge surface, as shown in Figure 3 [3,4]. Therefore, it is necessary to have a high degree of preferred orientation of hexagonal planes along the fiber axis if a high modulus is desired. To improve the orientation of graphite crystals various kinds of thermal and stretching treatments are carried out with accurate control.

Carbon fibers have been made from a number of precursors, but are now mainly produced from polyacrylonitrile (PAN), pitch and rayon. The conversion of PAN to carbon fiber is a process involving spinning and stretching of the precursor. Pitch–based fibers are made by a liquid crystal spinning process. These fibers possess particularly high modulus with moderate strength. Rayon–based fibers are processed much like PAN. Since the yield is low compared to PAN–based fibers, the fibers tend to be relatively expensive. All these carbon fiber fabrication processes involve the following essential steps regardless of the types of precursors used:

- A stabilization treatment in air at about 220°C to prevent the fiber from melting in subsequent high temperature treatments.
- Carbonization at about 1500°C that removes a great majority of monocarbon elements.
- Graphitization at about 2300°C in inert atmosphere to improve the properties of carbon fiber.

Most commercial carbon fibers are obtained from PAN, and there are high strength (Type I), high modulus (Type II) and ultra–high modulus (Type III) products available. Representative properties of these materials are included in Table 2. The properties of carbon fibers are a function of the composition of the precursor and of time–temperature profile of the above processes. In general, the higher the maximum processing temperature, the greater the extent of crystalline orientation parallel to the fiber axis, and thus the higher the fiber modulus. However, an increase in modulus is usually accompanied by a dramatic reduction in strength and ductility because of increasing sensitivity to flaws. An outstanding feature of carbon fiber is its resistance to fatigue and corrosion. It is also inert to temperature and humidity effects over a range of service temperature for most matrix resins. Cyclic fatigue resistance in fiber dominant failure mode is exceptional, to be discussed in Section 5.2 .

2.1.3 High Strength–High Modulus Organic Fibers

High strength–high modulus organic fibers, typically polymeric aramid fibers and ultra high molecular weight (UHMW) polyethylene fibers, have found applications in a broad range of composites. Figure 4 [5] shows the chemical structures of several high modulus organic fibers, and Table 2 compares properties of these fibers with those of glass and carbon fibers. Although the organic fibers invariably display low compressive strength (i.e. several orders of magnitude lower than the axial tensile strength) which often limits applications of these fibers to non–primary structures, their high specific mechanical properties coupled with high ductility and low thermal and electrical conductivity are often the material of choice. All these fibers exhibit excellent environmental and dimensional stability as well as low coefficient of thermal expansion.

As shown in Figure 4, the high strength–high modulus polymeric fibers mostly consist of chain polymers containing covalent carbon–carbon bonds. These polymers possess a high degree of aromaticity, high planarity and essential linearity in the chain backbone. There are now two generic routes to these fibers which include:

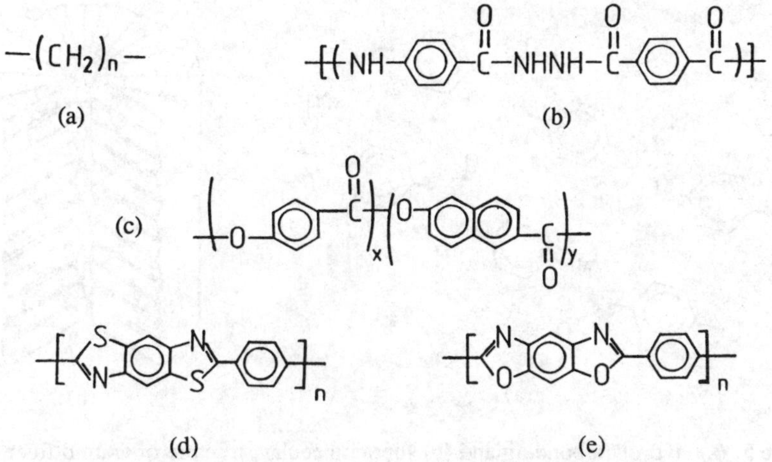

Figure 4. Chemical structures of high strength–high modulus organic fibers: (a) polyethylene; (b) aramid: poly(phenylene terephthalamide) (PPTA); (c) thermoplastic copolymer; (d) poly (phenylene benzobisoxazole) (PBO); (e) poly (phenylene benzobisthiazole) (PBZT). After Jaffe [5].

- The spinning of stiff, nematogenic polymers which easily form highly oriented structures in the solid states.
- The morphological manipulation of flexible, conventional polymers into highly oriented extended–chain fibers through complex processing.

Aramid Fibers

The aramid fibers are commercially produced by du Pont (Kevlar), AKZO (Twaron) and Teijin (Technora). These fibers are known to be spun from liquid crystal dopes through a dry–jet wet–spinning process. Chemically the Kevlar fiber is poly (phenylene terephthalamide) (PPTA) which is a polycondensation product of terephthaloyl chloride and p–phenylene diamine. The molecules form a planar array with interchain hydrogen bonding (Figure 5(a)), and the stacking sheets form a crystalline array of which bonding is rather weak. Electron microscopy and diffraction study on Kevlar fibers [6,7] show radially arranged axially pleated crystalline supramolecular sheets (Figure 5(b)). Kevlar fibers are circular in cross section with a smooth surface which exhibits little texture except for isolated scuffed fibrillar regions. They normally fracture by splitting into small fibrils (i.e. fibrillation) in the longitudinal direction without being broken transversely as shown in the scanning electron microphotograph (Figure 6 [8]) which is a direct manifestation of the microstructures and ductile nature of the fiber.

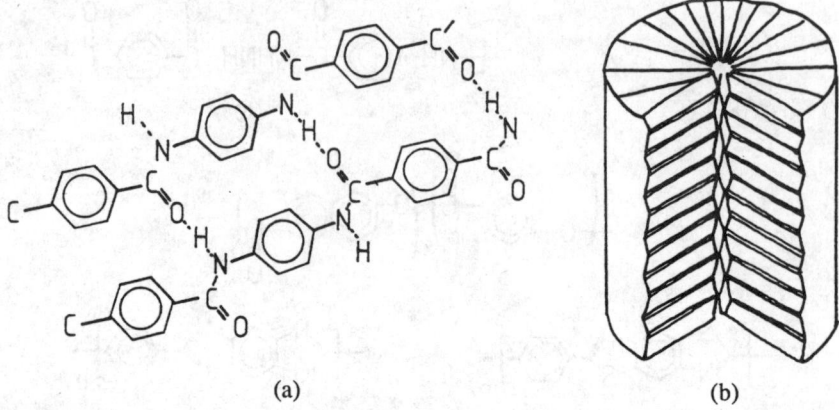

Figure 5. (a) Hydrogen bonding and (b) supramolecular structure of aramid fiber (Kevlar 49). After Dobb et al. [6,7].

The properties of Kevlar 29, Kevlar 49 and recently developed Kevlar 149 are about the same, the latter two having slightly better modulus than the first (Table 2). The difference between the high and low modulus variants of Kevlar fibers are purely structural. The low modulus type (Kevlar 29) is used primarily for tensile members such as ropes, cables, webbings, and ballistic cloth, while the Kevlar 49 and 149 are reinforcing fibers for high performance polymer matrix composites. These fibers have relatively high use temperatures (i.e. stable up to 300°C in the absence of hydrolysis agents) and very low creep, in addition to the high specific strength. To overcome the poor compressive performance of these fibers while fully exploiting their advantages, they are often used in hybrid constructions with glass and carbon fibers. They are degraded when exposed to visible or ultraviolet lights, which results in discoloration with accompanying loss of mechanical properties.

High Modulus Polyethylene Fibers

High modulus polyethylene fibers are produced via a gel–spinning process in which a low concentration solution of ultra high molecular weight (UHMW) ($M > 2 \times 10^6$) polyethylene is extruded to form a gel precursor fiber. This precursor fiber is subsequently hot drawn to produce a very highly oriented fiber with an extended chain fibrillar microstructure.

A number of commercial high modulus polyethylene fibers are available which include Spectra (Allied Signal), Dyneema (DSM/Toyobo) and Tokilon (Mitsui Toatsu). These polyethylene fibers are very light with a density of 0.97 g/cm^3. Their strength and modulus are slightly lower than those of aramid fibers but its specific tensile properties are the highest achieved with any organic materials. It has a very high elongation at break, two or three times higher than glass or aramid fibers. It is chemically inert and is particularly resistant to alkali environments compared to aramid fibers. Disadvantages of this fiber are low creep resistance and applications limited to low temperatures (up to 100°C) because of its low melting point.

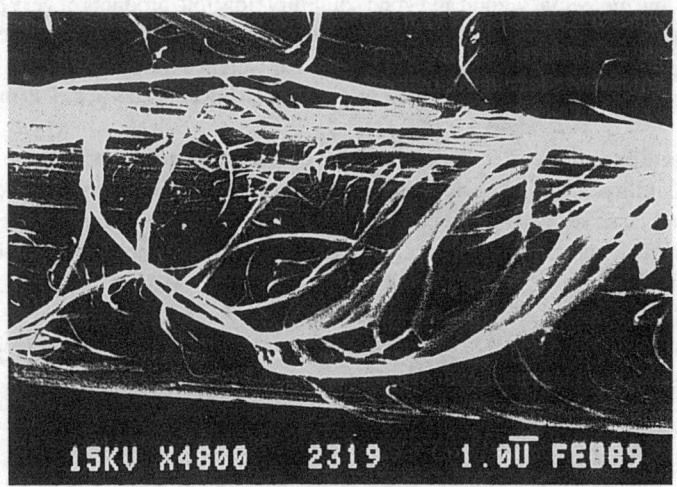

Figure 6. Scanning electron microphotograph of a fibrillated Kevlar 49 fiber. After Kim and Mai [8].

2.2. Matrix Systems

The matrix of an advanced composite material is the continuous phase which binds the fibers and transfers the stress to the fiber across the interface. The matrix is required to fulfill the following functions:
- To bind together the fibers and to protect their surface from damages during handling, fabrication and the service life of the composites.
- To disperse the fibers and maintain the desired fiber orientation and spacing.
- To transfer stresses to the fibers by adhesion and/or friction across the fiber–matrix interface when the composite is under load, and thus to avoid any catastrophic propagation of cracks and subsequent failure of the composites.
- To be chemically and thermally compatible with the reinforcing fibers.
- To be compatible with the manufacturing methods which are available to fabricate the desired composite components.

The matrix material has a primary influence on important mechanical properties inclusive of interlaminar shear strength, interlaminar fracture toughness in various failure modes,

compression and flexural strength; and it also dictates the processibility of the composite components. There is no single ideal matrix material present that satisfies all the requirements imposed by different applications. Polymers in general may be grouped into two broad classes:
- Thermosetting resins: These harden by a process of chemical cross–linking, whereby resins of low molecular weight and good solubility flow into products of very high molecular weight and limited solubility. The cross–linking is an irreversible process.
- Thermoplastics: These are already high molecular weight strong solids. They soften upon heating, and upon cooling regain their original mechanical properties. The process is reversible.

Thermosetting resins currently dominate the fiber composite industry due to their availability, availability of capital equipment, the existence of a large database and low material cost. Thermoplastic resins also offer other significant attractive features to be detailed later. Some of the advantages and disadvantages of these two different polymer types are summarized in Table 3 [9].

Table 3. Sumary of trades–offs of thermosets and thermolastics as composites matrix resins. After Hergenrother and Johnston [9].

Property	Thermosets	Thermoplastics
Formulations	Complex	Simple
Melt viscosity	Very low	High
Fiber impregnation	Easy	Difficult
Prepreg tack	Good	None
Prepreg drape	Good	None to fair
Prepreg stability	Poor	Excellent
Processing cycle	Long	Short to long
Processing temperature/pressure	Low to moderate	High
Mechanical properties: (–54 to 93°C, hot/wet)	Fair to good	Fair to good
Environmental durability	Good	Unknown
Solvent resistance	Excellent	Poor to good
Fracture toughness/damage tolerance	Poor to good	Good to excellent
Database	Very large	small

2.2.1.Thermosetting Resins

Among thermosetting matrix materials most common are epoxy, unsaturated polyester, vinylester, phenol–formaldehyde. Epoxy resins are by far most important polymeric matrix materials because of the good balance of properties obtained with the relative ease of handling and processing. Unsaturated polyester resins have maximum use temperature of around 100°C, vinyl–ester and epoxies extend the temperature range to about 150°C. There is an ever–increasing need for matrices with enhanced high–temperature capability which results in the introduction of a number of high performance resin systems. The underlying principle is to formulate a highly aromatic compound and to extend its thermal stability by incorporating further cross–linking groups with adequate processibility. These thermosetting resins include polyimide, bismaleimides and cyanate resins which offer use temperature up to 250°C. Moulding temperatures are around 300°C which may cause residual stress problem due to differential shrinkage when cooling from the process temperature. The remainder of this section discusses the chemistry of thermoset resins derived from epoxy resin and unsaturated polyesters. Important properties of several thermoset polymers are compared in Table 4.

Epoxy Resins

Epoxy resins are extensively used in advanced structural composites, particularly in the aerospace industries. Characteristics of epoxy resins useful for composite applications are high chemical and corrosion resistance, good mechanical and thermal properties, outstanding adhesion to various substrates including advanced fibers, low shrinkage upon cure, flexibility, good electrical properties and ability to be processed under a variety of conditions.

There is a wide choice of systems available, consisting of epoxy resins and curing agents or hardeners. Epoxy resins range from low–viscosity liquids to high melting point solids, and they can be readily formulated to give suitable products for the manufacture of prepregs. The majority of commercial epoxy resins used in composite manufacture is characterized by a three–membered ring known as the epoxide group [10]:

$$R-[CH_2-CH-CH_2]_n \atop \diagdown O \diagup$$

The glycidyl ether and amines account for most of the epoxy resin used in composites. Although the epoxide group reacts with a wide variety of groups, only three categories of curing agents are widely used. These are amine, anhydrides and catalitic curing agent, the first being co–reactive. The co–reactive curing agents act as a co–monomer in the polymerization process while the last two catalitic curing agents function as an initiator for epoxy resin homopolymerization. The objective in curing reaction is to convert as many epoxide group as possible to a highly cross–linked network. Treatment with curing agents gives three–dimensional insoluble and intractable thermoset polymers via an irreversible process. The choice of curing agent is of paramount importance in epoxy reaction since it controls the type of curing reaction which in turn determines the time and temperature of the cure cycle, processing method and the final physical and chemical properties of the cured resin. Typical cure cycles of composite prepregs containing epoxy matrix are one hour at 120°C followed by post cure for two hours at 240°C.

Table 4. Typical properties of thermosetting resins used in composites.

Property	Epoxy	Polyester	Vinyl ester	Polyimides (PMR-15)
Density (g/cm^3)	1.1–1.4	1.2–1.5	1.12	1.43
Young's modulus (GPa)	3–6	2–4.5	3–3.5	3.2
Poisson's ratio	0.38–0.4	0.37–0.39	0.38	0.38
Tensile strength (MPa)	40–100	40–90	60–90	56
Elongation at break (%)	1–6	2	2–5	–
Compressive strength (MPa)	100–200	90–250	110–125	187
Coefficient of thermal expansion (10^{-6}/K)	60	100–200	80	50
Heat distortion temperature (°C)	130–250	50–110	100–140	350

Epoxy resins can be modified with a variety of different materials. For example, polymers are often added to the resin to improve such properties as fracture toughness and damage tolerance of the final cured composites [11,12], or to control the viscosity of the resin during the manufacturing process. Examples of polymer additives include acrylonitrile–butadiene –styrene copolymers and polyethersulfones or liquid rubbers such as carboxyl–terminated butadiene–acrylonitrile (CTBN) copolymers. These polymers often possess reactive end groups that are capable of addition reactions with the epoxy resin.

Polyester Resins

Polyester resin contains the unsaturated diester group:

$$R-[O-\underset{\underset{O}{\|}}{C}-CH=CH-\underset{\underset{O}{\|}}{C}-O]-$$

The polyester backbone polymer is synthesized by condensation polymerization of the acid/glycol combination. A variety of dibasic acids or anhydrides and diols can be condensed with the maleic anhydride, and the properties of cured resin matrix can be tailored to requirements by changing the ratio and nature of the components. By dissolving the unsaturated polyester in monomers like styrene, a low–viscosity can be obtained to impart good processibility.

Unsaturated polyester resins can be cross–linked at ambient or elevated temperature. This involves free radical copolymerization of the maleic anhydride derived unsaturated polyester with styrene monomer (or other reactive diluent). The curing reaction produces a cross–linked polymer in which the polyester chains are bridged by one or more styrene units and become incorporated in a long copolymer chain. One of the most important reasons for the popularity of unsaturated polyester resins in contact molding applications is their ability to cure rapidly and satisfactorily at room temperature, by decomposition of peroxide (e.g.

methylethyleketone (MEK) peroxide) in the presence of small amounts of transition metal ions.

Polyester resins at room temperature show generally good short–term resistance to common acids, bases and solvents. At elevated temperatures and for long exposure times, the chemical resistance depends on the structures of the resin. The most important degradation mechanisms due to the attack of water and other chemicals in polyester matrix composites, particularly glass fiber–polyester matrix systems which are the major application of this resin for composite manufacture, include hydrolysis of the resin itself, destruction of the fiber–matrix interface and physical damage from osmotic pressure effects or from swelling forces [13].

2.2.2. Thermoplastic Resins

Thermoplastic resins are increasingly more important as matrices of composites for high temperature applications. Thermoplastic resins contain high molecular weight (M > 15000) linear or branched molecules which flow upon application of heat and pressure, and usually soluble in a suitable solvent. They may be melted and re–formed into another shape and size. Chemical structures of several thermoplastic resins of high temperature applications are shown in Figure 7 [14]. A characteristic feature of these resins is the presence of a high proportion of aromatic rings, usually connected by a stable group which provide a high degree of chain rigidity and thus high glass transition temperature, T_g. In addition, these resins contain low aliphatic hydrogen which is necessary for good thermal stability at high temperatures. All these resins shown are amorphous, except polyphenylene sulfide (PPS), polyetherketone (PEK) and polyetheretherketone (PEEK) which are semicrystalline polymers.

The major reasons for increasing use of thermoplastics as matrices for composites are:
- Cost savings can be achieved from improved processing and handling compared to highly cross–linked epoxies, e.g. adaptability of thermoplastics to high rate processing, reduced scrap rate because of reformability, simple storage, easy handling of materials and indefinite shelf life without refrigeration.
- Mechanical properties of the composites can be improved, particularly the fracture toughness and moisture sensitivity, compared to epoxies.

Mechanical properties of the high temperature thermoplastic resins are summarized in Table 5. In general, tensile strengths are slightly better than or comparable to, while tensile moduli are marginally worse than structural epoxies. Ultimate strain capability of these resins is at least an order of magnitude higher than the epoxies: typically polysulfone (PS) and PEEK have ultimate elongations over 50% and 150%, respectively. The ductile behavior in tension and the high Izod impact strength values are both indicative of better toughness for the thermoplastics. Moisture absorption for these resins is also low (i.e. typically 2% at equilibrium in water immersion) compared to epoxies.

Table 5. Typical properties of thermoplastic resins used in composites, After Hancox [14].

Property	poly-carbonate	polysulfone	polyether-sulfone	polyether-etherketone	polyether-ketone	polyether-imide	polyamide-imide	polyphenyl-enesulfide	liquid crystal polymer (Vectra)
crystallinity	mainly A	A	A	semi-C	semi-C	A	A	semi-C	liquid crystal
density (g/cm^3)	1.2	1.25	1.37	1.26–1.32	–	1.27	1.4	1.36	1.37–1.4
Young's modulus (GPa)	2.3	2.5	3.2	3.2	–	3.3	4.8	3.5	9–15.2
tensile strength (MPa)	70	60–75	84	93	110	105	93	84	165–188
elongation at break (%)	120	50–100	40–80	50	–	60	17	4	1.3–3.0
compressive strength (MPa)	86	–	–	–	–	140	275	110	–
compressive modulus (GPa)	–	26	–	–	–	29	28	–	–
coefficient of thermal expansion (10^{-6}/K)	70	56	55	47	–	62	63	54	–5 – +75
heat distortion temperature (°C)	130	174	200	150	165	200	273	136	220
glass transition temperature (°C)	–	190	230	143	165	217	–	93	–
melting temperature (°C)	260	–	–	334	365	260	–	285	280
Izod impact energy (Jm^{-1})	120–850	70	76–84	83	–	50	53	21	45–530
water absorption in 24h at RT (%)	0.12	0.2	0.43	0.1	–	0.25	0.3	0.2	0.02–0.04

A = amorphous
C = crystalline
RT = room temperature

However, thermoplastic resins have poorer resistance to creep and chemicals than thermosetting resins. Amorphous thermoplastics are especially susceptible to environmental stress cracking with chemicals such as chlorinated solvents (used as paint strippers) and most highly polar organic solvents. In contrast, for semicrystalline polymers, like PEEK in which rigid rings are connected to chemically inert groups to impart high crystalline, high melting point polymers, the chemical resistance is excellent.

There are also potential disadvantages of using thermoplastics which replace thermosetting resins for composite manufacture. There requires a fundamental change in processing. In general, much higher processing temperatures are required to form into desired shapes. It is noted that processing temperatures of most thermoplastic resins start well above 300°C, and in some cases they extend to 400°C or above. The resulting heat distortion temperature (HDT) are not much higher than those of high–temperature cured epoxies. Apart from these economic disadvantages, there are other technical difficulties which include proper wetting of the individual fibers in melt impregnation of fiber bundles and complete removal of solvent prior to final consolidation in solvent impregnation process.

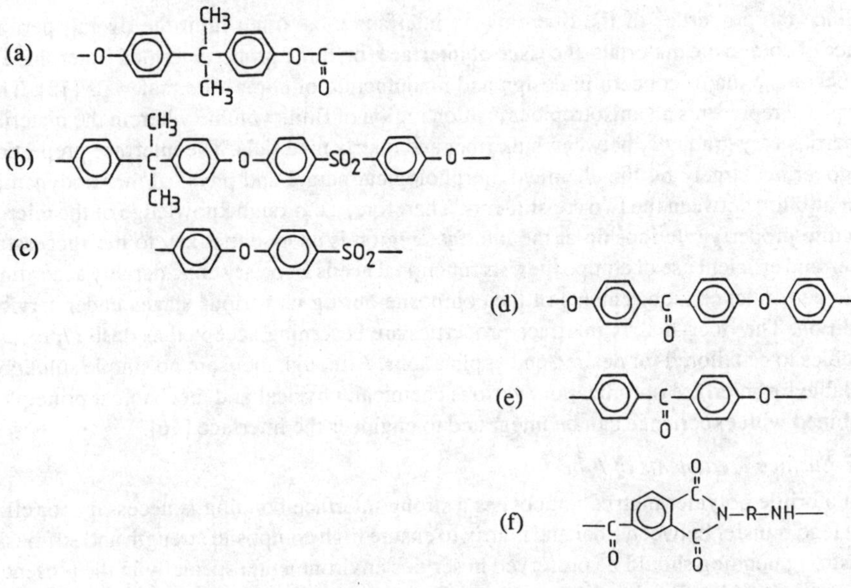

Figure 7. Basic structural repeat units of thermoplastic polymers: (a) polycarbonate (PC, Lexane); (b) polysulfone (PS, Udel); (c) polyethersulfone (PES, Victrex); (d) polyetheretherketone (PEEK, Victrex); (e) polyetherketone (PEK, Victrex); (f) polyamide imide (PAI, Torlon); (g) polyether imide (PEI, Ultem); (h) polyphenylene sulfide (PPS, Raton); (i) Liquid crystal polymer (Vectra). After Hancox [14].

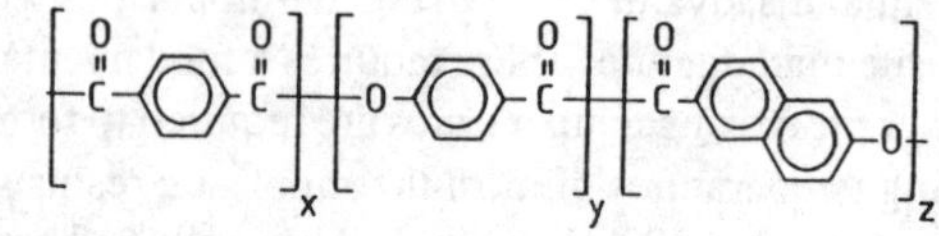

Figure 7. Continued.

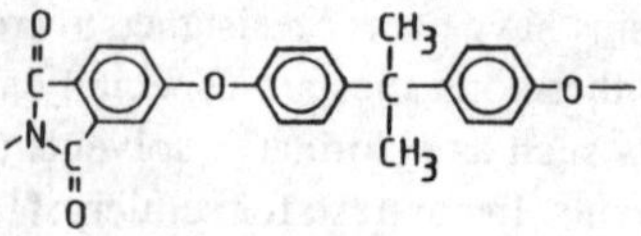

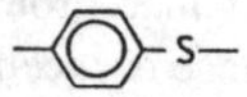

2.3. Interfaces

2.3.1 Introduction

Since the properties of the fiber–matrix interface most often limit the overall performance of composite materials, the issue of interface (or, more properly termed 'interphase') has become a major concern in design and manufacture of composite materials [15]. The interphase represents an anisotropic transition region of finite volume wherein the material properties vary gradually between bulk fiber and matrix materials. The interface properties are governed largely by the chemical/morphological nature and physical/thermodynamic compatibility between the two constituents. Therefore, a thorough knowledge of the microstructure property relationship at the interface region is an essential key to the successful design and efficient use of composites. Its functional needs also vary considerably according to the performance requirements of the composite during its various stages under service condition. The fiber–matrix interface properties are becoming accepted as design/process variables to be tailored for desired end applications. Although there are no simple solutions available for interface optimization, various chemical, physical and mechanical principles combined with experience can be integrated to engineer the interface [16].

2.3.2 Surface Treatments of Fibers

In a brittle polymer matrix composites a strong interface bonding is necessary for efficient load transfer between fiber and matrix to ensure high composite strength and stiffness. The strong bonding should be preserved in service environment, especially in the presence of moisture and heat, if the structural integrity of the composites is to be retained for the whole life of the components. There have been significant research efforts toward the modification of interface by means of surface treatment of fibers for this purpose. The methods of surface treatment and the exact mechanisms of improvement in the interface bond quality for a given fiber are different for different types of matrix materials as are the thermal and chemical compatibilities.

Glass Fibers

Coupling agents are normally applied on the surface of glass fibers. Its basic function is to form a chemical link with the polymer matrix. A typical coupling agent consists of a silane which contains functional groups compatible with the resins. A simple model for the action of silane coupling agent is schematically shown in Figure 8. The general chemical formula is shown as X_3Si-R which chemically react at one end (X group) with the glass surface and at the other end (R group) with the functional groups present in the polymer resin. Numerous experimental evidence exists to support the chemical reaction between glass fibers and resin matrix in the presence of a silane coupling agent [17]. Another important bonding mechanisms due to the coupling agents is the formation of interdiffusion and interpenetrating polymer network at the interface region. Several factors influence the structure of the coupling agent layers and subsequently the mechanical and physical properties of the composites. These include the silane structure in the treating solution and its organo–functionality, the drying conditions, the topology of the reinforcement, and the chemical composition of the surface.

It is generally accepted that glass fibers with silane coupling agents on their surfaces have stronger bonding than those without surface treatments, with most popular thermosetting resins such as epoxy, unsaturated polyester, vinyl ester and some thermoplastic resins. In particular, deterioration of interface bond quality and thus the gross mechanical performance in wet and hot environment can be reduced significantly when certain coupling agents are applied.

$$R-SiX_3 + H_2O \longrightarrow$$
$$R-Si(OH)_3 + 3HX$$

(a)

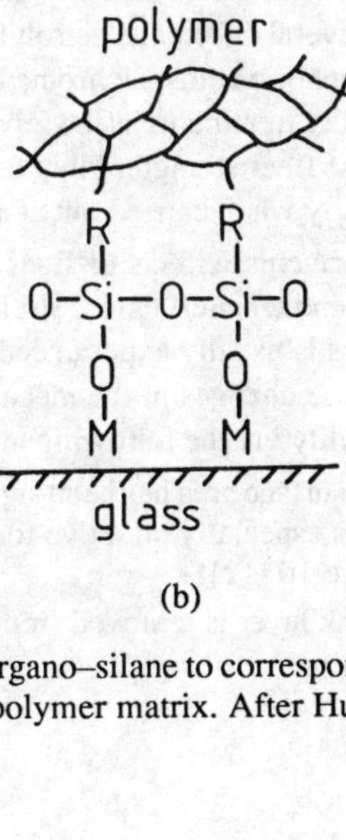

(b)

Figure 8. Functions of coupling agent: (a) hydrolysis of organo–silane to corresponding silane; (b) organo–functional R-group reacted with polymer matrix. After Hull [2].

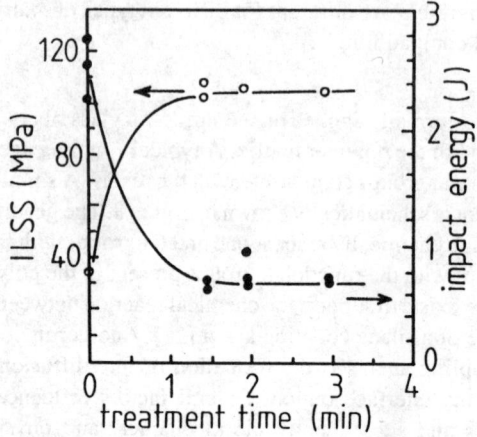

Figure 9. Effects of carbon fiber surface treatment level on interlaminar shear strength (○) and impact energy (●). After Goan [18].

Carbon Fibers

Surface treatment of carbon fibers can be grouped into oxidative and non–oxidative treatments. Non–oxidative methods involve plasma treatments and deposition of more active forms of carbon. There are two types of oxidative treatments: dry oxidation and wet oxidation. The dry oxidative treatments are carried out with air, oxygen or oxygen containing gasses like ozone and CO_2 at low or elevated temperatures. In the wet oxidative treatments, several different electrolytes have been widely used, including alkalis, nitric, sulfuric and phosphoric acids, dichromate permanganate. In general, the wet treatments are milder than the dry treatments as the former treatments do not cause excessive pitting and degradation of the fiber strength. Fiber damages are likely to occur in the dry oxidative treatments, particularly when carried out at a high temperature.

The principal effects of fiber surface treatments are to enhance the interlaminar shear strength and tensile/flexural strength of the composites, while a loss in the impact fracture toughness is usually experienced depending on the treatment level, as shown in Figure 9 [18]. These changes in the mechanical properties are attributed to the improved interface bond quality via the following modifications of the fiber surface:

- Fiber surface area has been significantly increased with associated variations in rugosity. This is especially effective for promoting mechanical anchoring at the interface region (Figure 10 [12]).
- A weak layer is removed from the fiber surface.

- There is an increase in the polar surface energy.
- Chemical modification takes place, and carboxyl, hydroxyl and carbonyl groups are produced on the fiber surface, which may promote chemical reaction with the functional groups present in the polymer resin.

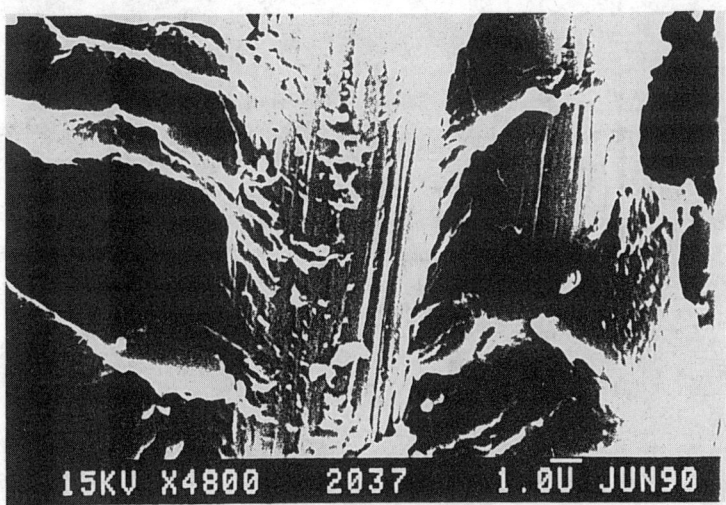

Figure 10. Scanning electron microphotograph showing well bonded carbon fiber–epoxy matrix interface. After Kim et al. [12].

2.3.3 Characterization of Interface Properties

A number of experimental techniques have been developed to characterize the interface/interlaminar properties in fiber composites. These technique may be grouped into two major means depending on the nature of specimens used and the scale of testing [14]: one involves testing of single fiber microcomposites wherein individual fibers are embedded in specially made matrix blocks; and the other uses bulk composite laminates to measure the interlaminar/intralaminar properties.

Test methods using microcomposites include the single fiber compression test, the fiber fragmentation test, the fiber pull–out test, the fiber push–out (or indentation) test and the slice compression test, as schematically shown in Figure 11 [19], with a variety of specimen geometries and scales involved. In these tests, the interface bond quality is measured in terms of the interface fracture toughness, interface shear bond strength at the bonded interface; and the interface frictional shear strength which is a function of the coefficient of friction and the radial residual fiber clamping stress at the debonded or frictionally bonded interface. Especially, the fiber pull–out test is one of the most reliable and direct test methods. It allows determination of the properties both at the bonded and debonded interfaces from the forces required to break the interface bond as well as to pull–out the fiber against the frictional resistance after complete debonding. All these micromechanical tests require a suitable theoretical model for proper evaluation of the experimental data. Further details of the theoretical analyses are given in Ref. [20–22].

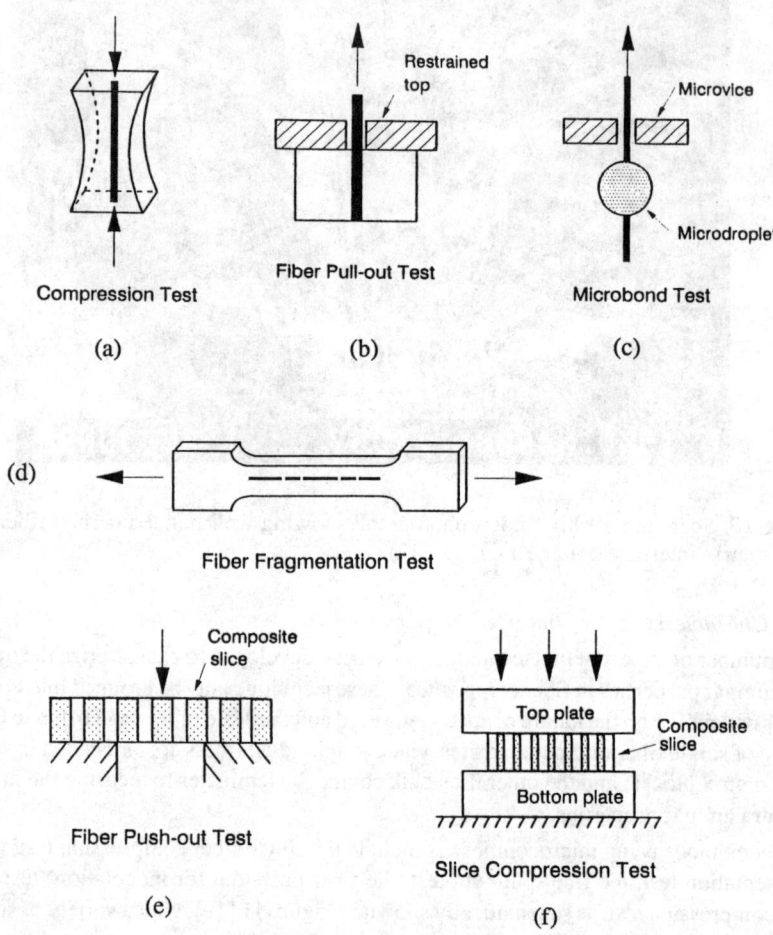

Figure 11. Various experimental configuration of single fiber composite test: (a) single fiber compression test; (b) fiber pull–out test; (c) microbond test; (d) fiber fragmentation test; (e) fiber push–out (or indentation) test. After Kim et al. [19].

The test methods in the second approach employ invariably laminated composites reinforced with continuous fibers. In addition to the short beam shear test (as specified in ASTM D2344) which measures the interlamniar shear properties, many different specimen geometries and loading configurations are available in the literature for the translaminar or in-plane strength measurements. These tests include Iosipescu shear test [23], the [±45°]$_s$ tensile test (ASTM D 3518), [10°] off–axis tensile test, the rail shear tests (ASTM D 4255), the cross–beam sandwich test, the in–plane lap–shear test [ASTM D3846], the thin walled tube torsion test and the transverse tensile test [19].

Testing with bulk composite materials always has a limitation in which the exact location and modes of failure must be consistent with the underlying principles of the test. Validity of the test must be based on the actual examination of the onset of failure during the test. In addition, both the interlaminar/intralaminar properties measured in these tests depend largely on the fiber volume content as well as the strength of the matrix relative to the bond strength at the fiber–matrix interface. In fact, failure may occur at the fiber–matrix interface, in the matrix, or a combination of these even in apparently interlaminar failure. Therefore, they cannot be regarded as giving the true values of the interface bond quality. The significance of these tests is that they provide some measure of the relative bond quality of different fiber, matrix and interface combinations.

2.3.4 Engineered Interfaces

There is now considerable evidence available which demonstrates the outstanding influences of interfaces on fracture toughness, strength and stiffness of fiber composites in various failure modes and loading geometries. A number of potential solutions have been suggested to improve the fracture resistance and damage tolerance of inherent brittle polymer matrix composites without sacrificing other important mechanical properties. There are two major approaches: one relies on the improvement of the intrinsic properties of composite constituents, and the other depends on suitable fiber–matrix interface and interlaminar control [15].

In the first approach, intrinsically tough matrices, including ductile thermoplastics and rubber modified thermosets, can be employed to improve the interlaminar fracture toughness of composites [12]. A tougher matrix is also effective for fracture resistance and tolerance against the low energy impact damages [24]. Different fibers can also be incorporated in a matrix material to form a hybrid fiber composite. For example, addition of glass or aramid fibers to brittle carbon fiber composites enhances the fracture toughness due to the failure mechanisms associated with the ductile fibers.

The second approach includes fiber coating with appropriate polymers, delamination arresters and promoters and reduction of shrinkage stresses in the matrix. One of the most effective methods among these is the application of polymer coatings, either fully [8] or intermittently [25], along the fibers. This method is simple to employ in various ways and facilitates a direct comparison of the mechanical properties of the composites with and without the fiber coating. The principal effect of altering the interface properties by fiber coating is to modify the failure modes and thus the potential energy absorption capability. The energy absorption of laminate composites in transverse fracture can also be increased by promoting controlled delamination or longitudinal splitting when the interlaminar bond strength is weakened.

3. Processing and Fabrication Technology

3.1. Introduction

The principal attractions of composite materials include their ability to be tailored so that their properties match the service requirements and the fact that a single composite component can replace an assembly of many components made from traditional metals. The versatility of composites extends to the number of processing possibilities. There are a number of manufacturing techniques developed for composite materials but there are a few general principles applied in selecting a most suitable processing method [26]:

- Select the cheapest and most reliable raw materials.
- Combine the maximum number of functions in a single part.
- Use the minimum number of processing stages.
- Use fast automated equipment, avoiding high labor content and extended resistance time in a costly plant.
- Use iterative on–line control which assures quality and minimizes post–fabrication inspection or testing.

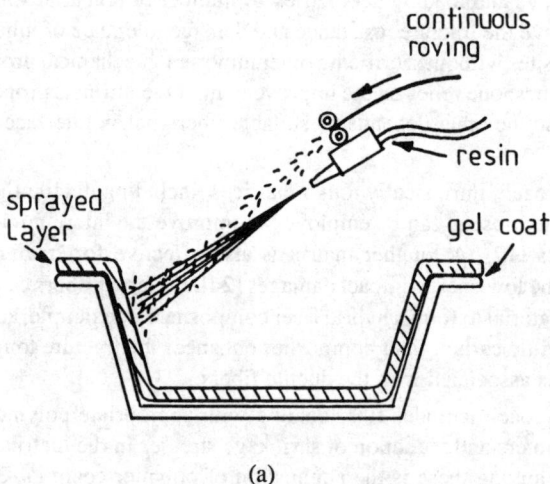

(a)

Figure 12. Direct roving processing technique: (a) spray–up; (b) filament winding; (c) pultrusion. After Bader [26].

The fabrication processes are classified here based on the technological choices available. In the direct roving processes, the dry reinforcements in the form of a spool of continuous tow (or roving) and chopped strand mats are converted directly to the component. The rest of the processes is grouped as molding process which use fibers already converted into web or sheet from. A separate section is also devoted to the techniques developed especially for thermoplastic matrix composites.

3.2. Direct Roving Processes

There are three basic processes, including spray up which is a variant of contact molding, filament winding and pultrusion. Figure 12 schematically show these processes.

3.2.1 Spray Up

The reins and chopped fiber rovings are sprayed simultaneously onto the mold surface. The reinforcement is fed into a special applicator head where it is chopped into short lengths, typically 25–50 mm and spray by an air jet onto the mold surface. Simultaneously another spray applies liquid resin so that both fiber and resin are delivered in the correct ratio at the mold surface. This process improves the degree of control of the process and increases the rate of production in comparison with hand lay–up. However, the process is capable of laying down only a random array of fiber which limit the fiber content to about 35% by weight. Consequently, properties will be inferior to those obtained from woven roving laminates. The main benefits of the process are realized in automated production units where rapid and uniform deposition of fiber–resin mix can be achieved.

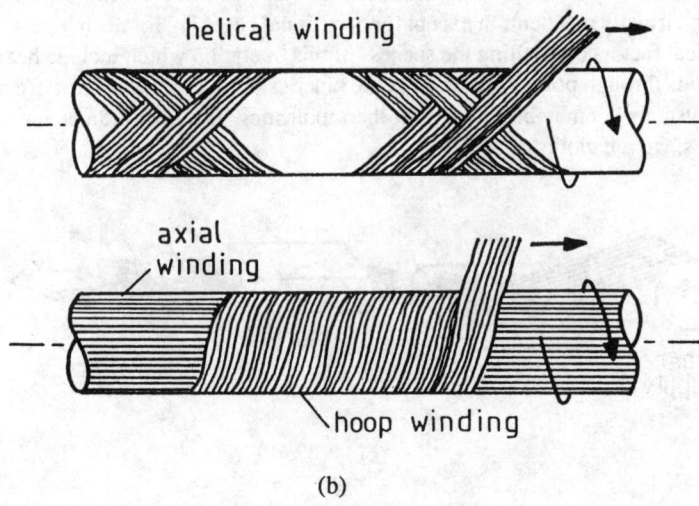

(b)

Figure 12. Continued.

3.2.2 Filament Winding

The continuous filament is wound onto a mandrel in layers of different angles to form the composite. There are two types of patterns normally used in filament winding: helical winding in which a constant angle is maintained; and biaxial winding where two or more separate winding angles (usually 0° and 90° to the rotational axis) are used. Helical winding is more common because the processing time is less, but biaxial patterns enable special properties to be obtained. In combination with either type of winding pattern, other materials may be a part of the completed structure such as mats or woven cloth, chopped fiber spray–up and thermoplastic or other liners. Techniques to apply the matrix resin include: (i) by spraying onto the mandrel; (ii) preimpregnating the roving in a bath of liquid (thermoset) resin; (iii) by using rovings which have been previously preimpregnated with a thermoplastic resin; (iv) using commingled yarns of fiber and thermoplastic resin and then continuously "welding" as it is applied to the mandrel. This process is most suited to tubular, axisymmetric shapes but modern computer design technology permits many variations.

3.2.3 Pultrusion

The process involves drawing the reinforcement through a bath of liquid resin and then directly and continuously through a heated die to produce a continuous section. The word "pultrusion" appears to be originated from "pulling" the fibers "through" a die. The process is suitable for simple section such as circular (e.g. fishing rods), tubes, channels, etc. Very good fiber alignment and very high fiber volume fractions can be achieved, which are required for excellent mechanical properties.

Pultrusion is a truly continuous process: with an automated cut–off saw, a pultrusion line can run with virtually no attention except for occasional checking for resin levels. There are several critical factors controlling the success of this operation which include heat transfer, flow of liquids through porous media and cure kinetics. This process is most often used for thermoset systems but may be adapted for thermoplastics. The fibers can be unidirectional roving, tapes, woven cloths or mats.

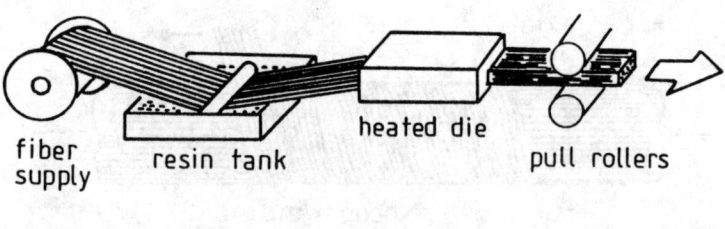

(c)

Figure 12. Continued.

3.3. Molding Processes

The common feature of the molding processes is that the reinforcement is first converted into an intermediate sheet or web form. This might be a woven cloth, a continuous or discontinuous random mat, a prepreg or sheet molding compound. The intermediate product means an added cost which can be minimized by simplifying the downstream process. The dry web must be impregnated with resin either before or during the final molding operation. This requires that the resin matrix be of sufficiently low viscosity to fully impregnate the fibers. The viscosity can be lowered by heating or dilution with a solvent, and impregnation may be assisted by pressure or vacuum. The molding processes have a number of variants, depending on the types of charges and reinforcement, viscosity of flow, use of vacuum, pressure and heat, the complexity and size of mold cavity, etc. Important features of several popular molding processes shown in Figure 13 [22,27] are discussed in the following.

3.3.1 Open Molding (or Contact Molding)

Open (or contact) mold is the oldest of the methods employed for fabricating fiber reinforced composite laminates. It is a versatile, slow and labor–intensive process. Although quality control relies almost entirely on the skill of the operator, since the process is fundamentally very simple, operators of low skill are often employed. In the process, a mold is prepared from a master pattern, and the reinforced plastic is applied to the mold to reproduce the shape of the original. The mold surface is treated with a mold release agent to prevent the molding from sticking to the mold. A pigmented resin, the so–called gel coat, is subsequently applied to the mold surface which will become the prime surface of the molding. A back–up lamina of a fine tissue is applied, which is followed by layers of chopped strand mat and/or woven roving and resin according to design specifications. After complete cure the mold is trimmed and removed from the master. This process is most commonly used with room temperature cure polyester resins and glass fibers, but can also be used for vinyl ester and suitable epoxy systems. The process time can be reduced by impregnating the fibers with liquid resin before being applied to the mold or by spraying the resin onto the mold at the same time as the chopped rovings.

3.3.2 Autoclave Molding

This process is widely used in the aerospace industries for production of high quality flat or curved panels from prepreg materials. The laminate is built onto a metal mold plate conforming to the shape of the panel to be produced. Both sides of the laminate are covered with a single layer of a fine peel–ply which will be stripped off after molding, leaving a clear smooth surface to the laminate. A perforated release film is stacked on the top surface of the laminate and if necessary bleeder or breather clothes will be added. The whole assembly is then covered with a non–porous vacuum bag which is sealed to the mold plate. Finally, the mold is loaded into the autoclave, and a vacuum is allowed to be drawn between the molding plate and the vacuum bag, while the pressure and temperature within the autoclave are separately controlled. The vacuum serves to continuously remove all volatiles that might have been trapped during the molding operation, reducing the incidence of porosity.

The combined effects of the vacuum and autoclave pressure allow a very uniform pressure over the entire surface of the molding. In many prepreg systems, the most critical processing requirement is to apply the molding pressure when the resin is at the correct state of cure (i.e. correct viscosity). The autoclave process is extremely versatile, but rather slow and the capital cost of the plant is relatively high. The consistent molding of high quality products makes the process very attractive in the high technology areas.

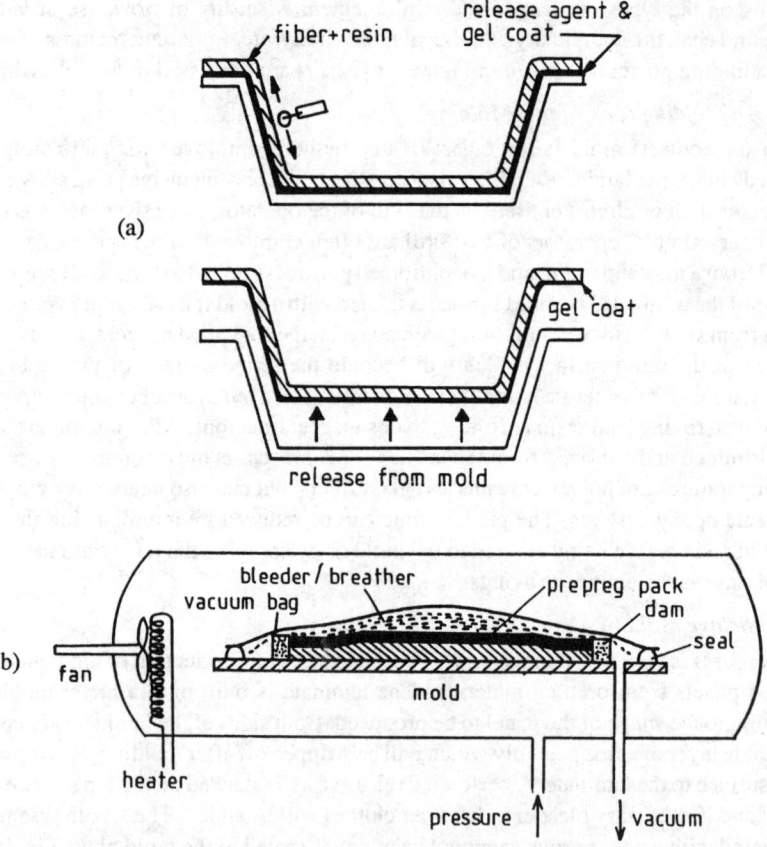

Figure 13. Molding processes of composite fabrication; (a) open molding; (b) autoclave molding; (c) sheet molding compound (SMC); (d) compression molding; (e) injection molding; (f) resin transfer molding (RTM). After Bader [26] and Strong [27].

3.3.3 Sheet Molding Compound (SMC) and Bulk Molding Compound (BMC)

SMC and BMC offer the automotive, appliance and equipment industries the capability for high volume production. The only fundamental difference between SMC and BMC materials is in the form of the feed stock. The SMC is in a sheet of about 6 mm thick, while BMC is usually in the form of a rope 20–50 mm diameter.

SMC is prepared by chopping continuous strand rovings onto a plastic film that has previously been coated with a resin paste. The paste and (glass) fibers are gently mixed together, and a sheet product is formed. Rolls of SMC are stored until the viscosity has increased to a predetermined level. BMC is prepared by thoroughly mixing chopped strands with a resin paste that can be used in bulk from or extruded into a rope for easier handling. BMC can be molded immediately after mixing or may be stored like SMC in order to increase the viscosity to specified levels.

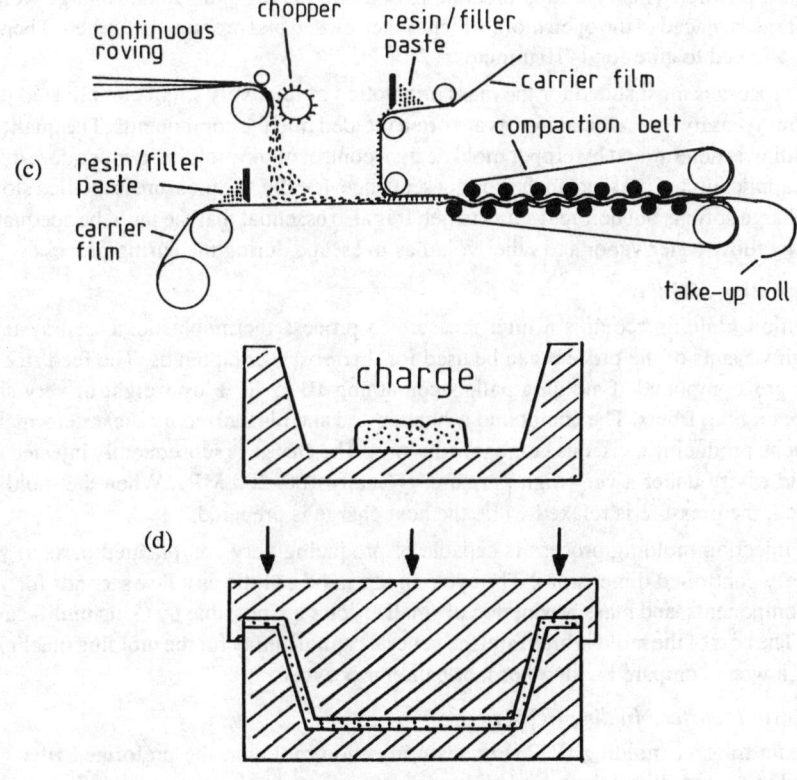

Figure 13. Continued.

For molding, both SMC and BMC are pre-weighted to specific size charges and placed into the mold prior to application of heat and pressure. The length of cure depends on temperature, resin, and part thickness. Unsaturated polyester or vinyl ester resins are the predominant polymers used in SMC and BMC. These compounds normally contain fillers to reduce cost and to control flow during molding. For automotive applications calcium carbonate is used extensively as a filler due to its low cost and ability to provide smooth molded surfaces.

3.3.4 Compression Molding

Compression molding is similar to the SMC process, and is usually applied to phenolic resins which require much higher consolidation pressure. A continuous pressure is maintained on the charge, and the dimensions of the molding are determined by the accurate control of the charge weight. The feedstock for typical fiber reinforced thermosets is a dry and coarse powder. This is usually precompressed into pellets of a suitable charge weight. This charge is placed in the open mold which is then closed and pressure is applied. Then the resin is allowed to cure for 1–10 minutes.

The process is most suited for the mass production of relatively small complicated parts using multi-cavity tools. Metal inserts are often molded into the components. The quality of the molding is determined by proper mold design, control of the molding temperature cycle, and the application of pressure in the correct sequence. Ideally, the pressure is applied slowly as the charge softens but before it starts to gel. It is also essential that the mold be adequately vented to allow water vapor and other volatiles to escape during the curing process.

3.3.5 Injection Molding

Injection molding requires a high pressure to process thermoplastics based systems, although variants of the process can be used for thermosets compounds. The feed stock is usually pre-compounded molding pallets containing 10 to 40 % by weight of very short glass (or carbon) fibers. The compound is hopper-fed and plasticized by shear deformation under heat, producing a viscous homogeneous mix. The charge is subsequently injected into the mold cavity under a very high pressure, typically 100–200 MPa. When the mold has solidified, the pressure is relaxed while the next charge is prepared.

The injection molding process is capable of producing very complicated parts to very accurately controlled dimensions. The cycle time can be as little as a few seconds for very small components, and mass production of small articles are possible by using multi-cavity molds. The cost of the mold is high, and the general capital outlay for the molding machine is also high when compared with other fabrication processes.

3.3.6 Resin Transfer Molding (RTM)

In resin transfer molding (RTM) or resin injection molding, the preformed fibers are enclosed in the mold, and the resin is subsequently injected to impregnate the fiber and fill the mold cavity. Low viscosity resins are preferred for this process to allow good penetration of the resin into the fiber preforms and thus to reduce entrapment of pores. The design of the mold is the most critical factor for successful RTM. The mold must be constructed so that the resin reaches all areas within the time allowed before the onset of gelation, without causing movement of the preformed fibers. The mold is often put under vacuum prior to injection of the resin to remove entrapped air from the reinforcement and to speed the RTM process. Very large (e.g. car body shell) and complex shapes can be made efficiently and inexpensively using this technique.

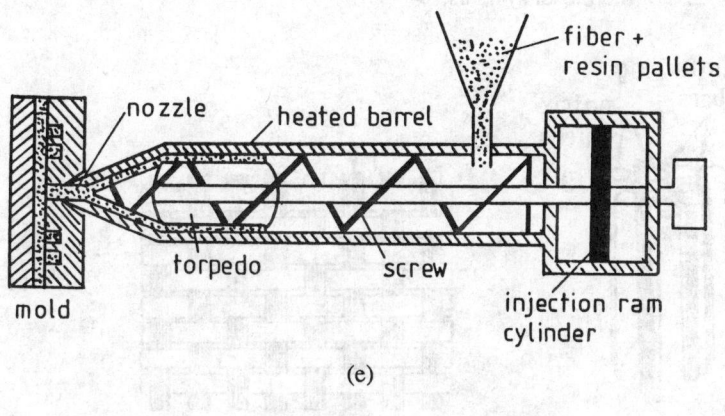

(e)

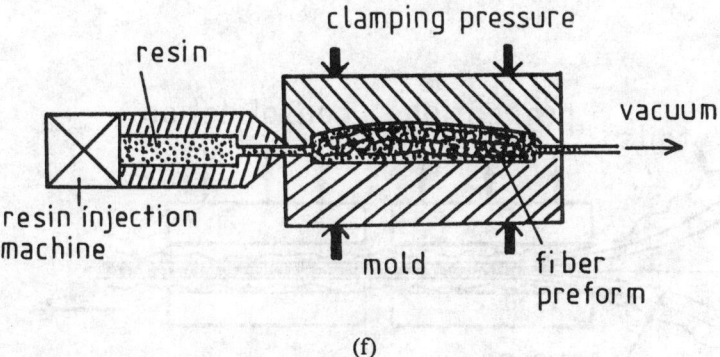

(f)

Figure 13. Continued.

3.4. Processing Techniques for Thermoplastic Matrix Composites

Thermoplastic matrix composites can also be processed by the same hand lay–up, autoclave molding and other processing technology (including injection molding and pultrusion) favored by thermosetting systems with some modifications. There are characteristic features that have to be taken into account in selecting proper processing methods for thermoplastic resins and their composites:

- The processing stage requires only heat and pressure. It is an entirely physical operation, and there are no hardener or curing agents (i.e. no chemistry) being involved.
- There are no constraints of lifetime in which the resin must be processed or molded.
- Thermoplastic resins lack the tack and drape which are considered to be indispensable to the process of thermoset systems.

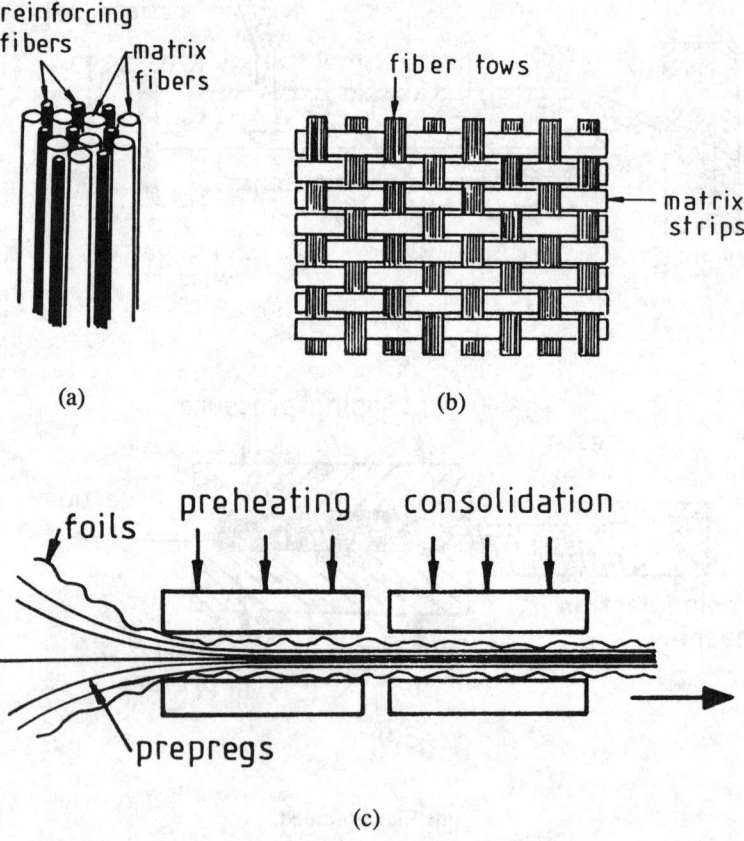

Figure 14. Processing techniques for thermoplastic matrix composites: (a) commingled yarn; (b) co–woven sheet; (c) step pressing; (d) roll forming; (e) hydro–rubber forming; (f) diaphragm forming. After Powell [28] and Cogswell [29].

A number of processing techniques have been developed especially suitable for these requirements, and two distinct stages of processes can be identified: impregnation processes and shaping processes. Film stacking, co–woven or commingled yarns, powder coating techniques are among the impregnation processes, while roll forming, extrusion, stamping and diaphragm forming are the shaping processes. These processes are schematically shown in Figure 14 [28,29] and are discussed in the following.

3.4.1. Film Stacking

For a long time film stacking was the standard technology for making thermoplastic composites. In this process, layers of reinforcing fibers are laminated between layers of thermoplastic polymer film. This laminate is fused under a pressure and heat to allow the resin to flow in between the fibers. A typical stack of reinforcement to be wetted out would be about fifteen fibers deep. Although this technique is still widely used as the most convenient method for small quantities, it has significant problems of improper wetting of fibers, particularly when the molten resins are highly viscous, and high concentration of fiber and resin rich regions in the finished product. Application of high pressure to infiltrate the resin between the fibers does not much help alleviate these problems because the high pressure also forces the fibers together.

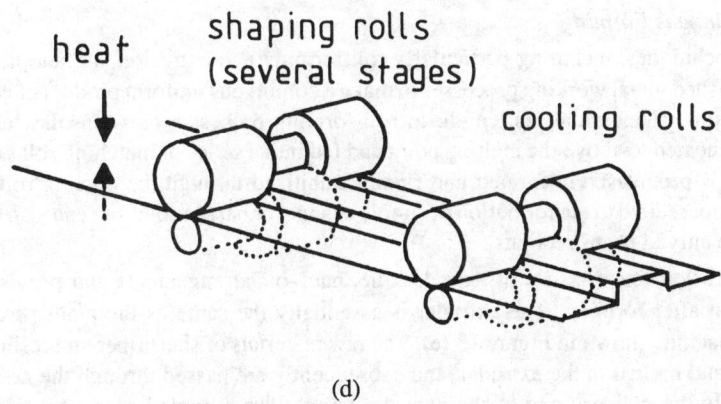

(d)

Figure 14. Continued.

3.4.2. Co–woven and Commingled Yarns

The inherent problems of wetting and insufficient intimacy between the composite constituents in the film stacking process can be in part eliminated by using the matrix material in the form of fiber or narrow tape, and co–weaving or intermingle it with the reinforcing fibers [30]. Upon melting, the matrix instantly coats the fibers. Beside improving drapability, such fiber forms of reinforcement and matrix resin can be knitted or braded together into complex three–dimensional structures, and thus can directly be consolidated without going through intermediate processing stages.

3.4.3 Powder Coating

It is also possible to impregnate the fibers with matrix resin in powder form. In this process, fiber rovings are passed through a fluidized bed of matrix powder and subsequently melted in a hot tube to wet the fibers. Spreading of the roving is a critical stage of the process and can be achieved by the use of electrostatics which also helps pin the powders to the fiber when using fine powders of the order of 5–10 µm.

3.4.4. Continuous Consolidation

Continuous consolidation of simple laminates in sheet form can be carried out by a variety of techniques. One of simplest batch processing is to operate a step pressing process in which a pre–assembled prepreg layers are fed through a heated press followed by a cold consolidating press. Tape laying is another simple continuous process where each layer of prepreg is melted and consolidated. To avoid uneven cooling across the laminate thickness which may generate residual stresses, a base plate is normally used which is heated to above glass transition temperature, T_g, of the resin.

3.4.5. Continuous Forming

These techniques, including particularly roll forming and extrusion, are adopted from well established metal working processes to make a continuous uniform product of constant cross–sectional shape along its length. In roll forming process a pre–consolidated sheet stock is preheated to above the melting point and fed into a series of matching roll sections. The shape is progressively formed and finally stabilized through the cooling rolls. This process is most suited to the formation of simple channel or hat sections, but can also be used to fabricate curved beam sections.

The extrusion line consists of extruder, die, haul–off arrangements and provision for stabilization after forming. The extruder is essentially the same as the main part of the injection machine shown in Figure 13 (e). The raw materials of short fiber and resin pallets are mixed and melted in the extruder, and subsequently are passed through the heated die connected to the delivery end of the extruder barrel. The extruded product swells upon leaving the die, allowance for which must be made in the design of the exit dimensions.

3.4.6. Stamping

Another successful technology adopted from metal forming is stamping of a preheated blank laminate in a cold tool. Various options for tooling have been employed, which include matched dies, rubber block and hydro rubber forming. This technique most suitable for products of simply folded shapes and uniform thickness, requiring minimum in–plane deformation during stamping. Since the whole process is extremely quick (i.e. typically 1 second or less) there is little time for large scale deformation.

3.4.7. Diaphragm Forming

Diaphragm forming is one of the most recent developments [31], in which prepreg layers are sandwiched between two constrained diaphragms. Pressure is then applied to one side to deform the diaphragms and thus to allow the laminate to take up the shape of the mold under heat. After conforming to the mold shape, the molding is cooled and the pressure is released to remove the diaphragm. This process is designed to produce large area structures with pronounced double curvature.

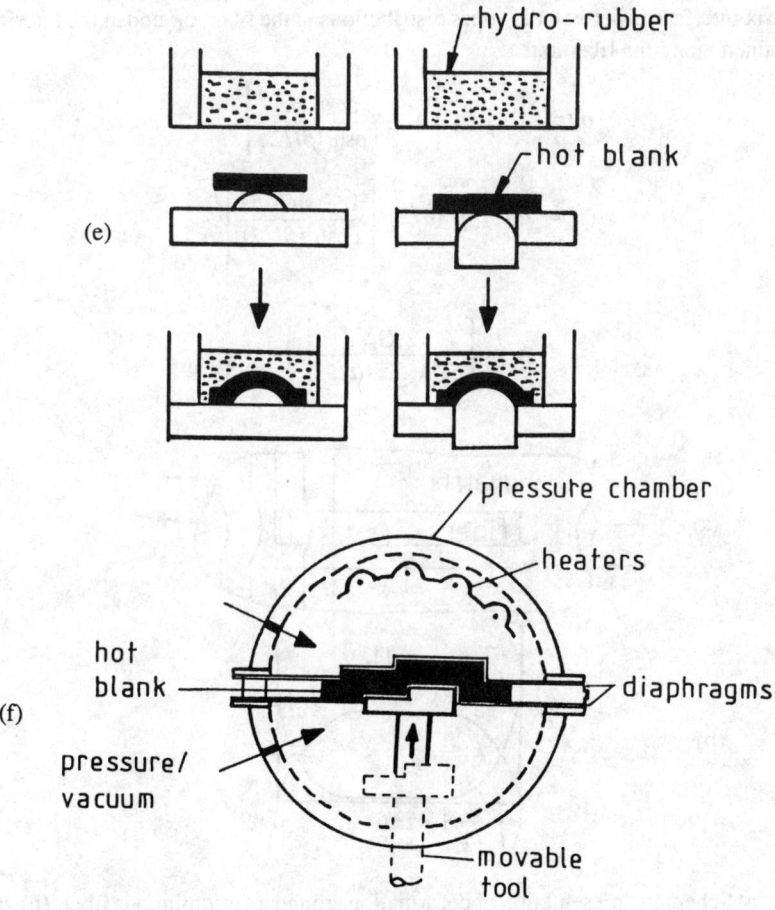

Figure 14. Continued.

4. Mechanics of Composites

4.1. Micromechanics of Stress Transfer

The stress transfer between fiber and matrix is an important phenomenon to understand the mechanical performance of a composite. Among many theoretical models developed to predict the mechanical response of a composite, the shear–lag model [32] has provided a sound basis for studying the micromechanics of stress transfer across the interface, the stress field near the broken fiber ends in particular. Consider a fiber of length, l, embedded in a matrix which is subjected to an axial strain ε_m as shown in Figure 15. Based on the assumptions that the fiber and matrix are elastic and isotropic, the bond at the infinitely thin fiber–matrix interface is perfect, the stress distributions in the fiber, σ_f, and at the interface, τ_i, are obtained along the fiber axis:

$$\sigma_f(z) = E_f \varepsilon_m \left[1 - \frac{\cosh \beta z}{\cosh (\beta l/2)} \right] \quad (1)$$

$$\tau_i(z) = E_f \varepsilon_m \beta (a/2) \left[\frac{1 - \sinh \beta z}{\cosh (\beta l/2)} \right] \quad (2)$$

where

$$\beta = \left[\frac{2 G_m}{E_f a^2 \ln (a/b)} \right] \quad (3)$$

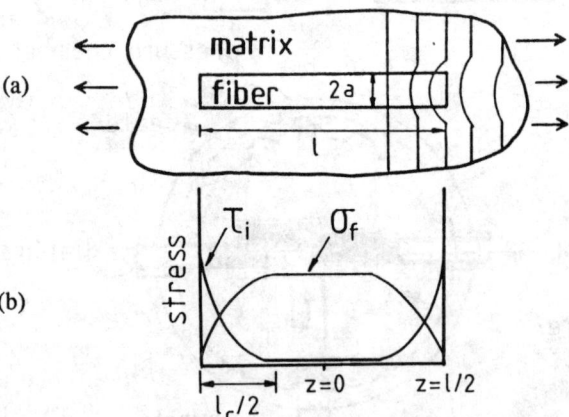

Figure 15. (a) Schematic presentation of deformation around a discontinuous fiber; (b) variation of fiber axial stress, σ_f, and interface shear stress, τ_i, according to Cox [32].

E_f is Young's modulus of fiber, G_m is shear modulus of matrix, and a and b are equivalent radii of fiber and matrix, respectively. The stress distributions in Figure 15 clearly show that the tensile stress is maximum in the center of the fiber while the shear stress is a maximum at the fiber ends. The implication is that there are regions near the fiber ends which do not carry the full load so that the average stresses in the fiber of finite length, l, is always less than that for a continuous fiber subjected to the same external loading. The average axial fiber stress, $\overline{\sigma}_f$, is given by:

$$\overline{\sigma}_f(z) = E_f \varepsilon_m \left[1 - \frac{\tanh(\beta l/2)}{\beta l/2} \right] \quad (4)$$

$\overline{\sigma}_f$ decreases as l decreases because a greater proportion of the fiber length is not fully loaded. To achieve the maximum stress the fiber length should be longer than a critical value, l_c (see Section 4.4 for critical transfer length).

4.2. Laminate Theory

A laminate is an ordered stack of plies or lamina. The most important class of non–isotropic bodies associated with simple laminate composite is orthotropic bodies which have three mutually perpendicular planes of material symmetry as in a unidirectional or bidirectional laminate. In an isotropic body, properties at any point are the same in every direction.

This section describes the methods which are used to calculate the elastic properties of laminates from the properties, orientation and distribution of individual laminates. This is important to predict the response of the laminate to external loads in the design of polymer composites which are mainly in the form of angle–ply laminates. The approach is based on laminate theory [33].

In an orthotropic laminate with bidirectional reinforcements, equal strength and stiffness values will result in the two directions. The Hook's law relationship in matrix form gives:

$$\begin{bmatrix} \sigma_1 \\ \sigma_2 \\ \tau_{12} \end{bmatrix} = \begin{bmatrix} Q_{11} & Q_{12} & 0 \\ Q_{12} & Q_{22} & 0 \\ 0 & 0 & Q_{33} \end{bmatrix} = \begin{bmatrix} \varepsilon_1 \\ \varepsilon_2 \\ \gamma_{12} \end{bmatrix} \quad (5)$$

where

$$Q_{11} = \frac{E_1}{1 - \nu_{12}\nu_{21}}, \quad Q_{22} = \frac{E_2}{1 - \nu_{21}\nu_{12}}$$
$$Q_{12} = \frac{\nu_{12} E_2}{1 - \nu_{12}\nu_{21}}, \quad Q_{33} = G_{12} \quad (6)$$

Therefore, orthotropic properties of the composite can be completely defined by four independent elastic constant E_1, E_2, ν_{12} ($= \nu_{21} E_1/E_2$) and G_{12}. The relations defined in Eq. 5 are related to the principal directions of the material. Consequently, because of material symmetry, the effects of the normal stress are independent of those of the shear stresses and hence the total effects can be obtained by superposition. If the laminate is loaded in such a way that the loading direction in the reference axes (x,y) is at an angle θ to the principal directions (1,2) as illustrated in Figure 16, elasticity theory shows that the stress–strain relation becomes:

$$\begin{bmatrix} \sigma_x \\ \sigma_y \\ \tau_{xy} \end{bmatrix} = \begin{bmatrix} \overline{Q}_{11} & \overline{Q}_{12} & \overline{Q}_{16} \\ \overline{Q}_{12} & \overline{Q}_{22} & \overline{Q}_{26} \\ \overline{Q}_{16} & \overline{Q}_{26} & \overline{Q}_{66} \end{bmatrix} = \begin{bmatrix} \varepsilon_x \\ \varepsilon_y \\ \gamma_{xy} \end{bmatrix} \quad (7)$$

The matrix $\overline{Q}_{ij}$ is called the transformed reduced stiffness matrix and the stiffness have the following values:

$$\begin{aligned}
\overline{Q}_{11} &= Q_{11}c^4 + 2(Q_{12} + 2Q_{66})c^2s^2 + Q_{22}s^4 \\
\overline{Q}_{12} &= (Q_{11} + Q_{22} - 4Q_{66})c^2s^2 + Q_{12}(c^4 + s^4) \\
\overline{Q}_{22} &= Q_{11}s^4 + 2(Q_{12} + 2Q_{66})c^2s^2 + Q_{22}c^4 \\
\overline{Q}_{16} &= (Q_{11} - Q_{12} - 2Q_{66})c^3s + (Q_{12} - Q_{22} + 2Q_{66})cs^3 \\
\overline{Q}_{26} &= (Q_{11} - Q_{12} - 2Q_{66})cs^3 + (Q_{12} - Q_{22} + 2Q_{66})c^3s \\
\overline{Q}_{66} &= (Q_{11} + Q_{22} - 2Q_{12} - 2Q_{66})c^2s^2 + Q_{66}(c^4 + s^4)
\end{aligned} \quad (8)$$

Note that $c = \cos\theta$ and $s = \sin\theta$. Eq. 8 can be inverted to obtain the strain–stress relations in the following general form:

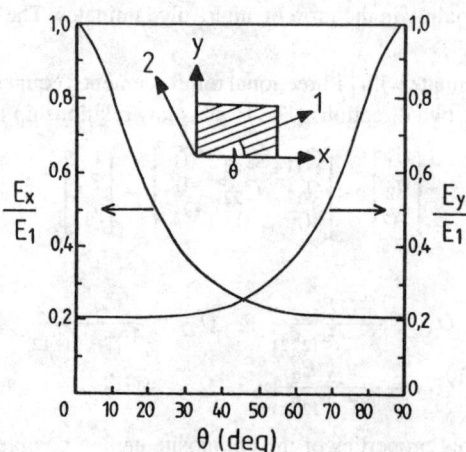

Figure 16. Normalized moduli of a glass fiber–polyester matrix lamina as a function of the angle θ.

$$\begin{bmatrix} \varepsilon_x \\ \varepsilon_y \\ \gamma_{xy} \end{bmatrix} = \begin{bmatrix} \bar{S}_{11} & \bar{S}_{12} & \bar{S}_{16} \\ \bar{S}_{12} & \bar{S}_{22} & \bar{S}_{26} \\ \bar{S}_{16} & \bar{S}_{26} & \bar{S}_{66} \end{bmatrix} = \begin{bmatrix} \sigma_x \\ \sigma_y \\ \tau_{xy} \end{bmatrix} \qquad (9)$$

in which $\bar{S}_{ij}$ are the transformed compliance constants and their relations to θ can be represented by a set of equations similar to Eq. 9. Therefore, along the (x,y) axes which are not aligned with the material principal directions, the elastic constants of the unidirectional lamina can be expressed as a function of elastic properties and the loading angle θ:

$$\begin{aligned}
\frac{1}{E_x} &= \frac{1}{E_1}c^4 + \left(\frac{1}{G_{12}} - \frac{2\nu_{12}}{E_1}\right)c^2s^2 + \frac{1}{E_2}s^4 \\
\frac{1}{E_y} &= \frac{1}{E_1}s^4 + \left(\frac{1}{G_{12}} - \frac{2\nu_{12}}{E_1}\right)c^2s^2 + \frac{1}{E_2}c^4 \\
\frac{1}{G_{xy}} &= 2\left(\frac{2}{E_1} + \frac{2}{E_2} + \frac{4\nu_{12}}{E_1} - G_{12}\right)c^2s^2 + \frac{1}{G_{12}}(c^4 + s^4) \\
\nu_{xy} &= E_x\left[\frac{\nu_{12}}{E_1}(c^4 + s^4) - \left(\frac{1}{E_1} + \frac{1}{E_2} - \frac{1}{G_{12}}\right)c^2s^2\right]
\end{aligned} \qquad (10)$$

Figure 16 plots the moduli, E_x and E_y, normalized with E_1 as a function of θ for a glass fiber–epoxy matrix composite, elastic constants of which are $E_1 = 40$ GPa, $E_2 = 8.2$ GPa, $G_{12} = 3.9$ GPa and $\nu_{12} = 0.26$ [2].

4.3. Strength of Unidirectional Continuous Fiber Composites

Unidirectional laminate can fail in many different modes depending on the external loading conditions. The fracture strength associated with these failure modes are quite different for different combinations of fiber and matrix with a range of fiber volume fraction, V_f. Some typical values of fracture strength for several most popular polymer matrix composites are given in Table 6.

4.3.1 Longitudinal Tensile Strength

The tensile failure of a fiber composite is a complex process which involves an accumulation of microstructural damage of various failure modes and extents depending on the external loading and environmental conditions. The exact failure mechanisms are further complicated by the large variety of different fiber–matrix systems available with a range of fiber volume fractions. In this section, a classical approximation of the rule of mixtures is adopted to account for composite axial strength.

By assuming equal strain in the fiber and matrix (i.e. $\varepsilon_f = \varepsilon_m$) within the range of elastic deformation, the longitudinal stress in the composite, $\sigma_{\parallel}$, under uniaxial loading can be expressed:

Table 6. Properties of unidirectional continuous fiber composites.

Property		E–glass –epoxy	S glass –epoxy	Type I carbon –epoxy	Type III carbon –epoxy	Kevlar 49 –epoxy
Density (g/cm^3)		2.1	2.0	1.5–1.6	1.63	1.38
Elastic moduli (GPa)						
	$E_\parallel$	45	55	145	220–290	80
	$E_\perp$	12	16	10	6.2–6.9	5.5
	G	5.5	7.6	4.8	4.8	2.1
	ν	0.2	0.26	0.25	0.25	0.31
Strengths (MPa)						
	$\sigma_\parallel^*$	1100	1600–2200	1240–2300	900–1200	2000
	$\sigma_\perp^*$	40	40	41	21	20
	τ^*	70	80	80	60–70	40
Coefficient of thermal expansion						
(10^{-6}/K)	longitudinal	6.3	3.5	–0.1	–0.5 to –0.8	–4.0
	transverse	30	29	25	27–30	60

$$\sigma_\parallel = \sigma_f V_f + \sigma_m(1 - V_f) \qquad (11)$$

where σ_f and σ_m are the stresses in the fiber and matrix, respectively. A similar expression can be obtained for the Young's modulus of the composite, E, by dividing Eq. 11 with the axial composite strain:

$$E = E_f V_f + E_m(1 - V_f) \qquad (12)$$

Now consider the deformation of a typical brittle fiber polymer matrix composite where the fiber fractures earlier than the matrix (i.e. $\varepsilon_f < \varepsilon_m$) and/or the matrix is allowed to flow plastically, as shown in Figure 17(a). Most advanced fibers employed in polymer matrix composites including glass, carbon and organic fibers behave elastically in tension up to the fracture strength. Epoxies and polyester resins and other thermoplastic matrix materials exhibit significant non–linear stress–strain curves. The matrix is unable to withstand the additional load transferred to it due to the fiber fracture. In this case, it is more appropriate to express the axial stress in the unidirectional composite as:

$$\sigma_\parallel = \sigma_f V_f + \sigma_m^\#(1 - V_f) \qquad (13)$$

where $\sigma^\#_m = (d\sigma_m/d\varepsilon_m) \times \varepsilon_f$, the matrix stress calculated at the moment of fiber fracture. It should be pointed out that to achieve the strengthening effects of the fiber reinforcement the composite must contain fibers more than a critical volume fraction. This is shown in Figure 17(b) where the lines A and B represent the strength–fiber volume fraction relationships for the composite and the matrix (in the absence of fiber), respectively. Strengthening cannot take place until $V_f > V_c$, and V_c is given by:

$$V_c = \frac{\sigma_m - \sigma_m^\#}{\sigma_f - \sigma_m^\#} \tag{14}$$

If the failure strain of the fiber is greater than that of the matrix (i.e. $\varepsilon_f > \varepsilon_m$, although this situation seldom occurs in polymer matrix composites), two different failure sequences can be envisaged depending on V_f. For a low V_f, the strength of the composite depends primarily on matrix fracture strength, since the fibers are unable to support the load and break. In contrast, when V_f is large the matrix takes up only a small proportion of the load because of the high modulus of the fiber so that when the matrix fractures the load transferred to the fibers is insufficient to cause fracture. Therefore, the load can be further increased until the fracture strength of the fiber is reached, then the composite strength is solely contributed by the fibers only.

The above simple analysis does not fully account for the fact that the strength of a brittle fiber is a statistical quantity which results from flaws being randomly distributed along the length. This means that the fiber strength depends on the fiber length, and is thus not a fixed quantity. In addition, there are large stress concentrations in the vicinity of a fiber breakage which may occur during fabrication and the early stage of loading. The stress concentration can be detrimental to the strength of unidirectional composites.

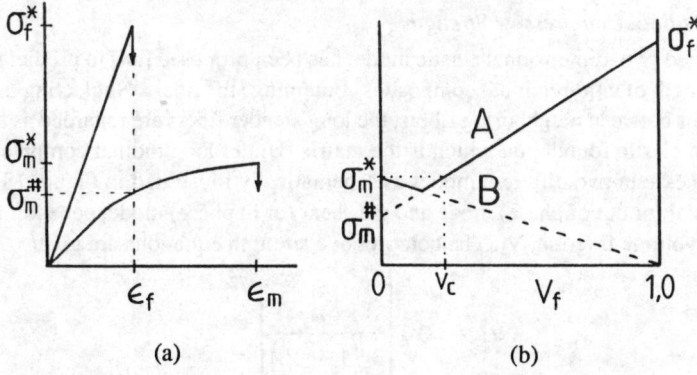

Figure 17. (a) Stress–strain response of fiber and matrix for $\varepsilon_f < \varepsilon_m$; (b) the relationship of the critical fiber volume fraction, V_c, to stresses in the fiber and matrix.

4.3.2 Transverse Tensile Strength

The low transverse tensile strength of unidirectional composites presents a major problem in the design of composite structures. Unlike the longitudinal tensile strength which is controlled almost entirely by the fiber strength, the transverse strength is governed by many factors including the properties of the composite constituents, interface bond quality, and presence and distribution of voids. One of the most characteristic features of the transverse strength is that it is usually less than the strength of matrix material measured in the absence of fibers, a clear indication of the negative effect of the reinforcing fiber. When bonding at the fiber–matrix interface is weak, the fibers can be regarded as cylindrical holes, the transverse strength of unidirectional composites is estimated as a function of matrix strength, σ_m^* [2]:

$$\sigma_\perp = \sigma_m^* \left[1 - 2\left(\frac{V_f}{\pi}\right)^{1/2}\right] \quad (15)$$

A strong bonding at the fiber–matrix interface improves the transverse tensile strength since the stress and strain magnification taking place in the matrix between neighboring fibers enhances the fracture strength. However, there also exist triaxial stress concentrations in the matrix, and the overall beneficial effect of strong interface bonding is not much significant.

An expression similar to Eq. 12 can also be derived for the elastic modulus of the lamina in the transverse direction, $E_\perp$, by assuming uniform stress in the fiber and matrix:

$$\frac{1}{E_\perp} = \frac{V_f}{E_f} + \frac{(1 - V_f)}{E_m} \quad (16)$$

4.3.3 Longitudinal Compressive Strength

A simplified two–dimensional elastic model has been proposed [34] to predict the compressive strength of unidirectional composites containing stiff fibers. Neglecting the effects of interactions between neighboring fibers, the long slender fibers are regarded as a column resting on an elastic foundation which is the matrix. Under longitudinal compression, the fibers will buckle in two different modes as schematically illustrated in Figure 18, namely the extensional (or out of phase) mode and the shear (or in phase) mode, depending mainly on the fiber volume fraction, V_f. The compressive strength equations are given:

$$\sigma_{\|c}^* = 2V_f \left[\frac{V_f E_m E_f}{3(1 - V_f)}\right]^{1/2} \quad (17)$$

$$\sigma_{\|c}^* = \frac{G_m}{1 - V_f} \quad (18)$$

for the extension mode and the shear mode, respectively. Comparisons with experimental measurements on several polymer matrix composites indicate that the predicted values based on the above equations are always an overestimate. In particular for Kevlar fiber–epoxy matrix composites the compressive yielding of the fiber resulted in formation of unfractured kink bands. In carbon fiber composites where the fiber–matrix bonds are strong, fibers are normally fractured in bending mode and the fracture processes in the matrix are dominated by shear. Therefore, it was noted that, in addition to the compressive failure due to buckling, the shear stress generated in the laminate may be sufficient to cause a shear failure mode. In this case, a lower estimate of compressive strength of the composite can be expressed by [2]:

$$\sigma^*_{\|c} = 2\left[\tau_f V_f + \tau^*_m(1 - V_f)\right] \qquad (19)$$

where τ_f and τ_m^* are shear strengths of the fiber and matrix, respectively. The factors which contribute to enhance the compressive strength of composites are high compressive strength, large diameter, high volume fraction and good alignment of the fiber, high matrix stiffness both in compression and shear, and strong bonding at the fiber–matrix interface.

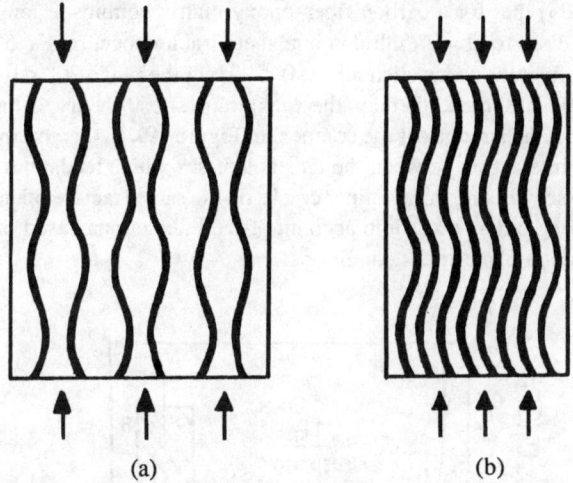

Figure 18. Compressive failure of unidirectional composites: (a) out–of–phase and (b) in–phase modes of buckling. After Hull [2].

4.3.4 Fiber Orientation Effects

When the unidirectional laminate is loaded at an angle θ to the fiber axis, Figure 19, the strength of the composite depends largely upon the failure mechanisms of the composite including longitudinal tensile fracture, intralaminar shear fracture and transverse tensile fracture. The three important stresses corresponding to these failure mechanisms are $\sigma_\|$, τ_m and $\sigma_\perp$. Based on the maximum stress theory, fracture is assumed to occur when the stress in the principal material direction reaches a critical value. i.e. for the three major failure mechanisms, respectively:

$$\sigma_\| = \sigma^*_\|, \quad \tau_m = \tau^*_m, \quad \sigma_\perp = \sigma^*_\perp \qquad (20)$$

The values of $\sigma_\parallel$, τ_m and $\sigma_\perp$ for a given applied stress, σ_θ, can be obtained by transformation:

$$\sigma_\parallel = \sigma_\theta \cos^2\theta$$
$$\tau_m = \sigma_\theta \cos\theta \sin\theta \qquad (21)$$
$$\sigma_\perp = \sigma_\theta \sin^2\theta$$

At angles between 0° and 90°, the failure strength predicted by the maximum stress theory depends on the relative magnitude of $\sigma_{\parallel c}^*$, τ_m^* and $\sigma_\perp^*$ for a given angle θ and is obtained from the smallest value of the following stresses:

$$\sigma_\theta = \frac{\sigma_\parallel^*}{\cos^2\theta}, \quad \sigma_\theta = \frac{\tau_m^*}{\cos\theta \sin\theta}, \quad \sigma_\theta = \frac{\sigma_\perp^*}{\sin^2\theta} \qquad (22)$$

It was found [35] that for a carbon fiber–epoxy matrix composite longitudinal tensile fracture occurred close to $\theta = 0°$, intralaminar shear fracture occurred at $5° < \theta < 20°$, and transverse tensile fracture occurred at $45° < \theta < 90°$. In the range $20° < \theta < 45°$, a mixed mode fracture was observed, thereby the maximum stress theory cannot be properly employed to predict the strength of the composite (Figure 19). In this intermediate range of angle, there is an interaction between the tensile stresses which leads to transverse cracks and the shear stresses causing intralaminar cracks on the same fracture plane. The Tsai–Hill criterion [36] among others takes into account of the interactions based on the maximum work theorem under a plane stress conditions:

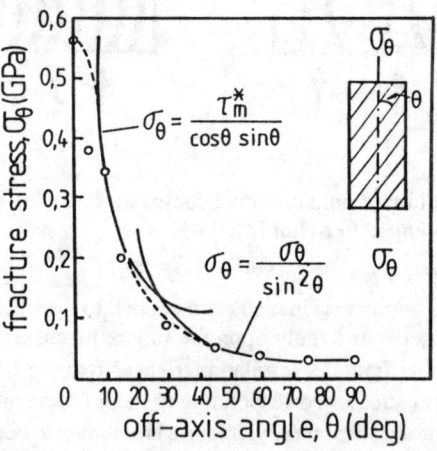

Figure 19. Orientation dependence of fracture strength of a unidirectional carbon fiber–epoxy matrix laminate ($V_f = 0.5$). Dotted line is the prediction based on Eq. 24. After Sinclair and Chamis [35].

$$\left[\frac{\sigma_\parallel}{\sigma_\parallel^*}\right]^2 + \left[\frac{\sigma_\parallel \sigma_\perp}{\sigma_\parallel^{*2}}\right] + \left[\frac{\sigma_\perp}{\sigma_\perp^*}\right]^2 + \left[\frac{\tau_m}{\tau_m^*}\right]^2 = 1 \tag{23}$$

Substituting σ_c^*, τ_m^* and σ^* with Eq. 22 and neglecting the second term, Eq. 23 gives the fracture strength in the intermediate angles:

$$\sigma_\theta = \left[\frac{c^4}{\sigma_\parallel^{*2}} + \left(\frac{1}{\tau_m^{*2}} - \frac{1}{\sigma_\parallel^{*2}}\right)c^2s^2 + \frac{s^4}{\sigma_\perp^{*2}}\right]^{-1/2} \tag{24}$$

Eq. 24 shows better agreement with the experimental results as shown in Figure 19.

4.4. Short Fiber Composites

Unlike continuous fiber composites the mechanical behavior of short fiber composites is often dominated by complex stress distributions associated with the fiber discontinuity. In particular, the local stress concentration near the broken fiber ends plays a critical role in determining the mechanical performance of these composites. One of the theories developed to predict the strength of short fiber composites is based on the modification of the simple rule of mixtures which was originally proposed for continuous fiber composites.

4.4.1 Aligned Fibers

Eq. 11 is derived based on the assumption that the tensile strain in the composite is uniform along the axial direction. When the fibers are discontinuous, there are differences in the strain between fiber and matrix near the fiber ends which induce a shear stress along the fiber axis as discussed in Section 4.1. Assuming constant interface shear strength, τ_i^*, the variation of fiber tensile stress, σ_f, is schematically illustrated in Figure 20 for different fiber length with respect to the critical transfer length [37]. The fiber critical transfer length, l_c, is defined as the minimum fiber length necessary to build up the fiber axial stress to the tensile strength σ_f^* and is given by:

$$l_c = \frac{\sigma_f^* d}{2 \tau_i^*} \tag{25}$$

Fiber lengths are not always constant and often show very wide variations, particularly those fabricated by molding in which there is considerable fiber breakdown into very small lengths during the processing operation. Taking the average fiber length, l, two different failure conditions are possible depending on the relative magnitude of l and l_c. If $l \geq l_c$, then the stress in the fiber reaches the fracture strength, and thus the full capacity of the fiber strength is attained over the central portion of the fiber. If $l < l_c$, the stress is not sufficient to cause fiber fracture when the matrix strength is reached. The average fiber strengths, $\bar{\sigma}_f$, are then given by:

$$\bar{\sigma}_f = \sigma_f^* \left(1 - \frac{l_c}{2l}\right) \quad (l \geq l_c)$$
$$\bar{\sigma}_f = \sigma_f^* \left(\frac{l}{2l_c}\right) \quad (l < l_c) \tag{26}$$

Therefore, the tensile strength of the composites containing aligned short fibers can be predicted by substituting the fiber stress, σ_f, with the average strength, $\bar{\sigma}_f$, in Eq. 13. For further treatment on the effects of fiber length variations, see Ref. [38].

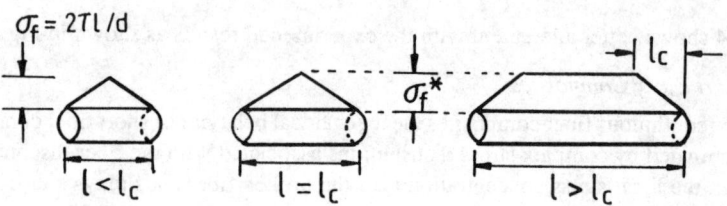

Figure 20. Stress transfer affected by fiber length, l, and its relations with the critical transfer length, l_c.

4.3.2 Random Orientation

Many composites contain short fibers of random orientation, particularly for those produced in contact mold, SMC, BMC and other mold processes. The strengths in these composites can be predicted by considering the effective lengths of the fibers which are projected on the plane of load application. As shown in Figure 21 where fibers are distributed randomly in a three–dimension space, only those fibers which lie parallel to the stress direction will give full reinforcement to the composite. Fibers randomly distributed with an angle, θ or ϕ, to the loading direction give only an effective reinforcement equivalent to a reduced fiber length. Assuming a total number of N fibers are oriented uniformly, the fiber efficiency factor, η, which is the effective fiber length in the loading direction divided by the average fiber length, is given:

$$\eta = \frac{N \int_0^{\pi/2} \int_0^{\pi/2} \cos\theta \cos\phi \, d\theta \, d\phi}{N \left(\frac{\pi}{2}\right)^2} = \left(\frac{2}{\pi}\right)^2 \tag{27}$$

If the fibers are randomly distributed in two dimension, the efficiency factor is given:

$$\eta = \frac{N \int_0^{\pi/2} \cos\theta \, d\theta}{N\left(\frac{\pi}{2}\right)} = \left(\frac{2}{\pi}\right) \tag{28}$$

The strength of a randomly distributed short fiber composites can thus be obtained from Eq. 13, assuming $l > l_c$:

$$\sigma_\parallel^* = \eta \, \sigma_f^* \left(1 - \frac{l_c}{2l}\right) V_f + \sigma_m^\# \left(1 - V_f\right) \tag{29}$$

In the above analysis, the fiber length is assumed constant in the average sense and the fiber fracture dominates the strength of the composite regardless of orientation of individual fibers. The possibilities of other failure mechanisms apart from tensile failure of fiber or matrix have not been taken into account.

An averaging technique has also been employed [39,40] to treat the strength of random fiber composites, based on three failure mechanisms, namely fiber fracture, matrix fracture in shear and matrix fracture in the transverse direction. The operative failure mechanisms are controlled by the angle between the fiber direction and the direction of applied stress as in continuous unidirectional fiber composites (see Section 4.3.4). Therefore, the strength for random fiber composites can be obtained by considering the angular strength dependence as a piecewise continuous function integrated over 90°:

$$\sigma^* = \frac{2}{\pi}\left[\int_0^{\theta_1} \sigma_{\theta 1} \, d\theta + \int_{\theta_1}^{\theta_2} \sigma_{\theta 2} \, d\theta + \int_{\theta_2}^{\pi} \sigma_{\theta 2} \, d\theta\right] \tag{30}$$

where $\sigma_{\theta 1}$, $\sigma_{\theta 2}$ and $\sigma_{\theta 3}$ are the strength values obtained when the individual failure mechanisms mentioned above, respectively, are operative.

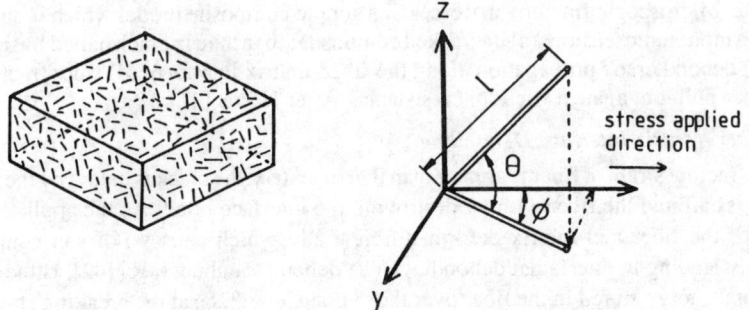

Figure 21. Determination of fiber efficiency in a randomly oriented short fiber composite.

5. Failure Analysis of Polymer Matrix Composites

5.1. Micromechanisms of Static Fracture

The fundamental understanding of how the fracture process initiates and progresses to final failure is a key to proper design of composite materials. Of particular importance is the local response of the fiber–matrix interface within the composite. To be able to avoid catastrophic fracture by enhancing fracture resistance and damage tolerance in composites, it is necessary to identify the basic failure mechanisms or origin of fracture toughness. It is well known that the fracture toughness of a composite is not simply the sum of the weighted contributions by the constituents, but is governed more importantly by the extent of energy absorption processes depending on the nature and bonding at the interface.

When a crack moves through a matrix containing fibers the following failure mechanisms may be expected to operate: matrix fracture, fiber–matrix interface debonding, post debonding friction, fiber fracture, stress redistribution, fiber pull–out etc. All these failure mechanisms apply to most composites reinforced with both short and continuous fibers, but it is not necessary for all these to operate simultaneously for a given system [15]. In order to distinguish the individual micromechanisms of toughening it is convenient to use a unidirectional fiber composite model where a crack travelling in the matrix approaches an isolated fiber, as schematically shown in Figure 22 [41].

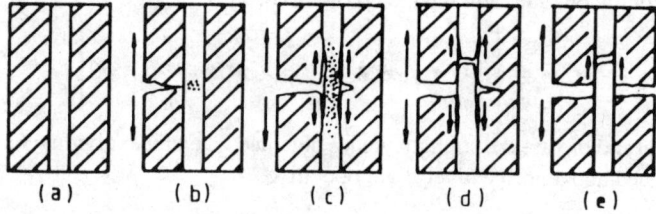

Figure 22. Microscopic fracture processes in a simple composite model which is subjected to a monotonic tension: (a) uncracked composite; (b) a matrix crack halted by the fiber; (c) debond crack propagation along the fiber–matrix interface; (d) fiber fracture; (e) fiber pull–out against frictional resistance. After Harris [41].

5.1.1 Fiber–Matrix Interface Debonding

If the fracture strain of fiber is greater than that of matrix, the crack is halted by the fiber or it may pass around the fiber without destroying the interface bond. As the applied load is increased, the fiber and matrix deform differentially which causes Poisson contraction eventually leading to interfacial debonding. The debond toughness, R_d [42], is the upper-bound total energy stored in the fiber over the debond length, l_d, at its breaking stress σ_f^*:

$$R_d = \frac{V_f \sigma_f^{*2} l_d}{2 E_f} \tag{31}$$

It has been pointed out [41] that implicit in the above equation is the separation of fibers from matrix in shear and the toughness contribution is a consequence of debonding rather than the debond process itself.

5.1.2 Post–Debonding Friction

After interfacial debonding has occurred, the fiber and matrix move relative to each other, resulting in energy dissipation due to post–debonding friction which is equal to the product of the frictional shear force and the differential displacement between the constituents [43]:

$$R_{df} = \frac{2 V_f \tau_{if} l_d^2 \Delta\varepsilon}{d} \tag{32}$$

where τ_{if} is the frictional shear strength at the interface, and $\Delta\varepsilon$ is the differential strain between fiber and matrix and d is the fiber diameter.

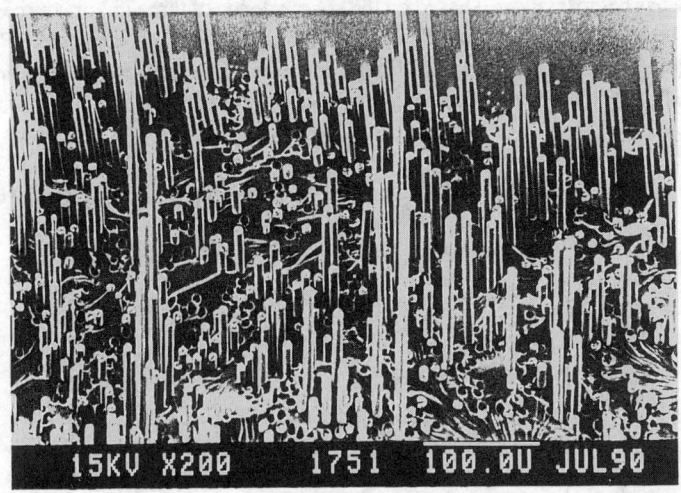

Figure 23. Fiber pull–out in carbon fiber–epoxy matrix composites. After Kim and Mai [46].

5.1.3 Stress Redistribution

When the fiber is loaded further after substantial debond extension, it may break at a weak point within the region near the main fracture plane. Upon failure the fiber instantly relaxes back and regain its original diameter. This leads to another source of toughness as the fiber strain energy is redistributed to the matrix [44]:

$$R_r = \frac{V_f \sigma_f^{*2} l_c}{3 E_f} \tag{33}$$

which is directly proportional to the debond toughness, R_d, given in Eq. 31.

5.1.4 Fiber Pull–out

The broken fiber end must be pulled out against the frictional grip of the matrix if total separation of the composite is to occur. This additional component of fracture toughness is the work done against sliding friction in extracting the broken fiber assuming a constant τ_f over a pull–out distance l_{po} [45]:

$$R_{po} = \frac{2 V_f \tau_{if} l_{po}^2}{d} \tag{34}$$

For most thermoset matrix continuous fiber composites, particularly those containing carbon fibers, the fiber fiber pull–out is a dominant source of fracture toughness (Figure 23 [46]).

5.1.5 Total Fracture Toughness Theory

A theory has been developed based on the co-existence of three major sources of fracture toughness, i.e. stress redistribution, R_r, fiber pull–out, R_{po}, and surface energy, R_s (inclusive of the toughness contributions due to fracture of fiber, R_f, and matrix, R_m, and interface debonding, R_i) [47]. Thus the total toughness, R_t, is given by:

$$R_t = R_r + R_{po} + R_s$$
$$= \frac{V_f \sigma_f^*}{\tau_{if}} \left[\frac{\sigma_f^* d}{6} \left(\frac{1}{4} + \frac{\sigma_f^*}{E_f} \right) + \frac{R_m}{2} \right] + \left(1 - V_f\right) R_m \tag{35}$$

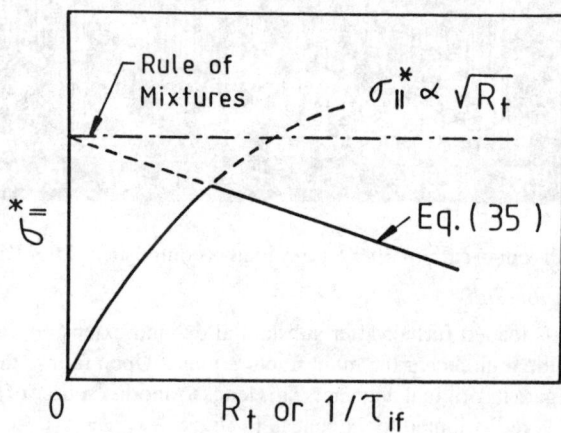

Figure 24. Relationship between composite strength, $\sigma_\parallel^*$, and total fracture toughness, R_t, or the inverse of frictional shear strength, τ_{if}. After Marston [47].

where

$$R_s = V_f R_f + \left(1 - V_f\right) R_m + V_f \left(\frac{l_d}{d}\right) R_i \approx V_f \left(\frac{l_d}{d} - 1\right) R_m \qquad (36)$$

In Eq. 36, R_f is neglected and R_i is approximated to R_m. Eq. 35 suggests that the total fracture toughness, R_t, varies linearly with the reciprocal of the interface frictional shear strength, τ_{if}, with a lower limit of $(1-V_f) R_m$ when τ_{if} approaches infinity. The variation of composite tensile strength, $\sigma_{\parallel}^*$, against R_t or $1/\tau_{if}$ is plotted in Figure 24, according to Eq. 35. For high τ_{if}, $\sigma_{\parallel}^*$ is determined assuming Griffith brittle fracture, i.e. $\sigma_{\parallel}^* \propto \sqrt{R_t}$, and for low τ_{if}, $\sigma_{\parallel}^*$ varies inversely with $1/\tau_{if}$. A further indication is that both high strength and high fracture toughness cannot be achieved simultaneously for a given combination of composite constituents, but these properties can be optimized.

5.2. Fatigue Failure Mechanisms

5.2.1 Introduction

Similar to other mechanical properties of fiber composites, the fatigue properties are also are anisotropic, i.e. dependent on the direction of loading relative to the fiber axis, and can be dangerously low in the transverse direction. This requires careful design and use of composites for application in fatigue environment, based on proper understanding of the mechanisms that govern the fatigue behavior. As for static fracture processes, fatigue damage in composite materials is associated with several major failure mechanisms including matrix cracking, debonding at the fiber–matrix interface, delamination and localized fiber breakage. These damage mechanisms depend greatly upon the relations between composite static failure strains and the matrix resin fatigue strain limits. The static failure strain of the composite is closely linked to the stiffness and failure strain of the fiber [48].

5.2.2 Unidirectional Continuous Fiber Composites

The fatigue damage in unidirectional composites when subjected to axial loading are characterized by three major mechanisms including fiber breakage, matrix cracking and interfacial shear failure. These mechanisms may operate simultaneously, but the predominant mechanism leading to failure may be effective in a limited range of the applied cyclic strain. This is illustrated in a fatigue life diagram, as given in Figure 25(a) [49] where the maximum strain (or stress) applied initially to a test specimen is plotted as a function of the number of load cycles.

The lower limit of the diagram is given by the fatigue limit (i.e. threshold strain) of the unreinforced matrix material, while the upper limit is given by the failure strain of the composite which is approximately equivalent to that of the fiber if fiber volume fraction is high. The latter mechanism is non–progressive since failure is not preceded by significant cycle dependent growth processes. The progressive damage mechanism is matrix cracking with associated interfacial shear failure which governs fatigue life. If fiber fracture strain is even lower than the fatigue limit of the matrix (e.g. high modulus carbon fiber–epoxy matrix system, Figure 25 (b)), the fatigue damage is totally suppressed by the highly stiff fiber [49].

When the loading cycle is inclined to the fiber axis, the predominant damage mechanism becomes matrix cracking and matrix shear along the fiber–matrix interface as in static loading (see Section 4.3.4). The static failure band in the fatigue diagram is lost and the fatigue limit drops with increasing off–axis angle. Therefore, the lowest limit is given by the strain for tensile debonding in the transverse direction at the off–axis angle of 90°.

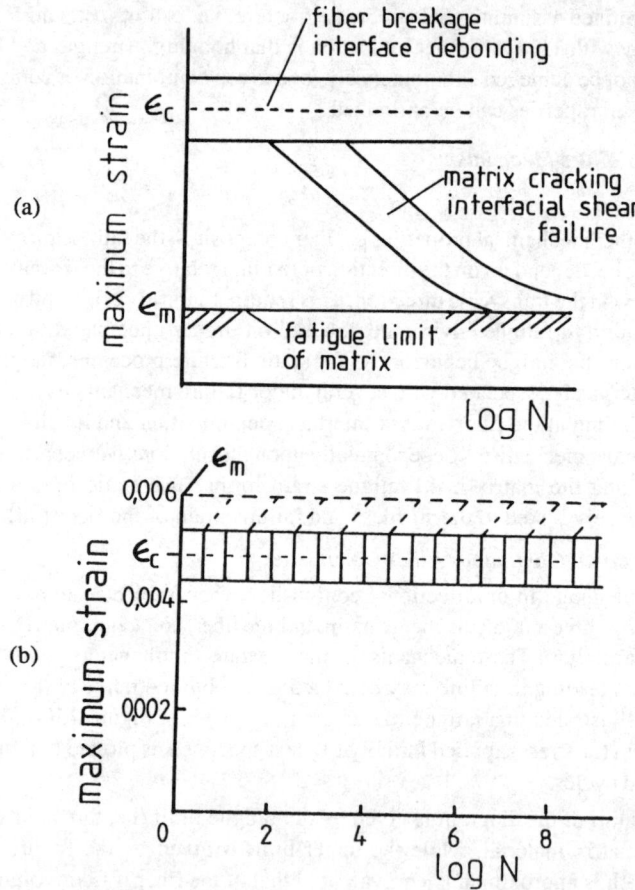

Figure 25. Fatigue life diagrams for unidirectional composites subjected to axial cyclic loading: (a) composites with $\varepsilon_c > \varepsilon_m$ (e.g. glass fiber–epoxy matrix); (b) composites with $\varepsilon_c < \varepsilon_m$ (e.g. high modulus carbon fiber–epoxy matrix). After Talreja [49].

5.2.3 Angle–plied Laminates

In the angle–plied laminates, the early stage of damage development is dominated by an array of primary matrix cracks in plies which are not aligned with the loading direction. The number of cracks increases monotonically with the number of cycles until a saturation density is reached which is called the characteristic damage state. This damage state is followed by initiation of cracks normal to the primary matrix cracks causing delamination. The final stage is dominated by fiber breakage and ultimate failure occurs when the locally damaged regions cannot support the maximum load.

Therefore, it can be summarized that the stiffness and failure strain of the fiber are key parameters in the determining fatigue performance of unidirectional composites in longitudinal loading. Matrix properties, failure strain and toughness in particular, also become important in the initial stage of fatigue loading when the laminate consists of angle–plies.

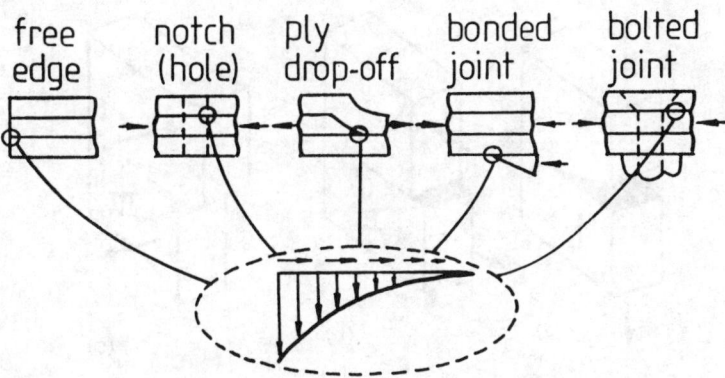

Figure 26. Source of delamination due to out–of–plane load from discontinuities in structure. After Wilkins [50].

5.3. Delamination

5.3.1 Interlaminar Fracture Modes

Delamination represents the weakest failure mode in laminated composites, and is considered to be the most prevalent life–limiting crack growth mode in most composite structures. Consequently, ever–increasing attention has been devoted to the understanding and characterization of this failure mode as well as to improve the durability against delamination. Delamination may be introduced during processing or subsequent service conditions. It may result from low–velocity impact, from eccentricities in structural load path, or from discontinuities in structures which induces a significant out–of–plane load, as shown in

Figure 26 [50]. In the absence of such discontinuities delamination can result from applied compressive loading, which may cause local or global buckling.

Delamination in composite structures seldom lead to immediate catastrophic failure. Instead, delamination occurring under in–plane loading normally induces local damages resulting in the loss of stiffness, local stress concentration and local instability. This delamination often leads to a redistribution of stresses which would eventually promote gross failure. In this context, delamination is indirectly responsible for the final failure of a composite. Composite structures in service are often subjected to complex three dimensional load paths. In general, a delamination will be subjected to a crack driving force with a mixture of mode I (opening), mode II (forward shear) and mode III (anti–plane shear) stress intensities, as schematically shown in Figure 27. Because delamination is constrained to grow between individual plies, both interlaminar tension and shear stresses are commonly present at the delamination front. Therefore, delamination is often a mixed–mode fracture process.

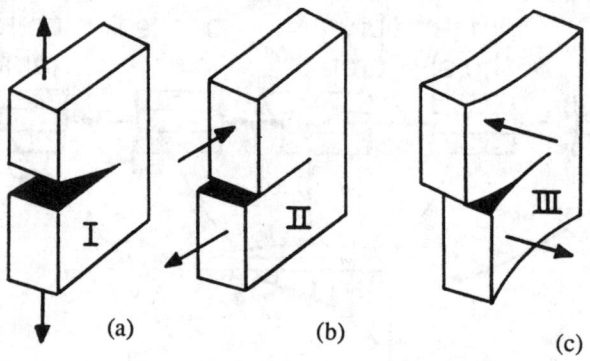

Figure 27. Modes of interlaminar crack propagation: (a) Mode I opening mode; (b) Mode II sliding shear mode; (c) Mode III tearing mode.

5.3.2 Interlaminar Fracture Testing

Mode I Opening Interlaminar Fracture Test (IFT)

Test methods based on fundamental mechanics concepts have been developed to evaluate the interlaminar fracture resistance of composites. Extensive research efforts have been devoted towards establishing a standard method using various interlaminar fracture tests (IFT), as shown in Figure 28. The double cantilever beam (DCB) specimens have been extensively used for measurement of the resistance to mode I interlaminar crack growth. It has been applied previously for testing adhesives where the specimen consists of a bond line between two metal arms to measure the strain energy release rate, G_{Ic} [51]. There are two basic configuration for the DCB specimens: the constant width specimen and the width tapered DCB (WTDCB) specimen [52]. The latter specimen geometry is designed to allow the strain energy release rate to be independent of the instantaneous crack length when the crack propagates under a constant load. Therefore, the crack length does not need to be monitored during the test. Its disadvantages are the cost for specimen preparation and the need for separate measurement of flexural modulus of the specimen.

The methods applied to evaluate the experimental data recorded during the DCB tests may be classified into two: compliance method and fracture energy method.

(i) Compliance Methods

The compliance methods are all based on the equation for strain energy release rate, G_{Ic}:

$$G_{Ic} = \frac{P_c^2}{2B} \frac{dC}{da} \tag{37}$$

where P_c is the critical load, B the specimen width and C the compliance and a the crack length. These methods require an equation for the relationship between C and a. A classical expression for this relationship is made by taking account of the strain energy due to the bending moment for a perfectly elastic and isotropic material:

$$C = \frac{8a^3}{EBh^3} \tag{38}$$

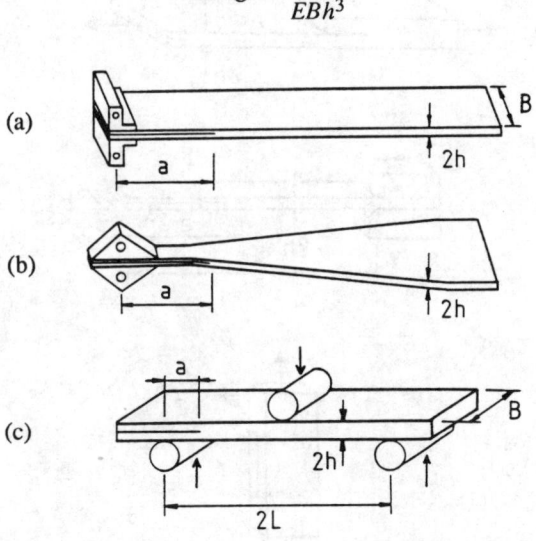

Figure 28. Interlaminar fracture tests: (a) double cantilever beam (DCB) test; (b) width tapered double cantilever beam (WTDCB) test; (c) end notched flexure (ENF) test; (d) cracked lap shear (CLS) test; (e) end loaded split (ELS) test; (f) edge delamination tensile test.

Therefore, the strain energy release rate in Eq. 37 becomes:

$$G_{Ic} = \frac{12 P_c^2 a^2}{EB^2 h^3} \quad (39)$$

Eq. 39 also applies to the width tapered DCB specimens where a/B is constant so that G_{ic} can be determined directly from the critical load P_c.

However, composite components are often made consisting of orthotropic laminates. In addition there are a number of factors which cannot be properly accounted for in the elastic beam theory as a consequence of various aspects of the practical DCB test. These include end rotation and deflection of the crack tip, effective shortening of the beam from large deflection of the arms, and a stiffening effect of the beam due to the presence of the end tabs bonded to the specimens. All these factors cause the values of the apparent elastic modulus E calculated from Eq. 35 to be varying. Therefore, a number of different analytical equations have been proposed in various forms to ascertain the correction factors in interpreting the experimental data, among which Williams and co–workers [53] have presented one of the most rigorous analyses.

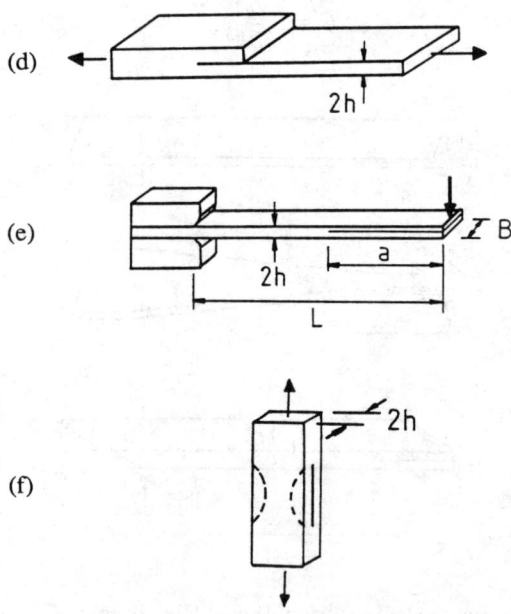

Figure 28. Continued.

Another approach developed based on an empirical compliance calibration [54] appears to avoid certain problems of correction factors. The compliance, $C (= \delta/P$, which is reciprocal of stiffness) is obtained from the loading–unloading experiments. based on the following empirical equation:

$$C = k\, a^n \tag{40}$$

A least squares line fit of a plot of C versus crack length a in log–log form allows the parameter k and n to be determined. Therefore, G_{Ic} can be written as:

$$G_{Ic} = \frac{P_c\, k\, n\, a^{n-1}}{2B} \tag{41}$$

(ii) Area Method

An alternative to the compliance methods is a direct fracture energy measurement technique. In this method, the crack extension is related directly to the area, ΔA, enclosed between the loading and unloading paths for extension of a known crack length, da, as shown in Figure 29. The mode I strain energy release rate is defined as

$$G_{Ic} = \frac{\Delta A}{B da} = \frac{1}{2B}\, \frac{P_1 \delta_2 - P_2 \delta_1}{a_2 - a_1} \tag{42}$$

Therefore, an average value of G_{ic} is determined by measuring the force P and displacement δ for a series of crack extensions. This method is still valid even if a non–linear elastic load–displacement response is observed, in contrast to the classical analytical compliance method where only the linear elastic response can be treated. However, the crack propagation has to be stable. If large unstable crack jumps with associated precipitous load drops are prevalent, the above expression becomes no longer valid due to the kinetic energy involved in the energy dissipation.

For the above reasons, interpretation of DCB test data should always be carried out in conjunction with examination of the fracture surface. Further complication is often encountered by the presence of fiber bridging. Fiber bridging occurs in mode I DCB test as a result of crossing the main delamination crack plane. There are two major reasons for this phenomenon. First, misalignment of fibers across the delamination plies or intermingle of plies results in no distinct interlaminar layer for crack propagation. Secondly, defects in plies adjacent to the main delamination plane becomes part of the main crack, fiber bundles of which bridge the crack. In both cases, a weak bond at the fiber–matrix interface and ductile fibers are favored which promote long debonding without being broken.

When fiber bridging occurs, the observed fracture toughness values during crack propagation are higher than the initiation value. An increase in fracture energy is required to debond the larger surface area generated by the bridged fibers and eventually fracture the bridged fibers. If fiber bridges are broken at roughly the same rate as that at which they are created a stable propagation value of G_{Ic} is obtained. The contribution of fiber bridging to fracture energy is also found dependent on specimen thickness [55,56]. It is noted that fiber bridging will significantly influence the compliance methods and fracture energy technique described above to determine G_{Ic}. A detailed discussion on this topic is given in Ref. [57].

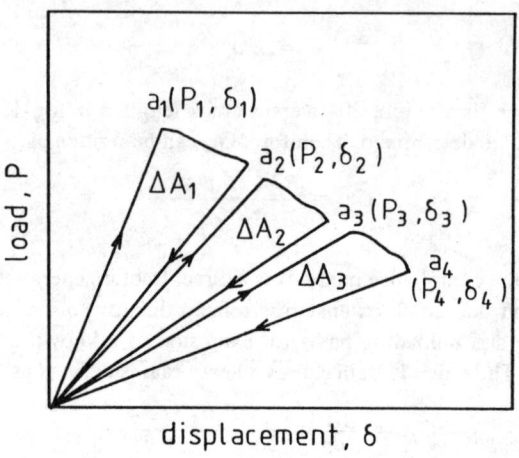

Figure 29. Area method for G_{Ic} determination.

Mode II Shear IFT Test

The pure mode II interlaminar fracture testing can be performed using the end notched flexural (ENF) specimen [58] (Figure 28(c)), which is essentially a three–point flexure specimen with an embedded through–width delamination located at the laminate mid–plane where the interlaminar shear stress is maximum. The major difficulty in designing a pure mode II specimen is in preventing any crack opening without introducing excessive friction between the crack faces [59]. A state of almost pure shear prevails at the tip of the mid–plane delamination [60]. The compliance method given by Eq. 37 can also be employed here to determine the mode II strain energy release rate, G_{IIc}. The relationship between C and a is much more complicated for the ENF test than for the DCB test, but may be found based on the classical beam theory [61]:

$$C = \frac{2L^3 + 3a^3}{8\ EBh^3} \qquad (43)$$

where L is the half span length. Therefore, the expression for G_{IIc} is obtained for small values of $(E/G_{IL})(h/a)^2$, where G_{IL} is the interlaminar shear modulus:

$$G_{IIc} = \frac{9a^2 P_c^2}{16\ EB^2 h^3} \qquad (44)$$

In general, interlaminar fracture toughness in mode II may also be derived from mixed mode tests. For example, cracked lap shear (CLS) test [62] which was originally developed for testing adhesively bonded metallic joints, the end notched cantilever beam (ENCB) test [63], the cantilever beam enclosed notch (CBEN) test, the end loaded split laminate (ELS) test [64], see Figure 28. The details of the expressions for the strain energy release rate from these tests are not treated here as the stress states are much more complicated than in pure mode II ENF test [65].

5.3.3 Free Edge Delamination

The problem of delamination along the straight free edge of composite laminates taking place under an in–plane uniaxial load (Figure 26) has attracted significant interests since this fracture phenomenon always acts as a precursor to final failure of laminate composites. Furthermore, the presence and growth of edge delamination may result in the progressive reduction of their stiffness and strength. The free edge delamination is attributed to the existence of interlaminar stresses which are highly localized in the neighborhood of free edge. The magnitude and distribution of the interlaminar stress components vary widely and depend upon the laminate lay–up, stacking sequence, properties of the composite constituents and the nature of loading [66].

Under tensile loading delamination is normally preceded by a number of transverse cracks, particularly in the 90° plies. Because of the presence of these cracks, the location of delamination is not unique as in the case of compressive loading which invariably results in gross buckling of the laminate. The path of delamination along the axial direction varies widely and depends upon the size and location of transverse cracks, type of laminates, and the composite constituents. The delamination threshold stress under which no delamination occurs appears also to vary according to the length or density of transverse cracks.

After delamination, the stiffness and strength are reduced depending upon the delamination area and type of loading. Without taking into account the effects of transverse cracking, the rule of mixtures analysis has been applied to predict the residual modulus, E_r, of a partially delaminated specimen [67]:

$$E_r = \left(E^* - E\right)\frac{A_d}{A^*} + E \qquad (45)$$

where E and E^* are the effective Young's moduli before and after delamination. A_d is the delaminated area and A^* is the total interface area. The strain energy release rate, G_e, associated with edge delamination growth has also been analyses. Form finite element analyses, it was noted that once the delamination progresses beyond a distance equal to a few ply thickness from the edge, G_e reaches a constant value. Based on the rule of mixtures and laminated plate theory, the constant G_e is derived as a function of the applied tensile strain, ε_c, and the laminate thickness 2h.

$$G_e = \varepsilon_c^2 h\left(E - E^*\right) \qquad (46)$$

6. Concluding Remarks

Polymer matrix composites have experienced remarkable growth in the past decades for numerous applications in aerospace, transportation, sporting goods, construction and household consumables industries. The most significant use of advanced composites came in the aerospace industry, where the weight savings gained from high specific strength and stiffness are a major benefit. Much of this chapter is concerned about the relation between the structures of composite constituents and the properties of the fiber, matrix and composites made therefrom. There are many manufacturing routes available for composites. It is important to recognize the profound effects that the manufacturing methods and processes have on the final properties of composites due to their direct influence on the microstructures. As a consequence of inhomogeneous and anisotropic nature, composites exhibit unique fracture behavior where cracks seek the least resistance path to propagate along the weak fiber–matrix interface and/or laminar interfaces. Fatigue and delamination have always been the limiting factor for damage tolerance design. The ability to anticipate and quantify these failure modes using various analytical techniques will play an important role in understanding the mechanical response and fracture phenomena in composite materials.

References

1. K.K. Chawla, *Composite Materials* (Springer–Verlag, New York, 1987).
2. D. Hull, *An Introduction to Composite Materials* (Cambridge University Press, Cambridge, UK. 1980).
3. L.S. Singer, in *Concise Encyclopedia of Composite Materials*, ed. A. Kelly (Pergamon press, Oxford, 1989) p. 47.
4. S.C. Bennett and D.J. Johnson, in *Proc. 5th London Carbon and Graphite Conference*, Vol. 1 (Society for Chemical Industry, London, 1978), p. 377.
5. M. Jaffe, in *Concise Encyclopedia of Composite Materials*, ed. A. Kelly (Pergamon press, Oxford, 1989) p. 129.
6. M.G. Dobb, D.J. Johnson and B.P. Saville, *J. Polym. Sci. (Polym. Phys. Ed.)* **15** (1977) 2201.
7. M.G. Dobb, D.J. Johnson and B.P. Saville, *Phil. Trans. Roy. Soc. Lond.* **A294** (1980) 483.
8. J.K. Kim and Y.W. Mai, *J. Mater. Sci.* **26** (1991) 4702.
9. P.M. Hergenrother and N.J. Johnston, *Polym. Mat. Sci. Eng. Proc.* **59** (1988) 697.
10. L.V. McAdams and J.A. Gannon, in *Polymers–High Performance Polymers and Composites*, ed. J.I. Kroschwitz (John Wiley & Sons, New York, 1991) p. 258.
11. A.C. Garg and Y.W. Mai, *Composites Sci. Technol.* **31** (1988) 179.
12. J.K. Kim, C. Baillie, J. Poh and Y.W. Mai, *Composites Sci. Technol.* **43** (1992) 283.
13. H. Ishida and J.L. Koenig, *Polymer Eng. Sci.* **18** (1978) 128.
14. N.L. Hancox, in *Concise Encyclopedia of Composite Materials*, ed. A. Kelly (Pergamon press, oxford, 1989) p. 134.

15. J.K. Kim and Y.W. Mai, *Composites Sci. Technol.* **41** (1991) 333.
16. J.K. Kim and Y.W. Mai, in *Structure and Properties of Fiber Composites*, Materials Science and Technology, Series Vol. 13, Vol. ed. T.W. Chou (VCH Publishers, Weinheim, Germany, 1993) p 239.
17. E.P. Plueddemann, in *Interfaces in Polymer, Ceramic and Metal matrix Composites*, Proc. ICCI–II, ed. H. Ishida (Elsevier Science Publishing, New York, 1988) p. 17.
18. J.C. Goan, T.W. Martin and R. Prescott, in *Proc. of 28th SPI/RP Annual Technical Conf.* (Society of Plastics Industries, New York, 1973) paper 21–B.
19. J.K. Kim, L.M. Zhou and Y.W. Mai, in *Handbook of Advanced Materials Testing*, ed. N.P. Cheremisinoff (Marcel Dekker, New York 1994) in press.
20. M.R. Piggott, in *Composite Applications: The Role of Matrix, Fiber and Interface*, ed. T. Vigo and B. Kinzig (VCH Publishers, Weinheim, Germany, 1993) p. 221.
21. Y.C. Gao, Y.W. Mai and B. Cotterell, *J. Appl. Math. Phys.* **39** (1988) 550.
22. L.M. Zhou, J.K. Kim and Y.W. Mai, *J. Mater. Sci.* **27** (1992) 3155.
23. S. Lee and M. Munro, *J. Composite Mater.* **24** (1990) 419.
24. J.K. Kim, D.B. Mackay and Y.W. Mai, *Composites* **24** (1993) 485.
25. Y.W. Mai and F. Castino, *J. Mater. Sci.* **19** (1984) 1638.
26. M.G. Bader, in *Processing and Fabrication Technology*, Delaware Composite Design Encyclopedia, Vol. 3, eds. L.A. Carsson and J.W. Gillespie (Technomic Publication, Lancaster, PA. 1990) p. 89.
27. A.B. Strong, in *Handbook of Composite Reinforcements*, ed. S.M. Lee (VCH Publishers, New York, 1993) p. 310.
28. P.C. Powell, *Engineering with Polymers*, (Chapman and Hall, London, 1983).
29. F.N. Cogswell, *Thermoplastic Aromatic Polymer Composites* (Butterworth–Heinemann, Oxford, 1992).
30. M.E. Ketterer, NASA Contract Report CR–3849 (1984).
31. P.J. Mallon and C.M. O'Bradaigh, *Composites* **19** (1988) 37.
32. H.L. Cox, *Brit. J. Appl. Phys.* **3** (1952) 72.
33. R.M. Jones, *Mechanics of Composite Materials* (McGraw-Hill, New York, 1975).
34. B.W. Rosen, in Fiber Composite Materials (ASM, Metals Park, OH. 1965) Chap. 3.
35. J.H. Sinclair and C.C. Chamis, in *Proc. of the 34th SPI/RP Annual Technical Conference* (Society of Plastics Industries. New York. 1974) Paper 22–A.
36. S.W. Tsai, NASA Contract Report CR–71 (1974).
37. A. Kelly and W.R. Tyson, *J. Mech. Phys. Solids* **13** (1965) 329.
38. T.W. Chou, *Microstructural Design of Fiber Composites* (Cambridge University Press, Cambridge, UK. 1993).
39. L.H. Lees, *Polymer Eng. Sci.* **9** (1969) 213.
40. P.E. Chen, *Polymer Eng. Sci.* **11** (1971) 51.
41. B. Harris, *Metal Sci.* **14** (1980) 351.

42. J.D. Outwater and M.C. Murphy, in *Proc. of the 24th SPI/RP Annual Technical Conference* (Society of Plastics Industries, New York, 1974) Paper 11–D.
43. A. Kelly, *Proc. Roy. Soc. Lond.* **A 319** (1970) 95.
44. J. Fitz–Randolph, D.C. Phillip, P.W.R. Beaumont and A.S. Teleman, *J. Mater. Sci.* **7** (1972) 289.
45. A.H. Cottrell, *Proc. Roy. Soc. Lond.* **A 282** (1964) 2.
46. J.K. Kim and Y.W. Mai, *Composites Sci. Technol.* **49** (1993) 51.
47. T.U. Marston, A.G. Atkins and D.K. Felbeck, *J. Mater. Sci.* **9** (1974) 447.
48. O. Konur and F.L. Matthews, *Composites* **20** (1989) 317.
49. R. Talreja, *Proc. R. Soc. Lond.* **A 378** (1981) 461.
50. D.J. Wilkins, NASA CP–2278 (1983) p. 67.
51. E.J. Ripling, S. Mostovoy and R.L. Patrick, *Mater. Res. Standards* (1964) p. 129
52. S.S. Wang, *AIAA Journal* **22** (1984) 256.
53. S. Hashemi, A.J. Kinloch and J.G. Williams, *Composites Sci Technol.* **37** (1990) 137.
54. J.P. Berry, *J. Appl. Phys.* **34** (1963) 62.
55. A.J. Russell and K.N. Street, in *Proc. 4th Int. Conf. on Composite Materials*, eds. Hayashi et al. (Japan Society of Composite Materials, Tokyo, 1982) p. 279.
56. Y.J. Prel, P. Davies, M.L. Benzeggagh and F.X. de Charentenay, *ASTM STP 1012* (Americal Society for Testing and Materials, Philadelphia, USA 1989) p. 251.
57. X.Z. Hu and Y.W. Mai, *Composites Sci. Technol.* **46** (1993) 147.
58. A.J. Russell and K.N. Street, *ASTM STP 876*, ed. W.S. Johnson (Americal Society for Testing and Materials, Philadelphia, USA 1985) p. 349.
59. A Sela and O. Ishai, *Composites* **20** (1989) 423.
60. J.W. Gillespie, L.A. Carlsson, R.B. Pipes, R. Rothschilds, B. Tretheweyand A. Smiley, NASA Contract Report CR–176416 (1985).
61. L.A. Carlsson, J.W. Gillespie and R.B. Pipes, *J. Composite Mater.* **30** (1986) 594.
62. D.J. Wilkins, J.R. Eisenmann, R.A. Chamin, W.S. Margolis and R.A. Benson, *ASTM STP 775*, ed. K.L. Reifsnider, (Americal Society for Testing and Materials, Philadelphia, USA 1982) p. 168.
63. T.K. O'Brien, in *Toughened Composite Materials–Recent Developments*, NASA Langley Research Center, Hampton, Verginia (Noyes Publications, NJ, 1985) p. 14.
64. C.R. Corleto and W.L. Bradley, in *ASTM Second Symposium on Composite materials: Fatigue and Fracture*, Cincinnati, OH. 1987.
65. J.M. Whitney, in *Interlaminar Response of Composite Materials*, ed. N.J. Pagano, (Elsevier Science Publishers, Amsterdam, The Netherlands, 1989) p. 161.
66. R.Y. Kim, in *Interlaminar Response of Composite Materials*. ed. N.J. Pagano (Elsevier Science Publishers, Amsterdam, The Netherlands, 1989) p. 111.
67. T.K. O'Brien, *ASTM STP–775*, ed. K.L. Reifsnider (Americal Society for Testing and Materials, Philadelphia, USA 1982) p. 140.

Problems

Q1. Consider a composite containing short fibers aligned parallel to the intended loading direction. The fibers are 0.04 mm in diameter, possess a fracture strength of 2 GPa, and the shear strength at the fiber–matrix interface 10 MPa. When the fibers are 2 mm long:

(a) Compare the fiber length with the critical transfer length, l_c.
(b) What is the maximum stress transmitted to the fibers?
(c) What is the average stress felt by the fibers?
(d) If the strength of the matrix is 30 MPa and the composite is to have a strength of 50 MPa, what is the volume fraction of fiber for the composite, according to the rule of mixtures prediction?

Q2. The figure below shows an elastic fiber of diameter, d, and of infinite length embedded in a semi–infinite elastic medium by an adhesive of interfacial fracture toughness, R_i. l_f and l_d are the free fiber and initial debond lengths, respectively. What are the fracture load, P, and the corresponding stress, σ_f, in the fiber? If σ_y is the elastic limit in the fiber, what is the fiber diameter for transition from tensile yielding of the fiber to adhesive fracture?

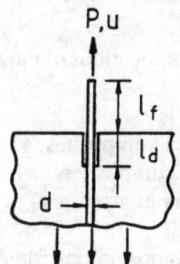

Q3. Why are high tensile strength and high fracture toughness mutually exclusive in brittle fiber–brittle matrix composites? What is the philosophy of the intermittent bonding theory to obtain high strength–high toughness in composites? What other methods are available to achieve this aim? (See Ref. [15,25]).

Q4. When pull–out toughness is the main contribution to the fracture resistance, one of the methods of improving the total toughness, R_t, is to have a rate sensitive coating on the fiber. The following figure is a graphical representation of the model.
    where   P = Fiber pull–out force
                   L = Embedded fiber length
                   d = fiber diameter
                   t = thickness of the coating material
                   u = fiber pull–out distance

Derive an expression for the maximum pull–out work, W_{po}, if the shear stress, τ, is a product of the coating fluid viscosity, η, and the velocity gradient across the thickness of the coating, dV/dz. z is the coordinate axis perpendicular to the fiber axis.
(Hint: Assume velocity of fiber pull–out is V_0 and $L = l/4$ where l is the finite length of the fiber in a composite.)

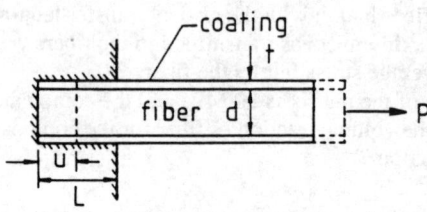

Also show that the optimum fiber aspect ratio for maximum pull–out work is given by:

$$\frac{l}{d} \leq \frac{\sigma_f^* \, t}{\eta \, V_0}$$

Q5. Give some typical commercial applications of the following advanced polymer matrix composites.

(a) Carbon fiber–polymer matrix composites.
(b) Glass fiber–polymer matrix composites.
(c) Aramid fiber–polymer matrix composites

(Hint: Refer to "Commercial Opportunities for Advanced Composites" *ASTM STP 704*, Ed. A.A. Watts, Americal Society for Testing and Materials, Philadelphia, 1979).

Q6. Determine the elastic moduli of orthotropic laminates in terms of the in–plane stiffness, A_{ij}.

Q7. Consider a 24–ply composite laminate, symmetrical and orthotropic, and all made of the same materials but with varying numbers of 0° and ±45° plies. The ordering of plies is immaterial but to maintain orthotropy there must be an equal number of +45° and –45° plies. The elastic constants of a carbon fiber–epoxy matrix lamina are:

$E_1 = 137.8$ GPa, $E_2 = 11.71$ GPa, $G_{12} = 5.51$ GPa, $\nu_{12} = 0.25$, $\nu_{21} = 0.0213$.

Calculate the elastic constants for the 24–ply laminates with 0° and ±45° plies given below.

0°	Number of plies +45°	−45°
24	0	0
16	4	4
12	6	6
8	8	8
0	12	12

Q8. A quasi–isotropic laminate is one in which the in–plane elastic properties are the same, i.e. $E_x = E_y$, $\nu_{xy} = \nu_{yx}$, and $G_{xy} = E_x/2(1 + \nu_{xy})$. Show that a 24–ply laminate with 8 plies oriented at each of 0°, +60° and –60°, or 6 plies at each 0° and +45°, –45° and 90° is quasi–isotropic with elastic constants E_x = 54.3 GPa, G_{xy} = 20 8 GPa and ν_{xy} = 0.305. Use the properties of lamina given in Q7.

Q9. The main hull molding for a small sailing yacht has been manufactured from layers of chopped–strand glass mat, and hand laminated with an isophthalic unsaturated polyester resin in a mold. The overall length of the molding is 10 m and it weighs 700 kg. The laminate generally varies in thickness from 5 to 10 mm, and each layer of mat contributes 1 mm to the thickness. The fiber volume fraction is 0.4. The in–plane Young's modulus is 10 GPa, and it is isotropic within the plane of lamination. The total area of the molding is approximately 40 m².

To improve the performance of the craft it is necessary to double the Young's modulus in the longitudinal direction of the hull, and to reduce the weight to 400 kg. The Young's modulus in the transverse direction must be lower than that in the original design.

Using the data in table below, design a modified laminate which will meet the new specification, can be laid up in the smae mold and carries the lowest possible cost penalty. (Taken from Lecture Notes on "Composite Materials Workshop: Manufacturing Technology and Design" M.G. Bader, University of Sydney, 1991).

Material	Form	V_f	E_1 (GPa)	E_2 (GPa)	Density (kg/m^3)	Cost ($/kg)
Glass	Chopped–strand mat	0.3	8	8	1590	6
Glass	Balanced cloth	0.5	16	16	1850	9
Glass	UD Woven roving	0.6	33	4	1980	9
Aramid	Balanced Cloth	0.5	35	35	1300	23
Aramid	UD Woven roving	0.6	76	9	1320	23
Carbon	Balanced Cloth	0.5	56	56	1500	38
Carbon	UD Woven roving	0.6	124	12	1560	38

UD = Unidirectional

Q10. A lightweight construction systems for agricultural building is based on square section steel tube, 40 mm square x 1 mm wall thickness, clad with steel sheet 1 mm thick, as shown below. It is proposed to replace this with a glass reinforced plastic (GRP) system of pultruded square tubes. The cladding is to be of GRP sandwich panels, consisting of 0.5 mm face sheets laminated from a balanced glass–cloth either side of a foam core. Properties of these materials given in the table.

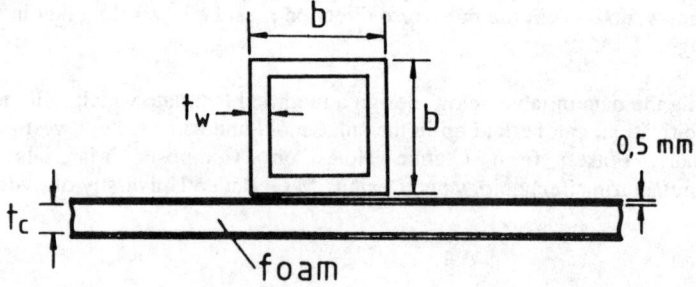

Determine the optimum tube dimensions, b and t, for minimum weight which will ensure that the tube are of at least equal bending strength and stiffness compared with the mild steel tubes to be replaced. Also determine the optimum core thickness for the face sheets which will give at least five times the stiffness of the 1 mm mild steel sheet. (Taken from Lecture Notes on "Composite Materials Workshop: Manufacturing Technology and Design, M.G. Bader" University of Sydney, 1991).

Material	Young's modulus (GPa)	Allowable tensile strength (MPa)	Density (kg/m^3)
Mild steel	200	150	7800
GRP (pultruded)	45	800	1600
GRP (cloth laminate)	28	500	1600
Foam core	0*	0*	100

* Assume that the core does not contribute to stiffness or strength of the samdwich but has "sufficient" compressive and shear strength.

Chapter 5
CERAMIC MATRIX COMPOSITES

*PRASHANT G. KARANDIKAR, TSU-WEI CHOU AND
AZAR PARVIZI-MAJIDI*
Center for Composite Materials, Department of Mechanical Engineering
University of Delaware, Newark, DE 19716

1. Introduction

Although ceramics are among the oldest and the most abundant materials known to mankind, are lighter than many structural materials and maintain their strength and environmental stability at very high temperatures; their use as engineering materials has been limited due to their inherent brittleness and the resultant catastrophic failure.

Tensile behavior of monolithic borosilicate glass (BSG) is shown in Figure 1. The stress-strain curve is linear up to failure and the strain-to-failure is much lower than that of metallic materials. Tensile behavior of ceramics is similar to that of glass. A monolithic ceramic has a distribution of cracks (flaws). When the applied load exceeds a critical value, the most severe crack starts growing. Lack of plasticity in these materials does not allow the so-called R-curve behavior whereby the resistance to crack growth increases as the crack grows. Thus, the crack grows unstably, resulting in fast fracture or catastrophic failure.

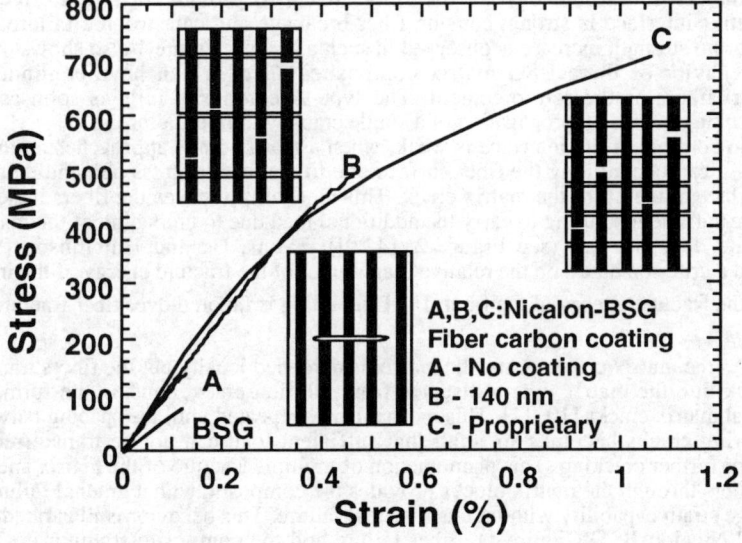

Fig. 1   Tensile stress-strain curves for borosilicate glass (BSG) and Nicalon fiber reinforced BSG (a) monolithic BSG (b) uncoated Nicalon-BSG (c) 140 nm carbon coated Nicalon-BSG (d) proprietary coating Nicalon-BSG [50]. Also shown are the damage modes observed.

Addition of a second phase material such as particles, whiskers, continuous or discontinuous fibers and phase transforming materials improves the crack growth resistance of ceramics and changes the failure mode [1-6]. The improved crack growth resistance of these materials may be due to one or more of the following mechanisms: crack bridging, crack deflection, crack blunting, fiber pull out and phase transformation induced stresses.

Incorporation of continuous fibers provides the highest increase in the toughness and strength of ceramics and imparts the desirable gradual failure mode [2-8]. The first continuous fiber reinforced ceramics (referred to as ceramic matrix composites, CMCs from hereon) were demonstrated in early 70's [7, 8]. In CMCs the idea is to make use of the higher failure strain of the fibers compared to the matrix failure strain. Matrix microcracking is the first damage mode observed in these composites. As the matrix fractures, if the fiber volume fraction is low, then fibers are unable to support the additional load transferred to them and the composite fails. In this case, the composite ultimate tensile strength (σ_{cu}) is given by

$$\sigma_{cu} = \sigma_{mu} V_m + \sigma_f' V_f \tag{1}$$

where σ_{mu} is the ultimate strength of the matrix, V_f and V_m are volume fractions of the fiber and the matrix respectively and σ_f' is the fiber stress at matrix failure.

If the fiber volume fraction is sufficiently high, the fibers can sustain the additional load due to matrix failure. But, since the fibers themselves are brittle, the matrix failure not only transfers additional load on the fibers but also provides a notch effect if the fiber/matrix interface is strong, causing fiber breakage and catastrophic failure. Only a limited or no strength increase is observed in such a system. Figure 1 also shows the stress strain behavior of three BSG matrix composites all of which have continuous SiC (Nicalon) fibers as the reinforcement. The type A composite fails as soon as matrix microcracking begins by propagation of a single crack with fiber failure.

In contrast, if the interface is weak, when a matrix crack approaches an interface, debonding can occur along the fiber surface due to the tensile stress perpendicular to the fiber surface at the tip of the matrix crack. This mechanism leaves the fibers intact. If the fibers are sufficiently strong to carry the additional load due to unloading of the matrix, the composite does not fail (see Figure 2, [4, 9]). Evans, He and Hutchinson [9] have proposed a criterion based on the relative magnitudes of the fracture energy of the interface, Γ_{ic} and the fracture energy of the fiber, Γ_{if}. Debonding is favored over fiber fracture if $\Gamma_{ic} / \Gamma_{if} < 1/4$.

As the matrix crack forms, the matrix is unloaded locally but the fibers transfer the load back into the matrix over a distance from the first crack, causing the formation of additional matrix cracks [10, 11]. This mechanism is repeated until the spacing between the neighboring cracks becomes so small that sufficient load can not be transferred to the matrix for further cracking. This phenomenon of multiple fracture of the matrix and sliding of the fibers through the matrix blocks provides the composite with a gradual failure mode with large strain capability without macroscopic failure. This behavior is illustrated by type B and C Nicalon-BSG (Figure 1). Fiber failure and the composite strength in this case depends on the interfacial shear strength / sliding resistance. If all the fibers have a unique strength σ_{fu}, then the composite strength is simply given by

$$\sigma_{cu} = \sigma_{fu} V_f \tag{2}$$

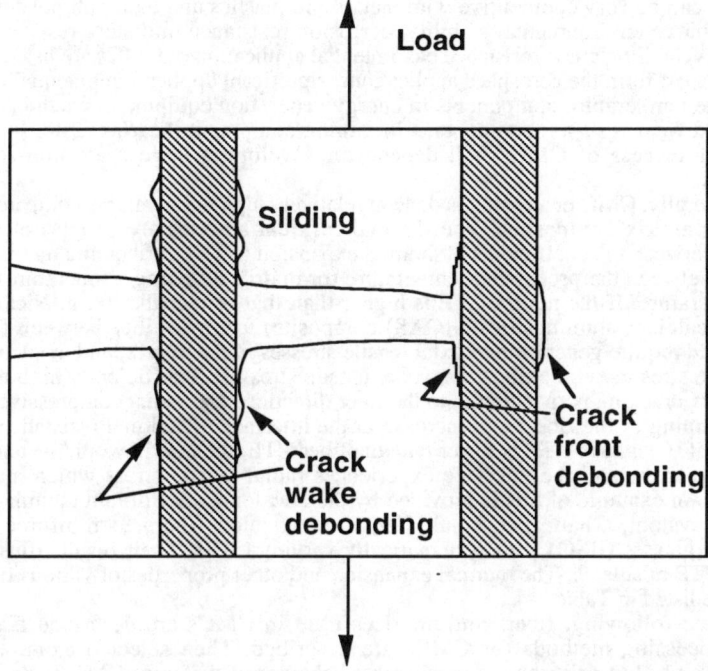

Fig. 2   Schematic of crack front and crack wake debonding [4].

In the absence of notch effect on the fibers, the critical volume fraction (V_f') which is necessary to prevent composite failure at matrix cracking is given by combining Equations (1) and (2).

$$V_f' \geq \frac{\sigma_{mu}}{\sigma_f - \sigma_f' + \sigma_{mu}} \qquad (3)$$

In reality, since the fibers are brittle, they show a strength distribution rather than a deterministic strength. Also, the notch effect on the fibers is a function of the interfacial strength/sliding resistance. Thus, the ultimate tensile strength of the composite cannot be predicted by Equation 2 and is determined by the fiber strength distribution and interfacial shear behavior.

The major impetus for development of CMCs has come due to their high strength at elevated temperatures. Currently, superalloys are used for making high temperature components of jet engines. These alloys are very heavy and there temperature capability is limited to about 1000 -1200°C. Ceramic matrix composite components based on glass-ceramics (up to 1200°C), silicon carbide (1400°C) and silicon nitride (1500°C) have the potential to provide the elevated temperature performance at much lower weights. The weight savings due to lower densities and higher efficiencies due to higher operating temperatures offered will reduce the fuel consumption significantly. At lower temperatures also CMCs can be very competitive with metals and plastics and their composites as they can offer unique environmental stability, corrosion resistance, moisture resistance, non-toxicity, recyclability, creep resistance etc. Potential application of CMCs are listed in Table 1 [12-18]. Apart form the aerospace applications, significant application potential exists for intermediate temperature components in energy generation equipment, chemical process industry and wear resistant components in mining equipment, bearings, etc. In fact, the commercial success of CMCs will depend on developing large scale non-aerospace applications.

Generally, CMC processing is done at relatively high temperatures compared to that of polymer matrix composites. The thermal residual stresses are proportional to the difference between the coefficients of thermal expansion (CTE) of fiber and matrix and the difference between the processing temperature (or matrix softening temperature) and the room temperature. If the matrix CTE is higher than that of the fiber (e. g. Nicalon fiber reinforced calcium aluminosilicate (CAS) composite), compatibility between fiber and matrix would require generation of axial tensile stresses in the matrix, and axial and radial compressive stresses on the fiber. The axial tensile stresses, if sufficiently high may give rise to matrix cracking perpendicular to the fiber direction. The radial compressive stresses lead to clamping of the fibers and increase in the interfacial frictional resistance. On the other hand, if the matrix CTE is lower than the fiber CTE, the matrix would be under axial compressive stress and the interface experiences radial tensile stress which may cause debonding. An example of this is provided by the Nicalon-LAS (lithium aluminosilicate) composite system. Chatterjjee et al. [19] used Nicalon fibers to reinforce various borosilicate glasses (BSG) with systematically varying CTEs. Their results illustrate the effects of CTE mismatch. The thermal expansion and other properties of various fibers and matrices are listed in Table 2-4.

In the following, fibers and matrices used in CMCs are described first. Then various processing methods for CMCs are described. Then selected examples from literature are used to describe the mechanical behavior of different CMC systems under tensile, compressive, shear, flexural, impact loadings at room temperature. Next, various techniques for the measurement of the fiber/matrix interfacial properties are described. Theoretical analysis of crack initiation, stress-strain behavior and modulus degradation (in unidirectional, cross-ply and woven composites) and ultimate strength (in unidirectional composites) under tensile loading is described next. The last three sections are devoted to the mechanical behavior at elevated temperatures and under fatigue and creep loadings. Majority of Sections 4 through 8 are excerpted from Reference [2].

2. Fibers and Matrices

A variety of fibers are available for reinforcing ceramic matrices; some of the common fibers and their properties are listed in Table 2 [20-23]. Fibers must offer higher strength and failure strain than the ceramic matrix. In addition, low mismatch between the coefficients of thermal expansion (CTEs) is extremely important. Residual thermal stresses in composites are proportional to the CTE mismatch ($\Delta\alpha$) and the difference between the processing/annealing temperature of the matrix (temperature at which matrix has sufficiently low viscosity) and room temperature (ΔT). In CMCs, ΔT can be as large as 500 to 1200°C. Therefore, it is extremely important to minimize $\Delta\alpha$. If the CTE of matrix (α_m) is larger than that of the fiber (α_f), residual axial tensile stresses develop in the matrix during cooling, causing matrix cracking. On the other hand, if $\alpha_m < \alpha_f$, fibers are under residual tension, which lowers the mechanical load carrying capacity of the fibers. Fibers can also shrink away from the matrix causing debonding. Low $\Delta\alpha$ also prolongs the life of the composite under cyclic thermal loading (e. g. during firing and cooling of a jet engine).

Table 1. Applications of Ceramic Matrix Composites [12-18].

Application	Materials	Status	Properties	Ref.
Cutting Tools	$TiC_{(p)}/Al_2O_3$ $ZrO_{2(p)}/Al_2O_3$ $SiC_{(w)}/Al_2O_3$ $SiC_{(w)}/Si_3N_4$	Current	Heat, corrosion and wear resistance, toughness	12
Valves, Seals, Bearings	$SiC_{(p)}/Al_2O_3$	Current	Heat, corrosion and wear resistance	12, 13
Piston Engines turbocharger rotor valve train components, camshafts	$SiC_{(p)}/Al_2O_3$ $SiC_{(w)}/Si_3N_4$	Current	Strength, toughness, stiffness, Heat, corrosion and wear resistance	12, 13
Adiabatic engines pistons, piston caps	Discontinuously reinforced ceramics	Developmental	Thermo-mechanical loads Heat, corrosion and wear resistance	12
Rocket engine turbo pump hot section components, nozzle extension, thrusters	C/SiC SiC/SiC SiC/Si_3N_4	Developmental, prototype	Extremely high temp., strength, toughness, stiffness, thermal shock, environmental resistance, reliability	14, 15
Space shuttle ablative tiles, nose cover	CMC	Current	Oxidation resistance, thermal shock	14
National Aerospace Plane (NASP) surfaces, leading edges structural parts propulsion system	C/SiC SiC/SiC SiC/Si_3N_4 SiC/Glass ceramic	Developmental	Oxidation resistance, thermal shock, Extremely high temp., strength, toughness, stiffness,	14
Turbine engines vanes, blades, disks, blisks, exhaust nozzle flaps, inner/outer flaps, center body, flame holder combustor, rotor	C/SiC SiC/SiC SiC/Si_3N_4 SiC/Glass-ceramic	Developmental, prototype	Extremely high temp., strength, toughness, stiffness, thermal shock	15, 16
Armor	$SiC_{(p)}/Al_2O_3$	Current	Toughness, light weight, wear resistance	17
Heat exchanger tube, Burner tube, Flue gas nozzle	$SiC_{(p)}/Al_2O_3$	Developmental		17
Wear products mining, chemical process power generation	$SiC_{(p)}/Al_2O_3$	Current	Wear resistance, toughness, chemical resistance	17
Fasteners	3-D C/SiC	Prototype	High temperature strength, pin shear, thread shear, bearing	15
Radiant furnaces Burner encapsulation tubes	Al_2O_3/Al_2O_3	Developmental	High temperature, environmental resistance	18

Subscript p and w indicate particulate and whisker reinforcements

Table 2. Properties of fibers for ceramic matrix composites [20-23].

Source	Type/Grade	Composition	Tensile Strength (MPa)	Tensile Modulus (GPa)	Density (g/cc)	Diameter (mm)	CTE (10^{-6} °C^{-1})	Use Temp (°C)	Elongation (%)
DuPont	FP	α-Al$_2$O$_3$	1380	370	3.90	20	5.7	1320	0.35
	FP-PRD-166	α-Al$_2$O$_3$+PSZ	2070	379	4.20	20	9.0	--	0.55
3M	Nextel-312	Mullite+14% B$_2$O$_3$	1750	154	2.70	11	3.5	1200	1.20
	Nextel-440	Mullite+2% B$_2$O$_3$	2100	189	3.05	11	4.5	1430	1.20
	Nextel-480	Mullite+2% B$_2$O$_3$	2275	224	3.05	11	4.5	--	1.00
ICI	Saffil	α-Al$_2$O$_3$	2000	300	3.30	3	--	1400	0.60
Sumitomo		γ-Al$_2$O$_3$+15% SiO$_2$	1800-2600	210-250	3.20	17	4.0	1250	0.90
Nippon Carbon	Nicalon CG	β-SiC+C+O	2520-3290	182-210	2.55	10-20	3.1	1200	1.50
AVCO	SCS-2	β/α-SiC on C	3450	407	3.05	140	--	--	0.85
	SCS-6	β/α-SiC on C	3920	406	3.00	143	4.2	1299	0.97
	SCS-9	β/α-SiC on C	2900	330	2.80	78	--	--	0.88
UBE	Tyranno	β-SiC+Ti+C+O	>2970	>200	2.37	8-10	3.1	1300	1.50
Dow Corning/	MPDZ	β-SiC+C+O+N	1750-2100	175-210	2.30	10-15	--	--	1.00
Celanese	HPZ	Si+N+O+C	2100-2450	140-175	2.35	10	--	--	1.00
	MPS	β/α-SiC+O	1050-1400	175	2.65	10-15	--	--	0.70
Union Carbide	Thornel T300	PAN, 92% C	3650	231	1.76	7	-0.6	--	1.40
	Thornel P100	Pitch, 99% C	2370	758	2.15	10	-1.45	--	0.32
Hercules	Magnamite AS4	PAN, 94% C	3587	235	1.80	8	--	--	1.53
	Magnamite HMU	PAN, 99.7%C	2760	380	1.84	8	--	--	0.70
Celanese	Celion 6000	PAN, 96% C	2758	234	1.76	7	-0.1	--	1.20

Table 3. Properties of glass and glass-ceramic matrices [24-26].

Type/Grade	Major Composition	Minor Composition	Flexural Strength (MPa)	Modulus (GPa)	Density (g/cc)	CTE ($10^{-6}°C^{-1}$)	Use Temp (°C)
7740 BSGlass	B_2O_3, SiO_2	Na_2O, Al_2O_3,	138	63	2.23	3.25	600
1723 ASGlass	Al_2O_3, MgO, CaO, SiO_2	B_2O_3, BaO	--	70	--	--	700
7930 HSGlass	SiO_2	B_2O_3	48	70	2.2	0.5	1150
LAS-I (GC)	Li_2O, Al_2O_3, MgO, SiO_2	ZnO, ZrO_2, BaO	103	86	2.5	1.5	1000
LAS-II (GC)	" + Nb_2O_5	ZnO, ZrO_2, BaO	--	--	--	--	1100
LAS-III (GC)	" + Nb_2O_5	ZrO_2	--	85	--	--	1200
MAS (GC)	MgO, Al_2O_3, SiO_2	BaO	110-170	120	2.6-2.8	2.5-5.5	1200
BMAS (GC)	BaO, MgO, Al_2O_3, SiO_2	--	83	103	2.6	2.5	1250
CAS (GC)	CaO, Al_2O_3, SiO_2	--	124	98	2.8	5.0	1350
Ternary Mullite	BaO, Al_2O_3, SiO_2	--	--	100	--	3.6	1500
Hexacelsian	BaO, Al_2O_3, SiO_2	--	--	--	--	6.5	1700

Table 4. Properties of matrices for ceramic matrix composites [27-30].

Matrix	Young's Modulus (GPa) 20°C 1400°C	Flexural Strength (MPa) 20°C 1400°C	Tensile Strength (MPa) 20°C 1400°C	Fracture Toughness MPa m1/2	Density (g/cc)	Oxidation weight gain (%/1000hr)	CTE (10^{-6} °C^{-1})
HP* Si_3N_4	250-325, 175-250	450-1100, 0-600	375, 150	2.8-6.6	3.1-3.4	0.05-2.5	3-3.9
S† Si_3N_4	195-315, --	275-840, 0-700	--, --	3-5.6	2.8-3.4	0.2	3.5
RS§ Si_3N_4	100-220, 120-200	50-300, 0-400	70-210, 140	3.6	2.0-2.8	0.1-2.6	2.5-3.1
HP SiC	430-450, 380	300-800, 175-575	200, 35-150	3-4	3.2-3.3	0.1-0.6	4.3-5.4
S SiC	375-420, 300-400	275-535, 240-450	--, --	2.5-6.5	3.0-3.2	0.01-0.25	4.4-4.8
RS SiC	350-375, 200-320	175-450, 70-450	77, --	--	2.9-3.1	0.06-0.16	4.3-4.4
CVD SiC	490	--, --	588, --	--	3.21	--	4.5
Al_2O_3	320-370	300-450, 150*	--, --	2-3	3.9-4.0	--	8-9
ZrO_2	150-250	700-1200, 500**	--, --	8-15	5.9-6.1	--	9-11

* - hot pressed, † - sintered, § - reaction sintered

CMC components will be exposed to high temperature during processing and service. Diffusion and chemical reactions between the fiber and matrix during these exposures can alter the properties of the interface and the composite dramatically. Therefore, compatibility between the fiber and the matrix is very important. Another important criterion is the ability of the fiber to retain its strength at elevated temperatures under service load. Temperature capability of the composite will be limited by the temperature capability of the fiber.

Current research has mostly focused on SiC-based fibers made by polymer precursor techniques (e. g. Nicalon fibers) and chemical vapor deposition techniques (e. g. SCS fibers). Although carbon fibers are available commercially at low cost and offer wide range and strengths and stiffnesses, their use in CMCs is limited due to the oxidation of these fibers at temperatures beyond 600°C in oxygen containing environments. Thus, oxidation resistant coatings may be necessary (similar to those in carbon/carbon composites).

Three major classes of matrices for CMCs include [24-30] glass and glass-ceramics, silicon-based ceramics (SiC and Si_3N_4) and oxide ceramics (Al_2O_3, ZrO_2). A wide variety of glass and glass-ceramics are available and the promising among these are listed in Table 3. Glass-ceramics have several advantages [26]: (1) their composition can be altered systematically by additions of a variety of components to produce matrices with tailored properties (CTE, compatibility with fibers, microstructure, etc.)., (2) they can be selectively doped to produce reaction and diffusion barriers at the fiber/matrix interface. (3) they can be processed at lower temperatures (than SiC and Si_3N_4) and pressures in the glassy phase and can be heat treated later to convert them to crystalline glass-ceramics with higher temperature capability. (4) they are oxidation resistant (5) their composites can be formed into shapes by viscous glass forming.

Important properties of oxide and silicon-based ceramic matrices are shown in Table 4. Monolithic SiC and Si_3N_4 have been the subject of intensive research in recent years with regards to developing engineering applications. These matrices have high strength, low CTE, good thermal shock resistance, and high oxidation resistance [27]. Si_3N_4 has significantly higher room temperature strength, lower CTE, and higher thermal shock resistance than SiC. But, SiC has several advantages over Si_3N_4: (1) SiC is not susceptible to slow crack growth, (2) SiC is less affected by impurities (3), SiC has much higher thermal conductivity and diffusivity, (4) SiC can retain strength at higher temperatures and is more oxidation resistant, and (5) SiC has very good creep strength.

3. Processing

Ceramic matrix composites can be fabricated by several methods such as chemical vapor infiltration (Figure 3), directed metal oxidation (Figure 4), slurry infiltration-hot pressing (Figure 5), sol-gel, polymer pyrolysis (Figure 6a and 6b) and reaction bonding. Some of these are derived from processes for polymer matrix composites (e. g. slurry infiltration-hot pressing), and some from processes for fabricating monolithic ceramics (e. g. reaction bonding). In general, due to high melting temperatures of ceramic materials fabrication has to be carried out at temperatures much higher than those used in processing of polymer matrix composites (PMCs). Fiber-matrix chemical compatibility (reaction at high temperatures) is important. Degradation of the reinforcement due to chemical and mechanical interactions is very much likely. Also, as emphasized in other sections, the conditions at interface after processing have to meet debonding and stress-transfer requirements. Thus, fabrication of CMCs involves lot more than consolidating the matrix and often a trade off must be made between matrix density, fiber strength retention and interface tailoring. In many cases reinforcement can be coated before incorporating into the matrix to provide a desirable interface.

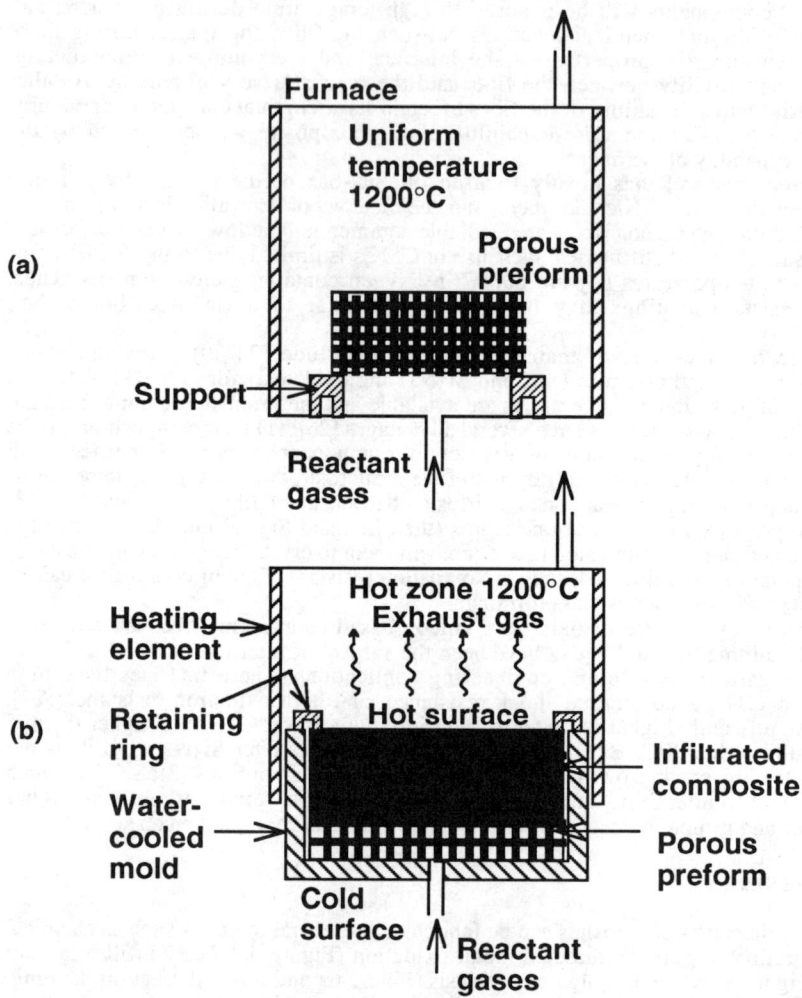

Fig. 3 Chemical vapor infiltration processes for fabriction of CMCs (a) isothermal (b) temperature and pressure gradient (forced) [31, 33].

3.1 Chemical Vapor Infiltration (CVI)

The CVI process [31-33] involves two phenomena: (i) mass transfer or penetration of vapor (gaseous reactants) into a porous preform (ii) chemical reactions to form and deposit the solid matrix on the fiber surface. As the matrix is deposited, the pores are filled. The chemical composition and purity of the matrix can be controlled by controlling the reactant gases and reaction conditions. SiC, Si_3N_4, Al_2O_3, ZrO_2, TiB_2 and TiC matrices can be deposited by this process although major development has been carried out for SiC matrix. For example, SiC matrix can be deposited by the following reaction:

$$CH_3SiCl_{3\,(g)} + \text{excess } H_{2\,(g)} \longrightarrow SiC_{(s)} + 3HCl_{(g)} + \text{excess } H_{2\,(g)}$$

Major advantages of the CVI process include ability to infiltrate complex shaped preforms to make near-net shape parts and ability to optimize interface through coatings deposited in the same reactor. Components such as turbine nozzle flaps, rotors, flame holders and combustors have been made of the SiC fiber fabric-SiC matrix system by CVI.

The CVI process can be isothermal/isobaric (Figure 3a) or may involve temperature and pressure gradients (forced CVI, Figure 3b)). In the isothermal/isobaric process the preform is set in a hot walled isothermal chamber through which the reacting gases flow. In this case mass transfer occurs mainly via diffusion. The deposition rates must be slow compared to the mass transfer rates to keep infiltration paths open and achieve uniform densification throughout the preform. As more and more matrix is deposited, deposition rates have to be reduced to prevent sealing of the infiltration paths. Depending on the processing conditions, a few hours to several weeks may be necessary to obtain 85-90% dense composites [31]. Typically, deposition temperatures range between 800 and 1100°C. This process can also be used to deposit interfacial coatings on the fibers before depositing the matrix.

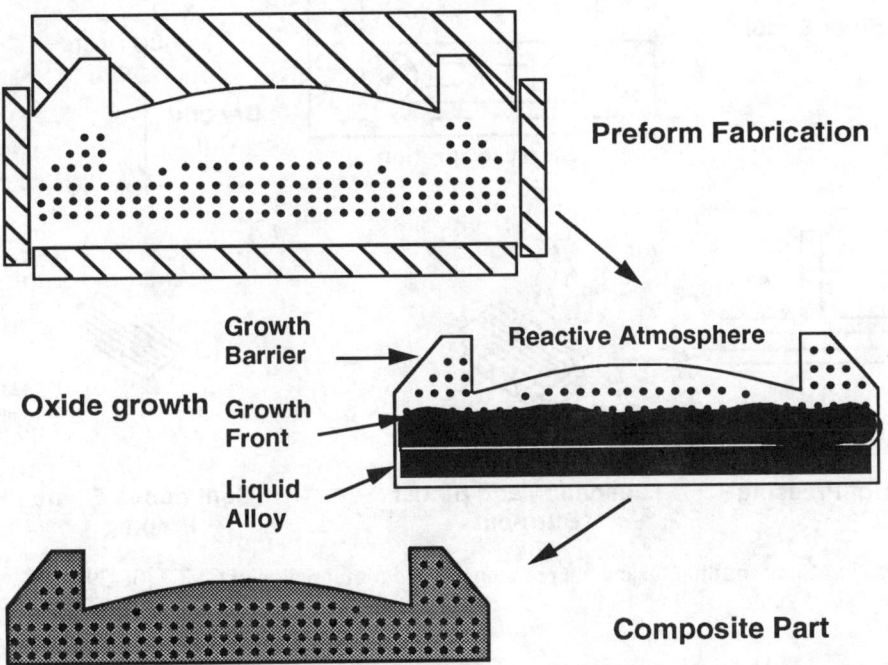

Fig. 4 Directed metal oxidation process for fabriction of CMCs [36, 37].

In the forced CVI process [33], reactants are forced under high pressure into one end of the preform maintained at low temperature. The other end of the preform is maintained at high temperature and the reaction products are suctioned out at the this end at low pressure. This prevents sealing of the entry point of the reactants and reduces the deposition times by an order of magnitude. The forced CVI however is less versatile than the isothermal CVI with respect to tooling and part geometries.

Tai and Chou [34, 35] have modeled the isothermal and forced CVI. In the isothermal CVI process model [34], pores in the fiber preform are simulated by cylindrical capillary tubes. The three components of the model are: diffusion of reacting gases into pore space, chemical reaction and film growth. Based on the model, optimum processing conditions for fabrication of SiC/Al_2O_3 composites could be predicted for different fiber preform geometries. Another model is developed to predict temperature and density distribution in the preform during forced CVI [35]. The model can also predict total time for the process and the vapor inlet pressure variations for maintaining constant flow rate of reactants.

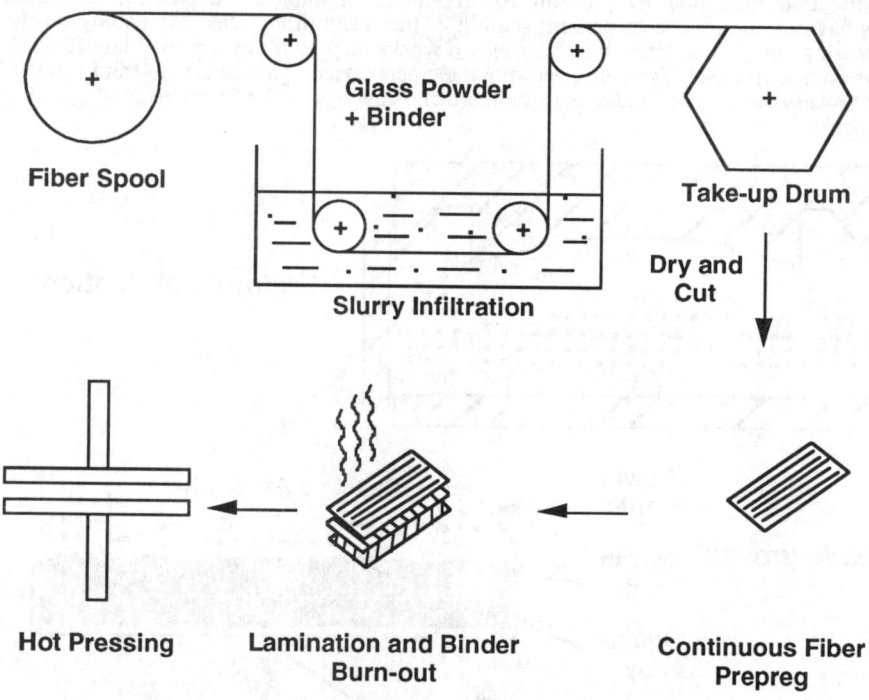

Fig. 5 Slurry infiltration and hot pressing method for fabriction of CMCs [38-39].

3.2 Directed Metal Oxidation

In the directed metal oxidation [36, 37] process (Figure 4) developed by Lanxide Corp., products of an oxidation reaction are directed to grow into a preform containing reinforcement. Reinforcements in the form of fibers, whiskers or particulates are first formed into a shaped preform. One end of the preform is then brought into contact with molten metal heated to appropriate temperature (900-1150°C for Al_2O_3). The gaseous reactants (oxygen, nitrogen etc.) are present on the other end of the preform. Alloying elements added to the melt (Si, Mg for Al) cause it to react with the gaseous reactants. The product of the reactions grows in the preform pore network away from the molten metal. The alloying additions also cause the metal to wick through the reaction product and reach the interface at which new reactions must occur. Thus, the process continues at an almost constant rate. A gas permeable growth barrier coating (e. g. calcium sulfate) is applied on the other end of the preform. This barrier allows penetration of gaseous reactants but terminates the growth of the reaction product maintaining the part shape. Although growth rates of the order of 1.5 inches/minute are achieved for Al_2O_3, slower rates are used to control the microstructure. A continuous interconnected matrix phase is formed in this process in the pore network of the preform. Some residual metal (3-15%) is also present in the final composite. Al_2O_3, AlN, TiN, ZrN, TiC, and ZrC matrix composites can be made by this process. The major advantages of the process are relatively low temperatures and versatility, and ability to make large, complex shapes, no shrinkage and high tolerances. Components made by this process include armor tile, heat exchanger tube, pump shaft sleeve, burner tube, desulphurization nozzle and wear plates (all SiC/Al_2O_3).

3.3 Slurry infiltration and hot pressing (viscous phase consolidation)

The slurry infiltration and hot-pressing method [31, 38, 39] is very similar to the method used for making polymer matrix composite prepregs and laminates. The first step in this process involves making a prepreg. This is done as shown in Figure 5. Fibers are spread over rollers and are passed through a slurry containing matrix powder, a solvent and a binder. These fibers are then wound on a plate or a drum to form a uniform layer. The layer on the drum is dried and cut to make handleable green tapes (prepregs). Prepreg layers are then stacked in desirable orientations in a mold. After burning out the binder (~600°C), the prepreg stack is hot pressed under inert atmosphere or vacuum to get composite laminate. The hot pressing temperature is selected so that the matrix powder becomes viscous and can flow and fill spaces between fibers under applied pressure. Typically, hot pressing temperatures are in the range 1200-1300°C, pressures are in the range 5-15 MPa and duration of processing is in the range of 15-30 minutes. High densification can be achieved in composites produced by this route leaving only a few percent porosity. This method is most suitable for glass and glass-ceramic matrices. It has been shown that through proper control of the process parameters, an in-situ carbon layer can be developed at the interface in the SiC fiber/glass systems resulting in tough, damage tolerant composites. The hot pressing temperatures required for more refractory matrices such as SiC, Al_2O_3 and Si_3N_4 are very high and cause degradation of the reinforcement. The glass-ceramic type of matrices can be formed by first creating a glass matrix at lower temperatures and then crystallizing it at a suitable higher temperature. This process is ideal for making flat plates or simple shapes (by forming techniques [39]) for materials development and fundamental characterization. Nicalon and carbon fiber reinforced borosilicate glass, calcium alumino silicate (CAS) glass-ceramic and lithium alumino silicate (LAS) glass-ceramic composites have been prepared by this method.

3.4 Sol-gel

A sol is a dispersion of fine solid particles (< 1 micron) in a liquid [40]. The particles remain suspended in the liquid indefinitely due to Brownian motion. A gel contains solid network structure with interspersed liquid. The sols can be aqueous-based or alcohol-based (using metal alkoxides). Fine particles in a sol have a very high surface area per unit volume compared to bulk materials. Due to fewer nearest neighbors, surface atoms have higher Gibbs free energy than those in the bulk. The particles therefore, have a tendency to agglomerate to reduce surface energy, and correspondingly, sols are thermodynamically unstable. Agglomeration can be reversible (flocculation) or irreversible (coagulation). For making solids by this method the gelation must occur in a controlled fashion. Therefore sols have to be kinetically stabilized by providing energy barriers to the agglomeration reactions by developing surface charge, adsorbed short chain polymers or solvation (hydration in aqueous systems). These barriers can be removed in a controlledfashion to allow gelation. This process can be used to make monolithic materials or matrices in composites.

To make a composite by this method [41-44], a fiber preform is first made. The sol is then forced into the preform under pressure. The gelation is then allowed to proceed by dehydration. The gels have to be calcined at higher temperatures (~600°C) to form a continuous oxide matrix. The following overall reaction shows formation of a metal (M) oxide matrix from a metal alkoxide $M(OR)_4$ by sol-gel

$$M(OR)_4 + 2H_2O = MO_2 + 4HOR$$

Significant shrinkage occurs during dehydration and calcination and the solid yield is low. The reinforcement constrains the shrinkage and causes cracking. Dehydration has to be carried out slowly to reduce cracking. Addition of solid particles to the sol (a combination of sol and slurry) can reduce the crack density and increase the solid content. Multiple cycles of infiltrations-gelation and calcination are necessary to overcome the low solid yield and to achieve significant densification. Even after multiple-step processing, the composite may have 20-25% porosity. Thus, a final hot pressing/sintering step (1200 to 1700°C) may be necessary to produce structurally viable composites. Examples of matrices that can be made by this method include SiO_2, ZrO_2, Al_2O_3, TiO_2, B_2O_3. Highly pure and homogeneous matrices or mixtures of oxide matrices can be produced by this method. Also, hot pressing or sintering temperatures for sol-gel derived composites are lowered than those required for composites made by the slurry infiltration technique described earlier. Sol can infiltrate a preform better than a slurry and mechanical damage to reinforcement by particles is minimized.

3.5 Polymer Pyrolysis

Polymer pyrolysis method [42, 45-47] depends on a pre-ceramic polymer which converts to a ceramic when pyrolyzed. For example, polycarbosilane and mithylvinylsilanes give SiC on pyrolysis and polysilazanes give Si_3N_4 on pyrolysis. The process is shown schematically in Figure 6a. The structures of various precursors are also shown in Figure 6b. First, the fibers are passed through a slurry containing the polymer, a solvent and a filler (powder of the desired matrix material) and are wound on a mandrel to make a prepreg. Layers of prepreg are stacked to create a desired laminate. Alternatively, a preform is made of the reinforcement (e. g. woven or braided) and the slurry is injected into it under pressure by techniques such as resin transfer molding or, the preform can be simply dipped in the slurry and the infiltration is aided by ultrasonic vibrations [47]. This step is followed by curing of the polymer (100-300°C). Then the composite is pyrolyzed at temperatures between 800-1200°C to convert the polymer into a ceramic. Significant gas evolution and shrinkage takes place during pyrolysis. Therefore, multiple cycles of infiltrations-curing-pyrolysis are necessary to obtain significant densification (~85%). A final hot pressing step may be performed to obtain fully dense composites [46]. Depending

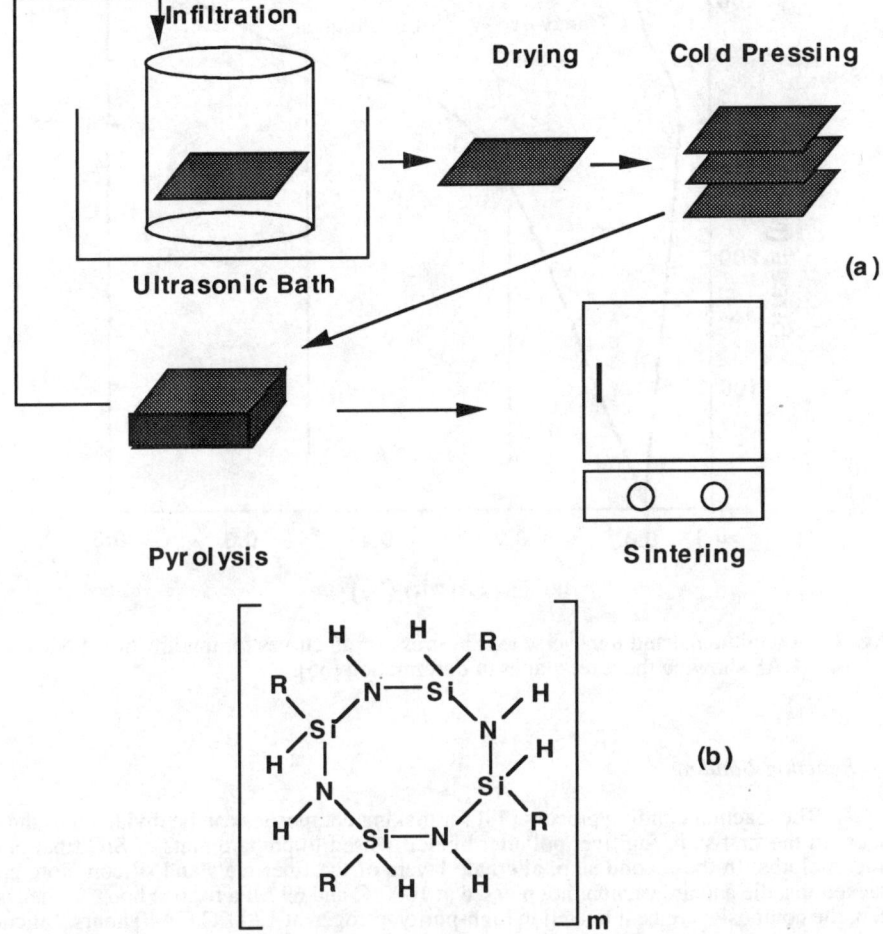

Fig. 6 (a) Polymer pyrolysis method for fabrication of CMCs [47] (b) structure of polysilazane preceramic polymer.

upon the pyrolysis temperatures, the products may be amorphous and may need to be crystallized by heat treatment at higher temperatures. The matrices derived by this method may not be stoichiometric. The pyrolysis atmosphere for polysilazane can be selected to produce Si_3N_4 (NH_3/N_2 atmosphere), or a mixture of SiC, C etc. (N_2 or Ar atmosphere) [46, 47]. Recently, Sato et al. have demonstrated that a carbon fiber reinforced Si-N matrix composite with 90-96% of theoretical density can be fabricated by using a perhydropolysilazane, as a pre-ceramic polymer [45].

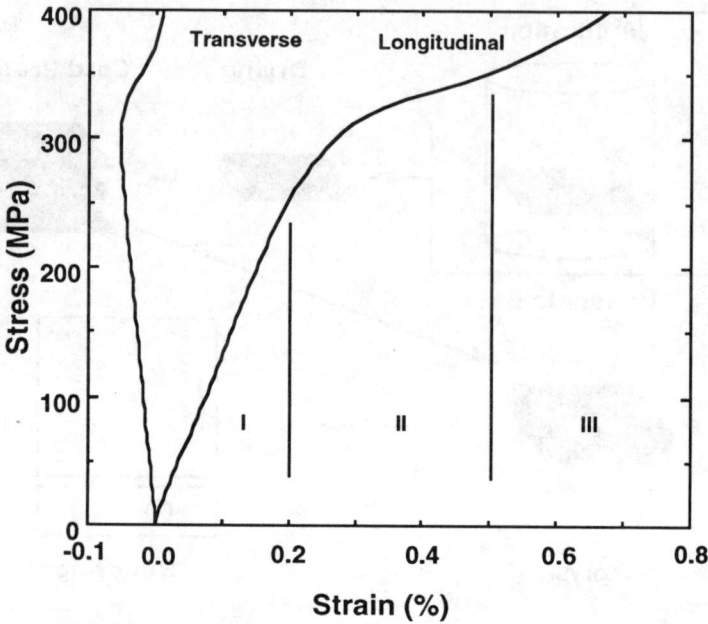

Fig. 7   Longitudinal and transverse tensile stress-strain curves for unidirectional Nicalon-CAS showing the three stages of deformation [55].

3.6 Reaction Bonding

The reaction bonding process [48] for making composites can be divided into three steps. In the first step, fugitive polymer binder is used to prepare mats of SiC fiber and silicon cloths. In the second step, alternate layers of the fiber mats and silicon cloth are stacked in a die and are vacuum hot pressed at 1000°C and 69 MPa for one hour. In the last step, the composite are heat treated in high-purity nitrogen at 1200°C for 40 hours. Silicon in the composite reacts with the nitrogen to form Si_3N_4 matrix. In the composite produced by this method, matrix has significant porosity (~30%). Eldridge et al. [49] have shown that the matrix can be fully densified by adding small amount of MgO to the silicon cloth and including a fourth step of hot pressing.

4. Static Mechanical Behavior at Room Temperature[*]

Behavior of ceramic matrix composites under different static loading conditions is outlined here using examples in the literature. The data available on tensile and flexural behavior are extensive. Data available on shear and compression are limited.

---

[*] Sections 4-8 were excerpted from Reference [2].

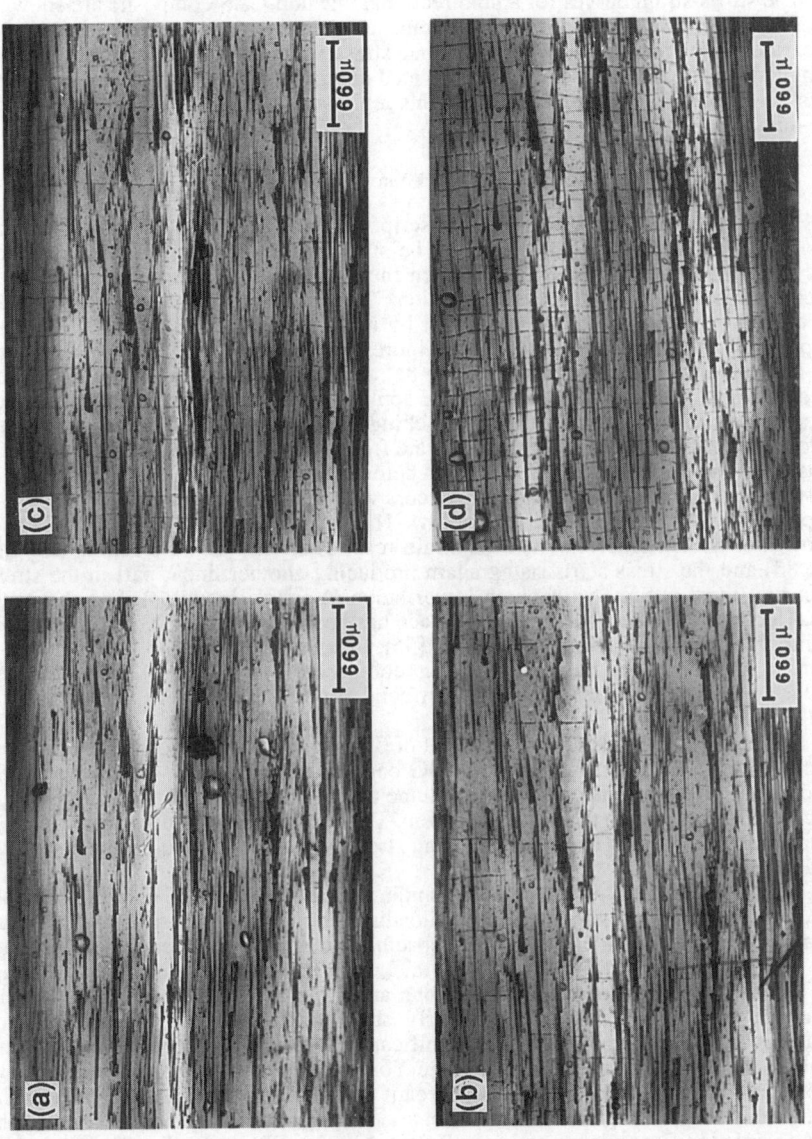

Fig. 8 Micrographs showing the sequence of cracking in unidirectional Nicalon-CAS composite at applied stress (MPa) or strain (%) levels (a) 157, 0.12 (b) 220, 0.17 (c) 283, 0.24 (d) 346, 0.53 [55].

4.1 Tensile

The stress-strain curves for a unidirectional Nicalon-CAS composite are shown in Figure 7. Both the longitudinal and the transverse strain responses are plotted. The nonlinearities in the longitudinal and transverse stress-strain curves are quite pronounced. The longitudinal stress-strain curve can be divided into three regions. In the first region, the curve is linear and the Young's modulus in this region can be predicted fairly well by the rule-of-mixtures.

$$E_1 = E_f V_f + E_m V_m \qquad (4)$$

where E denotes Young's modulus with subscripts 1, f and m denote composite (in fiber direction), fiber and matrix, respectively. The stress strain curve starts deviating from linearity at about 0.2% strain. The specimen then shows large strain increase without significant increase in the stress (second region). The process of damage development in unidirectional composites has been studied by many researcher [19, 50-55]. Damage development in Nicalon-CAS is shown in Figures 8a-d. Matrix cracking initiates in the matrix-rich regions of the composite (Figure 8a) at a longitudinal strain of 0.12%. These cracks do not span the entire thickness. As the applied strain increases, more cracks appear and the existing cracks grow. Although cracking can be detected at 0.12 % strain, the stress-longitudinal strain curve does not deviate from linearity until the strain reaches 0.2 %. Thus, cracking has to reach a critical level before the modulus is affected significantly. In the strain range 0.2 to 0.5 %, straining occurs without much increase in the stress and the slope of the stress-strain curve is very low. This region corresponds to multiplication and growth of matrix cracks. When the strain reaches around 0.5%, cracking saturates (Figure 8d) and the stress starts rising again producing another linear part in the stress-strain curve (third region) with much lower slope (Young's modulus) than the initial portion. Debonding at the fiber/matrix interface has been observed by X-ray radiography [51a] and transmission electron microscopy [56]. In some CMCs region II may not be as pronounced and clearly distinguishable (e. g. composite C in Figure 1). The additional strain due to matrix crack opening and interfacial slipping is probably generated more gradually in these systems.

Barsoum et al. [51b] studied the effect of fiber volume fraction on the matrix crack initiation stress in C-BSG and Nicalon-BSG composites. For the C-BSG system, the matrix cracking stress was independent of volume fraction up to volume fraction of 0.3 and then increased linearly with the volume fraction. The dependence was not as systematic in the Nicalon-BSG system. The matrix cracking stress was also found to decrease with the largest local fiber spacing.

The densities of matrix cracks in the unidirectional Nicalon-CAS composite and the corresponding reduction of the secant and unloading Young's moduli are shown in Figure 9 as functions of longitudinal strain (crack spacing can also be reported instead of crack density and for cracks extending through the thickness and width, it is the inverse of crack density). Similarly, Poisson's ratio reductions are superimposed on the crack density evolution in Figure 10. The graphs clearly show the correlation between moduli degradations and crack density. There is significant (30% for the unloading modulus and 55% for the secant modulus) reduction in the Young's modulus of the composite due to damage. The large difference between the secant modulus and the unloading modulus is due to the existence of a large permanent strain after damage development. The damage also leads to a very significant decrease in the Poisson's ratios. The unloading Poisson's ratio reduction is 55% and the instantaneous Poisson's ratio decreases more than 100%. The instantaneous Poisson's ratio becomes negative (-.025 to -0.05) at saturation damage as the transverse strain becomes positive. Nardone and Prewo [57] have attributed the positive transverse strain to the brooming due to fiber matrix debonding. From Figures 9 and 10 it becomes clear that the large strain produced in region II and the modulus decrease correspond to the onset and multiplication of matrix cracking and associated debonding (shown later in Section 4.1.2 with the help of a shear lag model). When the matrix cracking saturates, the stress-strain curve starts rising again but with much lower modulus as load is born mostly by fibers.

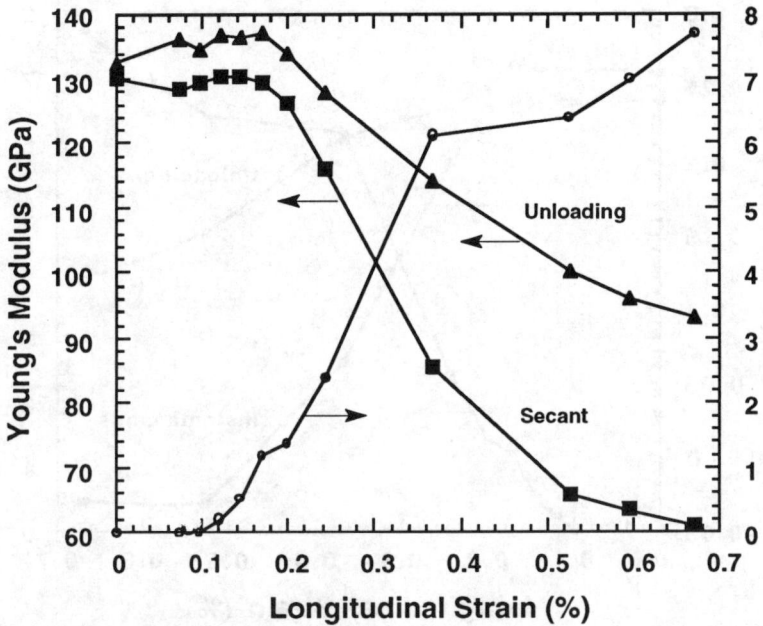

Fig. 9   Degradation of Young's modulus and evolution of crack density in unidirectional Nicalon-CAS as a function of applied strain [55].

Although, the failure strain of Nicalon fibers is of the order of 1.5%, Nicalon-based glass and glass-ceramic matrix composites fail at failure strains of about 0.8 to 1.0% [5, 52-55]. Prewo has shown that the strength of the fibers extracted from the Nicalon-LAS composites is lower than that of the virgin fibers [58]. Thus the lower failure strain of composites is at least partly due to degradation of the fibers during processing. This is explained in Section 3.1.4 based on strength theories.

Figure 11 shows the stress-strain behavior of HMU carbon fiber reinforced 7740 glass tested with the fibers at different angles to the loading direction [57]. It is clear that the composite loses its gradual failure mode at loading angles 10° and above. Loading at an angle of 10° to the fiber direction reduces the modulus from 169 to 140 GPa but the strength is reduced much more drastically from 785 to 171 MPa. Thus, ceramic composites cannot be used as unidirectional materials and it is necessary to manufacture laminates and composites with other 2-D and 3-D reinforcement architectures. The stress-strain behavior of HMU-7740 angle-ply laminates are also shown in Figure 11. Figures 12 and 13 show the variations of the composite elastic moduli and strengths, respectively as functions of angle for unidirectional and angle-ply composites. For both composites, modulus and strength decrease very rapidly with the loading angle. The strength decrease with angle is much more rapid under off-axis loading of unidirectional composites as compared to on-axis loading of the angle-ply laminate.

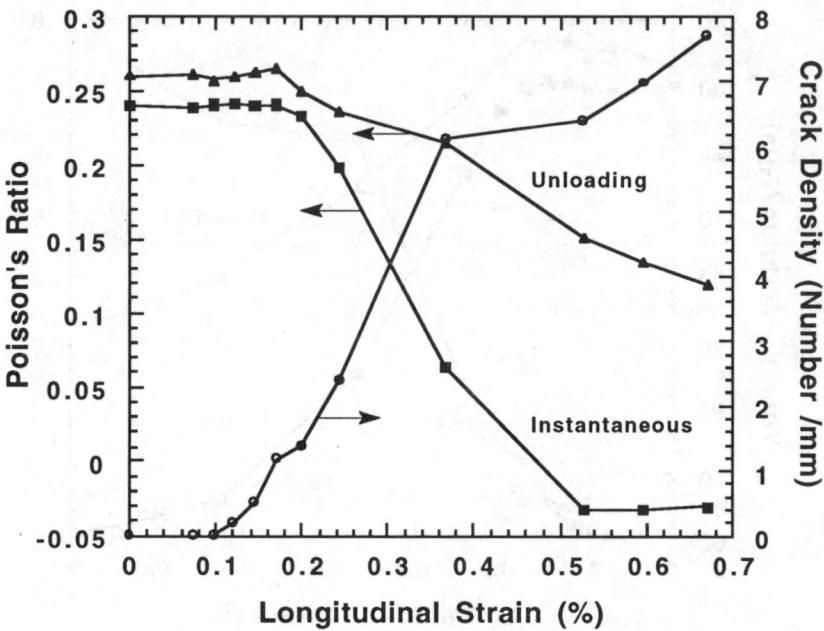

Fig. 10 Degradation of Poisson's ratio and evolution of crack density in unidirectional Nicalon-CAS as a function of applied strain [55].

The variation of modulus with the angle θ can be calculated by using the transformed reduced compliances (see, e. g. , Reference [59]) as

$$\frac{1}{E_\theta} = \frac{1}{E_1} \cos^4 \theta + \left(\frac{1}{G_{12}} - \frac{2\nu_{12}}{E_1} \right) \cos^2 \theta \sin^2 \theta + \frac{1}{E_2} \sin^4 \theta \qquad (5)$$

where E represents modulus, with subscripts 2 and θ indicate properties perpendicular to fiber direction and at an angle θ, respectively. G_{12} and ν_{12} denote the principal shear modulus and Poisson's ratio, respectively. Good fits between the experimental and predicted values were obtained for the unidirectional and angle-ply composites by using G_{12} values of 22 and 11 MPa, respectively (Figure 12). The following Tsai-Hill criterion [60] was used to predict the variation of the strength with angle for unidirectional composites.

$$\frac{1}{\sigma^2} = \frac{\cos^4\theta}{\sigma_1^2} + \left(\frac{1}{\tau^2} - \frac{1}{\sigma_1^2}\right)\cos^2\theta\sin^2\theta + \frac{\sin^4\theta}{\sigma_2^2} \qquad (6)$$

where σ_1 and σ_2 denote strengths of the unidirectional composite parallel and perpendicular to the fiber direction; σ denotes strength of the unidirectional composite in direction θ and τ is the lesser of the fiber-matrix interfacial shear strength or matrix shear strength. The following Tsai-Hill criterion [60] was used to predict the variation of the strength with angle for the angle-ply composites.

$$\frac{(\sigma_x^2 - \sigma_x\sigma_y)}{\sigma_1^2} + \frac{\sigma_y^2}{\sigma_2^2} + \frac{\sigma_s^2}{\tau^2} = 1 \qquad (7)$$

where σ_x, σ_y and σ_s denote stresses parallel to fibers, perpendicular to fibers and shear. Good predictions were obtained with $\tau = 32$ MPa for both unidirectional and angle-ply composites.

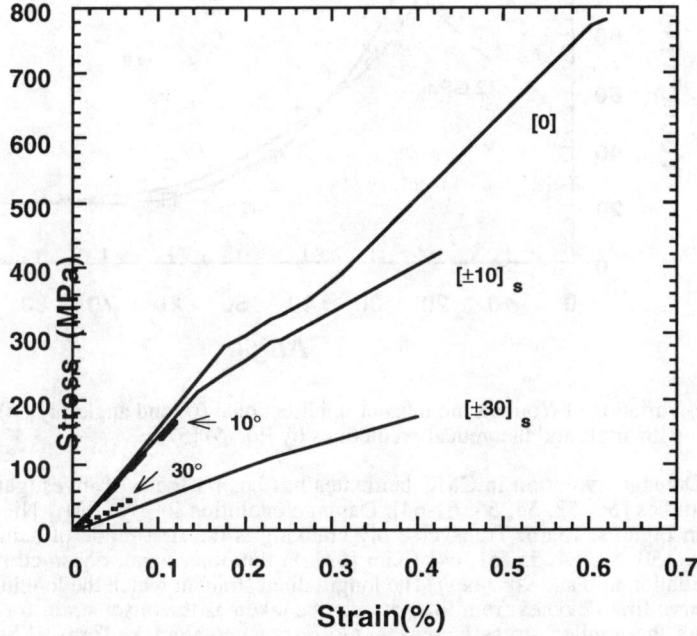

Fig. 11 Stress-strain curves for unidirectional HMU carbon-BSG composites loaded at various angle and HMU carbon-BSG angle-ply laminates [57].

The stress-strain behavior of unidirectional and cross-ply [55] Nicalon-CAS composites are shown in Figure 14. For stress-longitudinal strain curves of cross-ply laminates, the first deviation from linearity occurs at much lower longitudinal strains than for the unidirectional composite (0.2%). The average values (from 4 specimens) of longitudinal strains at which this deviation occurs are as follows: 0.027% for the [0$_3$/90$_3$/0$_3$] laminate, 0.035% for the [0/90/0/90/0/90/0/90/0] laminate, and 0.056% for the [0$_3$/90/0$_3$] laminate. Unlike the unidirectional composite, the first deviation leads to only kinking of the curves. The second deviation occurs at a longitudinal strain of 0.15 to 0.17% (for all cross-ply laminates) and leads to a very low slope region followed by a region with upward convexity.

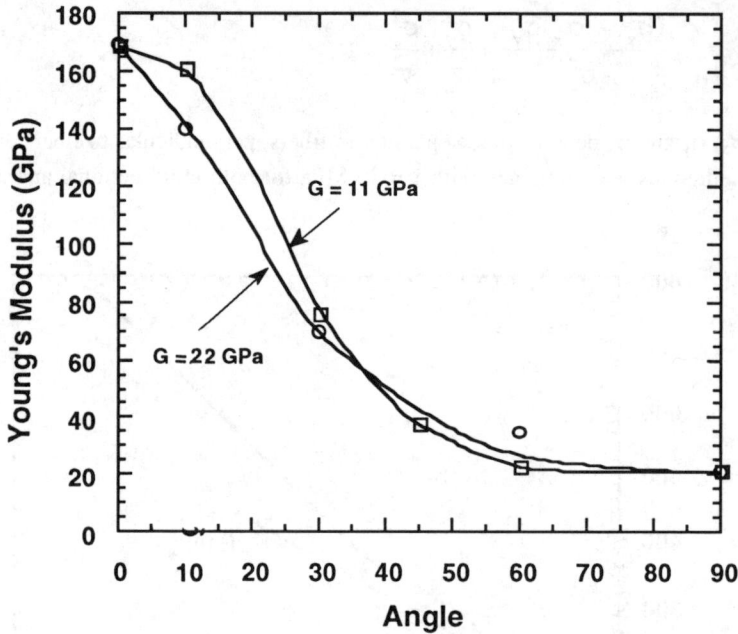

Fig. 12 Variation of Young's modulus of unidirectional (o) and angle-ply (□) HMU-BSG with angle and theoretical predictions by Eq. (5) [57].

Damage evolution in CMC laminates has been a focus of investigation in some recent studies [50, 52, 54, 55, 61-64]. Damage evolution in [0$_3$/90/0$_3$] Nicalon-CAS is shown in Figures. 15a-d. Transverse ply cracking is the first mode of damage in these laminates [50, 52, 54, 55, 61, 64] (Kim [63] on the other hand, observed simultaneous crack initiation in 0 and 90° plies). The longitudinal strain at which the longitudinal stress-strain curve first deviates from linearity can be taken as the onset strain for this damage mode. As the applied stress increases, more transverse cracks form. When the strain reaches 0.12%, some of the transverse ply cracks extend into the 0° plies as matrix cracks. With further loading, more matrix cracks appear in the 0° plies between the initial cracks. Apart from transverse cracks, a limited number of longitudinal splits connecting two adjacent transverse cracks are also found in middle of the 90° plies. For the Nicalon-CAS cross-ply laminates in Reference 17, longitudinal splitting in the 90° plies was found to be more extensive. No evidence was found for the occurrence of longitudinal splits in the 0° plies of the cross-ply laminates.

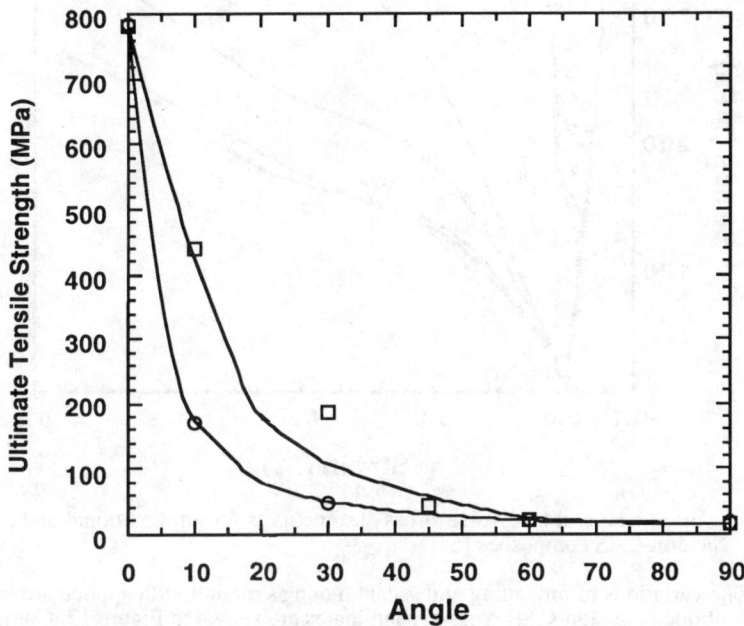

Fig. 13 Variation of ultimate tensile strength of unidirectional (o) and angle-ply (□) HMU-BSG with angle and theoretical predictions by Eqs. (6, 7) [57].

In Nicalon-CAS cross-ply composites transverse crack initiation strain decreases as the thickness of the 90° layer increases [52, 54, 55]. The densities of the matrix cracks and transverse cracks in Nicalon-CAS cross-ply laminates are shown in Figure 16 along with the densities of the matrix cracks in the unidirectional composite as functions of longitudinal strain. By comparing the initiation strains, evolutions, and saturation densities of the matrix cracks in the unidirectional composite and the matrix cracks in the 0° plies of the cross-ply laminates, it can be concluded that the presence of the transverse cracks in the cross-ply laminates has only a limited effect on the matrix cracks. The matrix cracks initiate at higher strains than the transverse ply cracks. The saturation density of the matrix cracks is higher than the saturation density of the transverse ply cracks in all cross-ply laminates. The saturation transverse crack density decreases as the thickness of the transverse layer increases [52, 54, 55]. The dependence of the initiation strain and the saturation crack density on the 90° layer thickness is similar to that observed in cross-ply polymer matrix composites and is explained on the basis of the constraint effect of the 0° plies [65]. Although reducing the thickness of the 90° layer postpones the initiation of transverse cracks, it leads to higher saturation transverse crack density. This trade-off must be considered when designing laminates.

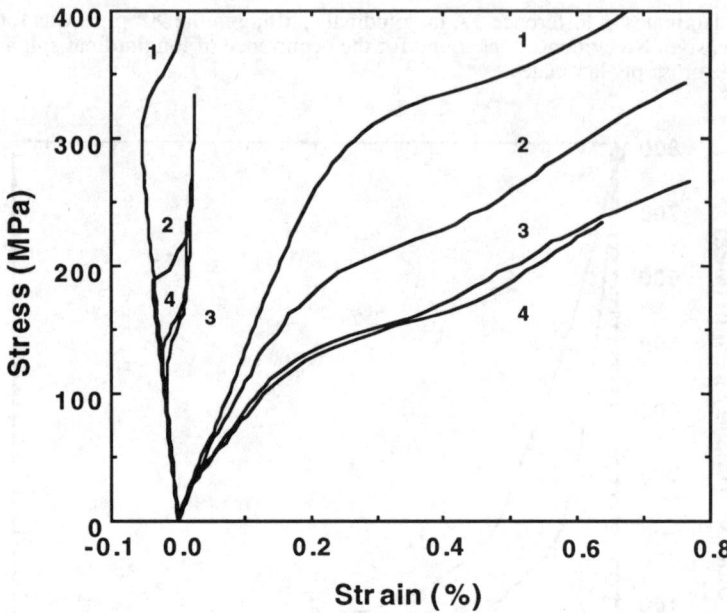

Fig. 14 Longitudinal and transverse stress-strain curves for unidirectional and cross-ply Nicalon-CAS composites [55].

The variations of unloading and secant Young's moduli with applied strain for the above mentioned Nicalon-CAS cross-ply laminates are shown in Figure 17. Compared to unidirectional composite, the modulus changes in cross-ply laminates are even more significant (45 to 55% for the unloading moduli and 60% to 69% for the secant moduli) due to the additional damage mode viz. transverse ply cracking. Experimentally determined Poisson's ratios (average from the two specimen faces) for the cross-ply laminates as functions of applied strain are shown in Figure 18. Again, the damage leads to a very significant decrease in the Poisson's ratios. The unloading Poisson's ratio reduction is 66% for the $[0_3/90/0_3]$ laminate, and 76% for the $[0_3/90_3/0_3]$ and the [0/90/0/90/0/90/0/90/0] laminates. The instantaneous Poisson's ratio reduces by over 100% for the cross-ply laminates. Thus, the percentage reduction of the major Poisson's ratio is higher than the percentage reduction of the longitudinal Young's modulus for all four configurations.

Tensile stress-strain behaviors of two-dimensional (2-D) woven, three-dimensional (3-D) two-step braided and 3-D four-step braided SiC-SiC composites are shown in Figure 19. Braided and woven composites are generally fabricated by depositing the matrix in the fiber preform using the chemical vapor infiltration (CVI) process. These composites typically have much higher percentage of porosity compared to laminated composites fabricated by hot pressing or sintering. Two main types of porosities are observed in these materials. Porosity occurring between two yarns is called interyarn porosity. Porosity occurring between individual fibers within a yarn is called intrayarn porosity. The interyarn pores are usually much bigger than the intrayarn pores and exhibit sharp angles. The size and shape distribution of porosity (particularly, interyarn) is dependent on preform geometry as the channels available for infiltration depend on the preform geometry. In the case of the 3D two-step composites, the fiber structure induces longitudinal pores spread

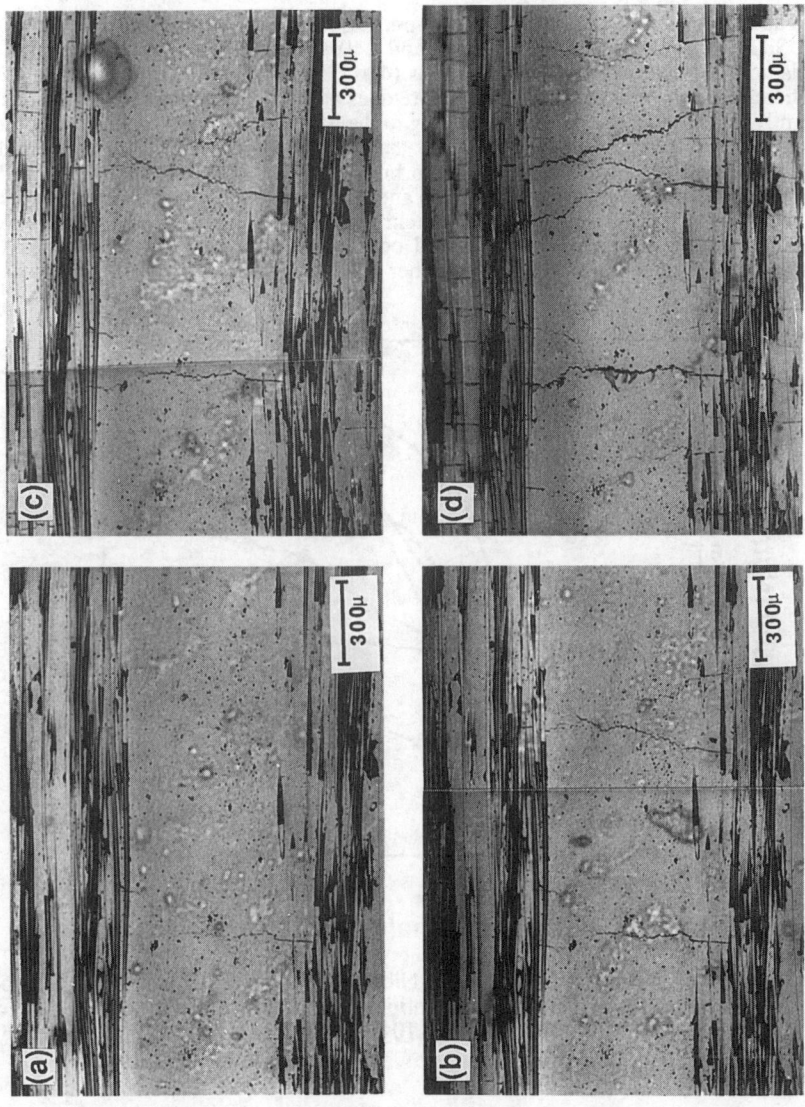

Fig. 15 Micrographs showing the sequence of cracking in [0₃/90₃/0₃] laminate at applied stress (MPa) or strain (%) levels (a) 68, 0.09 (b) 113, 0.16 (c) 141, 0.22 (d) 214, 0.55 [55].

along fibers and yarns. In contrast, in the 2D woven and the 3D four-step braided composites, the spaces created by the crossing of the yarns give rise to a more complex-shaped intrayarn porosity with sharp angles.

The Young's moduli for the 2-D woven, 3-D two-step braided and 3-D four-step braided SiC/SiC composites are 230, 260 and 240 GPa respectively. This composite system is unique because the matrix modulus (400 GPa) is twice that of the fibers (200 GPa). Therefore change in the fiber architecture does not affect the composites modulus of significantly. The yield strength, corresponding to the beginning of the non-linear part the curves, was of the order of 60 and 70 MPa for the 2D woven and the 3D four-step braided composites and 90 MPa for the 3D two-step braided composite. The sharp angles that characterize the interyarn pores in 2-D woven and 3-D four-step braided composites result in high local stresses which lead to matrix cracking and yielding at a lower applied stress. The higher density of the two-step braided composite compared with the two other composites could also partially explain the higher yield strength.

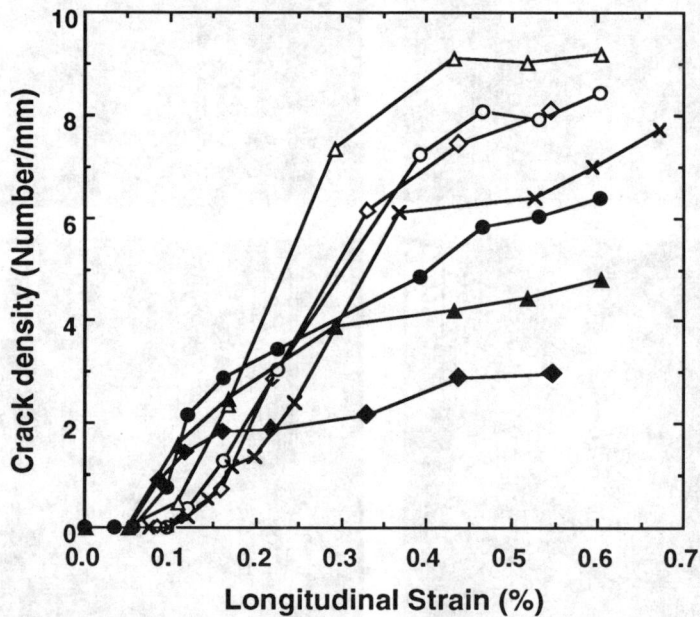

Fig. 16 Evolution of crack density for the Nicalon-CAS composites in Fig. 14 (open symbols indicate 0° ply cracks, solid symbols indicate 90° ply cracks; × - unidirectional, ○ - [0$_3$/90/0$_3$] , ◇ - [0$_3$/90$_3$/0$_3$] , ▲ - [0/90/0/90/0/90/0/90/0]) [55].

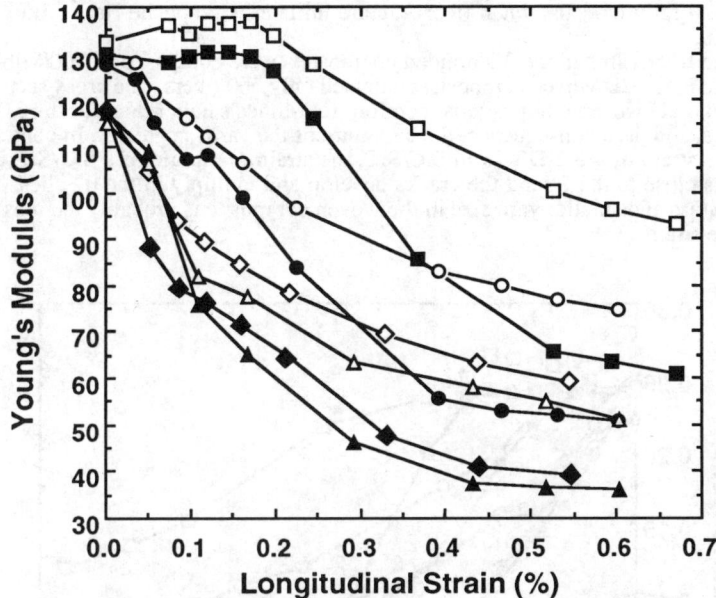

Fig 17 Variation of the longitudinal Young's modulus with applied strain for the Nicalon-CAS composites in Fig. 14 (open symbols indicate unloading modulus, solid symbols indicate secant modulus; □ - unidirectional, ○ - [0$_3$/90/0$_3$] laminate, ◇ - [0$_3$/90$_3$/0$_3$] laminate, △ - [0/90/0/90/0/90/0/90/0] laminate) [55].

The sequence of damage development in 2-D woven, 3-D two-step braided and 3-D four-step braided SiC-SiC composites is shown schematically in Figures 20a,b and c. In woven composites, cracks perpendicular to the applied load initiate from the sharp corners of the interyarn pores. These cracks then propagate into the transverse yarns. With increasing stress more transverse cracks appear in the transverse bundles. On further loading, these cracks grow into the longitudinal bundles leading to fiber fracture and composite ultimate failure. Unlike Nicalon-CAS cross-ply laminates described earlier, crack density in the longitudinal yarns is not higher than that in the transverse yarns. Composite final failure strain is only 0.2%.

In 3-D two-step braided composites, first cracking also initiates from interyarn pores. These cracks are arrested by the longitudinal yarns. Transverse cracking then develops within the transverse yarns (which are the braider yarns). At higher loads, the transverse crack in the transverse bundles and the interyarn regions grow into the longitudinal yarns. The longitudinal yarn cracking multiplies. The saturation density of longitudinal yarn cracks is higher than the saturation density of the transverse yarn cracks. Final failure occurs by fracture of the fibers in the longitudinal yarns. The composite shows final failure strain of 0.5%.

In 3-D four-step braided composites again, cracking initiates from the interyarn pores. These cracks propagate into the braiding yarns (which are at an angle to the loading axis). The cracking in the braiding yarns is found to be perpendicular to the fiber direction within the yarn rather than the loading directions. Cracking in the braiding yarn multiplies and the final failure occurs when fiber fracture initiates. Composite final failure strain is again 0.5%.

The fiber yarns in the 3D braided composites were composed of 2000 fibers each, while those in the 2D woven composite contained only 500 fibers. The cross-section of the yarns for the 2D woven composite is therefore four times smaller than for the 3D braided composites, and, as a consequence, the curvature of the yarns is greater. In contrast to the low failure strain of the 2-D woven SiC/SiC, the strain to rupture of a SiC/SiC cross-ply laminate is close to 0.5%, and the cracks develop and multiply in the 0° plies. The high yarn curvature and smaller yarn size in the woven composite is probably the reason for its low failure strain.

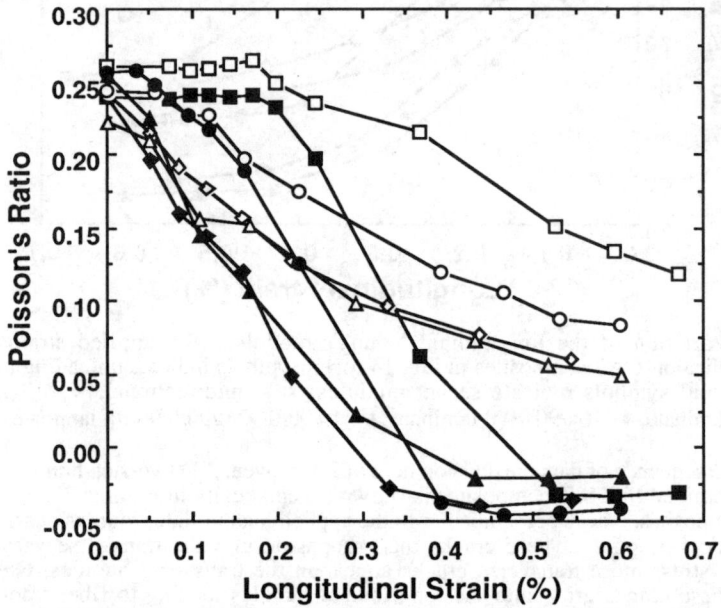

Fig 18 Variation of the major Poisson's ratio with applied strain for the Nicalon-CAS composites in Fig. 14 (open symbols indicate unloading values, filled symbols indicate instantaneous values; □ - unidirectional, ○ - [0₃/90/0₃] laminate, ◆ - [0₃/90₃/0₃] laminate, ▲ - [0/90/0/90/0/90/0/90/0] laminate) [55].

The response of the transverse strain versus stress was linear elastic for the 2D woven and the 3D two-step braided composites and non-linear for the 3-D four-step braided composite. The explanation for this lies in the fact that the cracks in the first two composites developed perpendicular to the loading direction while for the 3D four-step braided composite they were oriented at an angle to the loading direction and perpendicular to the direction of the fiber in the yarn. The cracks at an angle to the loading direction give rise to a component of the strain in the transverse direction.

The evolution of cracking in woven SiC/SiC composites was studied by Chen [67]. Two materials, one with standard density and one with higher density (lesser porosity) were studied. The composite failure strains were low (0.14 and 0.2% for high density and standard density composites respectively).The evolution of the inter-yarn cracking, intra-yarn cracking, longitudinal yarn cracking and number of fiber breaks in the longitudinal yarn were documented. The inter-yarn crack density was found to be higher in the high density material. The intra-yarn (90° yarn) crack density was same in standard and high density composites reaching a level of 2 cracks/mm. The cracking in the 0° yarns was dependent on composite density. The density of 0° yarn cracking was 0.5 and 2.0 cracks/mm respectively, for the standard and high density composites. None of the three crack systems showed saturation at composite ultimate failure. At a comparable stress level, the number of fiber breaks observed in a 45 mm gage length were higher in the higher density material.

Although the strengths of the laminated, woven and braided composites are not as high as that of unidirectional composite, they have reasonable strength in more than one direction and therefore are more useful in the applications where multidirectional loads are involved. The woven composites are made from fabric and therefore can be made to drape complex surfaces like missile nose cones or radomes. The 3-D braided composites can be made in near-net shape and in complex shape that cannot be made by lamination.

Significant amount of data on tensile behavior of various CMC systems are available in the literature. These data are listed in Table 5.

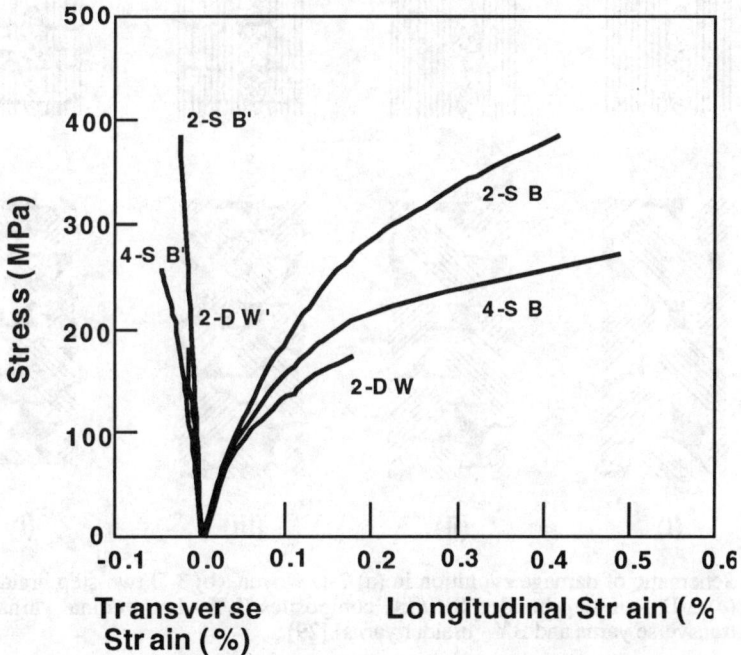

Fig. 19 Longitudinal and transverse tensile stress-strain curves for 2-D woven (2-D W), 3-D two-step braided (2-S B) and 3-D four-step braided (4-S B) SiC/SiC composites [66].

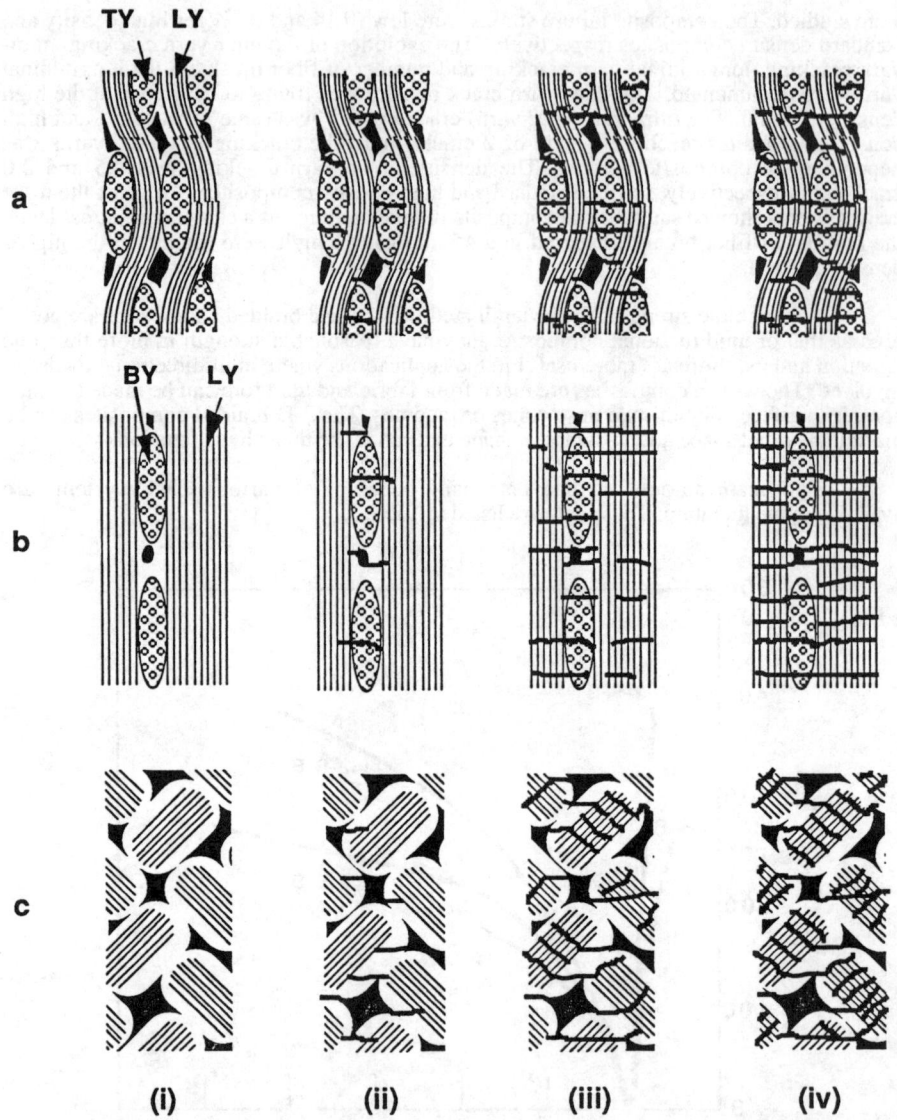

Fig. 20 Schematic of damage evolution in (a) 2-D woven, (b) 3-D two-step braided and (c) 3-D four-step braided SiC/SiC composites (LY - longitudinal yarns, TY - transverse yarns and BY - braider yarns) [29].

4.2 Compressive

Contrary to tensile loading, ceramics are not flaw sensitive under compressive loading as the applied load closes the flaws. Typically the compressive strength of monolithic ceramics is much higher than their tensile strength. Compressive and tensile stress-strain curves for unidirectional and cross-ply Nicalon-CAS composites are shown Figure 21 [68]. The longitudinal compressive stress-strain curves for the unidirectional composite and cross-ply laminate do not show any non-linearity as in the case of the corresponding longitudinal tensile stress-strain curves. The non-linearity in the tensile curve is associated with the creation and opening of the multiple matrix (and transverse) cracks. These damage mode do not arise under compressive loading. The existing flaws will be closed under compressive loading rather than being converted into cracks. The Young's moduli for the four configurations under compressive loading were 25-50% higher than those obtained under tensile loading.

The unidirectional composite shows a compressive strength of 1.4 GPa, which is almost three and a half times the tensile strength (400 MPa). The laminated composites show compressive strength twice that of their tensile strength. In polymer matrix composites, the axial compressive strength of unidirectional composite is much lower due to the low shear stiffness and yield strength of the polymer matrix. The ceramic matrices on the other hand, do not show yielding and have very high shear stiffness. As a result, the matrix dominated compressive strength is very high. Also, unlike fiber tensile strength, the fiber compressive strength may not deteriorate (due to flaw insensitivity) during processing of CMCs resulting in the high compressive strength of the composite.

Effect of strain rate on the compressive behavior of unidirectional and cross-ply Nicalon-LAS-II composites was investigated by Lankford [69]. It was found that the compressive strength of unidirectional and cross-ply composites increased dramatically above strain rate of 2 S^{-1} but compressive strength of $[\pm 45]_s$ laminate was not significantly affected by strain rate. The 0/90 and $[\pm 45]_s$ laminates show much lower compressive strength than the unidirectional composite because matrix cracking can be initiated in the laminates at lower stresses (by indentation of the matrix by the fiber) allowing fiber buckling and fiber tow kinking.

Wang et al. [70] on the other hand, have reported that in 2D cross-weave carbon-SiC composites, the compressive strength was lower than the tensile strength. This effect may be related to the inherent high porosity in composites processed by the chemical vapor infiltration technique. Compressive failure occurred by buckling of fiber bundles and matrix fragmentation. Low compressive strengths have been reported for a variety of 2D carbon-ceramic composites made by CVI by Rossignol et al.[71]. For these composites, the elastic modulus was found to increase as the porosity content decreased.

Phillippe et al. [66] reported compressive behavior of 2-D woven, 3-D two-step braided and 3-D four-step braided SiC-SiC composites. The compressive stress-strain curves are shown in Figure 22. (corresponding tensile stress-strain curves for these were discussed in the Sec. 2.1 and are shown in Figure 19). The compressive stress-strain curves for all the three composites were linear to failure and the woven composite shows highest compressive strength.

Table 5a. Tensile properties.

Composite	$\alpha_f-\alpha_m$	V_f	E_{11}	UTS	ν_{12}	$\sigma_{MC}(0)$	$\sigma_{PL}(0)$	$\sigma_{TC}(90)$	S_0	S_{90}	Ref.
C-pyrex [0]	-3.2,4.8	0.3	121	413(0.5)	0.23	(0.03)	--	--	0.3	--	50
[0/90]$_{3s}$	-3.2,4.8	0.31	75	320(0.6)	-0.04	--	--	--	--	--	50
SiC-Pyrex[0]	-0.1	0.34	95	726(1.29)	0.14	--	--	--	0.063	--	50
[0/90]$_{3s}$	-0.1	0.34	69	343(0.8)	0.06	--	--	--	--	--	50
C-7740 [0]	-4.0	0.42	190	530	--	123(0.07)	350(0.2)	--	--	--	51
HMU-BSG[0]	--	0.43	169	785(0.6)	0.18	--	290	--	--	--	57
[±10]$_s$	--	0.42	160	441(0.41)	0.28	--	185	--	--	--	57
[±30]$_s$	--	0.42	76	187(0.51)	0.90	--	15	--	--	--	57
[±45]$_s$	--	0.43	38	44(0.38)	0.67	--	5.9	--	--	--	57
[±60]$_s$	--	0.42	23	24(0.51)	0.28	--	3.4	--	--	--	57
[0/90]$_s$	--	0.40	82	300(0.48)	0.017	--	43	--	--	--	57
SCS6-7761[0]	1	0.4	175	436	--	117	--	--	--	--	19
SCS6-7740[0]	0.1	0.4	172	293	--	143	--	--	--	--	19
SCS6-9741[0]	-1.3	0.4	131	223	--	--	--	--	--	--	19
SCS6-7052[0]	-1.6	0.4	156	293	--	40	--	--	--	--	19

$\alpha_f-\alpha_m$ - difference between fiber and matrix CTEs, V_f - fiber volume fraction, E_{11} - Young's modulus (GPa), UTS - ultimate tensile strength (MPa), ν_{12} - principal Poisson's ratio, $\sigma_{MC}(0)$ - matrix cracking stress [MPa] (strain [%]) for 0° ply, $\sigma_{PL}(0)$ - proportional limit stress [MPa] or strain (%), $\sigma_{TC}(90)$ - transverse cracking stress [MPa] (strain [%]) for 90° ply, S_0 and S_{90} - saturation crack spacing in 0° and 90° plies, respectively (mm)

Table 5b. Tensile properties.

Composite	$\alpha_f - \alpha_m$	V_f	E_{11}	UTS	ν_{12}	$\sigma_{MC}(0)$	$\sigma_{PL}(0)$	$\sigma_{TC}(90)$	S_0	S_{90}	Ref.
Nic-LAS I [0]	--	0.46	133	455(0.33)	--	--	--	--	--	--	58
Nic-LAS II [0]	--	0.46	134	758(0.97)	--	--	381(0.28)	--	--	--	58
Nic-LAS IIC [0]	--	0.46	130	664(0.86)	--	--	439(0.34)	--	--	--	58
Nic-LAS II [0]	--	0.46	128	670(0.90)	--	--	345(0.27)	--	--	--	58
Nic-LAS IIC [0]	--	0.46	136	680(1.03)	--	--	391(0.29)	--	--	--	58
Nic-LAS IIIC[0]	3	0.5	120	530	--	290	--	--	0.4	--	56
[0/90]$_s$	3	0.5	110	276	--	145	--	--	--	--	56
Nic-LAS III [0]	0	0.5	129	557	--	238	--	--	0.12	--	56
Nic-LAS III[0]	2.7	0.39	130	510	0.21	0.15	--	--	--	--	63
[90]	2.7	0.39	20	5.8(0.04)	--	--	--	--	--	--	63
[0/90]$_s$	2.7	0.39	86	216	--	--	--	0.08	--	--	63
Nic-LAS [0]	--	0.40	124	570(0.75)	--	--	372	--	0.3	--	136b
Nic-LAS III [0]	2.4	0.39	132	700	--	195(0.15)	306(0.73)	--	--	--	53
Nic-1723A [0]	-2	0.58	153	--	--	201(0.13)	480(0.31)	--	--	--	53
Nic-1723B [0]	-2	0.43	136	390	--	161(0.11)	286(0.2)	--	--	--	53
Nic-CAS [0]	-1.1	0.4	133	--	--	132(0.1)	211(0.17)	--	--	--	53
Nic-BMAS [0]	1.2	0.43	146	500	--	214(0.14)	254(0.17)	--	--	--	53
Nic-AS [0]	1	0.50	138	555	--	240	--	--	0.09	--	56
Nic-sodalime[0]	-5	0.50	120	348	--	< 0	--	--	0.07	--	56
Nic-SiO$_2$ [0]	4	0.50	110	66	--	--	--	--	--	--	56

Table 5c. Tensile properties.

Composite	α_f-α_m	V_f	E_{11}	UTS	ν_{12}	$\sigma_{MC}(0)$	$\sigma_{PL}(0)$	$\sigma_{TC}(90)$	S_0	S_{90}	Ref.
Nic-CAS [0]	-1.3	0.34	128	400(0.78)	0.24	(0.12)	--	--	0.14	--	52
[0/90]s	-1.3	0.34	120	173(0.63)	0.24	(0.10)	--	(0.03)	0.15	0.9	52
[0₂/90₄]s	-1.3	0.34	101	107(0.65)	0.20	(0.14)	--	(0.05)	0.16	0.25	52
[0/90₃]s	-1.3	0.34	110	146(0.63)	0.22	(.08)	--	(0.05)	0.18	0.25	52
Nic-CAS [0]	-1.0	0.35	131	344(0.75)	--	(0.13)	--	--	0.12	--	54
[0₃/90/0₃]	-1.0	0.35	129	332(0.9)	--	(0.13)	--	(0.05)	0.11	0.28	54
[0₃/90₂/0₃]	-1.0	0.35	128	253(0.77)	--	(0.13)	--	(0.025)	0.11	0.70	54
[0₃/90₃/0₃]	-1.0	0.35	126	169(0.4)	--	(0.13)	--	(0.012)	0.11	0.80	54
Nic-CAS [0]	-1.0	0.35	130	398 (.68)	0.24	(0.12)	(0.2)	--	0.13	--	55
[0₃/90/0₃]	-1.0	0.35	128	344(.76)	0.24	(0.12)	(o.17)	(0.056)	0.12	0.16	55
[(0₂/90₂)2/0]s	-1.0	0.35	118	267(.77)	0.22	(0.12)	(0.17)	(0.035)	0.11	0.21	55
[0₃/90₃/0₃]	-1.0	0.35	118	235(.64)	0.24	(0.12)	(0.15)	(0.027)	0.12	0.34	55
Nic-CAS [0]	--	0.40	139	266	0.25	--	--	--	--	--	62
[0/±45/90]s	--	0.40	115	126	0.23	--	--	38(.034)	--	--	62
[0/90]2s	--	0.40	122	121	0.22	--	--	35(.028)	--	--	62
[±45]2s	--	0.40	105	93	0.25	--	--	38(.037)	--	--	62
Nic-CAS [0]	-0.8	0.40	134	420	0.25	(0.1)	--	--	--	--	63
[90]	-0.8	0.40	124	35(0.029)	--	--	--	--	--	--	63
[0/90]s	-0.8	0.40	127	205	--	--	--	(0.032)	--	--	63
[0/±45/90]s	-0.8	0.40	115	--	--	--	--	--	--	--	63

Ceramic Matrix Composites 281

Table 5d. Tensile properties.

Composite	α_f-α_m	V_f	E_{11}	UTS	ν_{12}	$\sigma_{MC}(0)$	$\sigma_{PL}(0)$	$\sigma_{TC}(90)$	S_0	S_{90}	Ref.
Nic-CAS [0]	--	0.37	124	334(0.7)	0.24	--	--	--	--	--	68
[±45/0]2s	--	0.37	105	283(0.9)	0.17	--	--	--	--	--	68
[0/90₃]s	--	0.37	89	210(0.9)	0.16	--	--	--	--	--	68
[±45]3s	--	0.37	81	109(0.1)	0.13	--	--	--	--	--	68
Nic-CAS [0]	-1.9	0.39	121	393(0.84)	0.18	100-180	--	--	0.036	--	117
[90]	-1.9	0.39	112	55(0.05)	--	--	--	--	--	--	117
SiC/SiC [0] A	--	0.45	270	535(0.43)	--	--	295	--	--	--	141
SiC/SiC [0] B	--	0.45	245	479(0.40)	--	--	200	--	--	--	141
SiC/SiC (2D-w)	--	0.43	211	207(0.2)	0.17	--	84	--		--	67
SiC/SiC (2D-w)	--	0.40	230	180(0.18)	0.16	--	60	--	--	--	66
SiC/SiC (3D-2b)	--	0.40	260	390(0.42)	0.17	--	85	--	--	--	66
SiC/SiC (3D-4b)	--	0.40	240	270(0.5)	0.22	--	70	--	--	--	66
C/SiC (2D-w)	--	--	--	450(1.0)	--	--	--	--	--	--	70
C-Mullite	--	0.59	210	448(0.24)	--	--	(0.06)	.	--	--	140b
SCS6-RBSN[0]	--	0.24	175	576(0.7)	--	--	195	--	3.47	--	51
SiC-RBSN[0]	--	0.3	193	682(0.45)	0.21	227(0.11)	0.11	--	0.8	--	73
[90]	--	0.3	69	27(0.03)	0.08	--	--	--	--	--	73
[0₂/90₂]s	--	0.3	124	294(0.6)	0.12	127(0.1)	0.1	--	0.8	--	73
[±45₂/-45₂]s	--	0.3	78	88(0.2)	0.36	75(0.1)	0.1	--	--	--	73

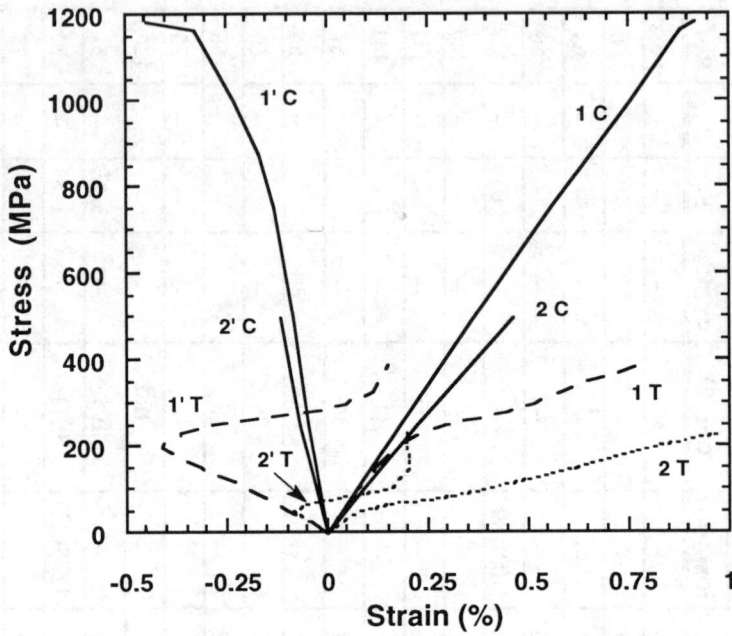

Fig. 21 Comparison of tensile and compressive longitudinal and transverse stress-strain curves of Nicalon-CAS [68]: 1 - [0], 2 - [0/90]$_{3s}$, C - compressive, T - tensile, unprimed - longitudinal and primed - transverse.

4.3 Flexural

Flexural tests are easier to perform at room as well as elevated temperatures than tensile tests. Hence they are used widely for material evaluation. But the stress state in a specimen subjected to flexural load is more complex than that in a tensile specimen. The stress in the specimen varies gradually from tensile on one side to compressive on the other. Also, specimen is subjected to interlaminar shear stresses. The relative magnitudes of the tensile and shear stresses depend on the span to depth ratio of the specimen. Prewo [58] studied the flexural (four-point-bending) behavior of Nicalon-LAS composites (span to depth ratio 25). The tensile and compressive moduli were found to be identical. The cracking in the specimen began on the side subjected to tensile load and progressed gradually towards the compressive side with increasing load. The cracks however, did not reach the compressive side as they were diverted by the fiber/matrix interface producing large deflection and delamination-type failure. Typically, for CMCs the flexural strength is found to be higher than the tensile strength [58].

Thus, the flexural response depends on tensile, compressive and shear behaviors. Due to damage on the tensile side alone, the neutral axis of the specimen shifts towards the compressive side [72a]. The specimen can undergo large deflection before failure. Thus, the simple beam theory cannot be used in calculations of strength.

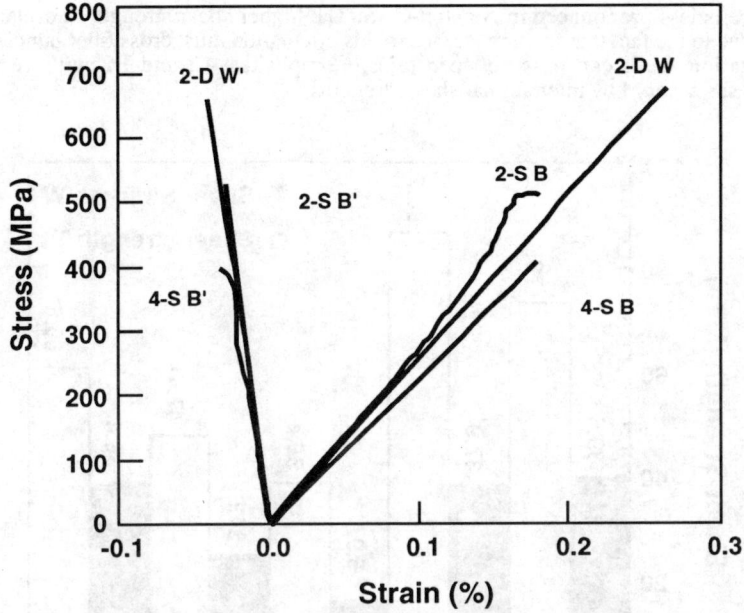

Fig. 22 Compressive stress-strain curves for 2-D woven (2-D W), 3-D two-step braided (2-S B) and 3-D four-step (4-S B) braided SiC/SiC composites (same materials as in Fig. 16) [66]. Primes indicate the transverse strain response.

4.4 Shear

Interlaminar and in-plane shear properties of composites are extremely important for predicting laminate behavior under tensile, compressive and flexural loadings. Four different techniques can be used to determine interlaminar shear strength [73-75]: short beam shear, four point flexural test, Iosipescu test and double notch shear test. Proper attention has to be paid in each of the methods to achieve uniform and pure shear stress in the gage section with desirable failure modes. In-plane shear properties can be determined using three methods: [+45/-45] laminate test, 10° off-axis test, and the rail shear test [73].

Bhatt and Phillips [73] studied the interlaminar shear behavior of SiC/reaction bonded silicon nitride (RBSN) composites by using double notch tests. Notches were cut such that the shear fracture plane was either parallel to the plane of hot pressing (designation W) or perpendicular to the plane of hot pressing (designation T). The interlaminar shear strengths were found to be 40 and 100 MPa in the W and the T orientations, respectively. Karandikar et al. [74] studied the effect of interface design on the double notch shear strength of Nicalon-BSG composites. Three Nicalon-BSG composites were studied with no interface coating (CG2), 25 nm interfacial carbon coating (A2) and 140 nm interfacial carbon coating (B2). The shear strengths were also evaluated for Nicalon-CAS composites and Nicalon reinforced epoxy (FRP). The shear strengths are shown in Figure 23. In all the composites, the interlaminar strength (orientation W) is lower than the splitting strength (orientation T). In particular, in the CG2, A2 and B2 composites, the interlaminar shear strength is almost half of the splitting strength. The difference is less pronounced in Nicalon-CAS. The higher shear strength in orientation T may be due to the fact that fracture surface in this orientation must cross fiber bundles while in orientation W, it can pass between plies. Composites A2 and B2 with very weak interface show very low interlaminar shear strengths.

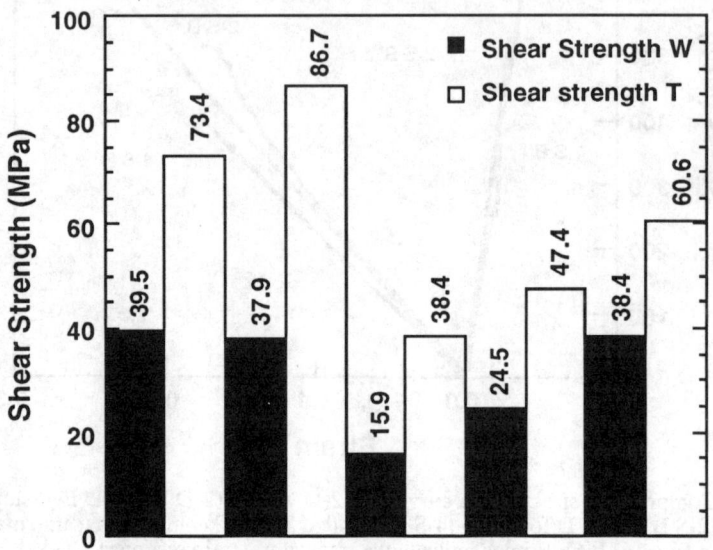

Fig. 23 Effect of interface design on double-notch shear strengths of Nicalon-fiber composites [74].

Fang and Chou [75] studied the effect of specimen design on the interlaminar shear strength of 2-D woven C/SiC, 2-D woven SiC/SiC and $[0/90]_{2s}$ Nicalon-LAS-II. Due to the often limited supply of CMC materials for testing, the possibility of using sub-sized double notch shear specimens (compared to the ASTM standards) was investigated. The interlaminar shear strengths were found to decrease as the notch distance decreased; however, the shear strength increased again when the stress fields around the notches interact. Shear strength data on different CMCs are summarized in Table 6.

4.5 Impact and Fracture Properties

Impact resistance of ceramic matrix composites has been found to be many times as high as monolithic ceramics [5a, 8, 50]. In the case of unidirectional Nicalon-CAS [5a] the notched and unnotched impact resistance is found to be five and fifty times that of monolithic silicon nitride. While the unnotched impact resistance of cross-ply Nicalon-LAS was only slightly higher than that of silicon nitride, the notched impact resistance was thirty times that of silicon nitride. The composite fracture surface under impact was jagged and showed fiber pull out while that of monolithic silicon nitride was flat. Similar results have been reported for C/LAS system [8].

Habib et al [50] attempted to measure the fracture toughness of unidirectional and cross-ply C-pyrex and Nicalon-pyrex composites. The unidirectional composites were found to be notch insensitive. The crack did not propagate in a self similar manner and was diverted by interfacial debonding. The cross-ply laminates were found to be more notch sensitive and showed interfacial debonding and transverse fracture. The C-pyrex cross-ply laminates were found to be more notch sensitive than Nicalon-pyrex. The fracture toughness values of 14 and 20 MPa m$^{1/2}$ were obtained for C-pyrex and Nicalon-pyrex respectively. The toughness values increased linearly with fiber volume fraction. Although these toughness numbers are not strictly valid, they reflect on the crack growth resistance of CMCs as compared to monolithic ceramics which have fracture toughness less than 5.

Fracture behavior of 2-D woven C/SiC and SiC/SiC composites was studied by Stull and Majidi [72b,c] using SENB and CT specimens of different sizes. The interface in the C/SiC system was weaker than the interface in the SiC/SiC system. The extent of matrix microcracking, fiber debonding and pull out was much higher in the C/SiC system and as a results the crack growth resistance was much higher. A specimen size independent R-curve could be obtained in the SiC/SiC system (R = 1.8 kJ/m2) but not in the C/SiC system. The relative specimen sizes needed to characterize purely material dependent R-curves were found to be proportional to the degree of irreversible damage.

5. Fiber Matrix Interface

Given the importance of the fiber-matrix interface, it is imperative to analyze the interface structure microscopically and measure the properties of the interface which affect the mechanical behavior and control them for optimizing composite performance at ambient and elevated temperatures. The term interface is used in a broad sense here to refer to the region between the fibers and the matrix which has a finite thickness and distinct structure and chemical composition.

In the silicate based composites a carbon-rich interface was produced during the processing of the composite via reaction between the Nicalon fiber and the matrix [76-81]. Figure 24 shows formation of such a layer in Nicalon-CAS composites [54a]. This interface layer prevents the formation of a strong bond between the SiC fiber and the glass-ceramic matrix. When a matrix crack approaches this layer, the stresses at the crack tip lead to fracture within this carbon layer along the fiber surface which blunts the matrix crack preventing fiber failure. It is proposed [77] that the interphase layer is formed by the following chemical reactions during processing of these composites.

$$SiC (s) + O_2 (g) = SiO_2 (s) + C (s) \tag{8}$$

$$SiC (s) + 2CO (g) = SiO_2 (s) + 3C (s) \tag{9}$$

The rate of this reaction is controlled by the rate of diffusion of O or CO through the reaction layer. Although initially the reaction layer thickness increases with time and temperature [78], it eventually starts decreasing again and the reaction layer may completely disappear [81]. Thus, proper control of the hot pressing is required to fabricate composites with reproducible interface layer.

The formation of C rich layer in the Nicalon-silicate system, however, is fortuitous and it may not be feasible to achieve the desirable interface layer in every composite system (e. g. SiC/SiC) through processing. Also, the intrinsic or processing-induced interface is unlikely to be stable at elevated temperatures. Therefore attempts have been made to deposit a desirable interface layer coating on the fibers before incorporating them in the matrix [74, 82-85]. Care must be taken in such situation to prevent alteration of this interface during processing as it often involves high temperatures and pressures.

Table 6a. Interlaminar shear properties.

Composite	V_f	Technique	Dimensions	L/D	ND	τ_{IL}	Ref.
C/Pyrex [0]	0.229	3PB	2.5x4x10	--	--	47	44
	0.233	3PB	2.5x4x10		--	44	44
	0.293	3PB	2.5x4x10		--	51	44
	0.402	3PB	2.5x4x10		--	63	44
	0.514	3PB	2.5x4x10		--	71	44
	0.595	3PB	2.5x4x10		--	18	44
C/LAS [0]	0.396	3PB	3.2x6.4x25.4	4	--	50.3	45
	0.396	3PB	3.2x6.4x25.4	5	--	46.9	45
	0.396	3PB	3.2x6.4x25.4	6	--	40.9	45
SiC/Pyrex [0]	0.340	3PB	--	10	--	28	50
C/Pyrex [0]	0.300	3PB	--	10	--	39	50
SiC/LAS [0/90]		4PB	3x4x30 (IS=6,OS=25)	--	--	20	61
SiC/RBSN [0]	0.300	DNS-W	6x6x27	--	19	40	73
	0.300	DNS-T	6x6x27	--	19	100	73
Nic-BSG-CG[0]	0.390	DNS-W	2x5x18	--	6	37.9	74
Nic-BSG-CG[0]	0.390	DNS-T	2x5x18	--	6	86.7	74
Nic-BSG-A[0]	0.440	DNS-W	2x5x18	--	6	15.9	74
Nic-BSG-A[0]	0.440	DNS-T	2x5x18	--	6	38.4	74
Nic-BSG-B[0]	0.450	DNS-W	2x5x18	--	6	24.5	74
Nic-BSG-B[0]	0.450	DNS-T	2x5x18	--	6	47.4	74
Nic-CAS [0]	0.350	DNS-W	2x5x18	--	6	38.4	74
Nic-CAS [0]	0.350	DNS-T	2x5x18	--	6	60.6	74
Nic-Epoxy [0]	0.600	DNS-W	2x5x18	--	6	39.5	74
Nic-Epoxy [0]	0.600	DNS-T	2x5x18	--	6	73.4	74

3PB - three point bend (short beam shear test), 4PB - four point bend (short beam shear test), IS and OS inner and outer span in 4PB, DNS - double notch shear with fracture plane parallel (W) and perpendicular (T) to hot pressing plane, dimensions in mm, L/D - span to depth ratio, ND - notch distance, τ_{IL} - interlaminar shear strength (MPa)

Table 6b. Interlaminar shear properties.

Composite	V_f	Technique	Dimensions	L/D	ND	τ	Ref.
Nic-CASII[0/90]	0.350	DNS-W	6.4x10.2x20.3	--	5.1	40	75
Nic-CASII[0/90]	0.350	DNS-W	6.4x10.2x20.3	--	7.6	32	75
Nic-CASII[0/90]	0.350	DNS-W	6.4x10.2x20.3	--	10	46	75
C-SiC PW	0.450	DNS-W	3.2x10.2x20.3	--	7.6	26	75
C-SiC PW	0.450	DNS-NT-W	3.2x10.2x20.3	--	7.6	19	75
SiC-SiC PW	0.400	DNS-W	3.2x10.2x20.3	--	7.6	53	75

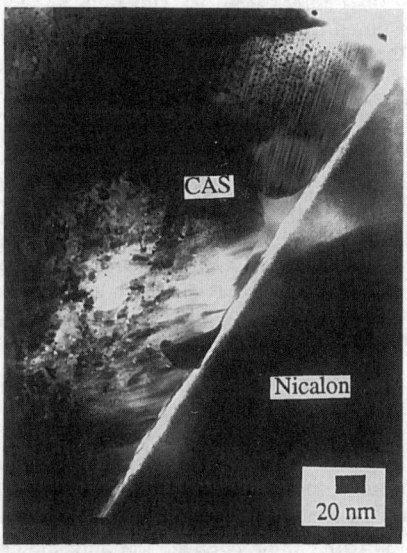

Fig. 24 TEM thin foil micrographs of SiC-CAS composites showing a light-colored structureless interface region (carbon) between the fiber and the matrix [54a].

Results of Monthioux and Cojean [83, 84] further show that not only the chemical composition of the interface but also its microstructure affects the composite performance. Six different types of interfacees were obtained in SiC/SiC composites systematically by depositing pyrolytic carbon on the fiber surface in various conditions. Correspondingly, four different stress-strain responses were obtained. The failure strain of the composites changes significantly with the interface condition. Composite with no pyrolytic carbon coating showed brittle failure with very low failure strain. Composites with highly anisotropic carbon coating (with aromatic carbon oriented parallel to the fiber surface) showed high toughness and the highest failure strain. Composite in which the aromatic carbon layer is more misoriented or microporous showed intermediate failure strain.

Carpenter and Bohler [85] have demonstrated that unbonded and porous multilayer SiC coatings remain stable even after several hours of exposure at 1316 and 1482°C and show debonding.

Given the importance of the interface, it is necessary to quantify the mechanical properties of the interphase itself. The stress transfer at the interface may be through chemical bonding or through frictional stresses. Several methods have been used in the literature for determining interfacial properties: ACK model [10, 11], microdebonding test [86], fiber push in test [87, 88] (Figure 25a,b), fiber push out test [89, 90] (Figure 25c), single fiber pull out test [91-93] (Figure 25d), microcomposite test [94] (Figure 25e), block test [95] (Figure 25f), frictional heating measurements [96], and bundle pushout technique [97, 98] (Figure 25g). The single fiber pullout, microcomposite and the block test all require fabrication of a test piece that is not a part of the composite itself. The success of the method therefore depends on how well the test piece represents the actual composite. The frictional measurement involves a dynamic shear stress at the interface. The interface property data obtained by various methods are summarized in Table 7. These methods have been outlined in the following.

According to the theory proposed by Aveston, Cooper and Kelly (ACK), the interfacial stress τ_f is assumed to be constant and can be calculated indirectly from the saturation spacing, S, of the matrix microcracks [10, 11, 99] as given by

$$\tau_f = \frac{1.337 \, V_m \, \sigma_{mu} \, R}{2 \, V_f \, S} \tag{10}$$

where R is the fiber radius and σ_{mu} is the matrix cracking stress.

Marshall [87] proposed a fiber push-in technique to measure the interfacial shear stress. In this technique fibers in a thick section of the composites are pushed down using a Vicker's indenter (Figure 25a). The penetration of the fiber itself is neglected, the force applied and the displacement of the fiber is calculated from the measurements of the indentation made on the fiber and the matrix by the indenter. The shear stress at the interface is assumed to be constant. The interfacial frictional stress is given by

$$\tau_f = \frac{F^2}{4 \pi \, u \, R^2 \, E_f} \tag{11}$$

where $F = 2 a^2 H$, is the force applied and $u = (b-a) \cot \psi$, is the fiber displacement, H is the fiber hardness, 2b and 2a are the diagonals of the indentations on the fiber and the matrix, respectively, and ψ is the angle between the opposite edges of the indenter. This analysis also neglected the transverse expansion of the fibers which would increase the sliding load. The analysis applies well to composites in which the interfacial fracture energy is very low.

Later, Marshall and Oliver [88] modified the indentation technique by monitoring the load and the displacement directly (Figure 25b). The penetration of the fiber was also subtracted from the displacement and an interface with finite fracture energy 2Γ was taken into account. The extent of debonding was estimated by an energy balance analysis for incremental crack growth. The relationship between fiber displacement and the applied force is given by

$$u = \frac{F^2}{4 \pi \, \tau_f \, R^2 \, E_f} - \frac{2\Gamma}{\tau_f} \tag{12}$$

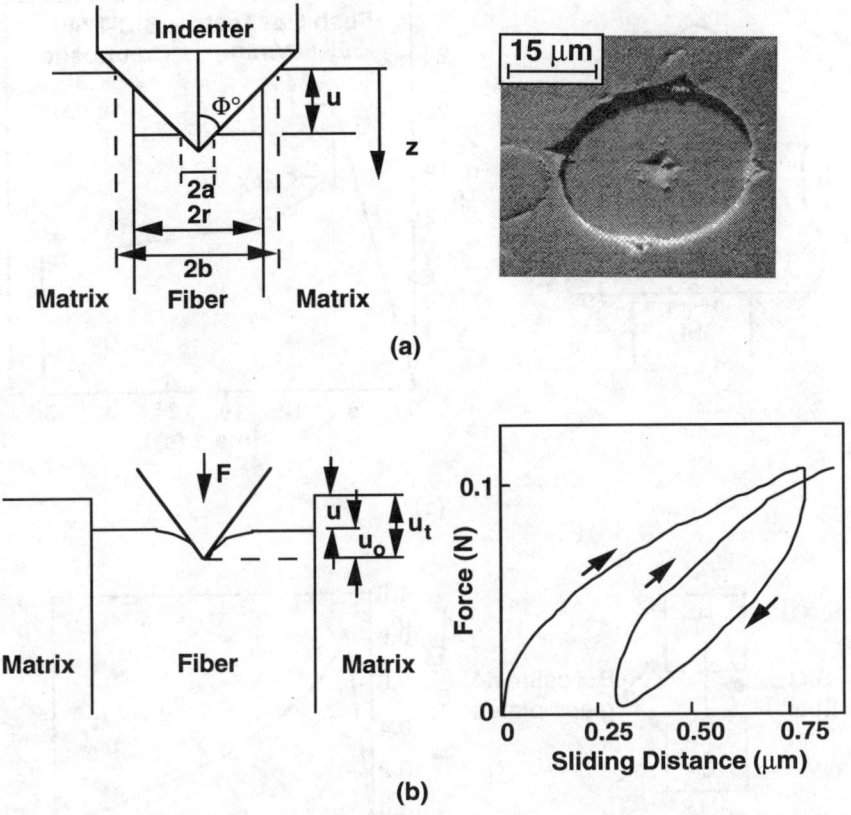

Fig. 25 Measurement techniques for fiber-matrix interfacial properties (a) Fiber push-in (indentation, [87]) (b) Fiber push-in (instrumented indenter, [88])

This analysis shows that the fiber displacement under pure sliding differs by a constant amount $2\Gamma/\tau_f$ from that under combined debonding and sliding. The τ_f and Γ can be calculated from the slope and the intercept of the plot of displacement vs force squared. The displacement of the fiber during unloading u_u and reloading u_r is given by the following two equations.

$$u_u = u_m [1 - (1-F/F_m)^2/2] \quad (13)$$

$$u_r = (u_m/2) [1 + (F/F_m)^2] \quad (14)$$

with the subscripts m denoting respective maximum values.

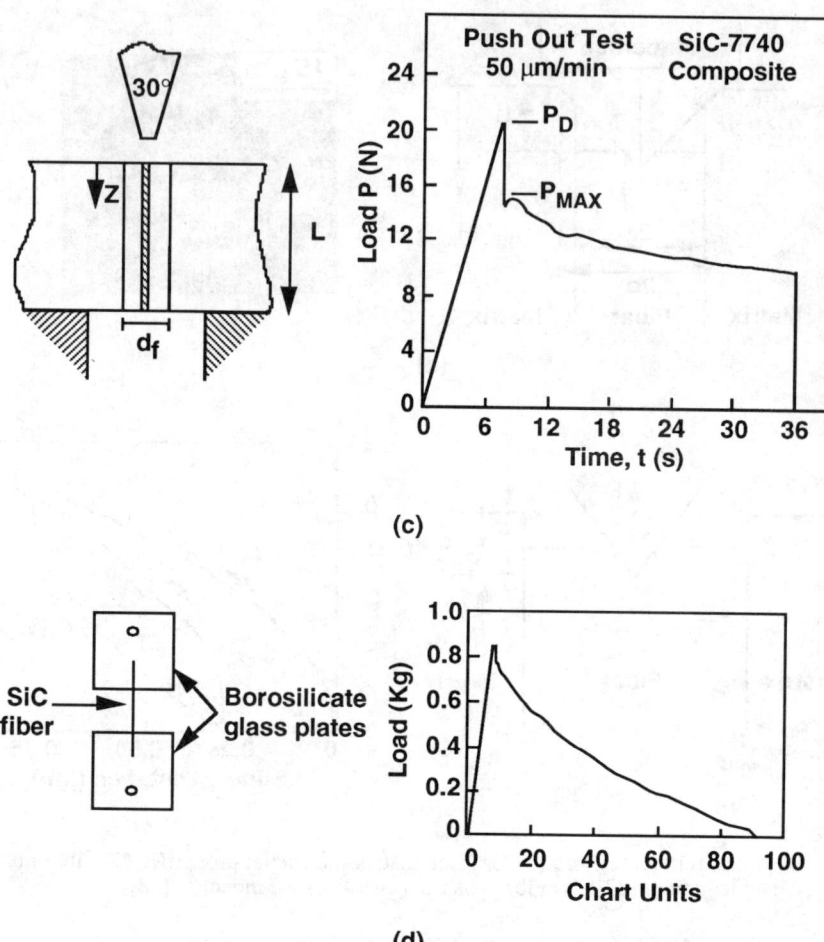

Fig. 25 Measurement techniques for fiber-matrix interfacial properties (continued) (c) Fiber push-out [105] (d) Single fiber pull-out [91].

The indentation method was later applied to push down fibers in thin specimens (thickness t) where the fibers can be pushed to an extent that they protrude from the other side of the test sample [89, 90, 105]. This technique is referred to as push-out technique (Figure 25c). A typical load displacement trace (displacement controlled tests) in these tests is also shown in Figure 25c. As the applied load reaches a critical level (P_D), debonding starts to occur and a sudden load drop to P_i may be observed. If the debonding is catastrophic, the load continues to decrease as the fiber is pushed out from the other side. If the debonding is progressive, after the load drop, load may increase to another peak value P_{max} (completion of debonding) and then decrease continuously. The interfacial debond strength is calculated from P_D and the frictional stress is calculated from P_i or P_{max}. The simplest of the analyses assumes a constant shear stress at the interface and the debond strength and the frictional stress are given by

$$\tau_{(d, f)} = \frac{P_{(D, i \text{ or max})}}{2 \pi R t} \tag{15}$$

In a load controlled test, at the onset of debonding, a displacement jump may be observed in the load-displacement curve in case of catastrophic debonding. The load at which this jump occurs will overestimate the frictional stress.

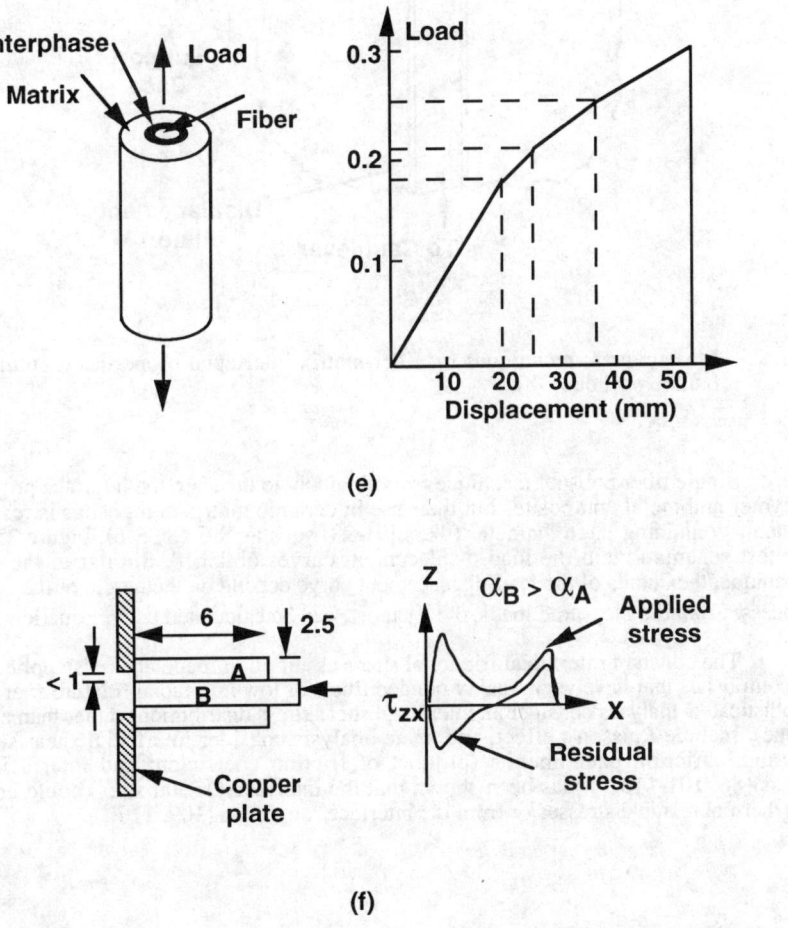

Fig. 25 Measurement techniques for fiber-matrix interfacial properties (continued) (e) microcomposite tensile test [94] (f) block test [95].

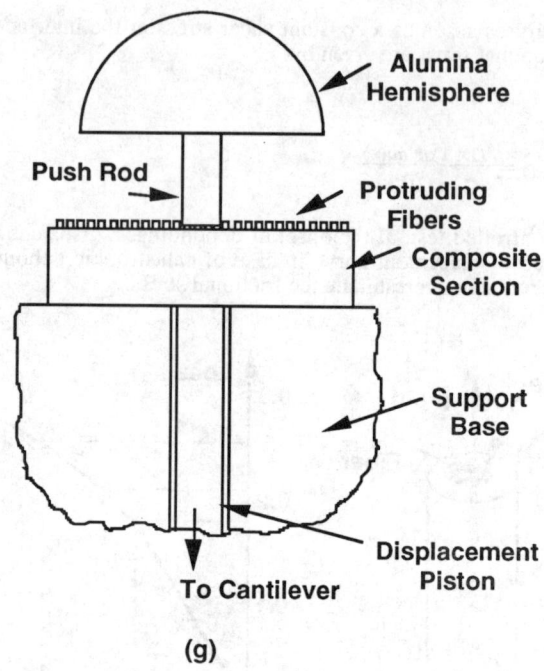

Fig. 25 Measurement techniques for fiber-matrix interfacial properties (continued) (g) bundle pushout [98].

Single fiber pull-out technique is used widely to measure the interface properties in polymer and metal composites but their use in ceramic matrix composites is restricted to systems containing large diameter fibers [91-93] such as SiC (SCS-6). Figure 25d shows the test schematic and the load displacement curves obtained. Similar to the push-out technique, the details of the load-displacement curve depend on the nature of the debonding process. From the measured loads, the τ_d and τ_f can be calculated using Equation (15).

The constant interfacial frictional stress assumption made above is applicable only to composites that have very weakly bonded fibers or low interfacial fracture energy. More sophisticated analyses consider an interfacial shear stress distribution (rather than a constant value), include Poisson's effect, and some analysis consider interfacial shear stress as a Coulomb friction phenomenon (product of friction coefficient and interfacial radial stress)[86, 101-110]. It has been shown that the interfacial radial stress could arise from the thermal residual stresses or from the interface roughness [109, 110].

Table 7a. Fiber-matrix interfacial properties.

Composite	$\alpha_f - \alpha_m$	V_f	I/F, nm	Tech.	τ_d	τ_f	μ	2Γ	G_{IIC}	σ_{MC}	UTS	Ref.
SCS6-7761	1	0.4	--	1,5	--	3.8,3.9	--	--	--	117	436	49
SCS6-7740	0.1	0.4	--	1,5	--	4.8,5.6	--	--	--	143	293	49
SCS6-9741	-1.3	0.4	--	1,5	--	5.3,7.1	--	--	--	--	223	49
SCS6-7052	-1.6	0.4	--	1,5	--	7.0,9.2	--	--	--	40	293	49
SiC-Pyrex	-0.1	0.34	--	1	--		--	--	--	--	726	50
C-Pyrex	-3.2,4.8	0.3	--	1	--	2.3	--	--	--	--	413	50
SiC-RBSN		0.24	--	1	--	5.0	--	--	--	--	576	51
Nic-CAS	-1.3	0.34	C,10	1	--	10.0	--	6	--	--	400	52
C-7740	-4	0.42	--	4	--	7	--	--	--	123	530	53
Nic-1723A	-2	0.58	--	4	--	7	--	--	--	201	--	53
Nic-1723B	-2	0.43	--	4	--	7	--	--	--	161	390	53
Nic-CAS	-1.1	0.4	--	4	--	7	--	--	--	132	--	53
Nic-BMAS	1.2	0.43	--	4	--	5.5	--	--	--	214	500	53

I/F, nm - type of interface and thickness in nanometers; Tech. 1 - crack spacing and ACK model 2 - push in 3 - push in load-displacement 4 - push out 5 - single fiber pull out 6 - microdebond 7 - microcomposite 8 - block test 9 - frictional heating 10 - bundle push out 11 - push out and shear lag; τ_d - interface debond strength (MPa), τ_f - interface frictional strength(MPa), μ - frictional coefficient, 2Γ - interface fracture energy (J/m^2), G$_{IIc}$ - interface mode II fracture toughness(MPa m$^{1/2}$), σ_{MC} - matrix cracking stress (MPa), UTS - ultimate tensile strength (*- bending strength)

Table 7b. Fiber-matrix interfacial properties.

Composite	$\alpha_f-\alpha_m$	V_f	I/F, nm	Tech.	τ_d	τ_f	μ	2Γ	G_{IIC}	σ_{MC}	US	Ref.
Nic-LAS III	2.4	0.39	--	4	--	3	--	--	--	195	700	53
Nic-CAS	-1	0.35	C, 10	1,2,4	--	14.4,12.4,12.3	--	--	--	(.13)	344	54
Nic-CAS	-1	0.35	C, 10	1,2	--	8.4,7.8	--	--	--	(.12)	398	74
Nic-LAS III C	3	0.5	C, 10	1	--	2	--	--	--	290	530	56
Nic-LAS III	0	0.5	--	1	--	7	--	--	--	238	557	56
Nic-AS	1	0.5	--	1	--	9	--	--	--	240	555	56
Nic-Sodalime	-5	0.5	--	1	--	12	--	--	--	<0	348	56
Nic-BSG CG	--	0.39	--	2	--	287	--	--	--	140	159	74
Nic-BSG A	--	0.44	C, 25	1,2	--	>5, 32.5	--	--	--	208	435	74
Nic-BSG B	--	0.45	C, 140	1,2	--	>8.6, 24.7	--	--	--	262	590	74
Nic-CAS	--	0.35	C,10	1,2	--	8.4, 7.8	--	--	--	118	398	74
Nic-1723	-.2	.35-.40	--	6,1	236	143, 6.9	--	--	--	--	--	86
Nic-LAS III	2.4	.35-.40	--	6	56	3.2	--	--	--	--	--	86
Nic-BMAS III	1.2	.35-.40	--	6	60	5.5	--	--	--	--	--	86
Nic-CAS I	-1.1	.35-.40	--	6	249	--	--	--	--	--	--	86
HMU-BSG	0	.35-.40	--	6	10.2	--	--	--	--	--	--	86
Nic-LAS III	2	0.5	C, 100	1,2	--	1.7, 2.5	--	--	--	300	620*	87
Nic-LAS III	3	0.5	C, 100	3	--	3.5	--	<.04	--	300	620*	88
SCS6-Oxide	--	0.25	C-rich	4	19.4	8	--	--	--	273	635	90

Table 7c. Fiber-matrix interfacial properties.

Composite	$\alpha_f - \alpha_m$	V_f	I/F, nm	Tech.	τ_d	τ_f	μ	2Γ	G_{IIC}	σ_{MC}	US	Ref.
SCS6-OxideHT	--	0.25	C-rich	4	7.6	5	--	--	--	285	740	90
SCS6-Oxide	--	0.25	?,1000	4	9.0	7.5	--	--	--	395	645	90
SCS6-OxideHT	--	0.25	?,1000	4	4.1	1.7	--	--	--	210	474	90
SiC-BSG	0.6	--	C-rich	5	--	3.6	0.72	--	--	--	--	92
SiC-Silica	5.6	--	C-rich	5	--	13.9	0.1	--	--	--	--	92
Nic-SiC	--	0.35	C-rich,10	7	--	2.3-6.1	--	--	.26-1	--	--	94
SiC-Pyrex	0.5	--	--	8	--	7	--	--	--	--	--	95
SiC-Pyrex	0.5	--	C	8	--	<1	--	--	--	--	--	95
Nic-CAS	--	0.35	C,10	9	--	3.5-5.0	--	--	--	--	--	96
SCS6-Pyrex	--	.1-.15	--	10	--	6.8	--	--	--	--	--	97
Nic-CAS	--	0.33	--	10	--	14-19	--	--	--	--	--	98
Nic-LAS III	--	0.5	--	3,11	--	0.2	0.1	--	--	--	--	104
SiC-RBSN	0.3	0.3	C-r,3000	4	19.1	14.7	0.27	--	--	--	--	105
SiC-BSG7740	0.3	0.03	C-r,3000	4	17.5	11.6	0.29	--	--	--	--	105
SiC-BSG7050	-1.0	0.4	C-r,3000	4	18.2	10.3	0.94	--	--	--	--	105
SiC-ZrO2	0	0.25	C-r,3000	4	39	16	--	--	--	217	287*	177
SiC-ZrO2	0	0.25	BN,1000	4	18	15	--	--	--	220	357*	177
SiC-ZrO2-w	0	.25, .2	C-r,3000	4	4.4	1.6	--	--	--	333	441*	177
Nic-BSG A1	--	0.47	O,Si,250	2	--	8.2	--	--	--	--	330	178

Table 7d. Fiber-matrix interfacial properties.

Composite	$\alpha_f-\alpha_m$	V_f	I/F, nm	Tech.	τ_d	τ_f	μ	2Γ	G_{IIC}	σ_{MC}	US	Ref.
Nic-BSG A2	--	0.27	O,Si,250	2	--	4.1	--	--	--	--	390	178
Nic-BSG B1	--	0.42	C-r,10	2	--	2.0	--	--	--	--	725	178
SiC-RBSN	--	--	C-r,3000	4	3.0	1.1	--	--	--	--	--	179
SiC-RBSN-LHp	--	--	C-r,3000	4	13.1	32.4	--	--	--	--	--	179
SiC-RBSN-HHp	--	--	C-r,3000	4	--	29.4	--	--	--	--	--	179
Nic-CAS	--	0.35	--	3	--	25	9.6	--	--	--	--	180
Nic-CAS H1	--	0.35	--	3	--	87	13.4	--	--	--	--	180
Nic-CAS H2	--	0.35	--	3	--	25	11.8	--	--	--	--	180
Nic-MAS	--	--	--	3	--	239	30.4	--	--	251*	504*	180
Nic607-MAS	--	--	C-r	3	--	48	12.4	--	--	665*	1168*	180
Nic6-BSG P1	--	--	C-rich	3	--	111	--	18	--	--	--	180
Nic6-BSG P2	--	--	C-rich	3	--	296	--	60	--	--	--	180
Nic-LAS III	--	--	--	4	--	1.1,8.3,1	.01-.1	--	--	--	--	181
Nic-CAS	--	--	--	4	--	25,19.8,16	.15-.2	--	0.32	--	--	181
Nic-1723	--	--	--	4	--	28,17,14	.15-.2	--	0.30	--	--	181

The debonding load P_d in a single fiber pullout test (fully bonded fibers) was related to the embedded length L by a shear lag analysis by Lawrence [100]. For a perfectly bonded case, the relationship is given by

$$P_d = \frac{\pi \, 2R \, \tau_d}{\alpha} \tanh(\alpha L) \qquad (16)$$

where α is a shear lag parameter dependent on the geometric configuration and elastic properties of the fiber and the matrix. According to shear lag analysis the maximum shear stress occurs at the point where the fiber emerges from the matrix. After its initiation, debonding may propagate without an increase in the applied load (i. e. unstable propagation) or on further loading (i. e. stable propagation).

After complete debonding, the pullout of a fiber is resisted by the frictional force at the interface. This force is a product of interfacial frictional coefficient (μ) and the normal stress on the interface. The Poisson's contraction (or expansion in push-out) affects the interfacial normal stress. Takaku and Arridge [101] proposed a shear lag analysis of the frictional pull-out problem taking into consideration the Poisson's contraction. The relationship between the frictional initial pull-out load P_i and the embedded length L is given by

$$P_i = \frac{\pi R^2 \sigma_0}{k} \left[1 - \exp\left(\frac{-2 \mu k L}{R}\right) \right] \qquad (17)$$

where $k = (E_m \, v_f)/[E_f(1 + v_m)]$, σ_0 is the residual compressive normal stress on the fiber and v_f and v_m are the Poisson's ratios of fiber and matrix, respectively. This analysis was extended to the push-out problem by Shetty [104] to give following relationship between the maximum frictional sliding load, P_{max}, and the embedded length

$$P_{max} = \frac{\pi R^2 \sigma_0}{k} \left[\exp\left(\frac{2 \mu k L}{R}\right) - 1 \right] \qquad (18)$$

For small friction coefficient μ, k and specimen thickness t or embedded length L and large fiber radius R, both Equations (17) and (18) reduce to Equation (15) which assumes a constant shear stress. Thus only under these conditions, the simple constant shear stress approximation should be used.

A microdebond technique was used by Grande et al.[86] to study the interface debond strength. In this technique, incremental loading is applied to a fiber using an indenter till fiber sliding is observed. A finite element analysis is used to calculate the interfacial shear stress distribution and the shear strength from the debond load. The finite element analysis showed that the maximum in the interfacial shear stress occurs at a distance 1-2 fiber diameters below the matrix crack (or specimen surface) and not at the matrix crack as predicted by shear lag analysis.

The different data analysis schemes for the pull-out and pushout tests have been reviewed in detail by Kerans and Parthasarathy [110]. They have suggested that the interface must be characterized in terms of interfacial fracture toughness rather than interfacial shear strength as the debonding occurs by propagation of a sharp mode II crack.

Their analysis shows that the loads P_D and P_{max} above cannot be directly used to calculate shear strength and frictional stress unless complete stress distributions are known. They derived the following relationships between the load point displacement δ and applied load P_a for the region between initiation and propagation of the debonding in pullout (for $P^* > P_a > P_{db} + P_r$) and pushout tests (for $P_a \leq P_{db} + P_r$), respectively.

$$\delta = \frac{L_0 P_a}{\pi R^2 E_f} + \frac{1 - 2\nu_f k}{2\mu k \pi R E_f} \left[P_{db} + P_r - P_a + (P^* - P_r) \ln\left(\frac{P^* - (P_{db} + P_r)}{P^* - P_a}\right) \right] \quad (19)$$

$$\delta = \frac{L_0 P_a}{\pi R^2 E_f} + \frac{1 - 2\nu_f k}{2\mu k \pi R E_f} \left[P_a - P_{db} - P_r + (P^* - P_r) \ln\left(\frac{P^* - P_a}{P^* - P_{db} - P_r}\right) \right] \quad (20)$$

where L_0 is the length of the fiber outside the matrix, P_r is the residual axial load in the fiber (product of fiber residual stress and its area), P_{db} is the load at which debonding starts occurring in the absence of axial residual load (not the value at which load drop occurs in load-displacement curve as in previous models), P^* is the pullout load at which Poisson's contraction due to axial stress cancels the residual normal stress resulting in no net friction condition and $k = [E_m V_f]/[E_m(1+\nu_f) + E_f(1+\nu_m)]$. The three unknowns, P_{db}, P^* and μ, can be calculated from three points on the experimental load-displacement curves. Alternatively, the parameters can be calculated by measuring the peak load in the pullout or pushout tests. The following two equations were derived for the peak loads P_m in the pullout (-) and pushout (+) tests

$$P_m = P^* + (P_{db} + P_r - P^*) \exp\left(\pm \frac{2\mu k L_c}{r}\right) \quad (21)$$

where L_c is the debond length at peak load. The four unknowns can be evaluated by measuring the peak load as a function of the specimen thickness or embedded length. The quantity P^* is the asymptotic limit of the peak load as a function of embedded length. The toughness of the interface, G_c in the absence of friction and residual stresses is given by [110]

$$G_c = \frac{(1 - 2\nu_f k) P_{db}^2}{4 \pi^2 R^3 E_f} \quad (22)$$

From the above models, the sequence of damage evolution in a unidirectional composite with chemical bond at the interface can be summarized as follows. When a matrix crack approaches a fiber, some debonding takes place and the crack passes around the fiber and is bridged by the fiber. The matrix is fully unloaded in the plane of the crack but the stress is transferred back into the matrix via interfacial shear stresses. Shear lag analysis or finite element analysis can give the shear stress distribution at the interface in the bonded region. The maximum shear stress occurs at the matrix crack plane (shear lag) or a few fiber diameters away from the matrix crack. When the interfacial shear stress reaches the shear strength (τ_d), debonding propagates. In the debonded region, the fiber can slide relative to the matrix and the stress transfer occurs through frictional stresses (τ_f). The frictional stress is a function of the friction coefficient (μ) and the radial stress on the interface (σ_N). The radial stress arises due to the thermal expansion mismatch between the fiber and the matrix and roughness of the debond surfaces. The debonding may propagate unstably or stably. Alternatively, the fracture mechanics approach can be used to characterize the onset of propagation of debonding. In this, debonding is considered as a mode II crack. Extension of this crack occurs when the strain energy release rate equals the toughness (or critical strain energy release rate) G_C.

Thus, a number of properties associated with the fiber matrix interface are deemed important for determining the behavior of the interface. The challenge is in measuring these through experiments which will simulate the actual process of matrix microcracking and ultimate failure of the composite.

Brun [111] has developed a push-out technique that can be used to measure the interfacial properties up to 1100°C. SiC fiber reinforced mullite and cordierite matrix composites were studied. The CTE of cordierite is lower than that of the SiC fiber while the CTE of the mullite matrix was higher than that of the SiC fiber. In cordierite matrix composite, the frictional stress at the interface increased gradually from 1.5 MPa at room temperature to 10.0 MPa at 1100°C. In mullite composite on the other hand, the frictional stress decreased from 10.5 to 9.5 MPa. The cordierite composite, has tensile residual stresses at the interface and matrix pulls away from the fiber. The contact between fiber and matrix is due only to partially engaged asperities and initial frictional resistance given by Equation (15) above is low. As the composite is heated, the fiber starts expanding and the asperities become fully engaged, increasing the frictional stress. An opposite effect is observed in the mullite system. This system has compressive residual stresses at the interface. These are reduced as the composite is heated and the frictional resistance decreases.

Interfacial property data of various CMC systems obtained by different techniques discussed above are summarized in Table 7.

6. Theoretical Analysis

6.1 Unidirectional Composites

6.1.1 Critical Stress for Matrix Cracking

The on set of matrix cracking in ceramic matrix composites is a very significant event as beyond this point the fibers can be subjected to environmental attack through the matrix cracks. Therefore, it is imperative to be able to predict the onset stress for matrix cracking theoretically. At the present time, several modeling approaches are available for predicting the critical stress of fiber-bridged cracking. These approaches, however, differ in their assumed initial crack state and the propagation mechanisms and can be divided into three categories.

The approaches of the first category (Aveston, Cooper and Kelly (ACK) [10,11] Budiansky, Hutchinson and Evans (BHE) [112] and Kuo and Chou [113]) consider the state of the composite before initiation of a crack and after complete propagation of the crack. The energy changes occurring during these processes are balanced to calculate the critical stress required for matrix cracking.

The approaches in the second category (Marshall, Cox and Evans (MCE) [114], McCartney [115] and Chiang, Wang and Chou [116]) assume a short starter crack and considered a self-similar crack propagation problem in the context of classical fracture mechanics. In this case, the composite is represented by an effective, homogeneous and isotropic medium that possesses a given fracture toughness. In addition, the fiber-bridged crack is replaced by one in which the cracked surface is acted upon by a certain surface traction distribution. The latter mimics the effects of the bridging fibers on the crack-tip stress field. Crack growth occurs when the stress intensity factor, K (or strain energy release rate, G) equals the critical stress intensity factor, K_c (or critical strain energy release rate, G_c).

From a physical viewpoint, the meaning of a short crack versus a long crack is illustrated in Figure 26. Here, three regions are identified with respect to the fiber-bridged crack. Region I, referred to as the *downstream region*, is sufficiently behind the crack-tip so that the stress and the strain fields are uniform with respect to the crack plane. In particular, the interface slipping length in this region is uniform along the crack plane. In region II, referred to as the *transient region*, the fields are complex on both sides of the crack-tip; the interface slipping length becomes shorter near the crack-tip. In fact, at the crack-tip, the slipping length vanishes because there is no crack opening at this point. Region III, the *upperstream region*, is sufficiently in the front of the crack-tip so that the

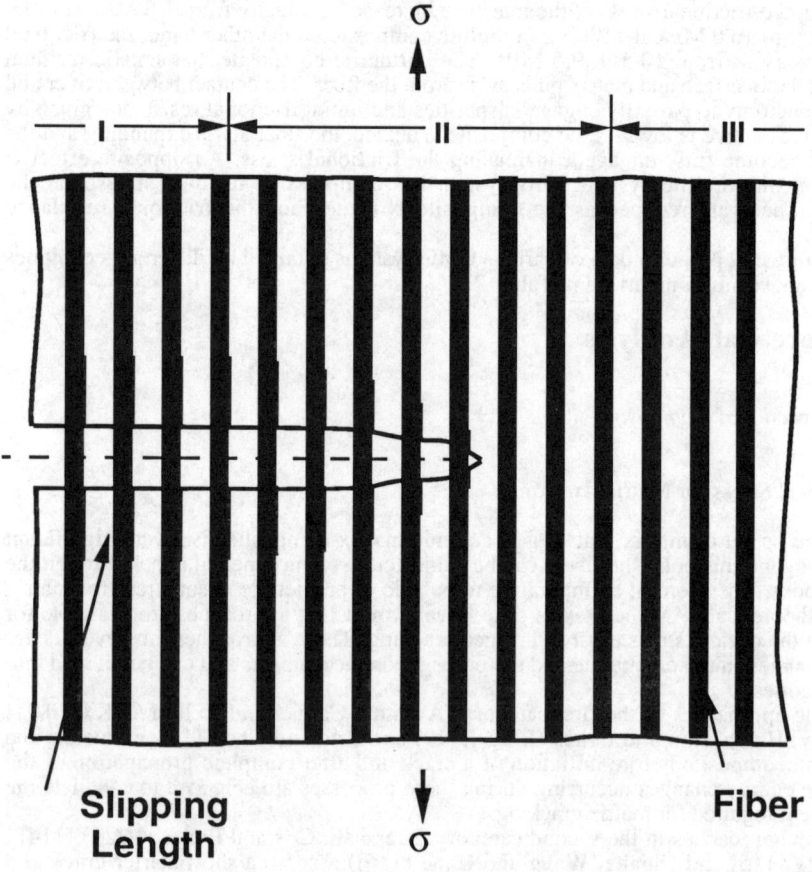

Fig. 26 The three regions associated with a fiber bridged crack [116].

stress and strain fields are again uniform. If a crack is long enough such that the stress/strain fields in region I are predominant, it can be regarded as a long crack. On the other hand, if a crack is located entirely within region II, it is regarded as a short crack. For a short crack, the slipping length is relatively short, especially near the crack-tip. Consequently, the effect of the matrix shear deformation above the slipping length may become important.

The third category includes the local flaw model [117]. In this model a matrix flaw and associated interface flaw is considered and a strain energy release rate for the growth of this crack is computed. Matrix cracking is assumed to initiate when the strain energy release rate equals the critical strain energy release rate.

The energy balance approach

The energy approach was first used by ACK [10,11] to predict the critical stress for matrix cracking. In this model, the bridging fibers can slip through the matrix against friction in the wake of the crack. The extent of fiber slippage depends on the frictional shear strength of the fiber/matrix interface. The energies involved in creating a matrix crack are the matrix surface energy (U_c), the energy required to debond all the fibers bridging a unit area of matrix crack (γ_{db}), the energy dissipated due to frictional fiber sliding (U_s), the fiber strain energy increment (ΔU_f), the decrease in matrix strain energy (ΔU_m), and the work done by external load (ΔW). The relationship among the energy terms is

$$U_c + \gamma_{db} + U_s + \Delta U_f \leq \Delta U_m + \Delta W \tag{23}$$

Assuming that the stress transfer from the fiber to the matrix occurs via constant interfacial shear stress, τ_f, ACK gave the following expressions for the energy terms.

$$U_c = 2\gamma_m V_m$$

$$\gamma_{db} = \frac{2\sigma_{mu} V_m G_{II}}{\tau_f}$$

$$U_s = \frac{E_f E_m V_m}{6\tau_f} \varepsilon_{mu}^3 \; \alpha \, R \, (1+\alpha) \tag{24}$$

$$\Delta U_f = \frac{E_f E_m V_m}{2\tau_f} \varepsilon_{mu}^3 \; \alpha \, R \, (1+\alpha/3)$$

$$\Delta U_m = \frac{E_f E_m V_m}{3\tau_f} \varepsilon_{mu}^3 \; \alpha \, R$$

$$\Delta W = \frac{E_f E_m V_m}{2\tau_f} \varepsilon_{mu}^3 \; \alpha \, R \, (1+\alpha)$$

where $\alpha = (E_m V_m / E_f V_f)$, γ_m is the matrix surface energy per unit area and G_{II} is the energy required to debond unit area of fiber/matrix interface. Substituting into Equation (23) and assuming zero debonding energy, the following expression is derived for the critical strain for matrix cracking.

$$\varepsilon_{mu} = \left\{ \frac{12\, \tau_f\, \gamma_m\, E_f\, V_f^2}{E_1\, E_m^2\, R\, V_m} \right\}^{1/3} \tag{25}$$

This equation indicates that the matrix cracking strain can be increased by reducing the fiber radius or increasing τ_f. For sufficiently low fiber radius or high τ_f, the matrix cracking can be totally suppressed. But, the composite will behave in a brittle manner with low work-of-fracture.

Aveston and Kelly (AK) [11] further extended this model by using non-constant shear stress at the fiber/matrix interface in the shear lag approach to relate the elastic displacements in the fiber and the matrix. The fundamental relationship governing load transfer between fibers and matrix is given by

$$\frac{dF}{dy} = \frac{2\, V_f\, \tau}{R} \tag{26}$$

where dF is the load transferred from fiber to matrix over distance dy. The shear lag analysis gives following expressions for the additional stress on the fibers ($\Delta\sigma$) due to matrix cracking and the shear stress at the interface as a function of the distance, y, from the crack plane.

$$\Delta\sigma = \Delta\sigma_o \exp(-\sqrt{\phi}\, y) \tag{27}$$

$$\tau = \frac{R}{2}\, \Delta\sigma_o\, \sqrt{\phi}\, \exp(-\sqrt{\phi}\, y) \tag{28}$$

where

$$\sqrt{\phi} = \left(\frac{2\, E_1\, G_m}{E_f\, E_m V_m}\right)^{1/2} \frac{1}{R\, [\ln(R_o/R)]^{1/2}}$$

$\Delta\sigma_o$ is the maximum value of the additional stress which occurs at the crack plane and R_o is the radial distance from the center of the fiber at which the displacement in the matrix is equal to the average displacement in the matrix.

If a perfect bond is assumed, the energy contributions to debonding and sliding are zero. The other energy terms can be calculated based on the stress fields given by the shear lag model. Carrying out the energy balance gives the following expression for the matrix cracking strain

$$\varepsilon_{mu} = \frac{2\, \gamma_m\, V_m}{\alpha\, R\, E_1} \left\{ \frac{2\, E_1\, G_m}{\varphi\, E_f\, E_m V_m} \right\}^{1/2} \tag{29}$$

Here, $j = \ln[\pi/(2\sqrt{3}\, V_f)]^{1/2}$ and G_m is the matrix shear modulus. The ratio of matrix crack initiation stress in the elastic case (Equation (29)) to that in the debonded case (Equation (25)) is given by $[\sigma_{mu}/3\tau]^{1/2}$.

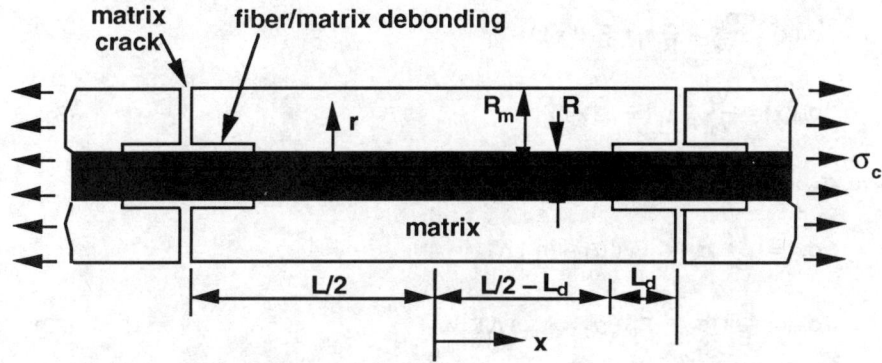

Fig. 27 Unit cell for the shear lag analysis of unidirectional composite [133].

The approaches followed by ACK and AK were generalized by Budiansky, Hutchinson and Evans (BHE) [112] and Kuo and Chou [113]. The strain energy terms were calculated using stress distributions given by a shear lag analysis. In these models, partial debonding was allowed and thermal stresses were taken into account. The analysis of Kuo and Chou is outlined here.

In the Kuo and Chou model, the fiber and matrix are modeled as concentric cylinders with radius R and R_m, respectively; the composite is composed of a series of repeating units bridged by the fiber (Figure 27). It is assumed that the normal stress in the fiber (σ_f) is a function of x only. Since the fiber failure strain is substantially higher than the matrix failure strain for most CMCs, the fiber is assumed to be intact during the development of matrix cracking. Since ceramic-matrix composites commonly possess stiff fibers and matrices, both the normal and shear deformations of the fiber and the matrix are considered. The crack spacing is denoted by L and the debonding length by L_d. By varying L and L_d, this model can encompass the cases of (1) perfect bonding, (2) partial debonding/sliding, and (3) complete debonding/sliding. In the debonded region, interfacial shear stress, τ is assumed to be a constant and equal to the interfacial frictional shear stress. In the bonded region, a shear-lag model is adopted to approximate the shear stress distribution in the interface.

The normal stress distributions in the bonded and debonded regions are as follows:

Bonded region : $|x| \leq L/2 - L_d$

$$\sigma_f(x) = \sigma_{fo} + \frac{\cosh \beta x}{\cosh \beta (L/2 - L_d)} \left(\frac{V_m}{V_f} \sigma_{mo} - 2 \frac{L_d}{R} \tau \right)$$

$$\sigma_m(x) = \sigma_{mo} - \frac{\cosh \beta x}{\cosh \beta (L/2 - L_d)} \left(\sigma_{mo} - 2 \frac{V_f}{V_m} \frac{L_d}{R} \tau \right) \quad (30)$$

Debonded region : $L/2 - L_d \leq |x| \leq L/2$

$$\sigma_f(x) = \frac{\sigma_c}{V_f} - \frac{2}{R}\tau_f\left(\frac{L}{2}-|x|\right)$$

$$\sigma_m(x) = \frac{V_f}{V_m}\frac{2}{R}\tau_f\left(\frac{L}{2}-|x|\right) \tag{31}$$

Here, σ_{fo} and σ_{mo} are the far-field stresses in the fiber and matrix, respectively:

$$\sigma_{fo} = \frac{E_f}{E_1}\sigma_c + E_f(\alpha_1 - \alpha_f)\Delta T$$

$$\sigma_{mo} = \frac{E_m}{E_1}\sigma_c + E_m(\alpha_1 - \alpha_m)\Delta T \tag{32}$$

where ΔT is defined as the ambient temperature minus the thermal stress-free temperature; normally, ΔT is negative; α is the thermal expansion coefficient with subscripts m, f and 1 indicating matrix, fiber and unidirectional composite, respectively and

$$\beta^2 = \frac{8}{R^2}\left(\frac{1}{G_f} + \frac{1}{G_m}\left(\frac{2}{V_m^2}\ln\frac{1}{V_f} - 3 - \frac{2V_f}{V_m}\right)\right)^{-1}\frac{E_1}{E_f E_m V_m}$$

It can be seen from Equations (30) and (31) that the stress distribution in the debonded region is governed by τ_f, and in the bonded region by the shear-lag constant β.

Based on this model, the following are the definitions of the terms involved in the energy balance

$$U_c = \gamma_m\, 2\pi(R_m^2 - R^2)$$

$$\gamma_{db} = G_{II}\, 4\pi R L_d$$

$$U_s = 4\pi R\tau\int_{L/2-L_d}^{L/2}(\Delta u_f - \Delta u_m)\,dx \tag{33}$$

$$\Delta U_f = \pi R^2\frac{1}{2E_f}\int_{-L/2}^{L/2}(\sigma_f^2 - \sigma_{fo}^2)\,dx$$

$$\Delta U_m = \pi(R_m^2 - R^2)\frac{1}{2E_m}\int_{-L/2}^{L/2}(\sigma_{mo}^2 - \sigma_m^2)\,dx$$

$$\Delta W = \pi R_m^2\sigma_c\int_{-L/2}^{L/2}\left(\frac{\sigma_f}{E_f} - \frac{\sigma_{fo}}{E_f}\right)dx$$

where Δu_f and Δu_m are, respectively, the displacement differences of the fiber and the matrix before and after matrix cracking. According to the definitions, all the energy terms in Equation (33) are positive. When the integrations are performed and the resulting energy expressions substituted into Equation (23), a general equation for the critical stress of matrix cracking can be achieved.

The general solutions of the critical stress were obtained numerically by Kuo and Chou, and closed-form solutions were given for two limiting cases. The first limiting case is perfect interfacial bonding, i.e., $L_d=0$. Also using $L=\infty$ to represent the initiation of matrix cracking, the cracking stress can be predicted as

$$\sigma_{cr1}^u = \left(\frac{4}{3}\beta\frac{\gamma_m E_1 E_f V_f}{E_m}\right)^{1/2} - E_1(\alpha_1 - \alpha_m)\Delta T \qquad (34)$$

The second limiting case is complete debonding and frictional sliding, i.e. $L_d=L_s$ and $G_{II}=0$, where L_s is the theoretical maximum sliding length determined by the condition that when the strain in the fiber is equal to that in the matrix, the sliding stops. In this case the cracking stress is given as

$$\sigma_{cr2}^u = \left(\frac{12\,\tau_f\,\gamma_m\,E_1^2\,E_f\,V_f^2}{R\,V_m\,E_m^2}\right)^{1/3} - E_1(\alpha_1 - \alpha_m)\Delta T \qquad (35)$$

The first part of this expression is identical to that given by ACK (Equation (25)), and the second part is due to the differential thermal expansions. Since ΔT is generally negative, in order to increase the critical stress, it is desirable that α_f be higher than α_m, which results in compressive axial thermal residual stresses in the matrix.

For practical composites, the degree of fiber/matrix debonding may lie in between the two limiting cases discussed above. For SiC/CAS, SiC/CAS and C/BSG systems the variation of the maximum sliding length L_s with bonding energy (γ_{db}) and interfacial frictional shear stress (τ_f) is shown in Figure 28. The numerical results indicate that for higher interfacial shear stress, shorter length is needed for the matrix stress to build up to the level at which the fiber strain equals the matrix strain decreasing L_s. On the contrary, L_s increases as the bonding energy becomes higher; this is due to the higher critical stress.

The effects of debonding length and debonding energy on the critical stress of matrix crack initiation are illustrated in Figure 29 for the above three composite systems. The predicted critical stresses are given with L_d/L_s varied from 0 to 1 and γ_{db} from 0 to $0.3\gamma_m$. The solutions of σ_{cr1}^u and σ_{cr2}^u are also indicated in the figures. σ_{cr1}^u is significantly higher than σ_{cr2}^u, indicating that matrix cracking with perfect bonding is unlikely to occur. Due to the fact that the curves for zero bonding energy are almost flat in the range of high degree of debonding, it is suggested that σ_{cr2}^u would be a good approximation for composites having negligible bonding energy. The experiment results are reasonably close to the theoretical predictions for zero bonding energy, also indicating that σ_{cr2}^u is a better approximation for the composite systems examined. It can also be seen that for a wide range of L_d/L_s, even a relatively small increment of γ_{db}/γ_m would result in a rather significant increase in the critical stress. The reason is that the interfacial debonding area, which depends on the debonding length, is much higher than the matrix cracking area. From the energy point of view, matrix cracking is likely to occur at the minimum point of each curve in Figure 29. This minimum point shifts to lower L_d/L_s values as γ_{db} increases. This decrease in debonding length is the result of increment in bonding energy.

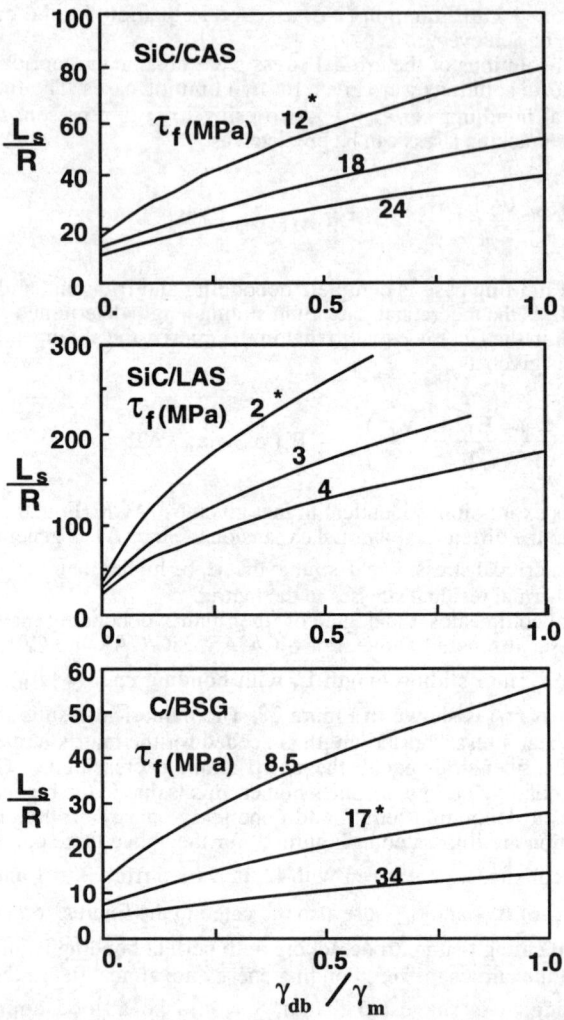

Fig. 28 Effects of interfacial frictional shear stress and bonding energy on sliding length L_s[96].

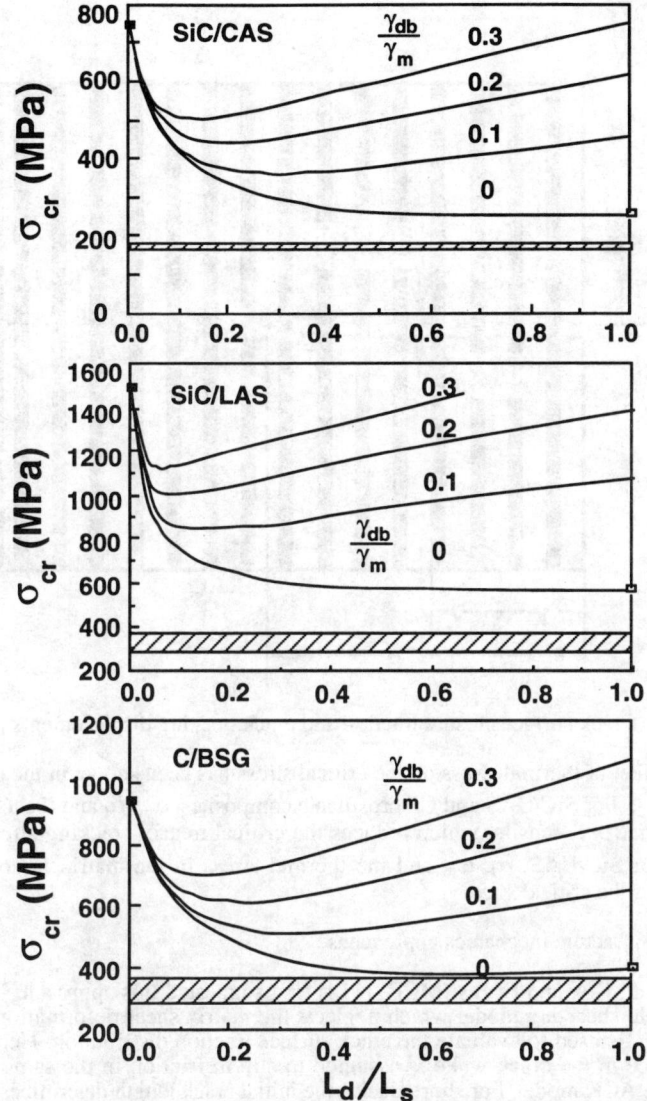

Fig. 29 Effects of bonding energy and debond length on critical stress for matrix crack initiation: solid squares: s_{cr1}^u; open squares: s_{cr2}^u. The shaded areas are used to indicate the experimental matrix cracking datas since the extent of debonding is not known [96].

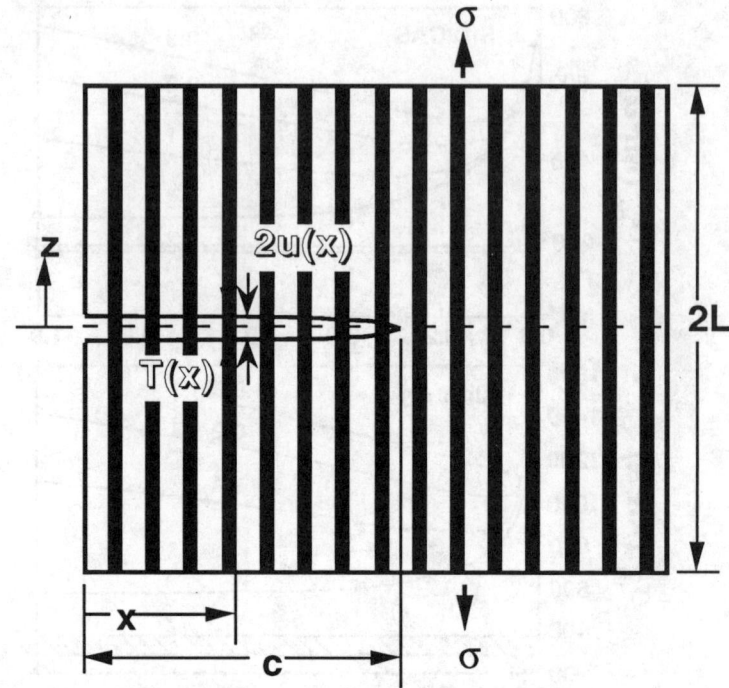

Fig. 30 Crack surface closure traction and crack opening displacements [116].

The effect of thermal stress on the critical stress has been shown in the expressions of σ_{cr1}^u and σ_{cr2}^u. For SiC/CAS and C/borosilicate composites, $\alpha_m > \alpha_f$ and the axial thermal stress in the matrix is tensile, which reduces the critical matrix cracking stress. On the other hand, for SiC/LAS, $\alpha_f > \alpha_m$ and the thermal stress in the matrix is compressive, which increases the critical stress.

The short crack-fracture mechanics approaches

Marshall, Cox and Evans (MCE) [114] first suggested this approach. In the MCE model, a simple shear-lag model which neglects the matrix shear deformation above the slipping region, is used to evaluate the crack surface traction distribution. Here the fiber-matrix interface in the crack wake is assumed to slip in friction in the same manner as assumed in the ACK model. For short cracks, the initial crack length determines the critical stress at crack propagation. Thus, a simple relationship between crack opening displacement (u) and crack surface traction (T) is obtained in the form $u \propto T^2$ (see Figure 30). Then at the crack-tip, u must vanish and so must T. This implies that the bridging fibers at the crack-tip carry no load.

McCartney [115] and Chiang, Wang and Chou [116] followed the approach suggested by MCE. But different definitions of critical stress intensity factor were used. Chiang, Wang and Chou also used a modified shear-lag model which takes into account the shear deformation in the matrix above the slipping region to derive the u-T relationship. In the MCE model, the composite strain is assumed to be the same as the matrix strain in the region immediately ahead of the fiber-bridged crack. Hence,

$$K_{IC}^c = K_{IC}^m (E_1/E_m) \tag{36}$$

This assumption is more valid in the upperstream region III (Figure 26). But, it is not obvious that the composite and matrix strains are the same in region II. Chiang, Wang and Chou further discounted the difference in the Poisson's ratios between the composite and matrix and adapted the following assumption

$$K_{IC}^c = K_{IC}^m (E_1/E_m)^{1/2} \tag{37}$$

McCartney proposed a fracture criterion of $\gamma_c = V_m \gamma_m$ or $G_c = V_m G_m$ at the crack-tip, where γ_c and γ_m are, respectively, the composite and matrix crack surface energies. This results in the following expression for K_{IC}^c:

$$K_{IC}^c = K_{IC}^m (V_m E_1/E_m)^{1/2} \tag{38}$$

The differences of these three fracture toughness expressions are quite significant. The MCE model gives highest values of K_{IC}^c and the McCartney model, lowest.

The MCE model predicts a discontinuity in the fiber stress at the end of fiber slipping length and also a decrease in the slip length of the fibers near the crack tip. Chiang, Wang and Chou have shown that the stress distribution becomes continuous when the shear deformation in the matrix is taken into account and the slip length distribution decays smoothly. Also, according to MCE and McCartney models the traction, T(X) approaches zero at the crack-tip while the Chiang, Wang and Chou analysis predicts a finite traction at the crack tip. Since the closure traction distribution T(X) determines the crack-tip stress intensity factor, K, the stress intensity factor predicted by the three models differ significantly.

Although each model predicts the ACK result for long cracks, the definition of long cracks also differs among the models. In the MCE model, a characteristic crack length is defined as

$$c_m = \left(\frac{\pi K_{IC}^c}{1.44 V_f^2 \omega} \right)^{2/3} \tag{39}$$

where K_{IC}^c is given by Eq. (37) and

$$\omega = \frac{8 (1-v^2) \tau_f E_f}{a \pi^{1/2} V_m E_m} \tag{40}$$

A long crack is defined as one whose length is equal to or greater than $c_m/3$.

Similarly, McCartney defines the characteristic crack length as

$$c_0 = (\sqrt{\pi} \, K_{IC}^c / \lambda^2)^{2/3} \tag{41}$$

where K_{IC}^c is given by Eq. (38) and

$$\lambda = 2 \frac{V_f}{V_m} \left(\frac{2 \pi \, \tau_f E_f E_c}{a \, E_m^2} \right)^{1/2} \tag{42}$$

A crack having a length of $5c_0$ or longer then meets the long crack condition.

The Chiang, Wang and Chou analysis determines the critical length numerically. For SiC/borosilicate, C/borosilicate and SiC/LAS composite systems, the Chiang, Wang and Chou analysis shows that the long crack condition is reached only after the crack has propagated to a length of many fiber diameters while the MCE and McCartney models show that the long crack condition may be achieved by a crack whose length is shorter than one fiber diameter.

The matrix cracking stresses predicted using the three models are shown in Figure 31, for SiC/LAS composites. The Chiang, Wang and Chou analysis predicts higher matrix cracking stress than the MCE and McCartney models in the region of short cracks, but all predictions approach the ACK results for long cracks. Note that, in both the MCE and McCartney models, the long-crack condition is reached at a very short length (on the order of one fiber diameter or less) in the borosilicate-based systems. In contrast, the Chiang, Wang and Chou analysis requires a much greater length.

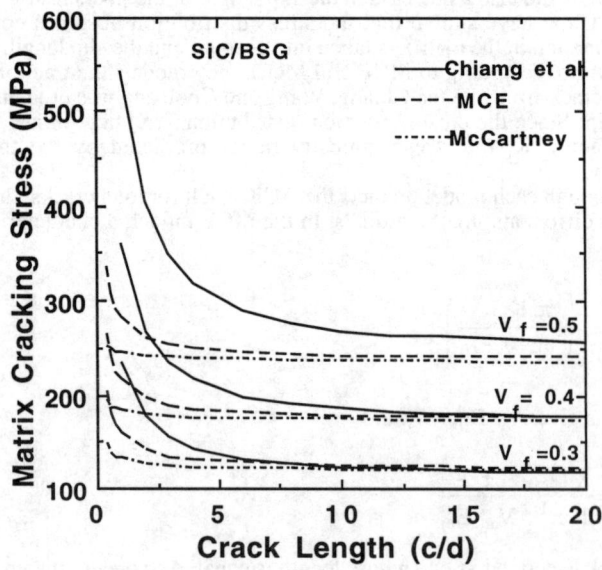

Fig. 31 Matrix cracking stress as a function of crack length: comparison of predictions by different models for Nicalon-BSG [116].

Local flaw model

The local flaw model addresses the crack initiation at a microscopic level. The unit cell in this model consists of a matrix flaw between two fibers. The interface is assumed to have a local flaw and interactions between the interfacial and matrix flaw are considered. The critical strain energy release rate criterion is used to predict the growth of the local flaw (crack initiation). Based on the sizes of the interfacial flaw, lower and upper bounds are predicted for matrix crack initiation. The experimental data for carbon-BSG system was found to lie within these bounds for a range of fiber volume fractions.

6.1.2 Stress-Strain Behavior and Modulus Reduction

Damage development in CMCs leads to non-linearity in the stress-strain behavior and degradations of the elastic moduli. Many theoretical analyses dealing with these phenomena have been proposed. The phenomenon of multiple cracking in brittle matrix composites was first addressed analytically by Aveston, Cooper and Kelly (ACK) [10] as follows. In the plane of a matrix crack, the matrix is fully unloaded. As we go away from the matrix crack, the load is transferred to the matrix via interfacial shear stresses. ACK assumed that the interfacial shear stress has a constant limiting value, τ_f. The stress in the matrix increases linearly (Figure 32a) away from the matrix crack. If the matrix strength is assumed to be single valued and equal to σ_{mu}, the stress in the matrix will build up to this value at a characteristic distance, x'. Consider a unit area of composite containing N fibers. Let A_m and A_f be the area fractions of the matrix and the fiber respectively. Then the load transferred to the matrix from the fibers is given by

$$\int_0^{x'} N\, 2\, \pi\, R\, \tau_f\, dx = \sigma_{mu}\, A_m \tag{43}$$

Then using the fact that A_m/A_f equals V_m/V_f and A_f equals $N \pi R^2$ the following expression can be derived for x'.

$$x' = \frac{V_m}{V_f} \frac{\sigma_{mu} R}{2\tau_f} \tag{44}$$

A new matrix crack will be created at a distance x' from the original crack (and no crack can be created at distances smaller than x'). However, if another matrix crack exists at a distance less than $2x'$ form the original crack, the stress can not be built up to the matrix cracking level between these two cracks and the spacing will remain higher than x' and less than $2x'$. Thus, the entire matrix will be traversed by cracks with saturation crack spacing between x' and $2x'$. Kimbler and Keer [99] have shown by statistical analysis that the saturation crack spacing is 1.337 times x'.

If the breaking strain of the matrix is single valued, the composite will be traversed by multiple cracks as soon as the first crack appears without further load increase. Due to the multiple cracking of the matrix, additional load is thrown on to the fibers resulting in additional strain. The additional strain in the composite varies between ($\alpha\, \varepsilon_{mu}/2$) and ($3\alpha\varepsilon_{mu}/4$) when the cracking is complete (corresponding to saturation crack spacing of $2x'$ and x' respectively). On further loading of the composite fibers carry all the load and stretch by slipping through the blocks of matrix resulting in a composite Young's modulus very close to $E_f V_f$. The final failure occurs when the composites stress reaches a value of

$\sigma_{fu}V_f$. Thus, the entire stress-strain behavior of the composite could be predicted approximately (Figure 32b).

Aveston and Kelly extended their shear lag analysis by considering non-constant interfacial shear stress as described in Section 4.1.1. Using this model they predicted the stress-strain behavior for portland cement reinforced by steel fibers (Figure 33). They calculated the shear stresses at the interface and showed that in practical ceramic composites, some debonding is bound to occur. The extent of debonding depends on the limiting fiber-matrix shear stress after debonding.

Karandikar and Chou [55] and Kuo and Chou [113] used the stress distributions given by the shear lag models (with and without thermal stresses, respectively) to compute the average strain in the fiber (and the composite) and hence the secant modulus.

$$\varepsilon_c = \frac{1}{L} \int_{-L/2}^{L/2} \left(\frac{\sigma_f}{E_f} + \alpha_f \Delta T - \alpha_1 \Delta T \right) dx \qquad (45)$$

where σ_f is given in Eqs. (30) and (31). Performing the integration gives the following equations for composite stress-strain relation and degraded secant modulus E_1' respectively.

$$\varepsilon_c = \frac{\sigma_c}{E_1'} \left(1 + \frac{E_m V_m}{E_f V_f} \frac{2L_d}{L} \right) + \frac{2L_d}{L} (\alpha_f - \alpha_c) \Delta T$$

$$+ \frac{2}{\beta L} \tanh\beta(L/2 - L_d) \left(\frac{E_m V_m}{E_1' E_f V_f} \sigma_c + (\alpha_f - \alpha_c)\Delta T - 2\frac{\tau_f}{E_f}\frac{L_d}{R} \right)$$

$$+ 2\frac{\tau_f}{E_f}\frac{L_d}{R}\frac{L_d}{L} \qquad (46)$$

$$E_1' = \frac{\sigma_c}{\varepsilon_c} \qquad (47)$$

It can be seen that the secant modulus approaches E_1' and $E_f V_f$ for no cracking ($L \to \infty$) and very dense cracking ($L \to 0$), respectively, if the thermal effect is not considered.

Both matrix crack spacing and debond length depend on the applied stress and can be calculated by the shear lag model assuming that matrix cracking strength is single-valued and the debond energy is well characterized. But in the actual composite, the matrix strength depends on the local microstructure (defects and fiber spacing) and debond energy is difficult to measure. Therefore, no attempt is made to predict the crack spacing and debond length as a function of applied stress.

The experimentally measured crack spacings for a unidirectional Nicalon-CAS composite are used instead and the debond length is varied systematically between zero and 8R as the saturation spacing was found to be of the order of 16R. No significant fiber fracture was observed at saturation spacing and therefore all the modulus decrease can be attributed to matrix cracking and debonding. The experimentally measured modulus values and the theoretical predictions are shown in Figure 34. It is clear that matrix cracking alone does not account for the modulus degradation. It can be seen that at high applied strain the experimental values approach the model predictions for debond length of 8R. The crack spacing at this stress level is about 16R. Thus, the entire interface is debonded and only frictional stress transfer occurs between the fiber and matrix. For Nicalon-CAS system, this conclusion was not changed significantly by including/excluding the thermal stresses [55, 113].

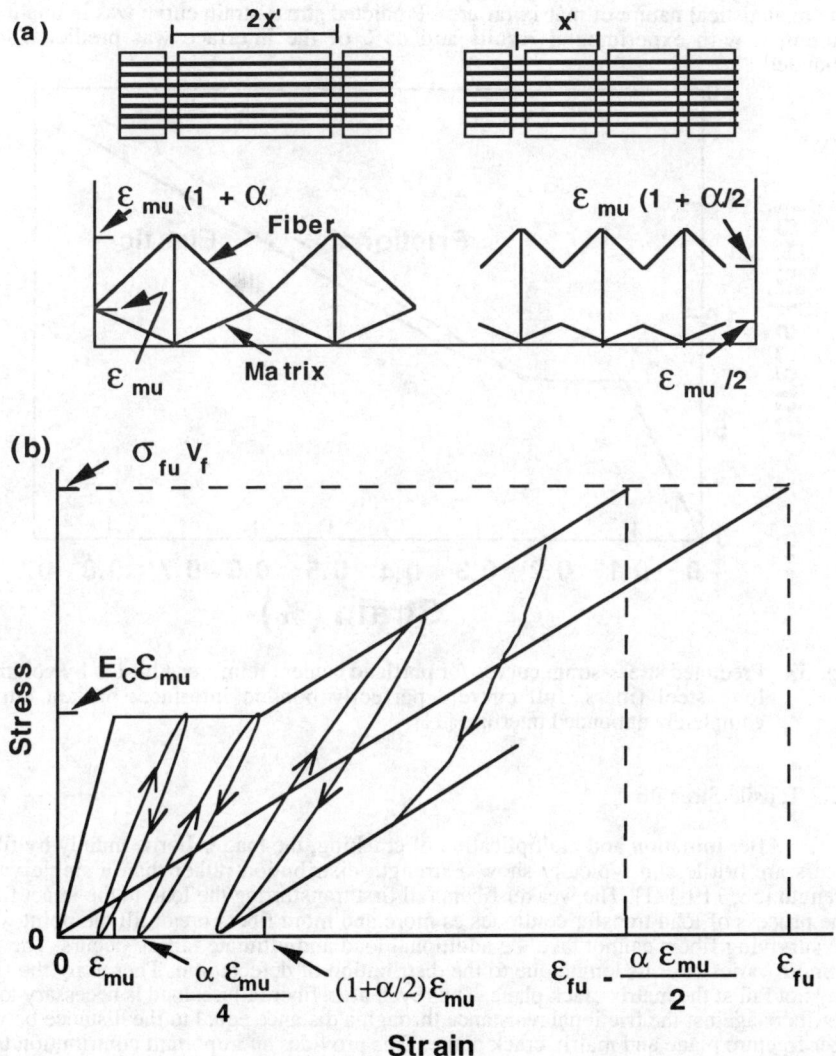

Fig. 32 (a) Fiber and matrix stress distributions by simple shear lag (b) Stress-strain curves expected for multiple fracture (matrix with single-valued failure strain) [11].

Lee and Daniel [118] used single valued matrix strength and interfacial shear strength and an iterative technique to predict the matrix crack spacing and debond length by a shear lag model. This model considered radial stress variations in the fibers and matrix but did not allow stress transfer by friction in the debonded area. While the experimental crack density increased gradually, the predicted crack density increased in a step manner (due to statistical nature of matrix failure). Predicted stress-strain curve was in qualitative agreement with experimental results and 85% of the interface was predicted to be debonded.

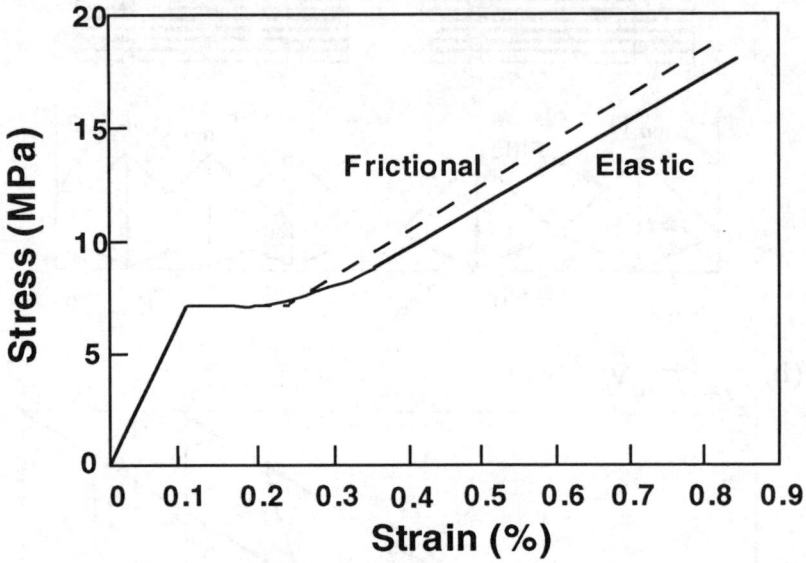

Fig. 33 Predicted stress-strain curves for portland cement reinforced by 1% by volume of long steel fibers: full curve - perfectly bonded interface, broken curve - completely unbonded interface [11].

6.1.3 Tensile Strength

After initiation and multiplication of cracking, the load is borne mainly by fibers. Fibers are brittle and typically show a strength distribution rather than a single valued strength [58, 119-121]. The weaker fibers fail first transferring the load to the intact fibers. The process of load transfer continues as more and more fibers break till the point where the surviving fibers cannot take the additional load and ultimate failure occurs. The fiber strength varies over its length due to the distribution of defects in it. Therefore, the fibers may not fail at the matrix crack plane. Thus even after fiber failure, load is necessary to pull the fibers against the frictional resistance through a distance equal to the distance between fiber fracture plane and matrix crack plane. This provides an important contribution to the toughness of the composite.

Prewo [58] applied the simple model given by Equation (2) in the introduction (average fiber strength multiplied by volume fraction) to predict the ultimate tensile strength of Nicalon-LAS composites. Strength measurements were done by tensile testing individual fibers carefully extracted from the composite. As noted previously, the strength of the extracted fibers was 30 to 40% lower than that of the as-received fibers. The agreement between the predictions and the experimental values was particularly good even though the strength variability of fibers was neglected.

Coleman [119] has shown that a Weibull distribution can be used to represent the strengths of a group of fibers. For long fibers the cumulative probability of failure, P_f, at applied stress, σ_f, is given by

$$P_f = 1 - \exp\left[-\left(\frac{l}{l_0}\right)\left(\frac{\sigma_f}{\sigma_0}\right)^m\right] \tag{48}$$

l is fiber length, l_0 is a reference length, σ_0 is the scale parameter and m is the Weibull shape parameter. The mean fiber strength, $\overline{\sigma}_f$, is given by

$$\overline{\sigma}_f = \sigma_0 \left(\frac{l}{l_0}\right)^{(-1/m)} \Gamma\left(1 + \frac{1}{m}\right) \tag{49}$$

where Γ denotes the gamma function.

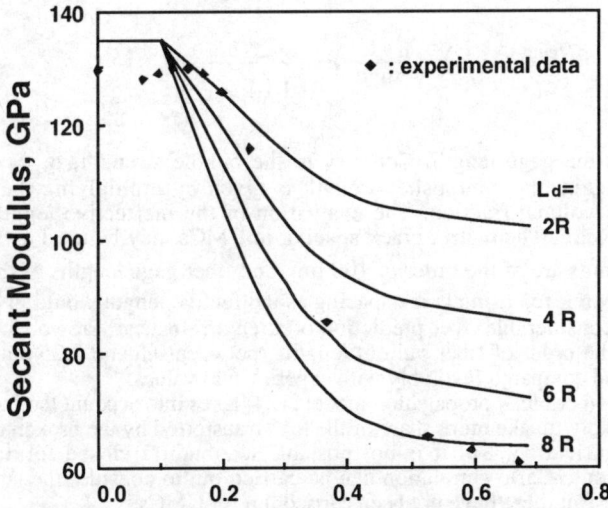

Fig. 34 Prediction of degradation of longitudinal Young's modulus with applied strain for unidirectional Nicalon-CAS - experimental values are shown by points and predictions at various levels of debonding are shown by lines [133].

If a loose bundle of fibers is loaded, the weaker fibers will break first transferring their load to unbroken fibers. The stress on the unbroken fibers is higher than the average bundle stress. The bundle as a whole will fail when the stress in the unbroken fibers reaches the ultimate strength. Coleman gives the following expression for the ratio of the bundle strength, σ_b, and the mean strength.

$$\frac{\sigma_b}{\bar{\sigma}_f} = \left(\frac{1}{m\,e}\right)^{\frac{1}{m}} \frac{1}{\Gamma\left(1 + \frac{1}{m}\right)} \tag{50}$$

The bundle strength is smaller than the mean strength and it reduces as the variation in the fiber strength increases (smaller m). Prewo [58], Bhatt and Phillips [73] and Davidge and Briggs [122] obtained the composite UTS by multiplying the bundle strength by the volume fraction in Nicalon-LAS, Nicalon-BSG and SiC-RBSN composites respectively. The UTS was found considerably underestimated for Nicalon-LAS while reasonably well predicted for Nicalon-BSG and SiC-RBSN.

The above equation gives the bundle strength of a loose bundle. It assumes that the load from broken fibers is evenly distributed among the unbroken fibers. But in an actual composite, matrix carries part of the load and also plays an important role in transferring the load to the unbroken fibers and parts of the broken fibers away from the break point. Also, the load will be transferred to adjacent unbroken fibers more effectively than fibers which are far away.

The cumulative weakening model [123, 124] considers the modified strains around the broken fiber ends. Around the broken end, there is a length of the fiber which does not support full load and is called ineffective length, l_c. The ineffective length depends on the interfacial shear strength. The following expression is given by Rosen for the ratio of the cumulative strength of the fibers to the mean strength.

$$\frac{\sigma_{cum}}{\bar{\sigma}_f} = \left(\frac{l}{l_c}\right)^{\frac{1}{m}} \left(\frac{1}{m\,e}\right)^{\frac{1}{m}} \frac{1}{\Gamma\left(1 + \frac{1}{m}\right)} \tag{51}$$

where l is the gage length. Contrary to the bundle strength, σ_{cum} exceed the mean strength. Again, the composite strength is given by multiplying the cumulative fiber strength by volume fraction. The evaluation of the ineffective length is however, not straight forward. The matrix crack spacing in CMCs may be used as ineffective length. Crack spacings are of the order of 100 μm, and fiber gage lengths of the order of 25 mm [55, 58]. Therefore, using crack spacing as ineffective length would give a large $(l/l_c)^{1/m}$ factor and considerable over prediction of strength. Instead, Prewo [58] used ineffective lengths of the order of fiber pullout lengths (between 0.5 and 0.25 cm). The predictions thus obtained compared favorably with experimental values.

The fiber break propagation model [124] takes into account the fact that the adjacent fibers are likely to take more share of the load transferred by the broken neighbor causing a stress magnification. But it is not possible to obtain a closed form equation for this strength. Monte-Carlo simulation can be carried out to consider the stress magnification effect [125]. But these have not been carried out for CMCs.

The above mentioned theories were not developed particularly for ceramic matrix composites. In ceramic matrix composites matrix microcracking and fiber/matrix interfacial sliding are present and modify the failure process significantly. A few models have been proposed recently specifically for CMCs [4, 126-132]. The approaches used for CMCs consider either failure due to a single crack or after formation of saturation multiple cracking. Statistical theories are used to predict the location of fiber failure away from the matrix crack plane and to predict the pullout distribution.

Thouless, Sbaizero, Sigl and Evans (TSSE) [126] considered multiple matrix cracking and stress transfer through constant frictional stresses. Once a fiber fails, it is assumed to be completely unloaded. In the TSSE model, the mean pull out length is given by

$$<h> = \frac{R \Sigma}{2 \tau_f (m+1)} \Gamma \left(\frac{m+2}{m+1}\right) \qquad (52)$$

where

$$\Sigma = \left[\frac{A_0 S_0^m \tau_f (m+1)}{2 \pi R^2}\right]^{\frac{1}{m+1}}$$

$$S_0 = \sigma_0 \left[\frac{2 \pi R L}{A_0}\right]^{\frac{1}{m}}$$

Also, L is the gage length and A_0 is a normalizing factor ($= 1$ m^2). The composite strength is given by

$$\sigma = V_f S \exp \left\{ -\frac{[1-(1-\frac{\tau_f d}{R S})^{m+1}]}{(m+1)[1-(1-\frac{\tau_f d}{R S})^m]} \right\} \qquad (53)$$

where d is the saturation crack spacing and

$$\left(\frac{R S}{\tau_f d}\right)^{m+1} = \left(\frac{A_0}{2 \pi R L}\right) \left(\frac{R S_0}{\tau_f d}\right) \left[1-(1-\frac{\tau_f d}{R S})^m\right]^{-1}$$

The model predicted the UTS of Nicalon-CAS composite fairly well but the mean pullout lengths predicted by Equation (52) significantly overestimate the pullout lengths in heat treated samples of Nicalon-LAS. This model was used by Cao et al. [56] to predict the UTS of six different Nicalon fiber-based composites by using UTS of one of the composites to scale the fiber strengths. Knowles and Yang [132] compared the predictions of Equation (53) with that of Equation (50) for TSSE's data. They showed that the TSSE model is essentially same as the bundle model with a non-uniform stress in the matrix. They also showed that the TSSE model predicts slightly higher UTS than the bundle model. The overestimation of the pullout lengths was attributed to the fact that the stress distribution was obtained by considering only a single matrix crack. TSSE also showed that mirror radii measurements on the failed fibers in composites could be used to estimate the strength of the fibers in-situ. But this approach requires assuming a fracture toughness for the fibers.

Cao and Thouless (CT) [127] followed the approach of TSSE to predict composite strength. The stress transfer was assumed to occur through constant frictional stresses (the fiber stress decreases linearly from the matrix fiber plane to a distance equal to half the crack spacing). According to the Cao-Thouless model, composite strength is given by

$$\sigma = V_f \Sigma \left[\frac{\Sigma R}{m (m+1) \tau_f L}\right]^{\frac{1}{m}} \exp(-1/m) \qquad (54)$$

The Cao-Thouless predictions for Nicalon-LAS composites agreed fairly well with experimental values.

Sutcu [130] has proposed that the load required for final fracture of a CMC is the load on the intact fibers plus the load required to pullout the fibers through average pullout length against the interfacial frictional stress. The composite strength is given by the maximum of S where S is given by

$$S = (1-P_f) \sigma_{max} V_f + P_f L_p^* 2 \tau_f V_f/R \tag{55}$$

where σ_{max} is the peak stress on the fiber, P_f is the cumulative fiber failure probability and L_p^* is the average pullout length. σ_{max} is increased gradually from 0 and corresponding values of P_f and L_p^* are computed. These are used in Equation (55) to to calculate S. The strength predicted by Equation (55) is really an upper bound on the strength and accordingly it over predicted the strength of Nicalon-LAS composites.

Curtin [131] has proposed a model in which the basic assumptions are similar to that proposed by TSSE with allowance for multiple failure of the fibers. Curtin has also proposed that the gage length is not simply the length of the specimen but depends on several materials parameters and is proportional to δ_c which is given by

$$\delta_c = \frac{\sigma_{fu}^* R L_o^{1/m}}{\tau_f} \tag{56}$$

where σ_{fu}^* is the stress required to cause one failure on an average in a fiber of length L_o (= stress at 63.2 % fiber failure or $\sigma_{fu}^* = \sigma_{fu} (\ln 2)^{-1/m}$, where σ_{fu} is the average fiber strength). According to this analysis the mean pullout length is given by

$$<L> = \frac{1}{4} \frac{\lambda_2(m)}{\lambda_1(m)} \delta_c \tag{57}$$

where λ_1 and λ_2 are slowly varying functions of m only. The composite strength, S is given by

$$S = V_f \sigma \left[\frac{2}{m+2}\right]^{\frac{1}{m+1}} \frac{m+1}{m+2} \tag{58}$$

where

$$\sigma = \left[\frac{(\sigma_{fu}^*)^m \tau_f L_o}{R}\right]^{\frac{1}{m+1}}$$

Curtin showed that his model could be used to predict the UTS of all different Nicalon-LAS composites reported by Prewo [58].

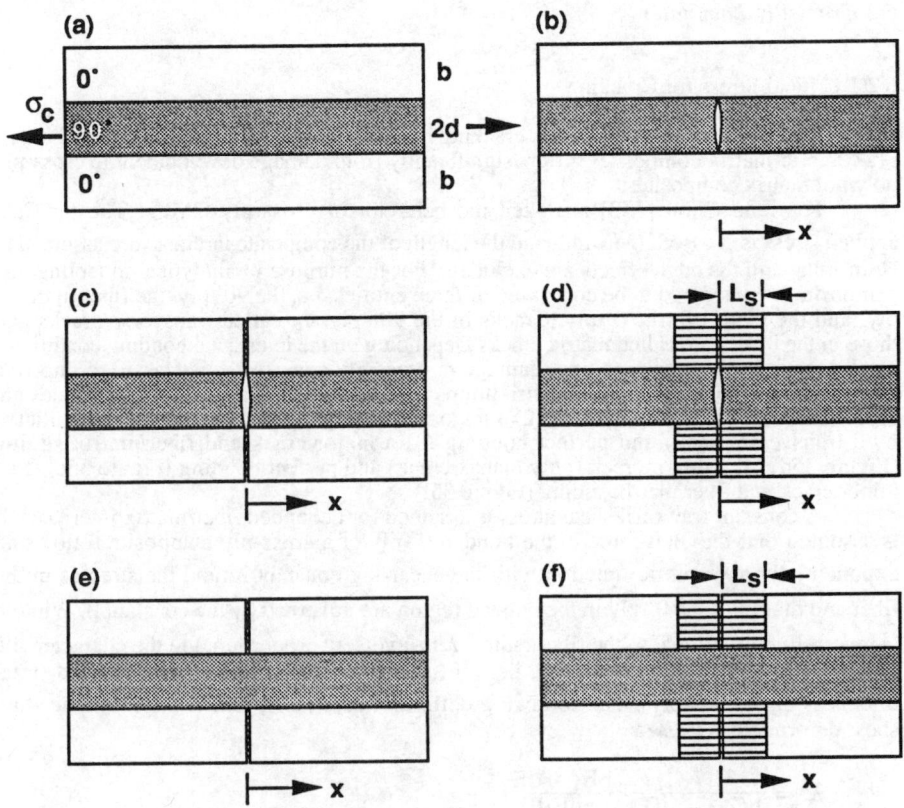

Fig. 35 Damage modes in cross-ply CMCs (a) No damage (b) a transverse crack (c) a major crack (a matrix crack initiated by a transverse crack) and perfect bonding (d) a major crack and fiber/matrix sliding (e) a minor crack (only matrix crack) and perfect bonding (f) a minor crack and fiber matrix sliding.

6.2 Cross-Ply Laminates

6.2.1 Critical Stress for Cracking

Due to the presence of matrix cracking in the 0° plies damage development in cross-ply ceramic-matrix composites differs significantly from damage development in cross-ply polymer matrix composites.

Kuo and Chou [133] analyzed the behavior of cross-ply CMCs. The far-field applied stress is σ_c. Both the width and the length of the composite laminate are assumed to be infinite, and the edge effects are excluded. For the purpose of analytical modeling, the composite is considered to be composed of three entities, i.e. the 90° ply, the fiber in the 0° ply, and the matrix in the 0° ply. Cracks in the 90° ply are called transverse cracks and those in the 0°-ply are called matrix cracks Depending on the interfacial bonding conditions, crack location, and crack spacing, damage in cross-ply composites is classified into five modes, and the solution of stress distributions is given for each mode. The five modes are (1) a transverse crack (Figure 35b) (2) a major crack (Figure 35c) (a matrix crack initiated by a transverse crack) and perfect bonding (3) a major crack and fiber/matrix sliding (Figure 35d) (4) a minor crack (only matrix crack) and perfect bonding (Figure 35e) (5) a minor crack and fiber matrix sliding (Figure 35f).

A constant frictional shear stress is assumed for debonded fiber/matrix interfaces. It is assumed that the total force in the bonded 0°-ply of a cross-ply composite follows an exponentially asymptotic function with the shear-lag constant λ, and the stresses in the fiber and matrix of the 0°-ply in the bonded region are governed by the constant β. While β is inversely proportional to the fiber radius, λ is inversely proportional to the characteristic lengths of the cross-ply composite, i.e. 2b (total thickness of the 0°-ply) and 2d (total thickness of the 90°-ply). The following definition of λ [118] which takes into account shear deformations is used

$$\lambda^2 = \frac{3G_{13}G_{23}}{bG_{23} + dG_{13}} \frac{bE_1 + dE_2}{bdE_1E_2} \tag{59}$$

where the subscript 3 stands for the thickness direction.

For the undamaged state and the five possible damage states, the stresses in the transverse ply, and fibers and matrix of longitudinal ply can be computed. These stresses are used to carry out the following energy balance involved in the initiation of transverse or matrix cracks in a cross-ply laminate.

$$U_c + \gamma_{db} + U_s + \Delta U_f \leq \Delta U_t + \Delta U_m + \Delta W \tag{60}$$

where ΔU_t is the strain energy decrease in the transverse ply, and U_c is the surface energy of the crack surfaces; the meanings of all the other energy terms are the same as those given for Eq. (23). The definitions of these energy terms are

$$\gamma_{db} = \frac{bV_f}{R} 2 L_s G_{II}$$

$$U_s = \frac{bV_f}{\pi R^2} \int_0^{L_s} 2\pi R \tau_f (\Delta u_f - \Delta u_m) \, dx$$

$$\Delta U_f = \int_0^S \left(\frac{bV_f}{2E_f} ((\sigma_f^f)^2 - (\sigma_f^i)^2) \right) dx \qquad (61)$$

$$\Delta U_t = \int_0^S \left(\frac{d}{2E_2} ((\sigma_t^i)^2 - (\sigma_t^f)^2) \right) dx$$

$$\Delta U_m = \int_0^S \left(\frac{bV_m}{2E_m} ((\sigma_m^i)^2 - (\sigma_m^f)^2) \right) dx$$

$$\Delta W = \int_0^S \left((b+d) \frac{\sigma_c}{E_f} (\sigma_f^f - \sigma_f^i) \right) dx$$

where σ denotes stress with subscripts m, f and t for matrix, fiber and 90°-ply, respectively and the superscripts i and f refer to the initial and final states, respectively. S is the length of the representing volume; Δu_f and Δu_m are, respectively, the displacement differences of the fiber and matrix of the 0°-ply before and after the occurrence of the crack. The definition of U_c varies with the cracking mode. U_c, γ_{db}, U_s, ΔU_f, and ΔW are always positive; the signs of ΔU_t and ΔU_m depend on the cracking modes. By using Eqs. (60) and (61), the critical stresses for crack initiation for the various cracking modes are discussed in the following. In order to simplify the analysis it is assumed that $G_{II} = 0$.

Numerical results for the prediction of critical stresses for matrix cracking have been obtained for damage states (1)-(5) denoted by σ_{cr1}^c, σ_{cr2}^c, σ_{cr3}^c, σ_{cr4}^c, and σ_{cr5}^c, respectively. The calculations are performed based upon the material properties of the SiC/CAS composite [55]. The effect of ply thickness on the critical stress for matrix cracking in the 0°-ply is shown in Figure 36. The experimental results of matrix cracking initiation are indicated by the solid circles; the stresses at which the crack density reaches 95% of saturated density are indicated by open circles. Matrix cracking develops in the stress range between the closed and open circles. The predicted σ_{cr3}^c and the experimental results follow the same trend. Thus, for Nicalon-CAS composites, damage mode (3) requires the lowest stress and is thus most probable i. e. initiation of matrix crack will occur from an existing transverse crack with accompanying fiber/matrix debonding and sliding. As d, the thickness of the 90°-ply, approaches zero, the solutions converge to two points, which coincide with the predicted critical stresses for a unidirectional composite (σ_{cr1}^u and σ_{cr2}^u). As the ratio d/b increases, σ_{cr2}^c and σ_{cr3}^c decrease, while σ_{cr4}^c and σ_{cr5}^c increase. The reason is that σ_{cr2}^c and σ_{cr3}^c are associated with major cracks in which higher d/b ratios mean higher stress concentrations. On the other hand, σ_{cr4}^c and σ_{cr5}^c are associated with minor cracks, and the 90°-ply shares the load with the fibers in the 0°-ply. Hence, the stress in the 0°-ply is reduced. For the composites examined, σ_{cr3}^c is the lowest, indicating that state (3) is the dominating matrix cracking mode. The fact that the theoretical predictions are higher than the experimental results may be due to the simplification of the model and the accuracy of the material properties used, especially τ_f and γ_m.

The effect of fiber volume fraction on the matrix cracking stress is illustrated in Figure 37. Again, σ_{cr3}^c is the lowest and the critical stress increase with the volume fraction. The effect of fiber modulus is illustrated in Figure 38, where E_c denotes the undamaged composite modulus. Here, σ_{cr2}^c/E_c and σ_{cr3}^c/E_c, unlike σ_{cr4}^c/E_c and σ_{cr5}^c/E_c, show trends of slight increase with E_f/E_m. The critical stresses for all damage modes were found to increase monotonically with V_f. Among the damage states, σ_{cr3}^c again is the lowest for the range of V_f studied.

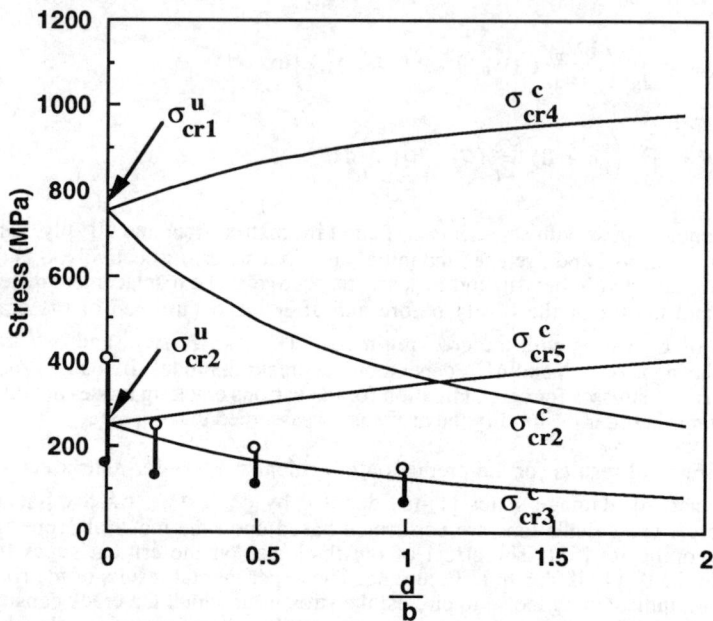

Fig. 36 Critical stress for matrix cracking in cross-ply CMCs as affected by the thickness of the 90° ply; solid circles - experimental crack initiation stress, open circles - experimental crack saturation stress [133].

6.2.2 Modulus Reduction

Karandikar and Chou [55] obtained the reduction in the Young's modulus of cross-ply CMCs by superposition of two different shear lag models. The transverse cracks in the 90° layer lead to redistributions of the stresses in the vicinity of the cracks. The effect of the presence of matrix cracking in the 0° plies on the stress distributions was neglected and the altered stress distributions in the 0° and 90° plies, $\sigma_1(x)$ and $\sigma_2(x)$ were computed by a shear lag analysis (see Figure 39) developed for cross-ply polymer matrix composites [134]. Based on the stress distributions, the reduced modulus of the cracked 90° layer E_2' is given by

$$E_2' = \frac{\overline{\sigma}_2}{\varepsilon_1} \tag{62}$$

where

$$\bar{\sigma}_2 = \frac{1}{l} \int_0^l \sigma_2(x) \, dx$$

$$\bar{\varepsilon}_1 = \frac{1}{l \, E_1} \int_0^l \sigma_1(x) \, dx$$

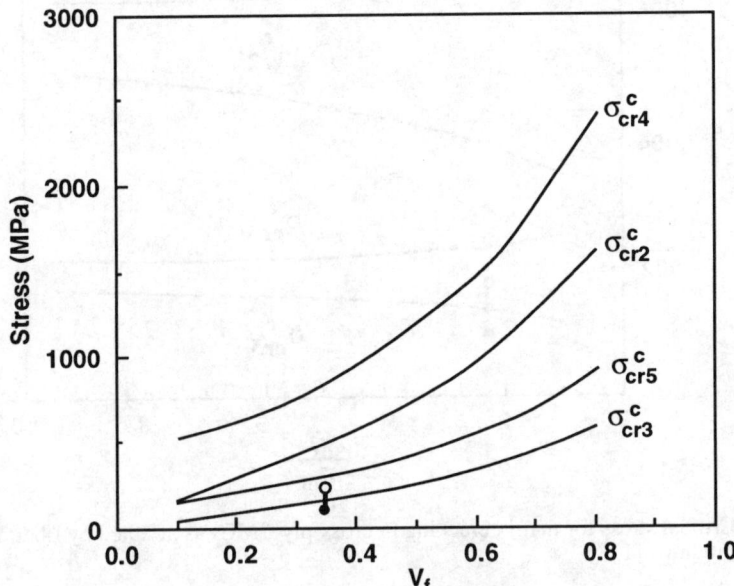

Fig. 37   Critical stress for matrix cracking in cross-ply CMCs as affected by fiber volume fraction [133].

The reduced modulus of the 0° plies was computed from Eq. (47), assuming that the cracks in the 0° plies and the 90° plies do not interact. The reduced moduli for 0° and 90° plies were input to the following rule-of-mixtures to calculate the modulus of the cross-ply laminate.

$$E_L' = E_1' \left(\frac{b}{b+d}\right) + E_2' \left(\frac{d}{b+d}\right) \qquad (63)$$

The experimentally determined secant Young's moduli (average from the two specimen faces) for the cross-ply laminates are shown along with the model predictions in Figure 40 for the $[0_3/90/0_3]$ Nicalon-CAS laminate. Similar to the unidirectional case (Figure 34), for the cross-ply laminate the experimental values at high applied strain compare very well with the model predictions for debond length of 60 μm. The saturation spacing of the matrix cracks in the 0° plies of all three cross-ply laminates is of the order of 125μm. Therefore, we can conclude again that almost the entire interface in the 0° plies is debonded. In other words, only frictional stress transfer occurs between the fiber and the matrix of the 0° plies at saturation damage.

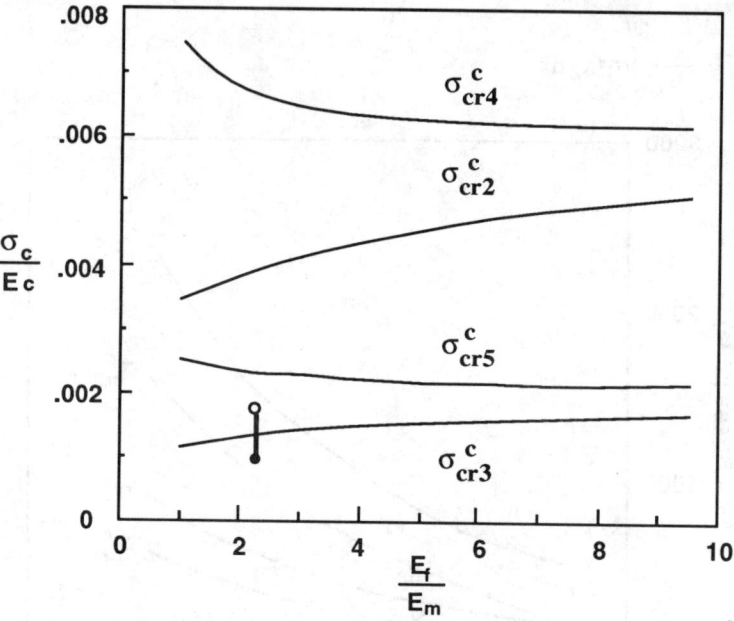

Fig. 38 Critical stress for matrix cracking in cross-ply CMCs as affected by fiber Young's modulus [133].

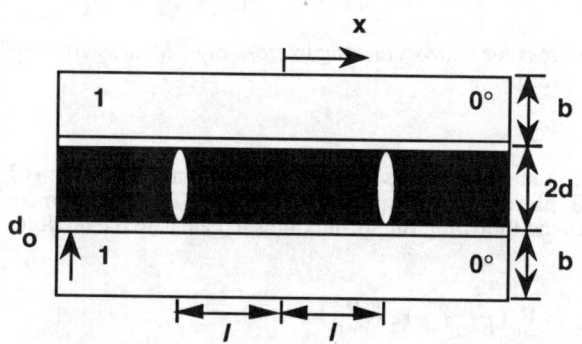

Fig. 39 Unit cell for the shear lag analysis of cross-ply laminates [55, 134].

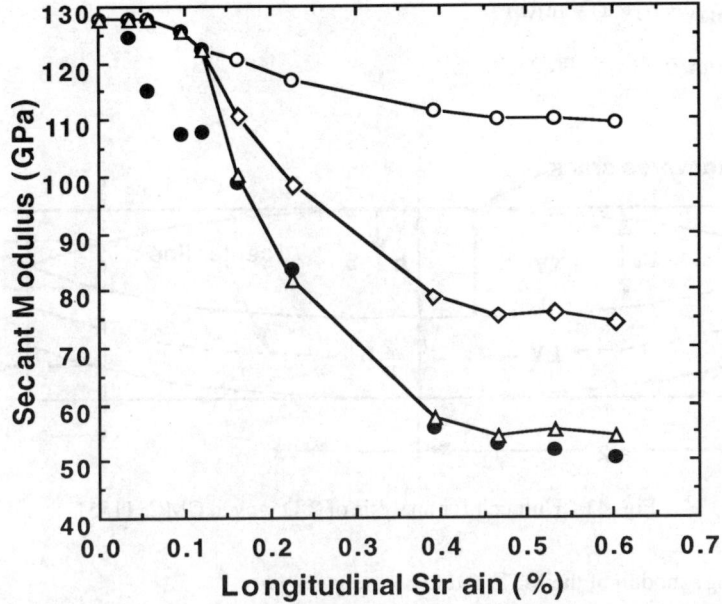

Fig. 40 Variation of longitudinal Young's modulus with applied strain for the [0₃/90/0₃] laminate (● - experimental values, ○ - predicted values $S_d = 0$, ◊ - predicted values $S_d = 30$, △ - predicted values $S_d = 60$) [55].

6.3 Woven Composites

6.3.1 Critical Stress for Matrix Cracking

Kuo and Chou [135] calculated the matrix cracking stress for woven composites by three approaches. In the first, the stress distribution obtained by shear lag analysis was used to carry out the total energy balance (TEB). In the second case, finite element analysis was used to obtain the stress distribution which was subsequently used to carry out the TEB. In the third approach classical fracture mechanics (CFM) approach was used. The strain energy release rate was calculated by the finite element analysis technique and was compared with the critical strain energy release rate to obtain the critical stress for cracking. These three approaches are summarized in the following.

The unit cell for the shear lag analysis of woven composites is shown in Figure 41. Since the volume occupied by the interyarn matrix is relatively small, the matrices in the upper and lower interyarn spaces (m_1 and m_2) are combined into the transverse yarn (TY) and longitudinal yarn (LY), respectively. Hence, the composite is considered to consist of two parts: a combined TY and m_1, denoted as a modified TY (MTY), and a combined LY and m_2, denoted as a modified LY (MLY). The thicknesses of the MTY and MLY are given by

$$h_{MTY} = h_{TY}(x) + m_1(x)$$
$$h_{MLY} = h_{LY} + m_2(x) \qquad (64a)$$

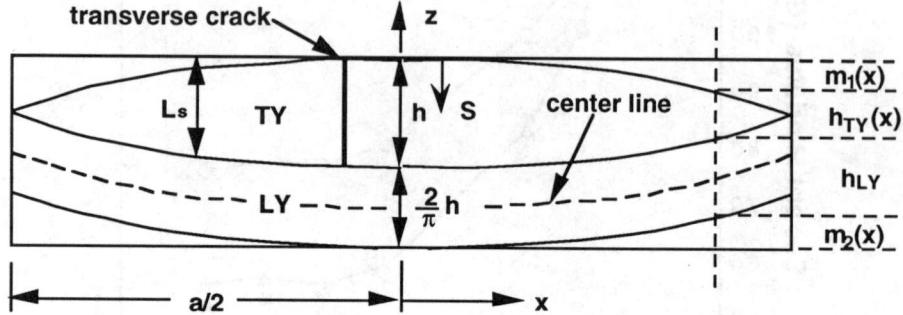

Fig. 41 Unit cell for analysis of 2-D woven CMCs [135].

The Young's moduli of the MTY and MLY are given by

$$E_{MTY} = \frac{E_2 h_T(x) + \eta E_m m_1(x)}{h_{MTY}}$$

$$E_{MLY} = \frac{E_x(x) h_L + \eta E_m m_2(x)}{h_{MLY}} \qquad (64b)$$

where η is the fraction of matrix content in interyarn region, and $E_x(x)$ is the Young's modulus of the LY in the x-direction. Due to fiber undulation, $E_x(x)$ can be expressed as

$$E_x(x) = \left[\frac{1}{E_1}\cos^4\theta_x + \left(\frac{1}{G_{12}} - \frac{2\nu_{12}}{E_1}\right)\cos^2\theta_x \sin^2\theta_x + \frac{1}{E_2}\sin^4\theta_x \right]^{-1} \qquad (65)$$

where θ_x is the angle between the LY and the x-axis.

It is assumed that the shear stress in the interface of the MTY and MLY is proportional to the difference between the average displacements of the MTY and MLY:

$$\tau = k(\bar{u}_{MTY} - \bar{u}_{MLY}) \qquad (66)$$

where k is a shear-lag coefficient, and $\bar{u}_{MTY}$ and $\bar{u}_{MLY}$ are the averaged displacements of the MTY and MLY, respectively. Using Equation (66), the equilibrium of MTY and MLY elements can be written as

$$\frac{d\sigma_{MTY}}{dx} = \frac{k}{h_{MTY}}(\bar{u}_{MTY} - \bar{u}_{MLY}) \qquad (67)$$

$$\frac{d\sigma_{MLY}}{dx} = \frac{k}{h_{MLY}} (\bar{u}_{MLY} - \bar{u}_{MTY}) \qquad (68)$$

Taking the derivative of Eq. (67), we have

$$\frac{d^2\sigma_{MTY}}{dx^2} = \frac{k}{h_{MTY}} \left(\frac{\sigma_{MTY}}{E_{MTY}} - \frac{\sigma_{MLY}}{E_{MLY}} \right) + \left[\frac{d}{dx} \left(\frac{k}{h_{MTY}} \right) \right] (\bar{u}_{MTY} - \bar{u}_{MLY}) \qquad (69)$$

The second term of the right-hand side of Eq. (69) depends mainly on fiber waviness; it is omitted in the following because the waviness is small in the case studied.

Since the total force carried by the MTY and MLY must be in equilibrium with the externally applied force, the following relation between σ_{MTY} and σ_{MLY} can be obtained:

$$\sigma_{MTY} h_{MTY} + \sigma_{MLY} h_{MLY} = \sigma_o h (1 + 2/\pi) \qquad (70)$$

where σ_o is the applied stress. Thus, Eq. (69) is rearranged as

$$\frac{d^2\sigma_{MTY}}{dx^2} - A(x)\sigma_{MTY} = -B(x)\sigma_o \qquad (71)$$

where $A(x)$ and $B(x)$ are given below

$$A(x) = \frac{k}{h_{MTY}} \left(\frac{1}{E_{MTY}} + \frac{1}{E_{MLY}} \frac{h_{MTY}}{h_{MLY}} \right)$$

$$B(x) = \frac{k}{E_{MLY}} \left(\frac{1}{h_{MTY}} + \frac{1}{h_{MLY}} \right)$$

It can be shown that far away from the cracks, σ_{MTY} approaches $B(x)\sigma_o/A(x)$, which is the solution for the no-crack case. By assuming a parabolic crack opening displacement, which is a consequence of the linear shear stress distribution in the yarns, and taking into account the shear deformations in both yarns, the following expression for the shear-lag coefficient, k, can be obtained:

$$k = \frac{3 G_{MTY} G_{MLY}}{G_{MLY} h_{MTY} + G_{MTY} h_{MLY}} \qquad (72)$$

where G stands for the shear modulus. Stress-free boundary conditions are prescribed for the crack surfaces. In addition, for the boundary condition at $x=\pm a/2$, it is assumed that the slope of the σ_{MTY} is zero. Therefore, for a set of n arbitrarily distributed transverse cracks, there are a total of n+2 boundary conditions, which are necessary for finding the stress distribution in the MTY from Equation (71). An iterative finite difference techniques is used for finding σ_{MTY} and σ_{MLY}.

Figure 42 presents the stress solutions for the MTY and MLY with a single crack at x=0 obtained by the shear-lag method. The results predicted for the no-crack situation are also indicated. At the crack surface, the stress in the MTY vanishes, and all the load is carried by the LY. Away from the crack surfaces, the stress in the MTY gradually increases through the action of the shear stress at the MTY/MLY interface. It can be seen that the solid curves approach the dotted curves, and after a distance of about two yarn thicknesses (2h), the stresses are almost unaffected by the crack. At the end points the predicted stresses are slightly different since the boundary conditions used for obtaining the two solutions (solid and dotted lines) are different.

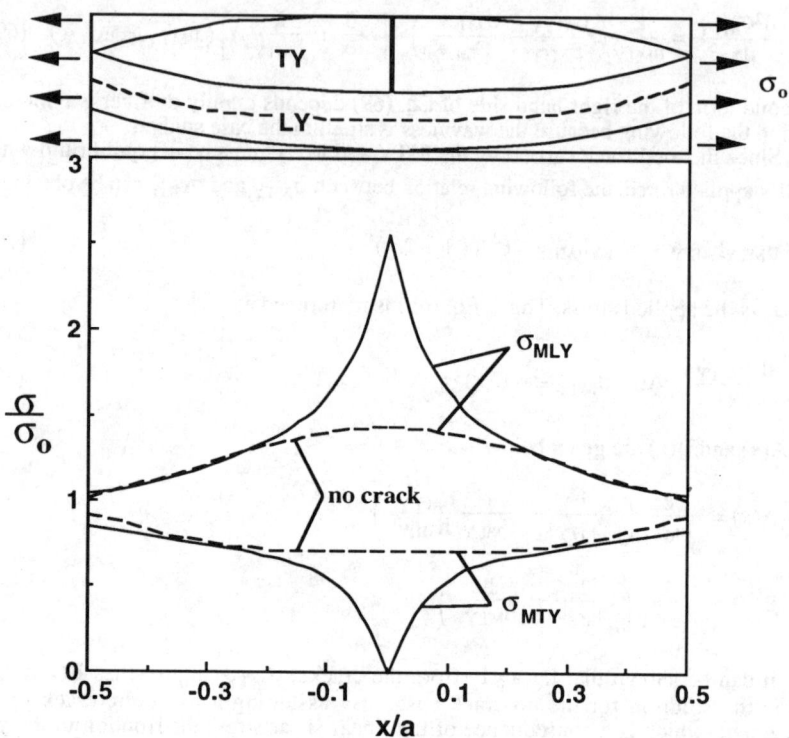

Fig. 42 Stress distribution in the unit cell for 2-D woven CMC before and after cracking [135].

Once the σ_{MLY} is found, the average composite strain can be obtained by averaging the strain in the MLY.

$$\varepsilon_o = \frac{1}{a} \int_{-a/2}^{a/2} \frac{\sigma_{MLY}(x)}{E_{MLY}(x)} \, dx \qquad (73)$$

Since σ_{MLY} is proportional to the applied stress σ_o, the composite stress-strain relation can be obtained. The strain energy stored in the unit cell is

$$U_c = \frac{1}{2} \int_{-a/2}^{a/2} \left(\frac{(\sigma_{MLY}(x))^2}{E_{MLY}(x)} h_{MLY} + \frac{(\sigma_{MTY}(x))^2}{E_{MTY}(x)} h_{MTY} \right) dx \qquad (74)$$

The strain energy released due to the creation of a crack is

$$\Delta U_c = U_c' - U_c'' \tag{75}$$

where U_c' and U_c'' correspond to, respectively, the stored strain energies before and after a transverse crack is formed. The strain energy release rate is given by

$$G = \frac{\delta(\Delta U_c)}{\delta s} \tag{76}$$

where s is the crack size. For the finite element analysis, the composite unit cell is composed of three parts: the LY, the TY, and the interyarn matrix. The element mesh has been so chosen that each single element contains only one material. The TY and interyarn matrix are isotropic, and the LY is transversely isotropic, with the fiber direction parallel to the yarn center line. Finer meshes are made for those areas with high stress gradients.

In the finite element calculation, strain energy release rate G was approximated by the following expression:

$$G \approx \frac{f_x \Delta x + f_y \Delta y}{2\Delta s} \tag{77}$$

where Δx and Δy are crack openings in the x and y directions, respectively, and f_x and f_y are the applied nodal forces (dimension N/m) for closing the crack; Δs is the crack increment. According to CFM, crack propagation from an initial flaw of length s occurs when G equals the critical strain energy release rate G_c.

$$G = G_c \tag{78}$$

and from TEB, the condition for the initiation of a crack of length L_s is

$$\int_0^{L_s} G(s)\, ds = G_c\, L_s \tag{79}$$

Thus, the critical strain for crack propagation from the CFM approach is a function of s, whereas the critical strain for crack initiation from the TEB method is single-valued.

The analysis shows that thinner fabrics have higher critical strains. It also shows that the crack initiation is more likely at the center of the transverse yarn (with respect to x direction) than away from the center. The predictions of the critical strains for crack initiation (TEB) and crack propagation (CFM) using the finite element and shear-lag methods are shown in Figure 43. The fracture toughness G_c of the transverse yarn is assumed to be 40 (N/m); no experimental data are available. Nevertheless, the relative positions of the curves in Figure 43 are not affected by G_c. As s increases, the curve predicted by CFM initially drops sharply, followed by more gradual variation. As mentioned, the critical strain predicted by the TEB is a constant, which lies in between the maximum and minimum values of the CFM curve. The critical strains predicted by the TEB using the finite element and shear-lag methods are very close.

Although the general behavior of crack initiation and growth can be modeled by the CFM and TEB approaches, the detailed behavior of an individual crack depends on the shapes and positions of intrayarn pores and can differ greatly from one location to another. This highly localized behavior cannot be fully understood without a micromechanics approach focusing on the fiber and matrix materials and on the intrayarn and interyarn pores.

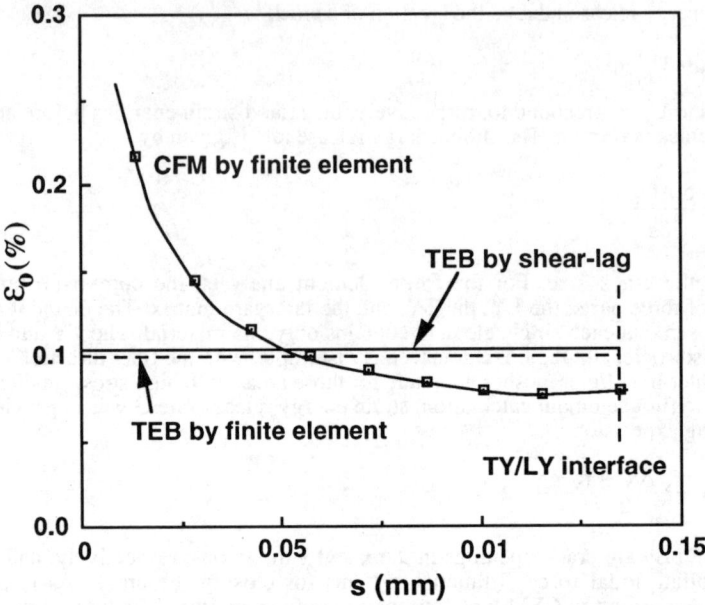

Fig. 43 Comparison of matrix cracking strain predictions for 2-D woven SiC/SiC obtained by classical fracture mechanics (CFM) and total energy balance (TEB) [135].

6.3.2 Modulus Reduction

The composite moduli without damage are calculated by considering Iso-Phase Model (IPM) and Random-Phase Model (RPM). In the IPM, all fabric layers are in the same relative position. The deformation, stress and strain distribution for all layers are identical. In the RPM, a vertical cross section contains fiber yarns with random locations. Therefore, it is reasonable to assume that the effective stiffness for any volume element with thickness dx (or dy) taken along the x (or y) direction is independent of position, and the composite strains can be expressed as $\varepsilon_x(x,y,z) \equiv \varepsilon_1$ and $\varepsilon_y(x,y,z) \equiv \varepsilon_2$.

In the case of IPM the axial Young's modulus and the Poisson's ratio of the fabric composite is given by

$$E_C = \frac{1-C_2^2}{h(a+b)C_1}$$

$$\nu_C = C_2 \tag{80}$$

where a and h are the major and minor axes lengths for the yarns (assumed to be elliptical) and b is the distance between to transverse yarns (see Figure 2 in Ref. [135]) and

$$(C_1, C_2) = \frac{2}{(a+b)} \int_0^{(a+b)/2} \frac{(1, Q_{xy}^{LY} A_{LY} + Q_{xy}^{TY} A_{TY} + Q_{xy}^{m} A_{m})}{Q_{xx}^{LY} A_{LY} + Q_{xx}^{TY} A_{TY} + Q_{xx}^{m} A_{m}} dx$$

Q_{xy} are the elements of the transformed reduced stiffness matrix and A represents area with subscripts LY, TY and m representing longitudinal yarn, transverse yarn and matrix, respectively.

Similarly, in the case of RPM the moduli are given by

$$E_C = \frac{C_3}{h(a+b)} (1 - v_c^2)$$

$$v_c = \frac{C_4}{C_3} \tag{81}$$

where

$$C_3 = \frac{\pi h a}{8} \left(Q_{11}(1 - \frac{3}{4}\phi^2) + (Q_{12} + 2Q_{66})\phi^2 + Q_{22}(1 + \frac{1}{4}\phi^2) \right)$$

$$+ \eta Q_{11}^{m} h \left(a + b - \frac{\pi}{4} a (1 + \frac{1}{4}\phi^2) \right)$$

$$C_4 = \frac{\pi h a}{8} \left(2 Q_{12} + \frac{1}{2}\phi^2 Q_{23} \right)$$

$$+ \eta Q_{12}^{m} h \left(a + b - \frac{\pi}{4} a (1 + \frac{1}{4}\phi^2) \right)$$

Q_{ij} are the coefficients of the reduced stiffness matrix and $\phi = \pi h/2(a+b)$. By neglecting Poisson's effects and the stiffness contribution of the interyarn matrix, i.e. high porosity content ($\eta \approx 0$), the composite stiffness can be further simplified as

$$E_C = \frac{\pi a}{8(a+b)} \left(E_{11}(1 - \frac{3}{4}\phi^2) + 2G_{12}\phi^2 + E_{22}(1 + \frac{1}{4}\phi^2) \right) \tag{82}$$

Application of above model in SiC/SiC composite shows that compared to polymer matrix composites, the stiffness reduction in ceramic matrix composites due to fiber undulation is much less pronounced. The effect of interyarn porosity is much more pronounced. The modulus decreases nearly linearly with increasing porosity.

The modulus of woven composites with cracked transverse yarn can be calculated by the shear lag approach and the FEM technique. Once the σ_{MLY} is found (by shear lag and finite difference technique or the finite element analysis) the average composite strain can be obtained by averaging the strain in the MLY.

$$\varepsilon_0 = \frac{1}{a} \int_{-a/2}^{a/2} \frac{\sigma_{MLY}(x)}{E_{MLY}(x)} dx \tag{83}$$

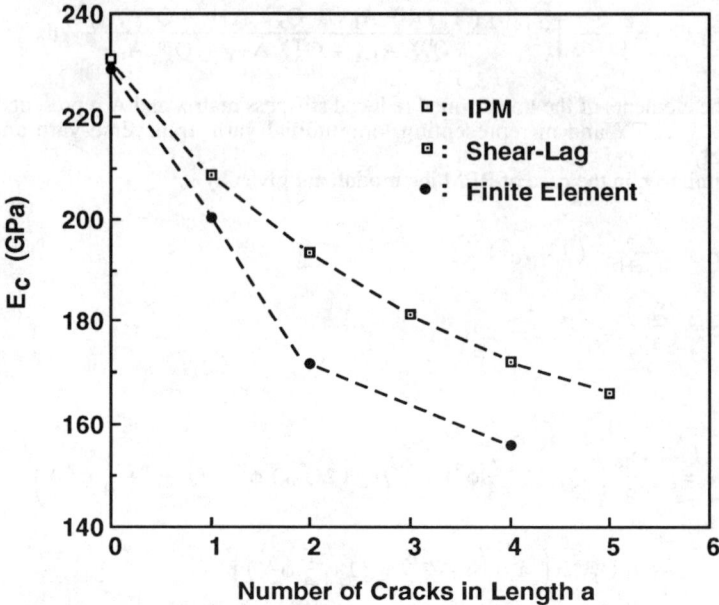

Fig. 44 Prediction of stiffness reduction in 2-D woven SiC/SiC due to cracking [135].

Since σ_{MLY} is proportional to the applied stress σ_o, the composite stress-strain relation can be predicted and the reduced composite stiffness can be calculated from the relation $E_c = \sigma_o/\varepsilon_o$. Predictions of modulus reduction (due to equally spaced transverse cracks in the length a) for SiC/SiC composite by the shear lag analysis and FEM are shown in Figure 44.

7. Static Mechanical Behavior at Elevated Temperatures

Property requirements for some very high temperatures and hostile environments can be potentially satisfied only by ceramic matrix composites. To realize this potential however, it is necessary to evaluate their performance at elevated temperatures and hostile environments. Another significant factor that must be taken into account is the fact that the matrix (and off-axis plies) undergo multiple cracking. If the component stress levels are such that multiple cracking is present, these cracks can provide passage for the hostile environments to reach the load bearing fibers. It has been shown that in cross-ply composites, the 90° plies undergo multiple cracking at very low strain levels. The component strain are very much likely to be higher than these levels. The passage of the environment is time dependent and adds time dependent factor to the strength of the composites, in addition to the inherent time dependent cracking of the fiber and the matrix (as found in the monolithic ceramics).

The efforts for such characterization are severely handicapped due to the limitations and complexities of present characterization techniques. Flexural testing has been used to evaluate elevated temperature mechanical response [58]. Although these data are useful for screening and demonstration purposes, their use in design is limited as flexural loading involves tensile, compressive and shear stress distributions in the specimens. Tensile tests at high temperature require special gripping techniques, specimen machining and large material availability. Measurement of strain at elevated temperature is also non-trivial. Nevertheless, limited tensile data have been reported recently [136, 138-140]. Since actual components are likely to undergo static or cyclic long term loading at elevated temperature, it is necessary to conduct long-term static fatigue and cyclic fatigue tests.

Prewo and Brennan [58] have reported flexural strengths of Nicalon-CAS, Nicalon-LAS-I and Nicalon-LAS-II composites up to 1100°C. For the LAS and LAS-I composite, in both air and argon atmospheres, composite flexural strength increased with temperature and the failure mode remained fibrous with significant pullout. On the other hand, for the LAS-II composites, in air, the flexural strength decreased above 700° C and the fracture surface lacked fiber pullout on the tensile side. Mah et al. [136] observed severe strength reduction for Nicalon-LAS-II composites but noted a strength increase at 1000°C. At 900°C flexural strength showed gradual degradation as the oxygen partial pressure in the test environment was increased. The LAS-II composites tested in argon showed flexural strength increase and retained the failure mode with fiber pullout till 1100°C [58, 136]. At room temperature, the LAS and LAS-I composites fail before appreciable microcracking. But LAS-II composites showed appreciable matrix cracking. Thus, in LAS-II composites matrix microcracks provide passage for the environmental attack of the interface and/or fibers. Luh and Evans [137] showed that the embrittlement of the Nicalon-LAS-II composite at 1000°C was due to twenty-fold increase in the interfacial shear stress.

The stress-displacement curves obtained by Mah et al. [136a,b] for Nicalon-LAS II at room temperature and 900°C are compared in Figure 45. At room temperature, the composites underwent multiple cracking and showed high strength. The fracture surface showed extensive fiber pullout. On the contrary, at 900 and 1025°C in air composites failed as soon as matrix microcracking began. The fracture surfaces of specimens tested at high temperatures showed complete brittle failure (lack of pullout) on the outside regions (150 µm from specimen surface) while the interior region retained fiber pullout. These observation suggests that the embrittlement occurs from outside towards inside as the environment transport occurs through the cracks. Jablonski and Bhatt [138] observed severe tensile strength degradation of SiC-RBSN composites at 1300 and 1500°C. Philippe et al. [139] have reported elevated temperature tensile behavior of 2-D woven SiC/SiC composites at 900 and 1300°C in air and argon. In argon atmosphere, while the strength was increased only slightly over room temperature strength, the strain to failure was increased substantially. Specimens tested in air showed strength reduction. The specimen tested in air at 1300°C showed a brittle zone (absence of pullout) of the order of 300 µm (from specimen surface).

Prewo et al. [140] characterized the tensile behavior of cross-ply Nicalon-LAS-III composite at 1000°C in air and argon (Figure 46). The composite strength was not degraded in argon atmosphere but it was severely degraded in air. The composite performance was still limited by the cracking in the 0° plies and not by the cracking of the 90° plies (which occurs at much lower stress).

Another approach that is taken to understand the elevated temperature performance of CMCs involves exposing the composites at elevated temperatures to different environments for certain duration and then testing them at room temperature (aging studies) [139, 141-145].

Most studies indicate that aging in inert atmosphere does not affect the composite strength and failure behavior [141-143] unless the temperature of exposure is higher than that which results in deterioration of the fiber itself by recrystallization, grain growth or decomposition. The composite strength / fracture strain may be enhanced as the reaction for formation of interfacial carbon layer, that began during hot pressing, continues [77].

Aging behavior in oxidizing atmosphere is more complex [141-143]. Filipuzzi et al. [141] compared the effect of aging in air on the mechanical behavior of unidirectional SiC/SiC composites with interfacial carbon layer thicknesses of 0.1 (material A) and 1.0 (materials B) μm. The as-received failure strains of both materials are same but the matrix cracking stress and the modulus is lower for material B. The tensile stress-strain curves after aging in air at 900 and 1400°C are shown in Figure 47a,b for materials A and B respectively. For a 900°C, 10 hour exposure, the failure strain of the composites is not affected but strength degradation is observed. The failure surface shows an outer brittle failure region. Composite B (thick interface) showed more strength degradation than composite A (thin interface). The extent of degradation increased significantly in both materials as the exposure time was increased from 10 to 100 hours. The area of the brittle region on the failure surface also increases. At 1400°C, failure strain for material A remained unaltered and strength was reduced. Increasing exposure time had only marginal effect. For material B, at 1400°C and 10 hrs composite failure strain was not affected but strength was degraded severely. Increasing the exposure time to 100 hrs leads to severe reduction of strain as well as strength. These observations suggest that (1) the thick interface material is more susceptible to degradation. (2) for material A, a long duration exposure at an intermediate temperature of 900°C is much more detrimental than that at 1400°C.

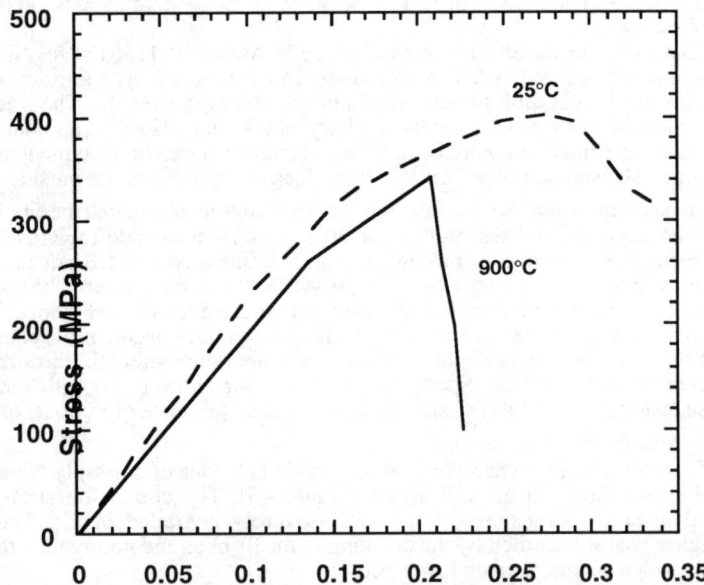

Fig. 45. Comparison of room temperature and 900°C tensile behavior of unidirectional Nicalon-LAS composite [136a, b].

Frety and Boussuge [143] have reported similar results on 2-D SiC/SiC composites: with and without a 5 μm interfacial C coating. At 800°C, both materials showed complete brittle failure and severe strength degradation. At 1400°C the material with no coating retained its room temperature properties. Bhatt [142] has reported that in SiC/RBSN composites heat treated between 600 to 1000°C, 35% of the strength was retained while for composites heat treated between 1200 and 1400°C 65% of the strength was retained.

The complex oxidation of these composites can be explained on the basis of active and passive oxidation of the interface [141-142]. During oxidation, the carbon interface oxidizes to carbon dioxide and carbon monoxide (gaseous products) leading to weight loss. On the other hand, the SiC fibers (or the Si_3N_4 matrix) oxidize to produce silica leading to weight gain. Silica also produces strong bond between the fiber and the matrix. These reactions are competitive. Low temperature and thicker carbon layer favor in-depth oxidation (active oxidation) of carbon over silica formation leading to severe interfacial degradation. Higher temperatures and thinner interfaces favor surface oxidation. The silica produced seals the pathway for diffusion of oxygen to carbon in the interior of the composite, and the oxidation of the carbon is prevented. This is termed as passive oxidation.

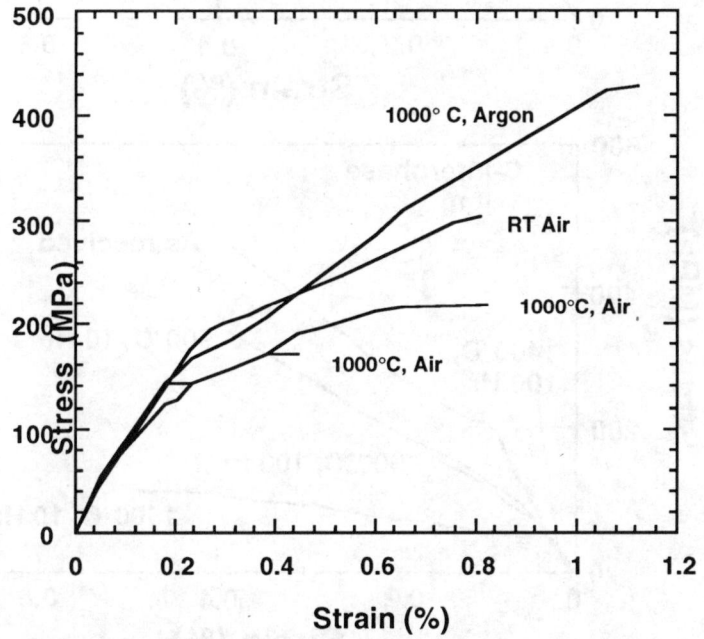

Fig. 46. Comparison of room temperature and 1000°C (air and argon atmospheres) tensile behavior of cross-ply Nicalon-LAS composite [140a].

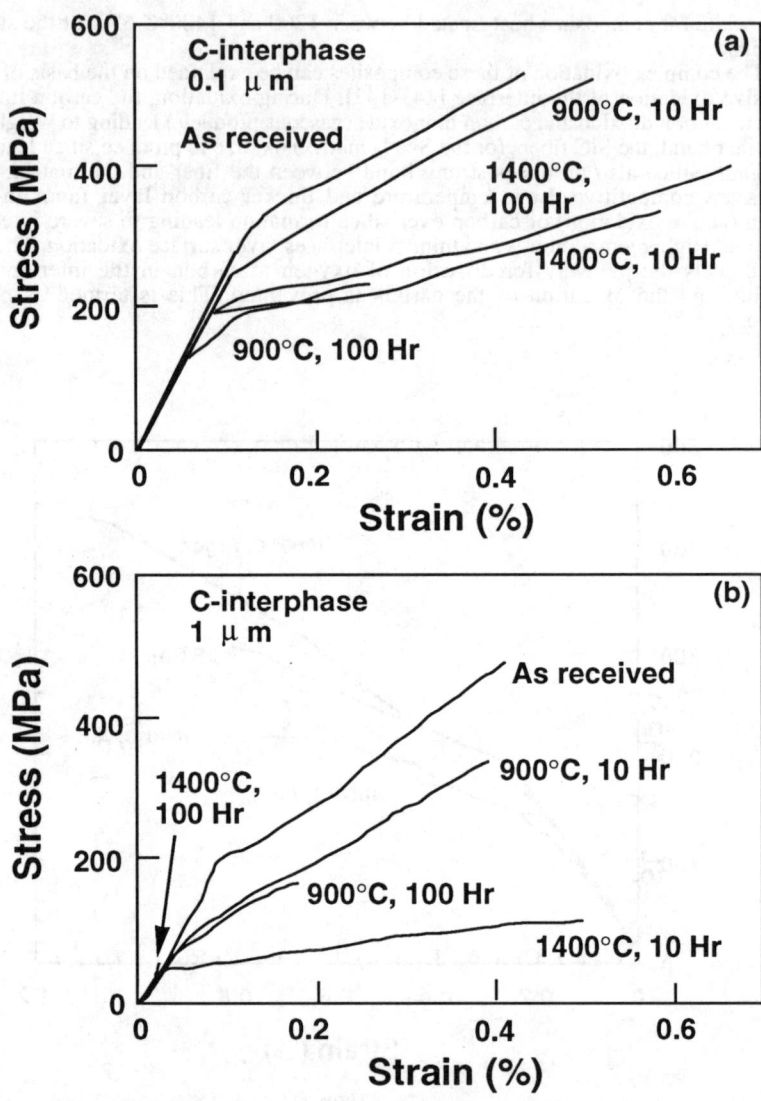

Fig. 47. Effect of ageing treatment on the stress-strain behavior of SiC/SiC composites with two different thickness interfaces [141].

8. Mechanical, Thermal and Thermomechanical Fatigue

A thorough review of fatigue behavior of CMCs is beyond the scope of this Chapter. But some important results have been summarized in the following. The behavior of many CMC systems under cyclic loading has been the subject of investigation of numerous recent studies [68, 96, 146-163]. Gradual modulus degradation in cyclic fatigue has been reported for the unidirectional and laminated ceramic composites at room temperature [68,146, 147,151,152,158, 160,161,163] and elevated temperatures [153, 154, 156]. It has been shown that the gradual damage growth accompanies modulus decrease in the CMCs under cyclic fatigue loading [68,152,158,159,162]. At maximum fatigue stress levels between matrix cracking stress and proportionality limit, the modulus in unidirectional composites was found to show recovery after initial decrease [151,158,160,162]. The modulus recovery was observed after saturation microcracking was reached and was attributed to the debris which may be wedged between crack faces, resulting in an increase in the frictional stress at the interface and realignment of debonded fibers with the applied stress. For the $SiC-Si_3N_4$ system tested at 1000°C and 1200°C [153, 154], gradual modulus decrease was found only for the specimens tested above the proportionality limit.

The results in the above studies also indicate the occurrences of fatigue failures and existence of fatigue limit for each of the materials investigated. At room temperature, the fatigue limit is higher than matrix crack initiation stress. Some studies show that the room temperature fatigue limit coincides with the proportionality limit associated with the non-linear behavior of the 0° plies [151,152,160] while other studies show the fatigue limit to be lower (10%) than this proportionality limit [146,161] and considerably lower (30%) than the proportionality limit at high loading frequencies [159,163]. At high temperatures, fatigue limit coincident with proportionality limit for unidirectional $SiC-Si_3N_4$ [153,154] and fatigue limit lower than matrix crack initiation stress for cross-ply Nicalon-CAS [152] have been reported. The relationship between residual strength of fatigued specimens with the monotonic strength is not clearly established but laminate geometry and applied maximum stress can be identified as parameters that affect the residual strength [151,152,160,161].

Karandikar and Chou [163] observed that gradual growth of matrix and transverse cracks in the Nicalon-CAS composites occurs under static fatigue as well as cyclic fatigue loadings. The data suggest that the crack growth is environmentally driven. The elastic moduli showed gradual degradation. The damage growth rate and the modulus reduction rate increased as the applied stress increased. The saturation densities of matrix and transverse cracks under fatigue loadings were found to be comparable to those under monotonic loading. A fatigue crack growth limit was observed for the matrix cracks and it is coincident with the matrix crack initiation strain (0.11%). Linear correlations between the moduli degradations and crack density could be used to predict the moduli degradations under cyclic fatigue loading using the experimentally measured crack densities. A logarithmic correlation was developed to predict the Young's modulus reduction under fatigue loadings in a limited applied maximum stress range.

Based on experimental observation and analysis of the experimental data two distinct regimes of modulus reduction were identified. The modulus reduction rate is constant in the first regime and decrease in the second. This behavior can be rationalized on the basis of a fatigue crack growth model. In the first regime, each matrix crack behaves like a steady state crack (Marshall et al. [114] McMeeking and Evans [164]). The stress intensity factor experienced by each crack is independent of the crack length which results in constant crack growth rate and modulus reduction rate. The modulus reduction rates in this regime can be related to the applied stress level and matrix fatigue properties. In the second regime of crack growth, the interactions between multiple cracks lead to reductions of the crack growth and modulus reduction rates.

Zawada and Wetherhold [165] investigated rapid thermal fatigue (TF) of unidirectional Nicalon-aluminosilicate glass composite with and without dead load. thermal cycling was done between 250-700 and 250-800°C. For the dead load case stress levels were 28 and 138 MPa (both below proportionality limit). It was found that the residual modulus was unaffected by thermal fatigue with or without load. However, residual flexural strengths after 500 cycles were lower in TF specimens. Residual flexural strength decreased as the dead load accompanying TF increased. Strength reduction was much more severe in TF range of 250-700. Better strength retention in the TF range 250-800 is attributed to the fact that at this temperature, matrix has lower viscosity than at 700°C and can flow to seal the surface cracks and prevent fiber degradation. This was confirmed by studies which showed more severe strength reduction after thermal aging at 650°C than at 800°C. This result is consistent with the one described previously for the aging of SiC/SiC composites.

Butkus et al. [166] studied isothermal (1100°C, 100 MPa) and thermomechanical fatigue (500-1100°C, 10-100 MPa) of a unidirectional Nicalon-CAS-II composite. The maximum stress 100 MPa was 40% of the 1100°C proportionality limit. In the thermomechanical fatigue (TMF), the load and the temperature increased in-phase. Tests were also done with added 60 second hold period at maximum load. All specimens survived 1000 cycles and did not show any modulus degradation. Strength under isothermal and TMF conditions were comparable to monotonic strengths at 1100°C. However, permanent strain in the composite was significantly higher than the monotonic specimen. The permanent strains were higher in the TMF specimens than in the isothermal specimens and were higher in specimens with added 60 second hold periods. These observations suggest that the fatigue behavior at 1100°C and 100 MPa is dominated by creep deformation.

Worthem and Ellis [167] compared in-phase and out-of-phase (180°) TMF behavior of a unidirectional Nicalon-CAS composite. In out-of-phase TMF, stress is maximum when the temperature is minimum and stress is minimum when the temperature is maximum. The temperature range was 600 to 1100°C. Tests were done at various maximum stresses (and zero minimum load) and in air or argon. In air in the maximum stress range 75 to 150 MPa, the out-of-phase TMF lives are 3 to 1 orders of magnitude less than the corresponding in-phase TMF lives. Also, isothermal fatigue lives and creep lives are found to be higher than corresponding TMF lives. In argon atmosphere however, no fatigue failures were observed up to 1000 cycles even when the stress was as high as 275 MPa in both type of TMF tests. These results clearly show that an out-of-phase combination of thermal and mechanical loads in air would severely limit the maximum allowed stress for components made of this material.

St. Hillaire and T. Erturk [168] studied thermal fatigue behavior of $[0/90_5/0_{1/2}]_s$ SCS-9 fiber (CVD SiC) reinforced silicon nitride composites. Thermal fatigue range was 500 to 1350°C and was provided by an impinging jet fuel flame to simulate the environment in a jet engine. A fatigue limit (no failure in 1000 cycles of 120 s each) of 125 MPa was observed and it was 70% of the room temperature proportional limit (185 MPa). Performance of SCS-6 fiber cross-ply laminates was found to be much inferior due to higher susceptibility to thermal gradients.

9. Creep

In crystalline materials, constant long term load at temperatures higher than recrystallization temperature (0.35-0.45 T_m, melting temperature) causes time dependent accumulation of strain called creep. The creep deformation can be broken down into three stages. In stage I, creep strain accumulates rapidly. Then the creep rate then starts decreasing and reaches a constant level. This constant strain rate creep is called steady state creep and this is stage II of creep. Finally, the creep rate starts rising again in stage III and leads to final failure. The extent of each stage depends on the stress level and the temperature. At very high temperatures or stresses, failure occurs in a very short period and the component cannot be effectively used. When the component has significant creep life, a large portion of it is in stage II. Then, the component life can be calculated based on the steady state creep rates ($\dot{\epsilon}$) in stage II. Therefore, creep research is often focused on determining steady state creep rates. In metallic and monolithic ceramic [169, 170] materials the following relation is found to hold

$$\dot{\varepsilon} = A \exp(-Q/RT) \sigma^n \tag{84}$$

where is the applied stress, Q is the activation energy for creep, R is the gas constant and T is absolute temperature, A and n are constants and n is called the creep exponent.

CMC components are expected to carry loads at elevated temperatures. Therefore, creep deformation must be accounted for when a CMC component is designed. Yet, significant database regarding creep of CMCs has not been generated. Abbe et al.[171] studied the flexural creep behavior of 0°-90° SiC/SiC composites made by CVI. The steady state creep rates at 1200°C and 1400°C at 70 MPa and 100 MPa under vacuum are listed in Table 8. The creep rate was not very sensitive to the applied stress. However, the composite creep rate was higher than the creep rate for corresponding monolithic matrix. Two main reasons for the poorer creep resistance of the composite are (a) matrix porosity (~20%) (b) poor stability of Nicalon fibers at 1200 and 1400°C.

Holmes [172, 173] has investigated the tensile creep behavior of SCS-6 SiC fiber (30 volume %) reinforced Si_3N_4 matrix composites. The matrix had 5% (by weight) Y_2O_3 and 1.25% MgO as sintering aids. Creep tests were carried out at stress levels above and below the proportional limit (84 MPa). The steady state creep rates are shown in Table 8. The steady state creep rate of the composite was found to be comparable to that of monolithic Si_3N_4 containing 4% Y_2O_3. Existence of matrix cracking beyond the proportional limit did not accelerate the creep rate.

Table 8. Steady state creep rates for ceramic composites / ceramics [171-173].

Material	Temperature °C	Stress (MPa)	Steady state Creep rate s^{-1}	Power law exponent
Nicalon/SiC (0°/90°)	1200	70	1.9×10^{-9}	1.7
	1200	100	1.1×10^{-8}	
	1400	70	1.1×10^{-8}	2.7
	1400	100	2.8×10^{-8}	
SCS-6 SiC/Si_3N_4 (0°)	1200	99	9.0×10^{-11}	
	1200	150	1.2×10^{-10}	6
	1200	175	4.0×10^{-10}	
SCS-6 SiC/Si_3N_4 (0°)	1350	70	2.5×10^{-10}	
	1350	110	1.0×10^{-8}	7
	1350	150	5.6×10^{-8}	
Si_3N_4 + 4% Y_2O_3	1370	100	4.0×10^{-9}	

Theoretical analysis of the creep behavior of short fiber reinforced CMCs [174-178] points out that since the creep rates of fibers and matrices in CMCs are comparable, composite creep depends on fiber and matrix creep rates. In addition, since the fiber/matrix interface is weak in CMCs, sliding at the interface takes place and affects the composite creep behavior significantly. In particular, the composite creep rate is very sensitively to the friction coefficient between the fiber and the matrix. In SiC fiber reinforced Al_2O_3 system higher fiber volume fraction leads to lower creep rates only at friction coefficients larger than 0.2.

10. Concluding Remarks

Significant experimental and theoretical knowledge has been generated on the behavior of continuous fiber reinforced ceramic matrix composites. The mechanical behavior of CMCs is affected by constituent properties, processing conditions, microstructure and interfacial properties. Many successful combinations of fiber and matrix systems have been identified and produced. Although the processing so far has been guided by heuristics, the database generated and the theoretical analyses available can definitely provide guidelines for future developments of CMCs. The selection of the fiber and matrix is governed by chemical compatibility at room as well as elevated temperatures and thermal expansion mismatch. The processing conditions must be optimized not only for complete densification but also for optimizing the interface microstructure and chemical composition.

The matrix microcracking stress and ultimate tensile strength are the two most important mechanical properties for design. Matrix microcracking stress also serves as the fatigue damage growth limit. The matrix microcracking stress and ultimate tensile strength are significantly affected by interface characteristics. Many interface characterization techniques exist, which measure various aspects of the interface properties (e. g. shear stress, fracture energy, fracture toughness, friction coefficient etc.). Although these properties are good for comparison and screening purposes, caution must be exercised in using these data in the analyses of the deformation processes. Increasing the interfacial strength can increase the matrix cracking stress but the composite may fail in a brittle fashion. Thus, interfacial shear strength must be optimized.

Due to low transverse strengths of CMCs two and three dimensional fiber architecture should be used. In such cases, matrix damage is created at much lower stresses than the unidirectional composite matrix cracking stress. The operating stress for a component is likely to be higher than that required to create damage in off-axis plies or yarns. Thus, any design with multi-directionally reinforced CMCs must take into account the presence of damage and its effect on thermo-elastic properties and environmental degradation.

Glass-based CMCs have limited temperature capability but can compete for use in applications involving temperatures up to 600°C. Glass-ceramic-based systems have shown promise with strength retention up to 1300°C. SiC and Si_3N_4-based systems are promising for applications involving temperatures even up to 1500°C.

Acknowledgment

The authors wish to express their gratitude towards NASA-Lewis Research Center (Dr. Gyekenyesi); NEDO, government of Japan; and RCAST, University of Tokyo (Professor T. Kishi) for providing the financial support for this work. Extensive and enlightening discussions with Drs. K. M. Prewo and W. Tredway (UTRC), Profs. N. Takeda and Y. Kagawa (University of Tokyo), Dr. Ichikawa (Nippon Carbon Co. Inc.) and Dr. Mark Headinger (DuPont Lanxide Composites Inc.) are also appreciated.

# References

[1] R. W. Rice, Ceram. Eng. Sci. Proc., 2 [7-8] (1981) 661-681.
[2] P. G. Karandikar, Tsu-Wei Chou and Azar Parvizi-Majidi, Chapter 11 "Mechanical Properties," in Handbook of Continuous Fiber Reinforced Ceramic Matrix Composites, Ed. R. H. Lehman, CIAC/CINDAS and Am. Cera. Soc. (1994) in press.
[3] A. G. Evans, J. Am. Cera. Soc. 73 [2] (1990) 187-206.
[4] A. G. Evans and D. B. Marshall, "The mechanical behavior of ceramic matrix composites," Acta Metall. 37 [10] (1989) 2567-2583.
[5] (a) J. J. Brennan and K. M. Prewo, J. Mat. Sci. 17 (1982) 2371-2383. (b) K. M. Prewo, J. J. Brennan and G. K. Layden, Ceramic Bulletin, 65 (1986) 305-313.
[6] K. M. Prewo, Ceramic Bulletin, 68 (1989) 395-400.
[7] D. C. Phillips, R. A. J. Sambell and D. H. Bowen, Journal of Materials Science, 7 (1972) 1454-1464.
[8] S. R. Levitt, Journal of Materials Science, 8 (1973) 793-806.
[9] A. G. Evans, M. Y. He and J. W. Hutchinson, J. Am. Cera. Soc. 72 [12] (1989) 2300-2303.
[10] J. Aveston, G. A. Cooper and A. Kelly, In *The Properties of Fiber Composites*, IPC Science and Technology Press Ltd., London (1971) 15-26.
[11] J. Aveston and A. Kelly, Journal of Materials Science, 8 (1973) 352-362.
[12] Buljan, S-T, Pasto, A. E., and Kim, H. J., Ceramic Bulletin, Vol. 68, No. 2, (1989) 387-394
[13] Maloney, L. D., Design News, March 13, (1989).
[14] Upadhya, K., Journal of Metals, May (1992) 15-18.
[15] Headinger, M., Preliminary Engineering Data Sheet, DuPont-Lanxide Composites Inc., Newark DE 19714.
[16] Gray, H. R. and Ginty, C. A., Journal of Metals, May (1992) 12.
[17] Ashley, S., Mechanical Engineering, July (1991) 44-49.
[18] Hindman, D. L., Snyder, J. E., Liang, W. W. and Schreiner, M. E., in High Tech Ceramics edited by P. Vincenzini, Elsevier Science Publishers B. V., Amsterdam, (1987) 2451-2461.
[19] A. Chatterjee, J. W. Moschler, R. J. Kerans, N. J. Pagano, and S. Mall, Ceramic Engineering and Science Proceedings, 10 [9-10] (1989) 1179-1190.
[20] J. Paresh, Ceram. Eng. Sci. Proc. 9 [7-8] (1988) 529-540.
[21] H. G. Showman and D. D. Johnson, Ceram. Eng. Sci. Proc. 6 [9-10] (1985) 1221-1230.
[22] F. Ko, Ceramic Bulletin 68 [2] (1989) 401-414.
[23] Azar Parvizi-Majidi, "Fibers and Whiskers - Chapter 2", *Materials Science and Technology: Volume 13 - Structure and Properties of Composites*, Vol. Ed. Tsu-Wei Chou, VCH Publishers, New York, (1993) 25-88.
[24] K. M. Prewo, J. J. Brennan and G. K. Layden, Ceramic Bulletin 65 [2] (1986) 305-313.
[25] I. W. Donald and P. W. McMillan, J. Mat. Sci. 11 (1976) 949-972.
[26] D. C. Larsen (Corning Glass Works), Submitted to National Academy of Science National Materials Advisory Board Committee on high Temperature Materials for Advanced Technological Applications (1987).
[27] N. Hecht, D. McCullum, et al., Ceram. Eng. Sci. Proc. 9[9-10] (1988) 1313-1332.
[28] V. J. Tennery (Ed.), Ceramic Materials and Components for Engines, The American Ceramic Society, (1989) 1480-1494.

[29] W. J. Lackey, D. P. Stinton, G. A. Cerny, A. C. Schaffhauser, and L. L. Fehrenbacher, Adv. Ceram. Mat. 2[1] (1987) 24-30.
[30] Mark Headinger, DuPont Lanxide Composites Inc., Private Communication.
[31] R. Naslain, "Thermostructural ceramic matrix composites: and overview." in Advanced Structural and Functional Materials, W.G. B Bunk Editor, Springer-Verlag (1991) 51-90.
[32] J. T. Hoyt and J. M. Yang, SAMPE Journal 27[2] (1991) 11-17.
[33] D. P. Stinton, A. J. Caputo and R. A. Lowden, American Ceramic Society Bulletin 65 [2] (1986) 347-350.
[34] Nyan-Hwa Tai and Tsu-Wei Chou, J. Am. Cera. Soc. 72 [3] (1989) 414-420.
[35] Nyan-Hwa Tai and Tsu-Wei Chou, J. Am. Cera. Soc. 73 [6] (1990) 1489-1498.
[36] G. H. Schiroky, A. W. Urquhart and B. W. Sorenson, Presented at the Gas Turbine and Aeroengine Congress and Exposition (1989), American Society of Mechanical Engineers, 89-GT-316.
[37] S. Ashley, Mechanical Engineer, [7] (1991) 44-49.
[38] R. A. Sambell, A. Briggs, D. C. Phillips and D. H. Bowen, J. Mat. Sci. 7, (1972) 676-681.
[39] K. M. Prewo, J. J. Brennan and G. K. Layden, American Ceramic Society Bulletin 65 [2] (1986) 305-313.
[40] C. W. Turner, Ceramic Bulletin, 70 [9] (1991) 1487-1490.
[41] J. A. Cornie, Y. Chiang, D. R. Uhlmann, A. Mortensen and J. M. Collins, Ceramic Bulletin, 65 [2] (1986) 293-304.
[42] J. R. Strife, J. J. Brennan, and K. M. Prewo, Ceram. Eng. Sci. Proc. 11[7-8] (1990) 871-919.
[43] H. K. Liu and A. P. Majidi, Ceram. Eng. Sci. Proc., 13 [9-10] (1992) 642-649.
[44] Ching-Li Hu and A. P. Majidi, Submitted to J. Am. Cera. Soc. (1994).
[45] K. Sato, T. Suzuki, O. Funayama, T. Isoda and T. Itoh, Ceram. Eng. Sci. Proc. 13 [9-10] (1992) 614-621.
[46] K. Nakano and A. Kamiya, Proceedings of European Conference on Composite Materials (ECCM) (1990) Bordeaux, France.
[47] Chih-Chin Lu, Tsu-Wei Chou and Azar Parvizi-Majidi, Proceedings of 39th International SAMPE Symposium, Ed. Drake et al., SAMPE (1994) 651-658.
[48] R. T. Bhatt, U. S. Pat. No. 4689188, (1987).
[49] J. I. Eldridge, R. T. Bhatt and J. D. Kaiser, Ceram. Eng. Sci. Proc. 12 [7-8] (1991) 1152-1171.
[50] F. A. Habib, R. G. Cooke and B. Harris, British Ceramics Transactions Journal, 89 (1990) 115-124.
[51] (a) G. Y. Baaklini and R. T. Bhatt, Ceramic Engineering and Science Proceedings, 12 [7-8] (1991) 1599-1615. (b) M. W. Barsoum, P. Kangutkar and A. S. D. Wang, Comp. Sci. & Tech. 44(1992) 257-269.
[52] A. W. Pryce and P. A. Smith, In *Proceedings of the 8th International Conference on Composite Materials*, Tsai, S. W. and Springer G. S. Edt., SAMPE, Covina, CA (1991) 24 A-1 to 24 A-10.
[53] R. Y. Kim and N. J. Pagano, J. Am. Cera. Soc., 74[5] (1991) 1082-1090.
[54] (a) S. W. Wang, PhD Dissertation, University of Delaware, (1990). (b) S. W. Wang and A. P. Majidi, J. Mat. Sci. (1992) 5483-5496.
[55] P. G. Karandikar and Tsu-Wei Chou, Composites Science and Technology 46 (1993) 253-263.
[56] H. C. Cao, E. Bischoff, O. Sbaizero, M. Ruhle, A. G. Evans, D. B. Marshall and J. J. Brennan, Journal of American Ceramic Society, 73 (1990) 1691-1699.
[57] V. C. Nardone and K. M. Prewo, Journal of Materials Science, 23 (1988) 168-180.
[58] K. M. Prewo, J. Mat. Sci. 21 (1986) 3590-3600.

[59] Tsu-Wei Chou, Microstructural Design of Fiber Composites, Cambridge University Press (1992).
[60] S. W. Tsai, in Fundamental Aspects of Fiber Reinforced Plastic Composites, Eds. R.T. Schwartz and H. S. Schwartz, Interscience, New York (1968) 3-25.
[61] O. Sbaizero and A. G. Evans, Journal of American Ceramic Society, 68 (1986) 481-486.
[62] S. Mall and R. Y. Kim, Composites 23 [4] (1992) 215-222.
[63] R. Y. Kim, Cera. Engg. and Sci Proc. 13 [7-8] (1992) 281-300.
[64] D. S. Beyerley, S. M. Spearing and A. G. Evans, J. Am. Cera. Soc. 75 [12] (1992) 3321-3330.
[65] J. E. Bailey, P. T. Curtis and A. Parvizi, Proc. Royal. Soc. Lond. A 366 (1979) 599-623.
[66] P. Pluvinage, A. Parvizi-Majidi and Tsu-Wei Chou, in Proceedings of the American Society for Composites 7th technical Conference, Technomic Publi. Co. (1992) 400-409.
[67] Azar Parvizi-Majidi and P. G. Karandikar, Proc. of 3rd Japan International SAMPE Symposium, Edt. Kishi et al. (1993) 586-593.
[68] B. Harris, F. A. Habib and R. G. Cooke, Proc. R. Soc. Lond. A 437 (1992) 109-131.
[69] J. Lankford, Composites, 18 [2], (1987) 145-152.
[70] Z. G. Wang, C. Laird, Z. Hashin, B. W. Rosen and Chian-Fong Yen, Journal of Materials Science, 26 (1991) 4751-4758.
[71] J. Y. Rossignol, J. M. Quenisset, H. Hannache, C. Mallet, R. Naslain and F. Christin, Journal of Materials Science, 22 (1987) 3240-3252.
[72] (a) N. Takeda, O. Chen, T. Kishi, W. Tredway and K. Prewo, Proceedings of the Int. Conf. Advanced Materials Mechanical Properties '90, Peragamon Press, Ed. T. Yokobori and T. Kishi, (1991) 791-800. (b) Kevin Stull and Azar Parvizi-Majidi, Ceram. Engg. Sci. Proc. 12 [7-8] (1991) 1452-1461 (c) Kevin Stull, CCM Research Report 92-34, University of Delaware (1992).
[73] R. T. Bhatt and R. E. Phillips, J. Comp. Tech. & Res. 12 [1] (1990) 13-23.
[74] P. G. Karandikar, Tsu-Wei Chou, A. Parvizi-Majidi and K. M. Prewo, Cera. Eng. Sci. Proc. 14 [9-10] (1993) 880-889.
[75] Nancy Fang and Tsu-Wei Chou, J. Am. Cera. Soc. 76 [10] (1993) 2539-2548.
[76] J. J. Brennan, in Tailoring Multiphase and Composite Ceramics, Eds. R. Tressler et al., Plenum Press, New York (1986) 549-560.
[77] R. F. Cooper and K. Chyung, J. Mater. Sci. 22 (1987) 3148-3160.
[78] J. Homeny, J. R. VanValzah and M. A. Kelly, J. Am. Cera. Soc. 73 [7] (1990) 2054-2059.
[79] L. A. Bonney and R. F. Cooper, J. Am. Cera. Soc. 73 [10] (1990) 2916-2921.
[80] M. H. Lewis and V. S. R. Murthy, Comp. Sci. & Tech. 42 (1991) 221-249.
[81] G. Qi, K. E. Spear and C. G. Pantano, Presented at the 1992 Fall Materials Research Society Meeting, Boston, MA.
[82] R. P. Boisvert, R. K. Hutter and R. J. Diefendorf, In *Proceedings of the 4th Japan-US Conference on Composite Materials*, Technomic Publishing Co., Lancaster, PA (1989) 789-798.
[83] M. Monthioux and D. Cojean, Proceedings of Fifth European Conference on Composite Materials (1992) 729-734.
[84] J. F. Despres and M. Monthioux, Proceedings of Fifth European Conference on Composite Materials (1992) 901-906.
[85] H. W. Carpenter and J. W. Bohlen, Ceram. Eng. Sci. Proc. 13 [7-8] (1992) 238-256.
[86] D. H. Grande, J. F. Mandell and K. C. C. Hong, Journal of Materials Science, 23 (1989) 311-328.

[87] D. B. Marshall, Journal of the American Ceramic Society 67 [12] (1984) C259-260.
[88] D. B. Marshall and W. C. Oliver, Journal of the American Ceramic Society 70 [8] (1987) 542-547.
[89] J. W. Laughner, N. J. Shaw, R. T. Bhatt and J. A. DiCarlo, Cera. Eng. Sci. Proc. 7 [7-8] (1986) 932.
[90] R. N. Singh, Cera. Eng. Sci. Proc. 10 [7-8] (1989) 883-893.
[91] U. V. Deshmukh and T. W. Coyle, Ceramic Engineering and Science Proceedings 9 [7-8] (1988) 627-634.
[92] U. V. Deshmukh, A. Kanei, S. W. Freiman and D. C. Cranmer, Mat. Res. Soc. Symp. Proc. 120 (1988) 253-258.
[93] D. C. Cranmer, U. V. Deshmukh and T. W. Coyle, In *Thermal and Mechanical Behavior of Metal Matrix and Ceramic Matrix Composites,* ASTM STP 1080, J. M. Kennedy et al. Edt., American Society for Testing and Materials, Philadelphia (1990) 124-135.
[94] J. Lamon, C. Rechiniac, N. Lissart and P. Corne, Proceedings of Fifth European Conference on Composite Materials (1992) 895-900.
[95] T. A. Parthasarathi, N. J. Pagano and R. J. Kerans, Cera. Eng. Sci. Proc. 10 [7-8] (1989) 872-881.
[96] Chongdu Cho, J. W. Holmes and J. R. Barber, Journal of American Ceramic Society 74 [11] (1991) 2802-2808.
[97] Y. Kagawa and K. Honda, Ceram. Eng. Sci. Proc. 12[7-8] (1991) 1127-1138.
[98] T. J. Mackin and F. W. Zok, J. Am. Ceram. Soc. 75 [11] (1992) 3169-3171.
[99] A. C. Kimber and J. G. Keer, J. Mat. Sci. Let. 1 (1982) 353-354.
[100] P. Lawrence, J Mat. Sci. 7 (1972) 1-6.
[101] A. Takaku and RGC Arridge, J. Phys.D: Appl. Phys., 6 (1973) 2038-2047.
[102] H. Stang and S. P. Shah, J. Mat. Sci. 21 (1986) 953-957.
[103] K. T. Faber, S. H. Advani, J. K. Lee, and J-T Jinn, J. Am. Cera. Soc., 69 [9] (1986) c208-209.
[104] D. K. Shetty, Journal of the American Ceramic Society 71 [2] (1988) C107-109.
[105] J. D. Bright, D. K. Shetty, C. W. Griffin and S. Y. Limaye, Journal of the American Ceramic Society 72 [10] (1989) 1891-1898.
[106] R. W. Goetteler and K. T. Faber, Composites Science and Technology 37 (1989) 129-147.
[107] Chun-Hway Hsueh, Mat. Sci. and Engg. A123 (1990) 1-11.
[108] J. R. Yeh, Journal of Composite Materials 25 (1991) 1158-1170.
[109] P. D. Jero, R. J. Kerans and T. A. Parthasarathi, Journal of the American Ceramic Society 74 [11] (1991) 2793-2801.
[110] R. J. Kerans and T. A. Parthasarathy, Journal of the American Ceramic Society 74 [7] (1991) 1585-1596.
[111] M. K. Brun, Journal of the American Ceramic Society 75 [7] (1992) 1914-1917.
[112] B. Budiansky, J. W. Hutchinson and A. G. Evans, Journal of Mechanics and Physics of Solids, 34 (1986) 167-189.
[113] W. S. Kuo and T. W. Chou, In *Metal and Ceramic Matrix Composites: Processing, Modeling and Characterization,* Ed. R. B. Bhagat, P. Kumar and A. M. Ritter. The Minerals, Metals and Materials Society, Warrendale, PA (1990) 311-318.
[114] D. Marshall, B. Cox and A. Evans, Acta Metallurgica, 33 (1985) 2013-2021.
[115] L. N. McCartney, Proc. Royal Society of London, A-409 (1987) 329-350.
[116] Yih-Cherng Chiang, A.S.D. Wang and Tsu-Wei Chou, J. Mech. Phy. of Solids 41 (1993) 1137-1154.
[117] A. S. D. Wang, X. G. Huang and M. W. Barsoum, Comp. Sci. & Tech. 44(1992) 271-282.

[118] J. W. Lee and I. M. Daniel, Composite Materials: Testing and Design (Tenth Volume), ASTM STP 1120, G. C. Grimes Edt., American Society for Testing and Materials, Philadelphia (1992) 204-221.
[119] B. D. Coleman, J. Mech. Phys. Solids 7 (1958) 60-70.
[120] P. W. Manders and T-W Chou, J. Reinforced Plastics and Comp. 2 (1983) 43-59.
[121] K. Goda and H. Fukunaga, J. Mat. Sci. 21 (1986) 4475-4480.
[122] R. W. Davidge and A. Briggs, J. Mat. Sci. 24 (1989) 2815-2819.
[123] B. W. Rosen, *Mechanics of Composite Strengthening*, Ch. 3, ASM, Metals Park, OH (1965).
[124] C. Zwebwn and B. W. Rosen, J. Mech. Phys. Solids 18 (1970) 189-206.
[125] H. Fukuda and K. Kawata, Fiber Science and Technology 10 (1977) 53-63.
[126] M. D. Thouless, O. Sbaizero, L. S. Sigl and A. G. Evans, Journal of American Ceramic Society, 72 (1989) 525-532.
[127] (a) H. Cao and M. D. Thouless, J. Am. Cera. Soc., 73 [7] (1990) 2091-2094. (b) P. G. Karandikar, Tsu-Wei Chou and Azar Parvizi-Majidi, Proc. of 3rd Japan International SAMPE Symposium, Edt. Kishi et al. (1993) 547-552.
[128] H. R. Schweitert and P. S. Steif, J. Mech. Phys. Solids 38 [3] (1990) 325-343.
[129] P. S. Steif and H. R. Schweitert, Ceram. Eng. Sci. Proc. 11 [9-10] (1990) 1567-1576.
[130] M. Sutcu, Acta. Metall. 37 [2] (1989) 651-661.
[131] W. A. Curtin, J. Am. Cera. Soc., 74 [11] (1991) 2837-2845.
[132] K. M. Knowles and X. F. Yang, Ceram. Eng. Sci. Proc. 12 [7-8] (1991) 1375-1388.
[133] Wen-Shyong Kuo and Tsu-Wei Chou, J. Am. Cera. Soc. (in press).
[134] H. Fukunaga, Tsu-Wei Chou, P. W. M. Peters and K. Schulte, J. Comp. Mat. 18 [7] (1984) 339-356.
[135] Wen-Shyong Kuo and Tsu-Wei Chou, J. Am. Cera. Soc. (in press).
[136] (a) T. Mah, M. G. Mendiratta, A. P. Katz, R. Ruh and K. S. Mazdiyasni, J. Am. Cera. Soc. 68 [1] (1985) C27-30. (b) T. Mah, M. G. Mendiratta, A. P. Katz, R. Ruh and K. S. Mazdiyasni, J. Am. Cera. Soc. 68 [9] (1985) C248-251.
[137] E. Y. Luh and A. G. Evans, J. Am. Cera. Soc. 70 [7] (1987) 466-469.
[138] D. A. Jablonski and R. T. Bhatt, J. Comp. Tech. Res. JCTRER 12 [3] (1990) 139-146.
[139] Philippe Pluvinage, Azar Parvizi-Majidi and Tsu-Wei Chou, Proceedings of the High Temperature Ceramic Matrix Composites-1 Conference, Edt. Naslain et al., Woodhead Publi. Co. (1993) 675-682.
[140] (a) K. M. Prewo, B. Johnson and S. Starrett, J. Mat Sci. 24 (1989) 1373-1379. (b) W. K. Tredway, K. M. Prewo, T. Isoda and M. Iwata, Presented at the 6th Japan Conference on Composite Materials, June, 1992.
[141] L. Filipuzzi, G. Camus, J. Thebault and R. Naslain, in Structural Ceramic-Processing, Microstructures and Properties, Eds. J. J. Bentzen et al., Riso National Lab., Roskilde, DK (1990) 283-289.
[142] R. T. Bhatt, J. Am. Cera. Soc. 75 [2] (1992) 406-412.
[143] N. Frety and M. Boussuge, Comp. Sci. Tech. 37 (1990) 177-189.
[144] F. Lamouroux, G. Camus and J. Thebault, Proceedings of the 5th European Conference on Composite Materials (1992) 499-504.
[145] K. M. Prewo and J. Batt, J. Mat. Sci. 23 (1988) 523-533.
[146] P. G. Karandikar and Tsu-Wei Chou, Ceramic Engineering and Science Proceedings 13 [9-10] (1992) 881-888 .
[147] P. G. Karandikar and Tsu-Wei Chou, Proc. Am. Society for Composites Seventh Technical Conference, Technomic Publishing Co., Lancaster, PA (1992) 695-704.
[148] K. R. Linger, Composite- Standards, Testing and Design, IPC Science and Technology Press, London (1974) 200-208.

[149] E. Minford and K. M. Prewo, in *Tailoring Multiphase and Composite Ceramics*, C. G. Patano et al. Edt., Plenum Publishing Co., New York (1986) 561-570.
[150] K. M. Prewo, Journal of Materials Science, 22 (1987) 2695-2701.
[151] L. P. Zawada, L. M. Butkus and G. A. Hartman, Ceramic Engineering and Science Proceedings, 11[9-10] (1990) 1592-1606.
[152] C. Q. Rousseau, In *Thermal and Mechanical Behavior of Metal Matrix and Ceramic Matrix Composites*, ASTM STP 1080, J. M. Kennedy et al. Eds., American Society for Testing and Materials, Philadelphia, 1990, 136-151.
[153] J. W. Holmes, T. Kotil and W. T. Foulds, In *Symposium on High Temperature Composites*, Technomic Publishing Co., Lancaster, PA (1989) 176-186.
[154] J. W. Holmes, Journal of American Ceramic Society, 74 (1991) 1639-1645.
[155] J. W. Holmes, Journal of Composite Materials 26 [6] (1992) 916-933.
[156] T. Kotil, J. Holmes and M. Comninou, J. Am. Cera. Soc., 73 (1990) 1879-1883.
[157] J. W. Holmes and S. F. Shuler, J. of Mat. Sci. Letters 9 (1990) 1290-1291.
[158] J. W. Holmes and Chongdu Cho, J. Am. Cera. Soc. 75 [4] (1992) 929-938.
[159] V. Ramakrishnan, J. W. Holmes and J. W. Jones, "High frequency fatigue of a fiber-reinforced glass-ceramic matrix composite," Submitted for publication.
[160] L. P. Zawada and L. M. Butkus, J. Am. Cera. Soc. 74 [11] 2851-2858 (1991).
[161] S. Mall and G. D. Tracy, J. Reinforced Plastics and Comp. 11 [3] (1992) 243-260.
[162] S. F. Shuler, J. W. Holmes and D. Roach, "Influence of loading frequency on the room temperature fatigue of a carbon fiber SiC-matrix composite," Submitted to the Journal of the American Ceramic Society.
[163] (a) P. G. Karandikar, Ph.D Dissertation, University of Delaware, May (1992). (b) P. G. Karandikar and Tsu-Wei Chou, J. Am. Cera. Soc. 76 [7] (1993) 1720-1728.
[164] R. M. McMeeking and A. G. Evans, Mechanics of Materials 9 (1990) 217-227.
[165] L. P. Zawada and R. C. Wetherhold, J. Mat. Sci. 26 (1991) 648-654.
[166] L. M. Butkus, J. W. Holmes, J. Am. Ceram. Soc. 76 [11] (1993) 2817-2825.
[167] D. W. Worthem and J. R. Ellis, Ceram. Eng. Sci. Proc. 14 [7-8] (1993) 292-299.
[168] G. M. St.Hilaire and T. Erturk, Ceram. Eng. Sci. Proc. 14 [7-8] (1993) 416-425.
[169] W. R. Cannon and T. G. Langdon, Journal of Materials Science, 18 (1983) 1-50.
[170] W. R. Cannon and T. G. Langdon, Journal of Materials Science 23 (1988) 1-20.
[171] F. Abbe, J. Vincens and J. L. Chermant, J. Mater. Sci. Lett. 8 (1989) 1026-1028.
[172] J. W. Holmes, J. Am. Ceram. Soc. 74 [7] (1991) 1639-1645.
[173] J. W. Holmes, J. Mat. Sci. 26 (1991) 1808-1814.
[174] J. R. Pachalis, J. Kim and T-W Chou, Comp. Sci. and Tech. 37 (1990) 329-346.
[175] J. R. Pachalis and Tsu-Wei Chou, Journal of Applied Mechanics 59 (1992) 27-32.
[176] Yuan Ruo Wang and Tsu-Wei Chou, J. Comp. Mat. 26 (1992) 1269-1286.
[177] R. N. Singh, J. Am. Cera. Soc., 73 [10] (1990) 2930-37.
[178] S. M. Bleay and V. D. Scott, J. Mat. Sci. 26 (1991) 2229-2239.
[179] J. I. Eldridge, R. T. Bhatt and J. D. Kaiser, Ceram. Eng. Sci. Proc. 12 [7-8] (1991) 1152-1171.
[180] A. M. Daniel and M. H. Lewis, Cera. Eng. Sci. Proc. 14 [7-8] (1993) 131-138.
[181] C. W. Lawrence and B. Derby, Cera. Eng. Sci. Proc. 14 [7-8] (1993) 139-146.

Homework problems

1. (a) What are the advantages and disadvantages of ceramics as a structural material?
(b) Discuss two methods of overcoming the disadvantages and explain the mechanisms involved.
(c) In what respects do ceramic matrix composites differ from polymer matrix composites?

2. Regularly spaced matrix cracking develops in ceramic matrix composites under tensile load. Assuming a constant shear stress τ at the interface, develop a relationship between minimum matrix crack spacing, x and matrix ultimate stress, σ_{mu}. Explain how the crack spacing depends on the fiber radius, fiber volume fraction and τ.

3. Explain/derive the expressions given by Aveston, Cooper and Kelly (ACK) for various energy terms (Eq. 24) in the derivation of matrix cracking stress by the energy balance approach.

4. What are the various methods for processing ceramic matrix composites? What are the advantages and disadvantages of each of the methods?

5. What criteria can be used in selecting ceramic matrices and the reinforcing fibers?

6. Explain the critical roll played by the interface in ceramic matrix composites. What properties of the interface can be used in the composite design? How are these measured?

7. Based on the data given calculate the interfacial shear stress for Nicalon-CAS composite: volume fraction 0.35, matrix cracking stress 117.5 MPa, fiber radius 7.5 µm. and experimental crack spacing 130 µm.

8. Using the data in Tables 1 and 2 calculate strengths of CG-I, A-I, B-I and Nicalon-CAS composites by average strength model, bundle strength model, Cao-Thouless model and Curtin model. Compare the calculations with the experiomental data and assess the suitability of the models for strength prediction (fiber gage length, L_0 = 25.4 mm; fiber radius 7.5 µm; $\sigma^*_{fu} = \sigma_{fu} (\ln 2)^{-1/m}$)

Table 1. Fiber strength data obtained by fracture mirror analysis and single fiber testing.

Compo-site	Fracture Mirror			Extracted Single Fiber[†]			As Recd. Single Fiber[†]		
	m	σ_{fu}	σ_0	m	σ_{fu}	σ_0	m	σ_{fu}	σ_0
CG-I	4.4	1887.8	2044.8	3.1	1472.5	1612.3	7.6	2986.5	3138.9
A-I	5.4	1967.7	2098.8	4.4	1425.3	1527.5	4.3	3196.4	3453.4
B-I	6.1	2327.6	2465.7	3.2	1196.2	1315.8	4.7	3145.1	3365.5
SiC-CAS	--	--	--	4.7	1448.0	1535.3	--	--	--

m, σ_0 - Weibull shape and scale parameters, σ_{fu} - average fiber strength, strength in MPa

Table 2. Data for strength and modulus calculations

Composite	V_f	τ_f MPa	Strength (Expt.) MPa
CG-I	0.39	1,10,50	158.8
A-I	0.44	1,10,50	558.1
B-I	0.45	1,10,50	585.1
SiC-CAS	0.35	12	445.8

9. Using the stress distributions given by shear lag analysis (Eq. 30-32), derive Eq. 47 for reduced modulus of a unidirectional CMC. Using the data in Tables 1-3 carry out modulus predictions at three levels (0%, 50% and 100%) of debonding. (E_m = 98 GPa, E_f = 195 GPa, G_m = 70 GPa, G_f = 80 GPa, α_m = 5x10^{-6}/°C, α_f = 4x10^{-6}/°C, ΔT = -1000°C)

Table 3. Data on SiC-CAS for modulus reduction calculations

σ (MPa)	Crack Spacing (μm)	Expt. E (GPa)
0.0	∞	130.0
94.6	∞	128.3
125.9	∞	129.5
157.4	4807	130.3
188.9	1923	130.4
220.3	854	129.5
251.8	727	126.3
283.3	418	116.1
314.8	163	85.6
346.4	157	65.7
377.7	143	63.4
409.2	130	61.0

10. A tubular component in an aircraft engine is to be made of SiC/SiC 2-D woven composite. The load and temperature cycles involved in one flight are as shown in Figure 1. What laboratory tests should be conducted to determine the number of flights such a component can survive? If the times required for the various parts of the loading cycle vary depending on the weather conditions, what sort of design curves would be required to calculate the residual life of the component?

Ceramic Matrix Composites 349

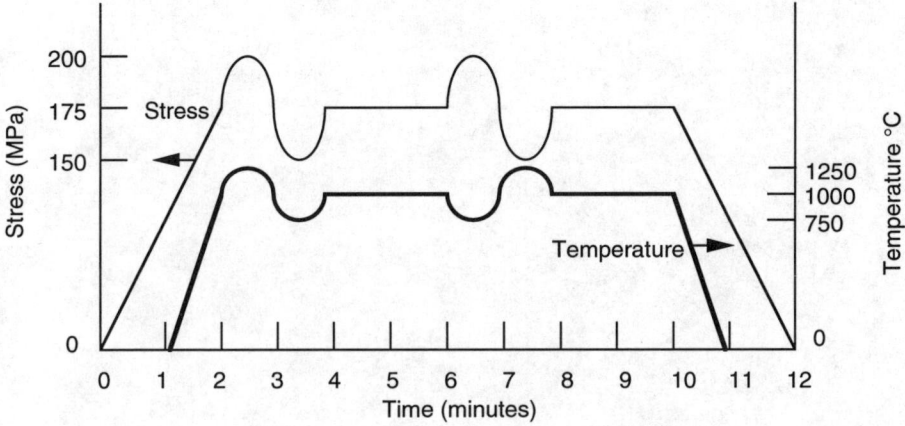

Figure 1. Temperature and stress cycles during one flight.

Chapter 6
MICROSTRUCTURE OF INORGANIC GLASSES

Robert H. Doremus
Materials Engineering Department
Rennselaer Polytechnic Institute
Troy, New York 12180-3590

1. Introduction

Most commercial inorganic glasses are uniform single-phase materials. Their optical clarity is important in many practical applications. Nevertheless second phases in glass can influence and even improve their properties. Glass-glass phase separation is the basis of the Vycor process, in which a soluble phase is leached from a borosilicate glass, leaving behind a glass porous on a very fine scale (5-10 nm), which is a useful adsorbent. This porous glass can be consolidated to a high silica glass by heating it to about 1000°C, which is much lower than the normal heating temperature of vitreous silica. Surface crystallization (devitrification) degrades glass properties, but uniform crystallization (glass ceramics) can strengthen glass and improves its chemical durability. Crystallization in optical components such as fibers and wave guides can seriously degrade their usefulness. Small particles in glass cause intriguing optical changes.

In this chapter the formation and influence of second phases, both glassy and crystalline, in glasses are discussed. Silicate glasses are emphasized because of their commercial importance; crystallization of fluoride glasses is included because they have potential optical applications as a result of their optical transparency in the infrared to a wave length of 8 μm. Nucleation and growth of second phases in a glass are first treated; then phase separation and crystallization are considered. Glass ceramics and their uses are then described, and finally the formation and optical properties of small metallic and semiconductor particles in glass are described.

2. Nucleation and Growth

2.1 Nucleation

The theory of nucleation of liquid drops from a vapor was developed by Volmer and Weber (1), and agrees well with experiments on nucleation in cloud chambers (2). Many correction factors have been proposed for the simple theory of Volmer and Weber, but they are unnecessary (2).

In the theory of Volmer and Weber the concentration N^* of critical nuclei is calculated from the fluctuation theory of Einstein (3) and the Gibbs result for the reversible work to form a critical nucleus (4). In this calculation one need not be concerned about the kinetics of growth or the disappearance of embryos of transforming material other than those of critical size. Furthermore only the chemical potential of embryos of critical size enters, so problems about embryos of smaller size are avoided. The critical assumption is that the material of the embryos has the thermodynamic properties of bulk material. See ref. 2 for discussion.

At constant temperature T the probability P of a fluctuation to form one thermodynamic state from another is

$$P = K \exp(W_R/kT) \qquad (1)$$

in which W_R is the reversible work done on the surroundings to provide the change in state, k Boltzmann's constant, and K a normalizing factor. Einstein derived Eq. 1 directly from Boltzmann's relation between P and entropy S

$$P = K' \exp(S/k) \qquad (2)$$

since at constant total energy (constant temperature $\Delta E = O$):

$$\Delta S = W_R/T \qquad (3)$$

This derivation is refined and discussed in depth by Tolman (5); Eq. 1 still results.

The probability to form a spherical embryo of critical radius r^* in a material containing N molecules per unit volume is N^*/N, where N^* is the number of embryos of size r^* per unit volume, and N is the normalizing factor, so

$$N^*/N = \exp(W_R^*/kT) \qquad (4)$$

in which W_R^* is the reversible work to form embryos of radius r^* from untransformed material. This process is reversible because the untransformed material is in equilibrium with embryos of radius r^*.

The reversible work to form a spherical critical nucleus, that is, one that is in unstable equilibrium with the surroundings, was found by Gibbs (4) to be:

$$W_R^* = -16\pi\gamma^3/3(\Delta P)^2 \tag{5}$$

in which gamma is the interfacial energy between the new and old phases, and ΔP is the pressure difference between the inside and outside of the critical nucleus. The derivation so far is valid for any combination of phases, although Volmer and Weber limited their derivation to drops nucleating from a vapor. For crystalline phases γ is an average over the surface facets in the critical nucleus. The volume free energy change ΔG_v for a transformation in which both phases are condensed is given by

$$\Delta P = \Delta G_v = \Delta H_e \, \Delta T / T_e V \tag{6}$$

in which ΔH_e is the heat of transformation at the equilibrium transformation temperature of T_e of a volume V of material, and $\Delta T = T_e - T$ is the undercooling, since T is the actual temperature of transformation. $\Delta T/T_e$ is a measure of the supersaturation. As an example, consider a pure crystal nucleating in a glass. Then T_e is the melting temperature of this crystal, ΔH_e is its heat of fusion at T_e, and T is the actual temperature at which nucleation takes place.

If nucleation is from a condensed phase, the rate of transport of atoms into the new phases influences the rate of nucleation I, which is given by

$$I = ZA*N* \tag{7}$$

in which N^* is taken from Eq. 4, A^* is the area of the critical nucleus, and Z is related to the rate of jumping of atoms across the interface between phases or to long-range diffusion from the matrix to the new phases. For either of these situations (2) the resulting nucleation equation has the form

$$I = B \exp(-Q + W_R^*)/kT \tag{8}$$

where B is a factor not much dependent on temperature and Q is the activation energy for transport. For transformations in glasses Q is usually found to be close to the activation energy for viscous flow. The nucleation rate of Eq. 7 has a maximum, because of two opposing tendencies. As the temperature decreases below the transformation temperature, the driving force ΔG_v increases, increasing the nucleation rate; at the same time the decreasing temperature leads to a lower rate of transport, eventually decreasing the nucleation rate.

The temperature dependence of W* from Eqs. 5 and 6 is approximately

$$-W^* = K/(\Delta T)^2 \qquad (9)$$

where K is a parameter not much dependent on temperature. Thus from Eqs. 7 and 8 one way to compare an experimental temperature dependence of a nucleation rate with theory is to plot log $I\eta$ as a function of $1/T(\Delta T)^2$, which should give a straight line; η is the viscosity.

The steady-state rate of nucleation requires some time to be established, especially in viscous glasses in which transport is slow. This process can be described by a time-dependent nucleation rate

$$I = I_o \exp(-\tau/t) \qquad (10)$$

in which I_o is the steady-state nucleation rate and τ is an "induction" time.

The preceding discussion applies to homogeneous nucleation, in which a second phase forms directly from another with the interfacial energy γ for a bulk phase. Homogeneous nucleation is usually characterized by formation of a large uniform number of small nuclei of second phase. In glasses foreign particles, surfaces, and defects (e.g. voids) can lead to heterogeneous nucleation with an effective interfacial energy lower than the bulk value. Often there is a spectrum of site potencies that lead to nucleation over a wide range of supersaturation.

2.2 Spinodal Decomposition

In the mechanism of nucleation described above, a small region of equilibrium phase grows to the critical size and beyond. In certain multicomponent mixtures an alternative mechanism is possible in which the phase transformation takes place uniformly throughout the transformation matrix instead of a discrete spots, and the composition changes gradually from the initial to the equilibrium value. This mechanism is called spinodal decomposition. In it there is a region of compositions in which d^2G/dC^2 is negative, in which G is the free energy of the system and C is its composition. In such a region there is no barrier to phase growth, since any small fluctuation leads to a reduction in free energy. The transformation proceeds by a continuous change in the compositions of the growing phases, whereas their extent remains constant. Table I contrasts the nucleation and spinodal mechanisms. An example of spinodal decomposition is given below in the section on phase separation.

Table 1. Comparison of Spinodal Decomposition and Nucleation and Growth Mechanism

Nucleation and growth	Spinodal decomposition
$d^2G/dC^2 > 0$	$d^2G/dC^2 < 0$
Small regions grow	Uniform transformation
Equilibrium composition of nucleus initially	Compositions change uniformly with time
Surface energy barrier	No surface energy barrier
Random morphology	Uniform morphology, characteristic wavelength

2.3 Growth

If the transport is controlled by an interface process, the rate of growth is constant as the new phase grows into a region of uniform composition. In diffusion controlled growth the rate of growth of a spherical particle of radius R is

$$\frac{dR}{dt} = \alpha \frac{D}{R} \qquad (11)$$

where α depends only on concentrations and D is a diffusion coefficient. The particle radius grows proportional to the square root of time.

At longer times, larger particles grow and smaller particles dissolve to decrease the total particle surface area. This "Ostwald Ripening", coarsening, or competitive growth is often observed in phase separated glasses, and occurs after the matrix in diffusion control is close to the equilibrium composition. The mathematical treatment of this process is given in refs. 6-8, see also ref. 2. The driving force is provided by the higher solubility of the smaller particles, resulting in transfer of material from them to the larger particles. The mean radius R of the particles as a function of time t is given by

$$R^3 - R_o^3 = 8\gamma C_e DVt/9KT \qquad (12)$$

in which R_o is the initial particle size, γ the interfacial energy, C_e is the solute concentration, V the molar volume of the particle, K the gas constant, T the temperature, and D the diffusion coefficient. Thus the mean particle radius increases as $t^{1/3}$.

3. Phase Separation

3.1 Introduction

Phase separation of sodium borosilicate glass is the basis of the Vycor process for making a 96% silica glass, as mentioned in the Introduction. Phase separation occurs in some other commercially important glasses, especially borosilicates and certain soda-lime silicates. Pyrex borosilicate glass is separated on a fine scale (9). As these glasses are heated in a temperature range above the glass transition temperature (perhaps from 600°C to 900°C) the scale of phase separation increases and can lead to degradation of properties.

Phase diagrams for glass-glass phase separation in silicates are discussed first, and then theories to explain this separation are considered. Kinetics of phase separation are then treated, and finally the influence of phase separation on properties is briefly discussed.

3.2 Phase Diagrams

Many binary liquid silicate systems show equilibrium phase separation; an example is the alkaline-earth series, with a greater tendency to phase separation in the series from MgO to SrO; BaO does not show equilibrium immiscibility. Many transition metal and rare earth oxides show liquid-liquid immiscibility with silica.

In binary alkali silicates there is no stable phase separation; however, these silicates are easily cooled to glasses without crystallization, and some of them show liquid-liquid phase separation below the liquidus temperature of crystallization, as shown in Fig. 1, which is called metastable phase separation. There is some indication, (15,16) of phase separation in binary potassium silicates with a boundary below that of Na_2O in Fig. 1.

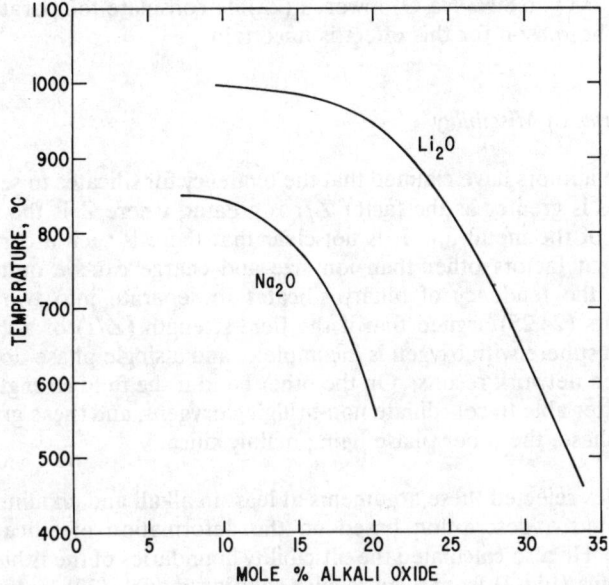

Figure 1. Phase separation diagrams for lithium[10-12] and sodium[13,14] silicates.

Douglas and coworkers studied phase separation in ternary systems soda-lime silicate (17), soda-lithia-silica (10), and soda-baria-silica (18). They found that the composition c of the immiscibility curves for several binary and pseudobinary (ternaries with constant silica) silicate mixtures was related to temperature by the equation

$$|c_c - c| = K(T_c - T)^n \tag{13}$$

where c_c and T_c are the composition and temperature of the maximum temperature (critical consulate point), K is a constant and n is a number usually between 1/4 and 1/2. This equation is of much theoretical interest, as described below.

Another important commercial ternary system is that of sodium borosilicates, already mentioned. Results of several studies on this system are summarized by Haller et al. (19,20). There is some questions about results in silicate rich mixtures, and the positions of tie lines.

Aluminum oxide has a strong tendency to reduce phase separation in silicates. For example addition of 1.8 mol% Al_2O_3 to a sodium borosilicate glass (70.5 mol%

SiO_2, 27.7% B_2O_3, 6.89% Na_2O) lowered (22) the consulate temperature from 754°C to 643°C. The reason for this effect is uncertain.

3.3 Theories of Miscibility

Some authors have claimed that the tendency for silicates to separate into two liquid phases is greater as the factor Z/r is greater, where Z is the ion charge and r the radius of the metal ion. It is not clear that there is such a correlation (23); I concluded that factors other than ion size and charge can be quite important in determining the tendency of binary silicates to separate into two liquid phases. These authors (24,25) argued that if the field strength (Z/r) of a cation is low its coordination sphere with oxygen is incomplete, and a single phase dominated by the silicon-oxygen network results. On the other hand if the field strength of a cation is high, it is better able to coordinate non-bridging oxygens, and these groups then form a separate phase, the other phase being mainly silica.

Charles rejected these arguments at least in alkali and alkaline earth silicates and substituted a description based on the deformation of silica chains in the network.(26) He also calculated the miscibility boundaries of the lithium and sodium silicate systems (Fig. 1) from a subregular solution model. (27)

McGahay and Tomozawa (28) calculated the critical temperatures of a variety of "ionic" systems, including silicate glasses, from an electrostatic model. They used Debye-Hückel theory for dilute solution to determine the thermodynamic properties of these systems. It is surprising that this theory, which breaks down for concentrated aqueous solutions, can be used for these dense solids. Perhaps the charges in them are highly screened.

The exponent n of the miscibility line equation (Eq. 13) has been derived from quite general considerations (29). One theory results from a three-dimensional Ising model as deduced from high temperature series, and the other from renormalization group methods. The former gave $n = 0.312$, the latter $n = 0.325$, and experiments (29) seem to be closer to at 0.325. The value of 0.5 found for a soda-lime glass by Douglas and Burnett was a rough estimate based on few data. The results of Simmons (29a) on lead borate glasses give a value of n of about 1/3. The relationship between statistical theories and the structural ones described above needs to be explored.

3.4 Kinetics

Hammel (30) measured nucleation rates of silica-rich particles in a glass containing 76 mole% SiO_2, 13% Na_2O, and 11% CaO. He measured the numbers of particles as a function of time after heating at different temperatures in the electron microscope. There was a transient in nucleation as described by Eq. 9; a constant nucleation rate was achieved only after a period of time.

To compare his results with nucleation theory, Hammel measured the critical nucleus size r* as a function of temperature. He made this measurement by holding a glass sample containing particles of uniform size in a temperature gradient and observing the temperature at which the particles neither dissolved nor grew. This temperature is the temperature at which the particles in the sample are of critical size. If ΔG_v is given by Eq. 6, a plot of 1/r* against temperature should be a straight line; Hammel found this relationship. For a comparison of the actual rates of nucleation measured with theory one must know W_R^* and therefore ΔG_v. Hammel calculated ΔG_v from the miscibility curve, and used Q = 395 kj/mol as estimated from growth experiments. The calculated values of nucleation rates from an equation such as Eq. 7 and the measured rates compare quite closely (23). This agreement, which is unusual, provides evidence that the nucleation was homogeneous, and supports the nucleation theory described in the last section.

Hammel found that the particle radius of the phase separating regions increased proportional to the square root of time, as expected for diffusion control. The activation energy of 395 kj/mole is close to that of 460 kj/mole for viscosity in this glass, suggesting that the diffusing units are related to the lattice units involved in viscous flow in this glass.

In a number of studies of later stages of particle growth a coarsening mechanism was followed (10,11,14,18,30), with the particle radius proportional to the cube root of time as expected from Eq. 11. A wide range of activation energies were found in these studies, mostly lower than for viscous flow.

In the above studies the volume fraction of particles was small and they grew as individual spheres. When the volume fractions of the two separating phases are not much different, the phases often separated into an interconnected "wormy" structures. Cahn (31,32) showed that the mechanism of spinodal decomposition described above leads to an interconnected structure of this sort in an isotropic system such as one containing two liquid phases. Cahn and Charles therefore suggested that this interconnected structure in glasses results from spinodal decomposition. Small-angle X-ray scattering also provides evidence that this structure results from spinodal decomposition in glass (33,37). It is also possible that the interconnected structure results from agglomeration of the two phases because they have close to the same volume fraction, and some experimental evidence supports this mechanism (23,38,39,40).

3.5 Influence of Phase Separation on Glass Properties

The viscosity of a phase-separated silicate glass is usually greater than that of a homogeneous glass of the same composition, because the separated silica-rich phase has high viscosity and dominates the flow behavior. The chemical durability of a glass can be increased by phase separation if the less durable phase is isolated from the corroding solution; if this phase is exposed to the solution, as in spinodal decomposition, the glass becomes less durable than if it were not separated. As Pyrex borosilicate glass is heated at 600°C, the scale of phase separation becomes larger, and the glass becomes less resistant to corrosion. (40a)

The electrical conductivity of a phase-separated silicate glass usually increases when the glass separates, if the higher conductivity phase is continuous. Other properties, such as strength, density, and refractive index also change with separation although not as much as viscosity and chemical durability.

4. Crystallization

4.1 Introduction

Crystallization or divitrification especially at the glass surface can lead to serious problems in glass manufacture because of resultant change in glass properties, such as viscosity and the coefficient of thermal expansion. Non-uniform crystals can develop stresses on cooling the glass and cause its fracture. Thus glass technologists have carefully avoided non-uniform crystallization by cooling glass rapidly through the temperature regime of high crystallization rates.

If crystals grow in the interior of a glass its optical properties are changed because of light scattered by the crystals, which can degrade desirable optical properties such as transparency. In an optical fiber for long distance transmission even a small amount of crystals can seriously degrade the optical transmission.

Translucent opal glasses were prized in antiquity for their colors and optical effects. Their translucency derives from light scattered by very small crystals in them, and they are often used as decorative glazes.

In this section the measurement of nucleation rates of crystals in glass and their comparison with theory are described first. Then nucleation agents and particles are discussed. Finally crystal growth rates are treated. Glass ceramics, which contain small uniform crystals, are discussed in the subsequent section.

4.2 Nucleation Rates

The best way to measure nucleation rates is to observe crystals directly in the optical or electron microscope after they have grown above the critical size. However, these critical sizes are extremely small, especially for homogeneous nucleation, usually only a few tens of nanometers or less, so the crystals must be grown to much larger size for easy observation in the optical microscope or even in the scanning electron microscope. In this method the glass is heated at nucleation temperature and then at a higher temperature to grow the particles to observable size. The relative nucleation and growth rates of crystals are shown schematically in Fig. 2.

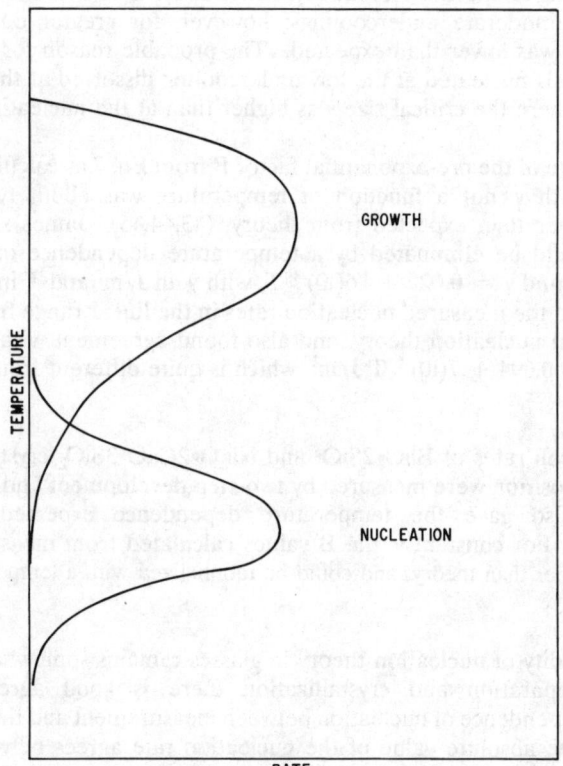

Figure 2. Schematic representation of nucleation and growth in a condensed phase.

If the difference between nucleation and growth temperatures is not too great, this method gives reliable counts of nucleated crystals. However, for larger undercooling, it is possible that some nucleated particles become subcritical at the higher temperature and dissolve. The nucleus of critical size $r^* = 2\gamma/\Delta P$, and ΔP is proportional to the undercooling, so for larger undercooling the critical nucleus size decreases.

Nucleation rates of $Li_2O \cdot 2SiO_2$ crystals in a $LiO_2 \cdot 2SiO_2$ glass have been measured by the two-step procedure, that is, by nucleating at a low temperature and growing the crystals to observable size at a higher temperature. (41-44) From Eqs. 7 and 8 one expects a plot of log $I\eta$ versus $1/T(\Delta T)^2$ to be linear, in which I is the nucleation rate, η the glass viscosity, T the temperature, and ΔT the undercooling below the melting temperature of the crystals. For these measurements this linearity was found for moderate undercoolings; however, for greater undercoolings the nucleation rate was lower than expected. The probable reason for this lowering is that some crystals nucleated at the low undercooling dissolved at the higher growth temperature, where the critical size was higher than at the nucleating temperature.

The value of the pre-exponential factor B from Eq. 7 as calculated from these experiments with γ not a function of temperature was about twenty orders of magnitude higher than expected from theory. (43,44,45) James showed that this discrepancy could be eliminated by a temperature dependence of the interfacial energy γ; he found $\gamma = 0.029 + 1.6(10)^{-4}$ T, with γ in J/m^2 and T in °K. Greer and Kelton analyzed the measured nucleation rates in the linear range from Refs. 41 and 42 with classical nucleation theory, and also found agreement with a temperature-dependent $\gamma = 0.094 + 7(10)^{-5}$ T J/m^2, which is quite different from that of James. (45a)

Nucleation rates of $BaO \cdot 2SiO_2$ and $Na_2O \cdot 2CaO \cdot 3SiO_2$ crystals in glasses of the same composition were measured by two-step development and direct counting, (43,44) and also gave the temperature dependence expected for moderate undercoolings. For constant γ the B values calculated from measured rates were again much higher than theory, and could be rationalized with a temperature-dependent interfacial energy.

The validity of nucleation theory in glasses remains somewhat uncertain. In both phase separation and crystallization there is good agreement for the temperature dependence of nucleation between measurement and theory, if all nuclei are counted; the absolute value of the nucleation rate agrees between theory and experiment for phase separation with constant γ, but agreement requires a temperature-dependent γ for agreement in crystallization experiments.

4.3 Nucleating particles and agents

In surface crystallization of glass, impurities are the most important source of nucleation. Dust particles adhere to the glass surface during processing, giving preferred nucleation sites. Alkalis in the glass can form crystals (e.g., Na_2CO_3) on the glass surface, enhancing nucleation.

Impurity particles can nucleate crystals in bulk glass. Platinum and noble metal particles are good nucleating agents. (47,48,23)

Certain dissolved oxides in glasses can enhance uniform nucleation rates in the glass. These "nucleation agents" are especially useful for making glass ceramics. Titanium dioxide is the most widely used nucleating agent. It is quite soluble in silicate glasses and lowers their viscosity. This agent has been used especially in lithium aluminum silicates. (47) Phosphorous pentoxide is also a successful nucleating agent in these glasses. (48) A large number of other oxides have been tried as nucleating agents, including refractory and transition metal oxides.

The mechanism by which these oxide nucleating agents increase the rate of nucleation is uncertain. In some cases it seems likely that they simply reduce the interfacial energy by preferential adsorption at the interface. Oxides with cations with high valence seems to be the most effective; perhaps these cations adsorb preferentially at the glass-crystal interface.

4.4 Growth

The rate of growth of a crystal from a melt or glass can be controlled by either heat flow or by a rearrangement process at the melt-crystal interface. A rough calculation suggests that heat flow will be important when crystal growth rates in a glass are greater than about 10^{-4} cm/s. Some maximum crystallization rates for oxides are given in Table 2. For the single-component oxides the rates are much slower than 10^{-4} cm/s., but for the binary glasses the rates are not far from this value, suggesting that heat flow may be an important factor in these growth rates.

Table 2. Maximum Crystallization Velocities of Glass-Forming Oxides

Material	Melting Point (°C)	Max. Cryst. Velocity (cm/s)	Temp. of Max. v (°C)	Refs.
Vitreous silica, SiO_2	1734	$2.2(10)^{-7}$	1674	49
Vitreous germania, GeO_2	1116	$4.2(10)^{-6}$	1020	50
Phosphorous pentoxide, P_2O_5	580	$1.5(10)^{-7}$	561	51
Sodium disilicate, $Na_2O \cdot 2SiO_2$	878	$1.5(10)^{-4}$	762	52
Potassium disilicate, $K_2O \cdot 2SiO_2$	1040	$3.6(10)^{-4}$	930	52
Barium diborate, $BaO \cdot 2B_2O_3$	910	$4.3(10)^{-3}$	849	53
Lead diborate, $PbO \cdot 2B_2O_3$	774	$1.9(10)^{-4}$	705	54

If the rate of crystal growth is determined by the rate of incorporation of molecules into the crystal (interface control), then these rates should be influenced by interface structure. Some theories of solid-liquid interface structure are reviewed in Ref. 55, and computer simulations of the interface described in Refs. 56 and 57. These studies support the importance of the entropy of fusion in determining the interface structure and its influence of crystal growth. Jackson (58) first emphasized the importance of the entropy of fusion in his factor α:

$$\alpha = \Delta H_f / RT_m \tag{14}$$

in which ΔH_f is the heat of fusion at the melting temperature T_m. When $\alpha < 2$, Jackson found a "rough" interface in which liquid molecules can transform to solid ones throughout the layer. For $\alpha > 2$, the interface is "smooth", so that the molecules from the liquid can only be incorporated into the solid at steps in the solid surface.

Empirically the factor α is valuable in distinguishing between types of crystal growth from the melt. For $\alpha > 2$, growth is found to be anisotropic and faceted whereas for $\alpha < 2$, growth is isotropic without facets. Furthermore, the simple kinetic model to be described below fits crystallization rates for $\alpha < 2$ but not for $\alpha > 2$.

As the temperature of a liquid is lowered under its freezing point its rate of crystallization first rises to a maximum and then decreases; this maximum can be understood qualitatively as follows. As the temperature difference from the freezing temperature becomes greater, the driving force for crystallization is increased, increasing the rate. However, the rate of motion of the molecules becomes slower as the temperature is lowered, decreasing their rate of incorporation into the crystals. The net effect of these two competing processes is a maximum in the growth rate.

Theories of varying complexity have been proposed to explain growth of crystals from the melt. (59-61) However, their validity is still a matter of considerable controversy.

A simple equation that is often compared to crystallization rates u from the melt is (Ref. 61a; Ref. 2, p. 116; Ref. 23, p. 91):

$$u = \frac{\Delta H_f (T_m - T)}{3\pi \lambda^2 \eta\, T_m} \tag{15}$$

in which ΔH_f is the heat of fusion of crystals of melting temperature T_m, T is the actual temperature, η the viscosity, and λ the thickness of the transition layer between crystal and glass, which is of atomic dimensions.

A convenient way to test the validity of the temperature dependence of Eq. 15 is to plot the product $u\eta$ as a function of temperature. Such a plot for crystallization and melting of germania (50) shows the linear dependence of $u\eta$ on

ΔT and the continuity between crystallization and melting rates. Similar plots result for silica (49) and P_2O_5 glasses (51). Values of $u\eta/\Delta t$ calculated from Eq. 14 and measured are compared in Table 3, using $\lambda = 0.3$ nm. The calculated values are ten to twenty times lower for SiO_2 and GeO_2 and about equal to that measured for P_2O_5. The discrepancies result from uncertainties in geometrical factors in Eq. 14 and in the value of λ. Nevertheless these comparisons suggest that the most important factors determining the crystallization rates in these oxides have been identified, especially for the temperature dependence.

Values of α in Table 2 are below two for SiO_2 and GeO_2 (rough interface) and above two for P_2O_5. Consistent with the above discussion, growth of SiO_2 and GeO_2 crystals is isotropic, whereas P_2O_5 is faceted.

For many other glass-forming systems with large values of α (four or greater), the temperature dependence does not fit Eq. 14, (61) nor are melting and crystallization rates similar. The reason for these deviations from the simple theory are not clear.

Table 3. Comparison of Measured and Calculated Rates of Crystallization in Glasses

Glass	Crystal	$u\eta/\Delta T$ in cm P/sec° K		Reduced Entropy of Fusion $\Delta H_f/RT_m$
		Measured	Calculated	
	Constant $u\eta/\Delta T$ at all undercoolings			
SiO_2	Cristobalite	0.14	0.008	0.46
GeO_2	Hexagonal GeO_2	0.21	0.02	1.3
P_2O_5	Tetragonal P_2O_5	0.035	0.036	3.1
17Na_2O, 12CaO 2Al_2O_3, 69SiO_2	Divitrite Na_2O.3CaO.6SiO_2	0.008		
	Constant $u\eta/\Delta$ at larger undercoolings only			
K_2O.2SiO_2	K_2O.2SiO_2	0.06		
Na_2O.2SiO_2	Na_2O.2SiO_2	0.05		3.7
Salol	Salol	$1.2(10)^{-4}$		>3.0

A suspension of colloidal particles about 0.2 mm in diameter serves as a hard sphere model of crystallization in simple liquids and glasses, and of the glass transition. (62) Such a suspension was studied as a function of volume fraction of the particles. Up to a volume fraction of 0.574, the simulated "crystllization" of these particles was analogous to classical nucleation and growth. Above this volume fraction, the "mechanism" of crystallization changed, and large asymmetric crystals formed. The authors identified the range of volume fractions from 0.574 to 0.581 with the glass transition. These particles with minimal interaction forces may be a reasonable model for molecular glasses such as glycerol, but it is not clear that they are valid models for network glasses such as silicates.

4.5 Combined Nucleation and Growth

The combined nucleation and growth of crystals in a glass can be measured from the total crystal volume or the heat given off as a function of time. If the crystals grow at a constant rate G, then the volume v of a spherical particle is $v = 4\pi G^3 t^3/3$. If the rate of nucleation I per unit volume is constant, then the volume fraction Wx of the glass that crystallizes is (ref. 2, p. 25.):

$$W_x = \int_0^t VI dt = \frac{4\pi G^3 I}{3} \int_0^t (t-\tau)d\tau = \frac{\pi G^3 I t^4}{3} \qquad (16)$$

To account for impingent of growing crystals the transformed volume fraction W becomes (46)

$$W = 1 - \exp(W_x) \qquad (17)$$

If the nucleation rate is not constant, an approximate equation for W is

$$W = 1 - \exp(Kt)^n \qquad (18)$$

where K is a constant and $n > 4$ if the nucleation rate increases with time and $3 < n < 4$ if it decreases. If nucleation is rapid and takes place on a constant number N of sites $n = 3$, because

$$W = NV = 4\pi NG^3 t^3/3 \qquad (19)$$

The kinetics of crystallization can be followed in a differential scanning calorimeter (DSC), which measures the temperature difference between a standard and a sample. In an isothermal measurement in the DSC, a phase transformation gives a peak in the output, which is a measure of the rate of transformation dW/dt as a function of time. Integration of the area under the curve for different times gives the fraction of transformation W as a function of time, and a plot of ℓn{ℓn(1-W)} versus ℓn time, gives a straight line with slope n. The value of K for each temperature can be plotted as a function of temperature to find the activation energy Q:

$$K = \nu \exp(-Q/RT) \qquad (20)$$

where ν is a constant, R is the gas constant, and T the absolute temperature.

The DSC can also be used to measure Q and n from nonisothermal (constant heating or cooling rate x) experiments. (63-66) A series of crystallization curves at different heating rates are measured. The temperature at the peak maximum T_p and the activation energy Q are related in the equation (66)

$$\ln(T_p^2/x) = \ln(Q/R) - \ln \nu + Q/RT_p \qquad (21)$$

Thus the activation energy Q can be found from a plot of $\ln(T_p^2/x)$ versus $1/T_p$ for different heating rates. The exponent n can be found from the width of the DSC curve δ at half the maximum temperature (64)

$$n = \frac{2.5\ RT_p^2}{\delta Q} \qquad (22)$$

The above treatment of nonisothermal DSC measurements is valid both for diffusion and interface controlled crystallization rates, as long as the temperature dependence of K is given by an Arrhenius relation. Effectively one assumes that the transformation is a function of W and temperature only; this assumption is sometimes called the "concept of additivity". In addition, Eq. 18 implies that growth is controlled by the transport-related interface process that is dominant at large undercoolings (see section on growth). At small undercooling, the growth rate increases as the temperature decreases (factor of ΔT), reaches a maximum and then decreases at larger undercooling, because of the influence of the reduction in transport (viscosity or diffusion in liquid), (see Fig. 2). It is only in this lower temperature range that Eq. 18, and consequently the above treatment of nonisothermal growth, are valid. If the DSC crystallization peak occurs close to the melting temperature, a more complicated temperature dependence must be used.

4.6 Fluoride Glasses

Crystals in fluoride glasses can degrade the optical properties of these glasses and reduce their value for many applications in fiber optics and as optical components. Crystallization in these glasses has been described in reports from conferences on halide glasses. (67-69) In this section, results of some studies at Rensselaer are discussed as an example of crystallization of these glasses. Compositions of some of the glasses studied are given in Table 4.

In zirconium barium based fluoride glasses, the main crystalline phases are $BaF_2.ZrF_4$ and $BaF_2.2ZrF_4$, both of which have α and β structural modifications. The heat of fusion of β-$BaZr_2F_{10}$ is estimated to be about 7800 j/mol from crystallization experiments with the DSC. (70) The melting temperature of this crystal is about 570°C, so α = 1.1. This result agrees with the nonfaceted morphologies of surface crystals, (71) although spherulites were observed by Neilson et al (72).

The crystallization of a ZBL glass isothermally and at different heating rates was studied in the DSC by Bansal et al (77). They found values of the activation energy Q and exponent n, calculated as described in the experimental section, within experimental error in the two methods. Thus one can use the nonisothermal method of calculating the kinetics parameters with confidence with the condition that Eq. 8 gives the correct temperature dependence. In these experiments, the coefficient n was within experimental error of three, suggesting that the particles are all of the same size (rapid initial nucleation) and grow at a constant rate. Bansal et al found a final crystallite density of about $5(10)^{15}/cm^3$. Nucleation was probably homogeneous to give such a high, uniform density.

Table 4. Acronyms and Batch Compositions of the Heavy Metal Fluoride Glasses

Glass	Glass Composition (mol%)						
	ZrF_4	BaF_2	LaF_3	AlF_3	LiF	NaF	PbF_2
ZB	65.0	35.0					
ZBL	62.0	33.0	5.0				
ZBLA	58.0	33.0	5.0	4.0			
ZBLLi	58.0	15.0	6.0	-	21.0		
ZBLN	58.0	15.0	6.0	-	-	21.0	
ZBLPb	58.8	31.4	4.8	-	-	-	5.0
ZBLAN	55.8	14.4	5.8	3.8	-	20.2	-
ZBLALi	50.7	20.7	5.2	3.2	20.2	-	-
ZBLALiPb	49.83	16.96	5.06	3.16	20.09	-	4.09

Measurements of kinetics from DSC experiments for a variety of fluoride glasses are summarized in Table 5 (from Ref. 67, p. 434). These results show agreement between isothermal and constant heating rate measurements. For the ZB, ZBL, ZBLPb and ZBLALi glasses, the exponent n is within experimental error of three, suggesting a constant number of particles growing at a constant rate (interface control). The ZBLA, ZBLLi, and ZBLALiPb fluoride glasses have n values significantly below three, suggesting some contribution of diffusion control in these glasses. Neilson et al (72) measured the growth in crystals in ZBLA glass directly, and found that the radius of the particles grew proportional to the square root of time, as expected for diffusion control.

The activation energies for crystal growth in Table 5 are consistent with the activation energies for viscous flow in these glasses. The activation energy for viscous flow at the low temperatures is in the range from 1000 to 1500 kj/mol, whereas at the higher temperatures, it is about 100 kj/mol. The activation energies for crystallization are between these two values, and are about what one would expect for viscous flow at the crystallization temperature if a smooth curve is drawn between viscosity data at the two temperature extremes.

At 390°C, surface crystals on a ZBL glass grew at a rate of about $7(10)^{-7}$ cm/s. The bulk crystallization velocity v as calculated from DSC measurements and X-ray measurements of R_f, was about $4(10)^{-8}$ cm/s, or more than an order of magnitude lower. These results can be compared with a crystallization rate at 390°C calculated from Eq. 14. Estimates of the required parameters are: ΔH_f = 7.8 kJ/mol, T_m = 560°, ΔT = 170°C, η = 2000 Pa-s, $\lambda = 3(10)^{-10}$ m. Then u is about $1.5(10)^{-8}$ cm/s, which is in surprisingly good agreement with the measured value.

Crystallization in the system of BaZnYb Th fluoride glasses shows a complex series of crystalline phases, including a BaYbTh fluoride phase that decomposes to $BaThF_6$ and YbF_3 phases as the temperature increases, and partial dissolution of a zinc fluoride phase. (73)

Table 5. Crystallization Parameters in Fluoride Glasses from Isothermal and Constant Heating Rate Measurements in the DSC

Glass	n		Q(kJ/mol)
	Isothermal	Constant Heating Rate	
ZB	3.1		374
ZBL	3.2	3.4	315
ZBLA		1.8	
ZBLALi	3.3	3.2	182
ZBLPb	2.9	2.8	320
ZBLALiPb	2.2	2.3	190

5. Glass Ceramics

Glass ceramics are glasses in which fine, uniform crystals are grown by special heat treatments. They are one of the few truly new materials developed over the past few decades, and can have superior properties, such as high strength and impact resistance, low coefficient of thermal expansion and hence good thermal shock resistance, low bulk and surface electrical conductivity, low dielectric loss and a range of optical properties from transparent through translucent to opaque.

To make a glass ceramic, a part is formed as a glass by traditional glass melting methods, and then the crystals are grown by controlled heat treatments to give the extent of crystallization desired. A variety of types of glass ceramics have been developed are available commercially; some types are described in Table 6, along with some uses. Among other important uses not listed in the table are bearings, heat exchangers, reactor control rods, seals, vacuum tube envelopes, substrates for electronic circuits, and capacitors.

McMillan (48) discusses these and other uses in his book entitled, <u>Glass Ceramics</u>, including many other aspects of these materials. Grossmann (74) has written a review of different types of commercial glass ceramics, including their properties and microstructures. Beall (75) has described new types of strong glass ceramics based on chain silicates such as enstatite; the microstructure of this material is similar to that of natural jade, $NaAlSi_2O_6$.

Table 6. Properties and Uses of Some Commercial Glass-Ceramics

Glass Type and Company	Crystals	Properties	Uses
Corning 9606	$2MgO \circ 2Al_2O_3 \circ 5SiO_2$ (Cordierite)	Strong, low expansion, transparent to radar	Radomes, nose cones for missiles
Corning 9608	β-Spodumene $Li_2O \circ Al_2O_3 \circ 4SiO_2$	Low expansion, low chemical reactivity, strong	Cookware
Owens-Illinois Cer-vit, Corning 9607, Schott Zerodur	β-Quartz	Very low expansion	Telescope mirrors
General Electric Re-X	$Li_2O \circ 2SiO_2$	Low electrical conduction, high strength	Insulators
Corning 9658	Mica	Fracture containment	Machinable
Slagsitall (Russia)	Gehlenite $2CaO \circ Al_2O_3 \circ SiO_2$	Wear resistance	Building materials

6. Small Particles in Glass

6.1 Introduction

Small particles of metals and semiconductors in glass can impart unusual optical properties to the glass. Small metallic particles absorb light and develop beautiful colors in the glass. These particles can also be used to study the optical properties of metals in transition from individual metal atoms to bulk metal, and the influence of particle size on the optical properties of the metal. Glasses containing small semiconducting particles such as cadmium sulfide show optical nonlinearity and a fast response time. Hence they have potential applications in high-speed optical switching, optical computing, and telecommunications fiber optic networks. If these semiconductor particles are coated with a metal, they should show enhanced nonlinearity.

In this section small metal particles are discussed first; then glasses containing semiconductor particles and finally coated particles are described.

6.2 Metallic Particles

Much information on growth and optical properties of metallic particles in glass is summarized in Weyl's monograph (76). Best known of glasses containing metallic particles is gold ruby, which has been known since the seventeenth century. Faraday (77) recognized that the color of gold ruby resulted from finely divided gold particles. This glass is made by dissolving gold in the glass melt, as an ion, in which state the gold is retained when the glass is cooled rapidly. (78) To form the gold particles the glass is reheated to an intermediate temperature. Certain agents, such as antimony oxide, in the glass aid nucleation of the particles, or they can be nucleated by ultraviolet, X-ray, or γ-radiation if a radiation-sensitive ion such as cerium is present. (78) The growth of the particles takes place by diffusion of gold atoms or ions to the particles. (79)

The color of gold ruby glass results from an absorption band at about 0.53 μm. This band comes from the spherical geometry of the particles and the particular optical properties of gold. (80) It is a "plasma-resonance" band, in which the free electrons in the metal are considered as a bounded plasma.

Many different measurements on gold and silver particles show that the mean free path lambda of free electrons in particles of radius R is (81-84):

$$\frac{1}{\lambda} = \frac{1}{\lambda_o} + \frac{1}{R} \qquad (23)$$

in which λ_o is the mean free path of free electrons in the bulk metal. This change in mean free path changes the effective optical properties of the metal, leading to a broader absorption band. In glasses containing gold or copper particles this broadening for very small particles removes the plasma absorption band, leading to glasses colored orange instead of the deep ruby red of glasses with larger particles.

Recent experiments on gold glasses show that Eq. 23 is valid for particles down to a diameter of one nm or less (85), confirming earlier work (81,86) These particles contain about thirty gold atoms. These results also reveal that the band-to-band transitions in gold are the same as in bulk gold, showing that a cluster of thirty gold atoms has the band structure of bulk gold. Some theories have suggested that the transition from atomic properties to bulk metal occurs for larger numbers of atoms than thirty. Preliminary experiments (85) suggest that clusters as small as ten to twenty gold atoms already have the electronic band structure of bulk gold. These gold particles in glass can be nucleated by radiation, so the particle size distribution is highly uniform, as confirmed by electron microscopy. These uniform particles are ideal for studying the optical properties of metal clusters as a function of their size.

The optical absorption of these small metallic particles in glass can also be used to determine the most reliable measurements of bulk optical properties. These measurements differ considerably from one investigator to another. The absorption of small gold, silver and copper particles agree most closely with the optical properties measured by Otter (87); the data of Roberts (88) on electropolished copper agree closely with those of Otter. Otter prepared a pristine bulk surface by melting and freezing in ultrahigh vacuum. More recent measurements of optical properties of these metals, mostly as thin films, are often chosen as the most reliable; the particle results show that the measurements of Otter and Roberts are the closest to the properties of bulk metals.

6.3 Semiconductors

Particles of the semiconductors cadmium sulfide, cadmium selenide, mixed cadmium sulfide-selenide, and cuprous chloride can be grown in glass. Many references to this work are given in Refs. 89-93. These materials are found to have non-linear optical properties. (89) In very small particles the optical absorption edge of these particles shifts with particle size (90), thus allowing wavelength tunability.

In cadmium sulfide the particle size changes the optical properties below a size of about 8 nm (90); this effect is sometimes called "quantum confinement". The growth of these particles in glass appears to be diffusion-controlled. (92)

A theoretical equation for the influence of particle size on the energy E of the absorption edge is (94)

$$E - E_g = \frac{h^2}{8\pi^2 \mu R^2} - \frac{1.8e^2}{\Sigma R} \quad (24)$$

in which E_g is the bulk energy (band gap), h Planck's constant, mu the reduced electron-hole mass, e the electronic charge, Σ the dielectric constant of the particles, and R their radius. The first term is the energy of electron and hole confinement and the second term is the Coulombic interaction energy. Some authors have neglected the second term in Eq. 24; however, a comparison between calculated and experimental results showed that it cannot be neglected. (92) Lippens and Lannoo calculated the change $E-E_g$ of the gap energy from the tight-binding approximation and a recursion method. For cadmium sulfide their results are close to linear on a plot of $E-E_g$ against $1/R^2$ for particle diameters from about 5 to 2.5 nm, but fall below this plot for smaller diameters.

6.4 Coated Particles

The nonlinear optical properties of the semiconductor particles described in the last section have not been strong enough for practical applications. If the particles have a coating of metal, they should have enhanced nonlinear properties (96,97). Thus we have grown a silver coating on cadmium sulfide particles, and reduced the surface of cuprous chloride particles to copper. These coatings were verified by the changes in optical properties and by electron microscopy.

Cadmium sulfide particles were grown and then coated with silver in a two-step process. (93) A potassium borosilicate glass contained about .5 wt% CdS and .01% silver. The glass was heated for 4 h at 725°C to grow cadmium sulfide particles of diameter 5 to 6 nm, as demonstrated by optical absorption and electron microscopy. Then the glass was heated at 550° for 4 h in hydrogen to grow silver on the cadmium sulfide particles. Only a few of the cadmium sulfide particles were coated to a size of about 15 nm in diameter, as shown by electron microscopy. The resulting absorption spectrum had a peak at a wave length of about 525 nm, shifted from the usual silver peak at 405 nm because of the cadmium sulfide core of the particles, as predicted from theoretical equations for coated particles. (96,98)

A high boria sodium borosilicate glass dissolved one mole% CuCl, and was clear as quenched. Particles of CuCl were grown in this glass at 450°C in air for one hour, giving an absorption edge at 380 nm as expected for these particles. Then the glass was treated in hydrogen at 500°C for one hour; it turned reddish and the absorption spectrum showed the absorption peak for copper particles (88) at 587 nm, but the shape of the peak was more symmetrical, unlike that for pure copper particles. The spectrum calculated for CuCl particles coated with copper agreed well with the measured spectrum. (93) A transmission electron micrograph showed particles 2 to 3 nm in diameter, and showed internal contrast as expected for CuCl coated with copper.

These examples show that semiconductor particles in glass can be coated with metals. The influence of the coating on the nonlinear optical properties of the particles awaits further measurements.

7. Problems

1. The viscosity of a liquid at its freezing temperature is a good indicator of the ease of forming a glass from the liquid; the higher the viscosity, the easier it is to form a glass. Explain.

2. The rate of nucleation of crystals from a bulk liquid is a maximum at a particular temperature below the freezing temperature of the liquid. If the liquid is divided into samll droplets, what happens to the temperature of maximum nucleation rate? Explain.

3. Explain why the rate of heat flow in the solid can influence the microstructure of a glass ceramic.

4. Explain why the crystallization model of colloidal particles may not be valid for silicate glasses.

5. Here are some measured maximum temperatures T_p at different heating rates x from a nonisothermal DSC measurement in a zirconium-barium-lanthanum fluoride glass. Calculate the activation energy of crystallization from them.

x, °K/min	T_p, °K
1.25	643
2.5	651
5.0	658
10.0	665
20.0	674

References

1. M. Volmer and A. Weber, *Z. Phys. Chem.* **A119** (1926) 277.
2. R. H. Doremus, *Rates of Phase Transformations* (Academic Press, San Diego, 1985) Chapter 4.
3. A. Einstein, *Ann. Phys.* **33** (1910) 1275.
4. J. W. Gibbs, *The Scientific Papers* Vol. I (Dover, New York, 1961) p. 219-274.
5. R. C. Tolman, *The Principles of Statistical Mechanics* (Oxford University Press, Oxford, 1938) p. 638ff.
6. G.W. Greenwood, *Acta Met.* **8** (1956) p. 243.
7. E. M. Lifshitz and V. V. Slyozov, *J. Phys. Chem. Solids* **19** (1961) p. 35.
8. C. Wagner, *Z. Electrochem.* **65** (1961) p. 581.
9. R. H. Doremus and A. M. Turkalo, *Science* **164** (1969) p. 418.
10. Y. Moriya, D. Warrington, and R. W. Douglas, *Phys. Chem. Glasses* **8** (1960) p. 19.
11. P. F. James and P. W. McMillan, *Phys. Chem. Glasses* **11** (1968) p. 377.
12. O. V. Mazurin and E. A. Porai-Koshits, Eds. *Phase Separation in Glass* (North Holland, Amsterdam, 1984).
13. R. J. Charles, *J. Amer. Ceram. Soc.* **47**, (1964) p. 559.
14. J. Zarzycki and F. Naudin, *Phys. Chem. Glasses* **8** (1967) p. 11.
15. Y. Kawamoto and M. Tomozawa, *J. Am. Ceram. Soc.* **64** (1981) p. 289.
16. J. M. Hyde and M. Tomozawa, *J. Noncryst. Solids* **109** (1989) p. 18.
17. D. G. Burnett and R. W. Douglas, *Phys. Chem. Glasses* **11** (1970) p. 125.
18. D. G. Burnett and R. W. Douglas, *Discuss. Faraday Soc.* **50** (1970) p. 200; *Phys. Chem. Glasses* **12** (1971) p. 117.
19. W. Haller, D. H. Blackburn, F. E. Wagstaff, and R. J. Charles, *J. Am. Ceram. Soc.* **53** (1970) p. 34.
20. G. Srinivasan, I. Tweer, P. B. Macedo, A. Sarkar, and W. Haller, *J. Noncryst.Solids* **6** (1971) p. 221.
21. O. V. Mazarin, M. W. Streltsina, and A. S. Totesh, *Phys. Chem. Glasses* **10** (1969) p. 63.
22. J. H. Simmons, P. B. Macedo, A. Napolitano, and W. K. Haller, *Discuss. Faraday Soc.* **50** (1971) p. 155.
23. R. H. Doremus, *Glass Science, 2nd Edition*, (Wiley & Sons, New York, 1994) p. 50ff.
24. B. E. Warren and A. G. Pincus, *J. Am. Ceram. Soc.* **23** (1940) p. 301.
25. E. M. Levin and S. Block, *J. Am. Ceram. Soc.* **40** (1957) p. 95, 113; **41** (1958) p. 49.
26. R. J. Charles, *Phys. Chem. Glasses* **10** (1969) p. 169.
27. R. J. Charles, *J. Am. Ceram. Soc.* **50** (1967) p. 631.
28. V. McGahay and M. Tomozawa, *J. Noncryst. Solids* **109** (1989) p. 27.

29. M. Levy, J-C. Le Guillou and J. Zinn-Justin, *Phase Transformations*, (Plenum Press, New York, 1982, see especially articles by D. Beyrens, p. 25 and J. V. Sengers, p. 95).
29a. J.H. Simmons, *J. Am. Ceram. Soc.*, **56** (1973), p. 284.
30. J. J. Hammel, *J. Chem. Phys.* **46** (1967) p. 2234.
31. J. W. Cahn, *Acta Met.* **9** (1961) p. 745.
32. J. W. Cahn, *J. Chem. Phys.* **42** (1965) p. 93.
33. J. Zarzycki and F. Naudin, *Comptes Rendus* **265** (1967) p. 1456; *J. Noncryst. Solids* **1**, (1969) p. 215; **5** (1971) p. 415.
34. G. F. Neilson, *Phys. Chem. Glasses* **10** (1969) p. 54.
35. M. Tomozawa, R. K. MacCrone and H. Herman, *Phys. Chem. Glasses* **11** (1970) p. 136.
36. G. B. Stephenson, *J. Noncryst. Solids* **66** (1984) p. 393.
37. R. Yokota and H. Nakajima, *J. Noncryst. Solids* **70** (1985), p. 343.
38. F. Naudin and J. Zarzycki, *Comptes Rendus* **266** (1968) p. 266.
39. T. P. Seward, D. R. Uhlmann and D. Turnbull, *J. Am. Ceram. Soc.* **51** (1968) p. 634.
40. J. F. MacDowell and G. H. Beall, *J. Am. Ceram. Soc.* **52** (1969) p. 17.
40a. M. Tomozawa and T. Takamori, *J. Am. Ceram. Soc.*, **62**, (1979), p. 370.
41. K. Matusita and M. Tashiro, *J. Noncryst. Solids* **11** (1973) p. 471.
42. P. F. James, *Phys. Chem. Glasses* **15** (1974) p. 95.
43. P. F. James, in *Nucleation and Crystallization in Glasses*, ed. J. H. Simmons, D. R.Uhlmann and G. H. Beall (American Ceramic Society, Columbus, Ohio, 1982) p. 1.
44. P. F., James, *J. Noncryst. Solids* **73** (1985) p. 517.
45. G. F. Neilson and M. C. Weinburg, *J. Noncryst. Solids* **34** (1979) p. 137.
45a. A. L. Greer and K. F. Kelton, *J. Am. Ceram. Soc.*, **74**, (1991) p. 1015.
46. M. Avrami, *J. Chem. Phys. I* **7** (1939) p. 1103; **8** (1940) p. 212; **9** (1941) p. 177.
47. S. D. Stookey and R. D. Maurer, in *Progress in Ceramic Science*, ed. J. E. Burke (Pergamon Press, New York, 1962).
48. P. W. McMillan, *Glass-Ceramics*, 2nd ed. (Academic Press, London, 1979).
49. F. E. Wagstaff, *J. Am. Ceram. Soc.* **52** (1969) p. 650.
50. P. J. Vergano and D. R. Uhlmann, *Phys. Chem. Glasses* **11** (1970), p. 30, 39.
51. R. L. Cormia, J. D. Mackenzie and D. Turnbull, *J. Appl. Phys.* **34** (1963) p. 2239.
52. A. Leontewa, *Acta Physicohem USSR* **16** (1942) p. 97.
53. J. A. Laird and C. G. Bergeron, *J. Am. Ceram. Soc.* **53** (1970) p. 482.
54. J. P. DeLuca, R. J. Eagan and C. G. Bergeron, *ibid* **52** (1969) p. 322.
55. D. W. Oxtoby and A. D. Haymet, *J. Chem. Phys.* **76** (1982) p. 6262.
56. G. H. Gilmer, *Science* **208** (1980) p. 355.
57. D. Frenkel and J. P. McTague, *Annu. Rev. Phys. Chem.* **31** (1980) p. 491.

58. K. A. Jackson, in *Growth and Perfection of Crystals,* ed. R. H. Doremus, B. W. Roberts and D. Turnbull, (Wiley & Sons, New York, 1958) p. 319.
59. J. W. Cahn, W. B. Hillig and G. W. Sears, *Acta Met.* **12** (1964) p. 1421.
60. K. A. Jackson, D. R. Uhlmann and J. D. Hunt, *J. Cryst. Growth* **1** (1967) p. 1.
61. D. R. Uhlmann, in *Nucleation and Crystallization in Glasses,* ed. J. H. Simmons, D. R. Uhlmann and G. H. Bell, (American Ceramic Society, Columbus, Ohio, 1982) p. 80.
61a. H. A. Wilson, *Philos. Mag.*, **50**, (1900), p. 238.
62. W. Van Megan and S. M. Underwood, *Nature* **362** (1993) p. 616.
63. H. E. Kissinger, *J. Res. Natl. Bur. Stand.* **57** (1957) p. 217.
64. J. A. Augis and J. E. Bennett, *J. Thermal Anal.* **13** (1978) p. 283.
65. K. Matusita and S. Sakka, *J. Noncryst. Solids* **38-39** (1980) p. 741.
66. N. P. Bansal, R. H. Doremus, A. J.Bruce and C. T. Moynihan, *J. Am Ceram. Soc.* **66** (1983) p. 233.
67. M. G. Drexhage, C. T. Moynihan and M. Robinson, eds. *Materials Sci. Forum* **19-20** (1987) p. 429.
68. M. Y. Yamane and C. T. Moynihan, Eds. *Materials Sci. Forum* **32-33** (1988) p. 185.
69. G. H. Frischat and C. T. Moynihan, Eds. *Materials Sci. Forum* **67-68** (1991) p. 187.
70. N. P. Bansal and A. J. Bruce, R. H. Doremus and C. T. Moynihan, *SPIE* **484** (1984) p. 51.
71. N. P. Bansal and R. H. Doremus, *J. Am. Ceram. Soc.* **66** (1983) C-132.
72. G. F. Neilson, G. L. Smith and M. C. Weinberg, *Mater. Res. Bull.*, **19** (1984) 577.
73. R. Garcia, R. H. Doremus, N. P. Bansal, S. H. Ko and T. Margraf, *J. Mat. Res.* **3** (1988) p. 989.
74. D. G. Grossmann, in *Encyclopedia of Materials Science and Engineering,* ed. M. B. Bever (Pergamon Press, Oxford, 1986) p. 1966.
75. G. H. Beall, *J. Noncryst. Solids* **129** (1991) 163.
76. W. A. Weyl, *Colored Glasses* (Society of Glass Technology, Sheffield, England, 1959).
77. M. Faraday, *Philos. Mag.* **14** (1857) p. 401, 512.
78. S. D. Stookey, *J. Am. Ceram. Soc.* **32** (1949) p. 246.
79. R. H. Doremus, in *Nucleation and Crystallization in Glasses and Melts,* (American Ceramic Society, Columbus, OH, 1967) p. 117.
80. G. Mie, *J. Chem. Phys.* **25** (1908) p. 377.
81. R. H. Doremus, *J. Chem. Phys.* **40** (1964) p. 2389.
82. R. H. Doremus, *J. Chem. Phys.* **41** (1965) p. 414.
83. R. H. Doremus, S.-C. Kao and R. Garcia, *Appl. Opt.* **31** (1992) p. 5773.
84. U. Kreibig and L. Genzel, *Surface Science* **156** (1985) p. 678.
85. P. G. N. Rao and R. H. Doremus, to be published.
86. R. H. Doremus and A. M. Turkalo, *J. Mat. Sci.* **11** (1976) p. 903.

87. W. Otter, *Z. Phys.* **161** (1961) p. 163.
88. S. Roberts, *Phys. Rev.* **118** (1960), p. 1509.
89. R. K. Jain and R. C. Lind, *J. Opt. Soc. Am.* **73** (1983) p. 647.
90. N. F. Borrelli, D. W. Hall, H. J. Holland, and D. W. Smith, *J. Appl. Phys.* **61** (1987) p. 5399.
91. L. C. Liu and S. H. Risbud, *ibid*, **68** (1990) p. 28.
92. V. Sukumar and R. H. Doremus, *Phys. Stat. Solidi* **b179** (1993) p. 307.
93. V. Sukumar, S.-C. Kao, P. G. N. Rao and R. H. Doremus, *J. Mater. Res.* **8** (1993) p. 2686.
94. L. E. Brus, *J. Chem. Phys.* **80** (1984) p. 4403.
95. P. E. Lippens and M. Lannoo, *Phys. Rev.* **B39** (1989) p. 10935.
96. A. E. Neeves and M. H. Birnboim, *J. Opt. Soc. Amer.* **B6** (1989) p. 787.
97. J. W. Haus, N. Kalyaniwalla, R. Inguva and C. M. Bowden, *J. Appl. Phys.* **65** (1989) p. 1420.
98. A. L. Aden and M. L. Kerker, *J. Appl. Phys.* **22** (1951) p. 1242.

Chapter 7
Microstructure and Properties of Superconducting Materials

C. S. Pande
Naval Research Laboratory
Washington, DC 20375-5320

Electrical resistance of many materials drop suddenly to zero when they are cooled below a certain temperature (T_c). This temperature is called critical temperature or superconducting transition temperature and these materials are called superconducting materials. Such materials are finding increasing use in the generation of high magnetic fields. Other applications are in lossless electrical power transmission and storage, in magnetic sensors, in electromagnetic radiation detectors and in high speed digital signal and data processing. In this chapter the microstructure and properties of these materials are briefly considered. For a more detailed coverage, the reader is referred to the references cited at the end of the chapter.

1. BRIEF HISTORICAL INTRODUCTION

1911: Dutch physicist H. Kamerlingh Onnes discovered that resistivity of mercury dropped suddenly to zero when it was cooled below 4.2K, the boiling point of liquid helium, which was used as the cooling agent. He named this phenomenon as superconductivity. (1)

1912: Onnes discovered that superconductivity was destroyed by the application of a sufficiently strong magnetic field or on the passage of sufficiently strong electrical current.

1913: Superconducting discovered in lead at 7.2K.

1930: Highest critical temperature in a pure metal discovered in Nb (T_c = 9.2K).

1933: Meissner and Ochsenfeld discovered that the magnetic field is expelled from the interior of the superconductor when it is cooled below T_c in the presence of a weak magnetic field. This is often referred as Meissner effect. (2)

1934: F. and H. London proposed a model to explain Meissner effect and showed that magnetic field does penetrate to a penetration depth of λ_L. (3)

1950: A phenomenological theory of superconductivity was proposed by Ginzburg and Landau. (4)

1957: Bardeen, Cooper and Schrieffer proposed a microscopic theory (BCS theory) of superconductivity. (5)

1962: Josephson discovered that supercurrents can travel through an atomically thin layer of an insulator. (6) this is the so called Josephson effect which forms the basis of superconducting electronics.

1986: Bednorz and Müller discovered the first high T_c superconductor ($T_c \sim$ 30K in LaBaCuO ceramic). (7)

1987: Wu, Chu, and co-workers discovered the superconductor $Y_1Ba_2Cu_3O_7$ with T_c=92K, the first superconductor to have a T_c above the boiling point of liquid nitrogen. (8)

1988: Bismuth and Thallium oxide superconductors discovered

1993: Mercury containing oxide superconductors discovered. Believed to be the superconductor with highest T_c. (For a good historical introduction see reference [9]

As mentioned before, in addition to zero resistance, superconductors exhibit Miessner effect i.e. they expel magnetic field from the interior as temperature is lowered below T_c. It should be noted that Meissner effect is not a consequence of zero resistivity, but another characteristic property of the superconducting state, when an external magnetic field is applied to the superconducting state. The transition from superconducting to normal state can be sharp (Type I superconductors) or broad (Type II superconductors). The strength of the magnetic field needed to bring out the transition is called the thermodynamic critical field H_c. With the exception of Vanadium and Niobium all the superconducting elements and most of their alloys are type I superconductors. Most of the practical materials are type II for reasons discussed in section 2.2. In these materials microstructure plays an important role.

2. THEORETICAL BACKGROUND

The superconducting state is characterized by a condensation of conduction electrons of a superconductor into a state of lower energy. Theoretical understanding of the superconductivity phenomena has been reached gradually. Certain results were obtained directly from thermodynamics. One of the most important is

$$C_s - C_n = \frac{T_c}{4\pi}\left(\frac{\partial H_c}{\partial T}\right)^2 \qquad (1)$$

where C_s and C_n are specific heats in the superconducting and normal state. H_c is a critical field, and T is temperature. Another relation obtained empirically is

$$H_c = H_o \left(1 - \frac{T}{T_c}\right)^2 \tag{2}$$

This relation relates the critical magnetic field H_c with critical temperature T_c ($H_o = H_c$ at T = 0). Results from thermodynamics are obtained by considering the transition between normal and superconducting state as a reversible thermodynamic transition. (See Further Reading.)

Other results have been obtained from London's phenomenogical equations. London (3) from physical considerations proposed that

$$J = \frac{cA}{4\pi\lambda_L} \text{ (in C.G.S. units)} \tag{3}$$

where J is current and A is the vector potential of the local magnetic field such that

$$B = \text{Curl } A$$

using Maxwell's equation viz

$$\text{Curl } B = \frac{4\pi J}{c}$$

we get
$$\lambda_L^2 \nabla^2 B = B \tag{4}$$

which accounts for Meissner effect, since B = constant is not a solution of the above equation unless the constant is zero. Consider the case when the current flows in y direction and the field B is in Z direction. The solution of the above equation is

$$B_2 = B_2^{(o)} \exp\left(-\frac{Z}{\lambda_L}\right) \tag{5}$$

i.e. field falls off exponentially inside the superconductor with a characteristic length λ_L. λ_L is called London Penetration length λ_L can be shown to be given as

$$\lambda_L^2 = \frac{mc^2}{(4\pi nq^2)} \tag{6}$$

where q is the particle charge, m its mass, n its concentration and c is the velocity of light. In Nb λ_L = 39 nm and in Pb λ_L = 37 nm.

Many other useful results have been obtained by the semi phenononogical model of Ginzburg and Landau proposed in 1950. (4) In this model superconductivity is described in terms of an order parameter Ψ. The model gave expressions for the kinetic energy of the superconducting electrons and described the magnetic behavior of the superconductors very well. A notable success was the prediction of Vortex state (see section 2.2) on the basis of this model. (10) However a basic mechanism of superconductivity was still lacking. A fundamental basis of superconductivity based on quantum theory was first given by Bardeen, Cooper and Schrieffer (5) which will now be briefly described.

2.1 BCS Theory

In spite of a wide diversity of materials, that are now known to be superconductors, there is a basic set of concepts based on BCS theory that forms the basis of our fundamental understanding of the superconducting phenomenon. What happens when a substance say Hg is cooled to below its T_c? The free energy change of the electron gas is only about 10^{-7} eV per atom, but the behavior of the electrons changes drastically. In normal state each electron scatters individually from lattice defects and impurities while in superconducting state, the electrons form a macroscopic quantum state, which involves pairing of conduction electrons of opposite spin by some inter electron attraction to form what is known as Cooper pairs and a condensation of these pairs to give a coherent many body system which are not scattered by lattice defects etc., hence zero resistivity. In BCS theory the interelectron attraction is provided by the interaction of the electron with the vibrations of the crystal lattice (phonons). Roughly speaking the distance between the two electrons of the Cooper pair is called the coherence length. Another length scale that is important is called penetration depth λ_L, which is the characteristic distance over which magnetic field changes in a superconductor. These lengths follow naturally from BCS theory.

BCS theory provides an expression for T_c as

$$T_c = 1.14 \; \Theta \exp\left(-\frac{1}{N(0)V}\right) \qquad (7)$$

where Θ is the Debye temperature, V is an attractive interaction and N(0) is the electron density of states. This equation is roughly obeyed by many superconductors. However when the coupling between the electrons is strong, a modification of this equation, the so called McMillan equation is used viz

$$T_c = \frac{\theta}{1.45} \exp\left[\frac{-1.04(1+\lambda)}{\lambda - \mu^*(1+0.62\lambda)}\right] \qquad (8)$$

where λ is electron phonon interaction parameter and μ^* describes the Coulomb repulsion and is of the order of 0.1. In strong coupling $\lambda \gg \mu^*$. This equation describes A15 materials very well but fails in case of high T_c materials. (See section 4.7)

The theory of type II superconductivity was given by Ginsburg, Landau, Abrikosov and Gorkov and is usually called GLAG theory. These researchers showed that the type I or type II behavior is determined by the ratio $\kappa = \dfrac{\lambda_L}{\xi}$ where λ_L is the London penetration depth and ξ is the coherence length.

Type I $\kappa < (2)^{1/2}$
Type II $\kappa > (2)^{1/2}$

For more details see for example reference (11), and those in further reading.

2.2 Role of Microstructure

In type II superconductors the magnetic field penetrates the material slowly at a value denoted by H_{c1} and continues up to a value denoted by H_{c2} at which point the material is transformed to normal (non-superconducting) state. The superconducting state between H_{c1} and H_{c2} is called mixed state. Type II behavior is shown in general by many alloys and compound superconductors. Abrikosov showed that the surface energy of the interface between the normal and superconducting state is negative in type II superconductors. The normal state therefore maximizes surfaces by forming flux tubes (called flux lines, fluxiods or vortices). It turns out that these tubes form a triangular lattice that can be observed.

The flux tubes are parallel to the magnetic field. The radius of the tube is of the order of coherent length. Magnetic field in the tube are generated by the supercurrent circulating around the tube. Each fluxiod carries a quantum unit of flux

$$\phi_0 = \frac{h}{2e} = 2.0 \times 10^{-15} \text{ Weber} \qquad (9)$$

At a field B the spacing and between the fluxiods is given by

$$d = \left(\frac{4}{3}\right)^{1/2} \sqrt{\frac{\phi_0}{B}} \qquad (10)$$

As the field is increased the fluxiods get closer and closer till at B_{C2} the cores of the fluxiods which are normal completely overlap and the whole specimen becomes normal i.e. non-superconductive.

Whenever a current J flows in a superconductor in mixed state, it experiences a Lorentz force F_L (per unit volume of the superconductor) given by

$$F_L = J \times B \qquad (11)$$

Force F_L acts in a direction perpendicular to both B (which is parallel to flux line) and J. This force can move the fluxiods, and create an electric field E and produce a resistance in the superconductor $\left(= \frac{E}{J}\right)$, unless the fluxiods are prevented from moving i.e. are pinned by some microstructural feature. The critical current J_c is that current that produces enough F_L to depin the fluxiods and make them move. Therefore the pinning force F_p is

$$J_c \times B = \text{Pinning Force.} \qquad (12)$$

It is thus clear that to optimize J_c one needs microstructure that can pin fluxiods strongly. Since high J_c is a prime requirement for most of the technological applications, the prime importance of suitable microstructure is obvious.

To obtain an expression of the pinning force (and hence J_c), in terms of microstructure one needs to obtain the local or 'elementary' pinning force between the fluxiod and the pinning center and then sum over the entire flux line lattice.

Flux pinning by a pinning agent such as a lattice defect can be due to several causes. Three main pinning interactions are elastic, core, and magnetic of which elastic is usually dominant. The normal core of a flux line is usually (very slightly) denser and stiffer than the superconducting matrix. A fluxiod thus produces a stress field which could interact with the stress fields of the pinning agents. This is the so called first order (elastic) interaction. The core interaction is due to a local change in the superconducting condensation energy at the pinning center, while the magnetic interaction occurs when the pinning center is larger than the penetration depth λ_L. The calculation of elementary pinning force has been performed for some simple cases.

Summation over such elementary pinning forces is still a matter of controversy. It is especially difficult if several mechanisms are operating, or if the system contains several classes of pinning agents. The reader is referred to references (12) and (13) for more detail.

2.3 Radiation Effects

As we have seen, both the normal and the superconducting properties of the superconductors are determined by the arrangement of atoms in the crystalline solid. This arrangement can be altered by the creation of various lattice defects using radiation and hence it is important to study irradiation effects. Further, it is necessary to ascertain the effect of radiation on superconducting magnets (for example those that might be used in fusion reactors, will be exposed to high energy neutron radiations) and other devices. Radiation is also a convenient way to introduce a desired microstructure and then study their correlation with basic properties of superconductors.

Radiation produces three major effects in a superconductor. There may be a change in critical current, and a change in transition temperature and thirdly there is usually an increase in normal state electrical resistivity.

In order to cause increase in critical current the size of the defects produced has to be of the dimensions of coherence length. Such defects could be displacement cascades, and dislocation loops, since they can act as pinning centers. The change in T_c depends very much on the nature of the material.

In ordered alloys such as A15 and high T_c superconductors, the T_c reduction could be quite significant, since these materials could be disordered. The changes in T_c in metallic superconductors are relatively small or insignificant. The change in resistivity usually depends on the total number of defects produced.

The nature of defects produced are (1) point defects, produced by almost all radiations, (2) cascades usually produced by high energy neutrons, protons and ions, (3) dislocation loops and voids, both of which usually are results of annealing, (4) antisite or disorder defects i.e. atoms occupying wrong positions without change of overall structure. Obviously this applies to ordered structures only. For further information see reference(14).

3. APPLICATION OF SUPERCONDUCTING MATERIALS

The following table (Table 1) gives a brief summary of some of the important applications and the corresponding material that is most often used for the purpose.

Table 1. Applications of Superconducting Materials

Applications	Superconducting Device	Material Used	Comments
Magnetic Resonance Imaging (MRI)	Magnets	Nb-Ti	
Fusion Reactors	Magnets	Nb-Ti, Nb$_3$Sn	Not suitable for very high magnetic fields
Superconducting Transformers	Wires	Nb-Ti	
Field Current Limiters		Nb-Ti	
Transmission Wires	Wires	Nb$_3$Sn, NbTi	
Energy Storage	Magnets	Nb Ti, Nb$_3$Sn	
Levitation Trains	Magnets Coils	Nb Ti	
Magnetic Measurements	SQUID	Nb$_3$Sn, High T_c Materials	SQUID stands for Superconducting Quantum Interference Device

Main limitations holding back a large scale use of superconducting materials are:

(1) Need to cool using liquid helium.
High T_c materials may eventually be coold by liquid nitrogen.

(2) AC losses (A superconductor showing no heat dissipation in direct current may still show a loss in alternating currents.)

(3) Instability of the system due to thermal, mechanical or radiation induced.

4. MICROSTRUCTURE IN VARIOUS SUPERCONDUCTING MATERIALS

In next sections, we describe some of the most important superconducting materials, their microstructure and their properties.

4.1 Niobium-Titanium Alloy Superconductors

A majority of superconducting magnets are constructed out of Nb-46-50 wt% Ti alloy (Nb-47 wt% T_i being most common) and the process is now commercially well established. All the superconducting magnets for the superconducting super collider (now cancelled) would have used this alloy system. The reason for this preference is the ease of fabrication and excellent mechanical properties [It should be noted that initially (1962) a different alloy of Nb, Nb 25% Zr was used. It had a higher T_c (11K vs 9.3K for NbTi) but was more difficult to process and had a lower B_{c2}. (B_{c2} for NbTi 11 Tesla at 4.2K, 14 Tesla at 2K.)]. If a higher B_{c2} is desired a partial replacement of Nb (up to 25% by wt) by tantlum can increase B_{c2} up to 13T at 4.2K). NbZr alloys have comparable critical current to NbTi which could be very high (for NbTi typically J_c ~3 x 10^9 Ampere/meter^2 at 5 Tesla and at 4.2K). Such a high J_c is achieved by developing processing procedures to produce strong pinning centers. As with some other practical superconductor, commercial NbTi superconductors is almost always produced in multifilamentary form, with up to 5000 filaments in a wire typically 0.5 to 2 mm in diameter. Below we describe the basic microstructure and properties of this important alloy. (Please note that magnetic field H and magnetic induction B are sometimes used interchangeably.)

4.1.1 Microstructure

The main flux pinning centers (15,16) in this material are α-titanium precipitates with hexagonal close packed structure and to a lesser extent dislocation cells introduced by plastic deformation. The α-titanium precipitation is usually carried out at 375°-420°C. ω phase precipitation can also occur during this heat treatment but is usually avoided by suitable cold work. Because though ω phase provides strong pinning it reduces wire drawability. (17) The single heat treatment mentioned above will produce a maximum amount of precipitate of about 10% (by volume). This is often not sufficient to produce the critical current the material is capable of. A series of strains given to the material followed by heat treatment can increase this volume fraction to up to 20%. Care has to be taken not to increase the size of the α precipates beyond their typical size which ranges between 80 nm to 200 nm, or to reduce the titanium content of the matrix below about 36 wt%. It should be noted that although initially the α precipitates are randomly distributed and roughly spherical in shape the precipitates are alongated along the drawing axes due to subsequent deformations, till by the final treatment, in modern practical wires they are highly distorted into densely folded sheets, until they are only about 1 nm thick, separated by only a few nm, and made up of long strands. (18,19)

Fig. 1 shows a transverse TEM micrograph of portions of one filament of Nb 46.5% 5 WT% Ti composite with a J_c of 3.2 X 106 A/meter^2 at 5 Tesla at 4.2K. For details see Lee and Larbalestier who also describe in detail the combination of heat treatment and deformation needed to produce the final microstructure.

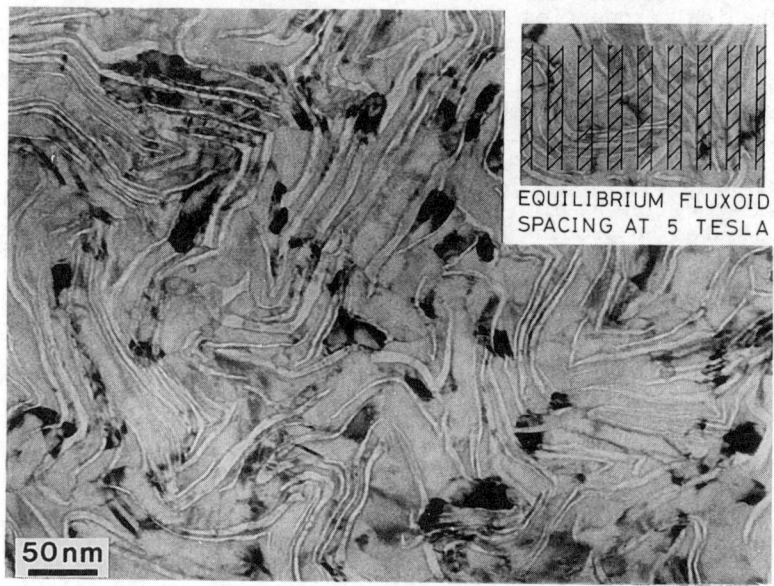

Fig. 1: Transverse transmission electron micrograph of portion of one filament of Nb46.5wt%T$_i$ composite. Equilibrium fluxiod spacing at 5 Tesla is also shown. Notice that this spacing is of the same order as the spacing of the microstructure. (Picture courtesy of Dr. Peter Lee.)

When the experimental pinning force F_p is plotted, as a function of the reduced field b $\left(b = \frac{B}{B_{c2}}\right)$ it is found that the data can be approximately represented as

$$\frac{F_p}{F_{pmax}} = b(1-b) \qquad (13)$$

Such a dependence suggests that the Fp is simply the sum of all elementary pinning forces. (20) However other functions have also been suggested.

Because of two competing mechanism of pinning due to precipitates and interfaces (grain boundaries and cell boundaries) it is difficult to separate the contribution of each mechanism. However it is now accepted that the pinning is nearly proportional to the volume fraction of precipitates, and is inversly proportional to the cell size or grain size. The proportion of the pinning centers depend on the processing conditions. It is found that at the optimum wire size, the density of the precipitates is about ten times the density of grain boundaries. (21)

Fabrication of NbTi conductors for practical applications are usually in the form of multifilametary filaments. The procedures for such fabrication is discussed in detail in reference (13).

4.2 A15 Superconductors

A15 superconductors are those with A15 type structure (sometimes also known as Cr_3Si structure or β-tungsten structure), (13) the most well known being Nb_3Sn, V_3Si (22) and V_3Ga compound superconductors. Another A15 superconductor, Nb_3Ge, had the record for highest transition temperature (23) before the advent of high T_c superconductors. About 76 A15 compounds are known to exist. Table 2 gives a list of some of the important A15 superconductors listed in order of their T_c. The typical compound A_3B is close to the stoichiometric ratio of 3:1. The A atoms (A = Nb, V, Mo, Ti, etc.) are on the faces of a cube, formed by B atoms (Ga, Sn, Si, Au, Pt, etc.) which occupy the bcc positions of the cube. (See Fig.2)

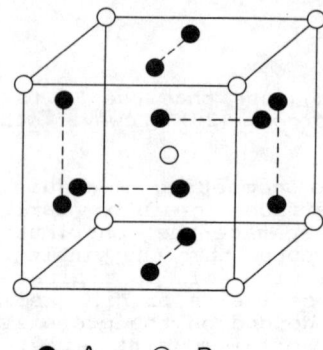

Fig. 2: A15 structure consisting of A (Nb, V, Mo, etc.) and B (Ge, Al, Ga, etc.) atoms. The A atoms form three orthogonal chains on the faces of the cube formed of B atoms which form a BCC sublattice.

Table 2
Critical Temperature of Important A15 Superconductors Listed in Order of Their Transition Temperature (T_c)

A15 Superconductor	T_c (K)	Comments
Nb_3Ge	23.2	Prepared by film deposition technique.
Nb_3Ga	20.7	
Nb_3Al	18.9	May be deficient in Al from 3 to 1 ratio.
Nb_3Sn	18.3	
V_3Si	17.1	
V_3Ga	15.9	
Mo_3Re	~15	May not have exact 3:1 ratio.
Nb_3Si	>11	Conjectered to have the highest T_c if suitably made.

A atoms form three othogonal chains in the crystal parallel to three <100> directions as shown in Fig. 2. The nearest neighbor distance along the chain is less than the distance of closest approach in a pure A crystal. For example in Nb_3Ge the distance between the Nb atoms along the chain is 0.2575 nm compared to 0.2858 nm in Nb bcc unit cell. (13)

4.2.1 Microstructure in A15 Material

The microstructure observed in A15 materials are grain boundaries, precipitates, radiation induced defects and a low density of dislocations. (24) Of these grain boundaries are the most important.

Grain Boundaries:

Grain boundaries are responsible to a great extent for the high value of critical current J_c observed in these materials. This relationship between J_c and grain size D has been studied by many workers. (25,26,27,28) Grain size can be altered by varying the growth conditions. It is then found that J_c increases linearly with the inverse of grain size sometimes reaching a maximum.

Fig. 3 is a picture of a superconducting Nb_3Sn tape coated with bronze. The central part of the micrograph shows Nb_3Sn grains roughly equiaxed and of size less than a micron. The J_c of the tape at 5T, and 4.2K exceeded $10^6 Amp/cm^2$.

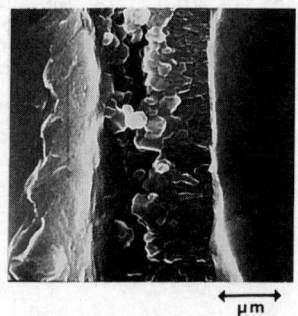

Fig. 3: Superconducting Nb_3Sn tape etched to expose Nb_3Sn grains in the center (Scanning electron micrograph).

Grain size can in fact be made as small as 30 nm in A15 materials by suitable annealing. Fig. 4 shows a picture of superconducting multifilamentary wire looking at one of the filaments.

V_3Ga grains have been formed by annealing CuGa and V sheets put together in the wire drawing process. The grains initially formed are columnar.

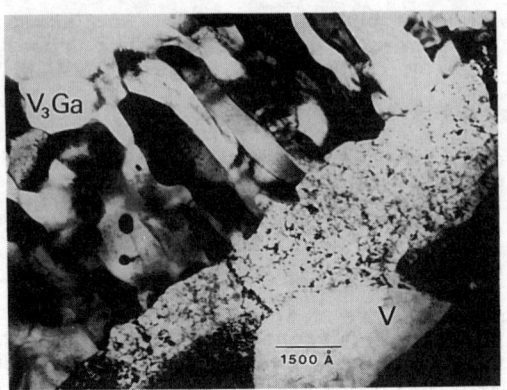

Fig. 4: Transmission electron micrograph of a filament of V_3Ga wire showing V_3Ga columnar grains.

So far there is no good model that provides a detailed mechanism by which grain boundaries pin flux line lattice. Since Kramer's model (29) is extensively used in analyzing J_c observations we provide a brief outline of this model. The basic premise in Kramer's model is that the depinning force is such that it exceeds the shear strength of the flux line lattice. Then the flux flow will occur by shear. Kramer then utilized the idea of the shearing of a real lattice. On such considerations he obtains a relation for flux pinning as

$$J_c B = \frac{1.3 \times 10^{10} \, B_{c2}^{5/2} \, b^{1/2}(1-b)^2}{\kappa^2 \left(1 - \frac{a_o}{D}\right)^2} \qquad (14)$$

a_o is the flux line lattice constant, D is grain size.

If $a_o \ll D$ the equation can be represented as

$$J_c^{1/2} B^{1/4} = \frac{1.14 \times 10^{15}}{\kappa} (B_{c2} - B) \qquad (15)$$

Therefore a plot of $J_c^{1/2} B^{1/4}$ vs B should give a straight line. This is the so called Kramer plot and the relation is usually obeyed. The intercept from this plot can be used to obtain B_{c2} and slope can provide a value of κ. Unfortunately though values of B_{c2} obtained are consistent, those for κ are not.

Kramers model has been critized by Dew-Hughes. (30) It certainly does not predict correct grain size dependence. We now briefly mention the model of Pande and Suenaga (31) where grain size dependence is considered in detail.

This model is based upon the following assumptions: (1) any grain boundary segment is regarded to be made up of a superposition of uniformly spaced dislocation walls spacing and other parameters are so chosen to satisfy all boundary conditions where on the average, the Burgers' vector are randomly distributed along all three axes; (2) grains are made up of such segments, the realtion between the grain size D, and grain segment, d, will depend on geometry and shape of the grain, but on the average $D \approx kd$ where k is of the order of 2; and (3) the flux pinning is primarily due to first order elastic interactions. Based on the above assumptions and after statistically summing the elementary pinning forces and expression for grain boundary pinning was obtained as

$$F_p(D) = \frac{P_o(1-b)^2}{a_o} \left(\frac{a_o}{D}\right) \sin^2\left(\frac{\pi}{\sqrt{3}\,k}\frac{D}{a_o}\right) \quad (16)$$

where $k \approx 2$, P_o is a constant independent of grain size D, b is the "reduced" magnetic field and a_o is the lattice constant of the flux line lattice which depends on the magnetic field. It should be emphasized that the above relation is based on several assumptions mentioned before and hence should be treated as very approximate. However the following implications of this result should be noted. Equation (16) shows that the flux pinning and hence critical current is approximately inversely proportional to grain size as expected from experiments. However, equation (16) also gives an oscillatory behavior in F_p as function of grain size. This can be seen by plotting F_p as a function of a_o/D. Such oscillations are not observed experimentally although experimental results do show large scatter in the data, especially in thin films. We belive that such an oscillatory behavior is due to the assumptions that all the grain are of the same size. In general, there is a distribution in grain sizes which should be taken into account. Pande [32] by his measurements of grain sizes in Nb_3Sn found that the distribution is roughly lognormal and can also be represented by a Rayleigh distribution

$$F(D) = \frac{\pi}{2}\frac{D}{\bar{D}} \exp\left[-\frac{\pi}{4}\left(\frac{D}{\bar{D}}\right)^2\right] \quad (17)$$

where $\bar{D}$ is the average grain diameter of the distribution. When this distribution is utilized, the pinning force becomes (33)

$$F_p(\bar{D}) = \frac{\pi}{4} \frac{P_o(1-b)^2}{a_o} \left(\frac{a_o}{\bar{D}}\right) \left[1 - \exp\left\{-\frac{4}{3}\frac{\pi}{k^2}\left(\frac{\bar{D}}{a_o}\right)^2\right\}\right] \quad (18)$$

This result is plotted in figure 5. It is seen that the oscillations are smoothed out but there is a peak in the F_p at a value of $\frac{a_o}{\bar{D}} \cong 1.83/k$. The exact location of the peak will depend on the value k used. Equation (18) leads to the conclusion that flux pinning cannot be indefinitely increased by grain refining. The formulation given above can also be used in principle to derive a magnetic dependence of flux pinning due to grain boundaries if P_o can be obtained using appropriate summation procedure.

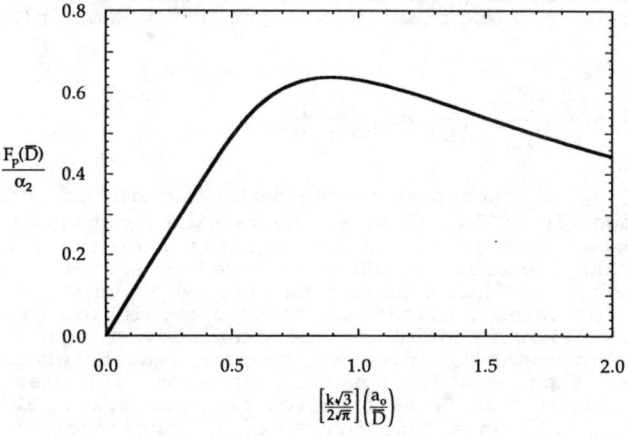

Fig. 5: A plot of Pinning Force F_p as a function of grain size in scaled variables. It is seen that F_p has a peak for a particular value of average grain size $\bar{D}$. α is a function of b only.

Precipitates

The precipitates play only a secondary role as pinning centers. Precipitates seems to refine grain size and hence indirectly raise critical current. There is some evidence (34) that Nb_5Ge_3 precipitates raise J_c in Nb_3Ge but a quantitative relationship between precipitate size, distribution, and field remains to be established.

Radiation Damage Defects

An interesting feature of A15 superconductors is the relation between its properties (especially T_c) with long-range crystallographic order, S. A convenient measure of S is the Bragg-Williams order parameter S defined as

$$S = \frac{p - r}{1 - r} \quad (19)$$

where p is the fraction of A atoms on A sites and r is the fraction A atoms in the compound A_3B. This parameter is such that for complete order S=1 (p=1) and for complete disorder S=0 (p=r). It is known that in A15 with high T_c, T_c is highest for S=1 and gradually decreases as S → 0. This is dramatically seen when the order parameter is gradualy reduced on irradiation. The effect of composition changes upon T_c can also be described in a similar fashion, by now one needs two order parameters S_A and S_B for A and B lattice sites respectively. When B is a transition element the sensitivity to long range order is much less.

Sweedler and Cox (35) give a relation between T_c and S as

$$T_c = T_{co} \exp\left[-\alpha\left(1 - \frac{S}{S_o}\right)\right] \quad (20)$$

where α is a constant, S_o is initial S value (~1) and T_{co} is the initial T_c value. S can be obtained from the radiation fluence ϕ by the relation where g is a constant

$$S = S_o \, e^{-g\phi} \quad (21)$$

Resistivity and specific heat also show one to one dependence on S, specific heat increasing with increasing S and resistivity decreasing with increasing S.

Apart from antisite defects (A atom on B site and B atom on A site) it has been shown that depending on the nature of radiation other defects are also created. High energy neutron or ions have been shown to create highly disordered regions (36) of the size 2 to 6 nm depending on the energy as shown in Fig. 6. Some dislocation loops are also observed. Disordered regions of the same order in size as coherent length may play significant role in determining both J_c and T_c of the irradiated A15 superconductors.

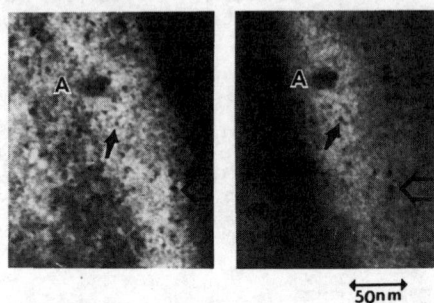

Fig. 6: Transmission electron micrograph of disordered regions (for example as pointed out by the arrows) in Nb_3Sn after high energy (energy = 1 MeV) neutron irradiation. The same region is imaged in two different reflections to reveal both disorder (right) and strain (left) in the disordered regions.

Dislocations

Most of the A15 materials are brittle and fail by cracking along grain boundaries. Hence it is difficult to produce dislocations by mechanical deformation at low temperatures. Some ingrown dislocations have been observed in Nb_3Sn and V_3Si. The Burgers vector (37) of these dislocations are of the type b = [100]. However dislocations play only a minor role in determining either superconducting or normal properties of A15 materials.

4.3 *Nitride and Carbonitride Superconductors*

Nitride and carbonitride superconductors all have cubic B1 (sodium chloride) structure. Around 1933 superconductivity was discovered in niobium carbide. In 1941 niobium nitride was discovered with a T_c ~ 16K, and in 1953 niobium carbonitride was discovered with a T_c of 17.8K, the highest in this class of superconductors. These superconductors are known to be strong coupling superconductors with a large electron-phonon interaction. The main drawback of these materials is the difficulty in preparation, since they are metastable. The preferred method is the thin film disposition technique including post-disposition nitriding and reactive sputtering. These materials have several very appealing features.

(1) The critical current can be quite large; over 2×10^6 A/cm^2 at 4.2K in zero field and over 10^5 A/cm^2 at 20 Tesla.

(2) They are relatively strain and radiation tolerant.

(3) They have relatively high Hc$_2$, in the range 20 to 25 Tesla.

(4) They are suitable for thin film electronics because of the above, and also because of their chemical stability.

The electron mean free path in these materials can be very large and hence coherence length can be small (4-5 nm). The pinning centers most responsible for high J_c are probably high density of point defects since some of them (eg NbN) is stable only at off stoichiometry. For additional information see reference (38).

4.4 Organic Superconductors

Although most of the organic solids are insulators there are a small number of organic compounds that behave like metals with high resistivity. Such organic charge transfer salts were first discovered in 1960s by researchers at DuPont. About thirty of them are known to exhibit superconductivity with transition temperature up to 12K (13K under pressure). Most if not all of these molecules contain sulfur or selenium atoms additions to the parent molecule tetrathiafulralene (TTF). These molecules, planar in nature, are stacked to form columns. Other molecular units connect these columns and act as donor or acceptors, the 'metal like' conduction is achieved by overlapping π orbitals. These superconductors are sometime thought of as one dimensional superconductors, and a theory of such superconductors by Little in 1964 aroused much interest. Not much information exists about defects in these superconductors. Further information about these superconductors can be obtained in references (40) and (41).

4.4.1 Doped C_{60} (Fullerene) as a Superconductor

It came as a great amazement to the scientific world that in addition to diamond and graphite, there exists a third form of element carbon, containing 60 (C_{60}) or 70 (C_{70}) carbon atoms. Kroto et al (42) proposed a structure viz truncated icosahedron formed from 20 regular hexagons and 12 regular pentagons for the structure of C_{60}. (Fig. 7) This structure is now well established and was called Buckminister Fullerene by them partly because their structure is similar to geodesic design of Buckminister Fuller. It is now popularly known as fullerene and sometime even as Buckyball.

The research in fullerenes advanced greatly when in 1990 Kratschmer et al (43) discovered a technique to make C_{60} in large quantities using a graphite arc in helium atmosphere. C_{60} crystallizes into an fcc structure at room temperature which could change to simple cubic at lower temperature. C_{60} has high electron affinity and so can be easily doped with alkali and alkaline-earth metals.

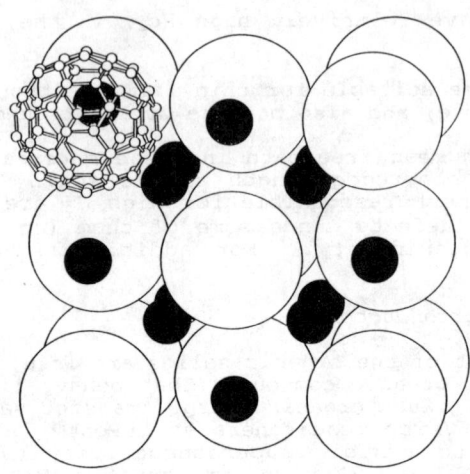

Fig. 7: Conventional cell for doped superconducting fullerene. A_3C_{60} (A = alkali). Large grey spheres are carbon molecules, smaller dark spheres are alkali atoms. Structure formed by small white spheres is the C_{60} structure prior to doping by A atoms. (Figure courtesy of Dr. Steven C. Erwin.)

Haddon et al (44) produced the first metallic fullerenes by such doping. Hebard et al (45) produced a fullerene superconductor (T_c = 18K) by using potassium dopant. (Fig. 7) T_c has been raised to 28K by doping with rubidium and further to 33K by mixed doping of rubidium and cesium. The stoichiometry of the superconducting phase can be written as A_3C_{60} where A is the alkali dopant or dopants. In case of K_3C_{60} it is known that the superconductor is of type II with λ = 240 nm and coherence length ζ = 2.6 nm. It appears that T_c is controlled by unit cell size. When superconducting transition temperature is plotted as a function of unit cell size for a variety of A_3C_{60} phases, it is found that T_c increases roughly with cell size. This result also suggests a pressure dependence on superconducting properties.

The exact mechanism for high temperature superconductivity in these materials is still being debated, but may turn out to be simpler than in high T_c oxide superconductors, because of the simplicity of the structure.

4.5 Chevrel Phase Superconductors

Chevrel phases can be represented as $M_yMo_6X_8$ where M is a metal, $0 \leq y \leq 4$, and X is usually sulfur or selenium. The structure was established by Chevrel et al in 1971, (46) and superconductivity was found in this class of ternary compounds in 1972. (47) The main attraction of these superconductors is relatively high value of critical field and hence they are suitable for ultra high field applications. It also bears some similarity with high T_c superconductors. Table 3 lists in order of their T_c values, some important Chevrel phase superconductors, their composition, and their critical field. It should be noted that Chevrel phase superconductors can be either stoichiometric or non-stoichiometric.

Table 3
Chevrel Phase Superconductors

Compound	Stoichiometric	T_c K	B_{c2} T	J_c
$PbMo_6S_8$	Yes	10-15	45-60	$5 \times 10^8 A/m^2$ at 10T and 4.2K
$SnMo_6S_8$	Yes	10-14	30-40	
$Cu_{1.8}Mo_6S_8$	No	10.8	14-16	
$YbMo_6S_8$	Yes	9	-	
$PrMo_6Se$	Yes	9	21	
$AgMo_6S_8$	Yes	8.5	-	
$LaMo_6S_8$	Yes	7.0	~6	

The ideal crystal structure of these ternaries consist of S (or Se) atoms at the corners of a cube with the vertices of Mo_6 octahedra lying on the faces of the cube. A network of four such cubes are loosely linked through a M atom (for example Pb).

These compounds are usually formed either by solid state diffusion or by thin film techniques. Single crystals have also been formed from melt. Chevrel phase superconductors are extremely brittle, and are at least as sensitive to radiation as A15 materials. In spite of all this several attempts have been made to make wires out of these materials.(48) The main microstructural features responsible for high critical currents appears to be grain boundaries and point defects.

4.6 High T_c Superconductors

High T_c superconductors are oxide superconductors. They should however be distinguished from the oxide superconductors discovered in 1960s which were based on Titanium and Niobium. The highest T_c of these oxides is 13.7K for $LiTi_2O_4$. High T_c superconductors are oxides of copper with perovskite structure.

Before their discovery superconductivity was unknown in copper oxides. In 1986 Bednorz and Muller (7) observed evidence of superconductivity with transition temperature over 30K in $La_{2-x}Ba_xCuO_4$. This material had perovskite structure. A substitution of Yttrium for Lanthanum led to the discovery of $YBa_2Cu_3O_{6+\delta}$ ($\delta < 1$) often referred to as YBCO or 123 because of the ratio of Y:Ba:Cu in this material. This was the first superconductor discovered with a T_c in the range 90-95K, (8) well above the boiling point of liquid nitrogen. This discovery was followed in 1988 by the discovery of Bismuth ($Bi_2Sr_2Ca_2Cu_3O_{10+\delta}$) and Thallium ($Tl_2Ba_2Ca_2Cu_3O_{10+\delta}$) based superconductors with even higher T_c. A large family of these superconductors with over 40 members are now known which could be represented by $A^1{}_xX_yA_{n-1}Cu_nO_{2n+\delta}$ n = 1,2,3...($\delta < 1$).

The crystal structure of 123 was found to be orthorhombic. For superconductivity the material must be deficient in oxygen. Typically the a b and c values in the orthorhombic unit all are a=0.382nm, b=0.389nm and c=1.168nm i.e. c axis is much larger than a, and b. a and b are approximately equal. They are exactly equal at high temperatures (about 750°C) i.e. the structure is then tetragonal.

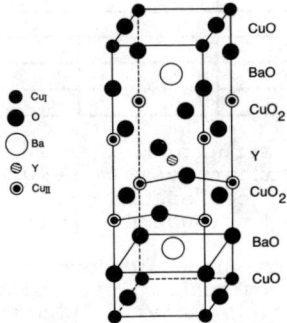

Fig. 8: Structure of $YBa_2Cu_3O_7$ showing the position of individual atoms, CuO planes (Cu_I) and Cu chains (Cu_{II}). In the superconducting state some of the oxygen atoms will be missing from the complete structure which could accommodate up to 9 oxygen atoms.

The 123 structure can also be considered as a stack of three simple perovskite cells as shown in fig. (8). Although the structure can take up to 9 oxygen atoms, the total number of oxygen atoms in the superconducting 123 range between 6 and 7. The properties of the superconductor depend sensitively on the oxygen content of the unit cell. Characteristic of the structure of 123, as of all high T_c oxide superconductor is the sheets of CuO planes and chains.

BISCO family of superconductors have also perovskite type structure with copper oxide planes and chains and orthogonal unit cells. Two most common members are $Bi_2CaSr_2Cu_2O_{8+\delta}$ ($\delta < 1$) commonly known as 2122 or 85K phase, and $Bi_2CaSr_2Cu_3O_{10+\delta}$ commonly known as 2223 or 110K phase. Thallium superconductors have similar structure. The highest T_c in this family is 125K. With the increasing number of CuO planes the T_c seems to increase in the oxide superconductor family. For details about the structure see reference (49).

The oxide superconductors allow a variety of substitutions and additions. In 123 structure the rare earth atom Yttrium can be replaced by most but not all of the rare earth elements. Copper can be partially substituted by iron, nickel, aluminum, silver, etc, though T_c is reduced by these substitutions. Ag seems to be least detrimental for this purpose. Another interesting partial substitution is that of lead for Bismuth. Lead seems to stabilize high T_c (2223) phase.

After the discovery of high T_c superconductors, they have been subjected to intense investigations to determine their structure using x-ray and neutron diffraction and lattice imaging and the microstructure present in them using transmission electron microscopy and related techniques. It is now known that the important defects in these superconductors include twin and grain boundaries, point defects and impurities, dislocations and stacking faults precipitates (second phase particles) voids and microcracks.

Relation between structure, microstructure and normal and superconducting properties have also been determined in many cases. Of course, as in conventional superconductors the microstructure present in oxide superconductors is expected to play a prominent role in determining the current carrying capacity of these materials

4.6.1 Twins

Planar defects (50,51) were discovered in $YBa_2Cu_3O_{7-\delta}$ ($\delta < 0.5$) almost immediately after the discovery of this superconductor using transmission electron microscopy. They were shown to be twin boundaries. Fig. 9 shows a bright field image with the electron beam parallel to 001 axis of the crystal. The

Fig. 9: Transmission electron micrograph of YBa$_2$Cu$_3$O$_7$ showing two set of twins approximately at right angles to each other. The twins are aligned along [110] direction and the interfaces are seen edge on. The orthorhombic a and b axes are found to alternate along these interfaces.

corresponding electron diffraction is shown in Fig. 10. In electron diffraction a small splitting of the h00 and 0k0 reflections can be seen. This information and the fringe contrast confirms that the planar defects are (110) twin interfaces. The twin formation is due to a tetragonal to orthorhombic phase transformation that this material undergoes below about 750°C. If a, b denote the lattice parameter (which are equal when the crystal is in tetragonal form), Pande et al (51) have shown that twin misorientation θ is given by

$$\Theta \approx 4\tan^{-1}\left(\frac{a-b}{a+b}\right)$$
$$\approx \frac{4(a-b)}{a+b} \quad \text{since } a \approx b. \tag{22}$$

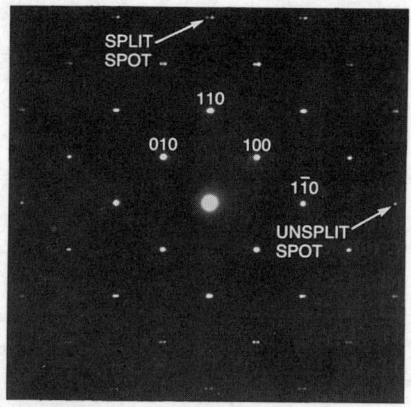

Fig. 10: Electron diffraction from the same region and orientation as in Fig. 9. Notice the splitting of spots along one direction.

Using the appropriate values of a and b, it is found that θ is about one degree in good agreement with the splitting observed in electron diffraction. Because of this twinning relation the twins give the alternate black and white contrast observed in TEM. Sometimes a fine tweed like structure is also observed instead. This structure has been shown to consist of fine (< 5 nm) orthorhombic domains each with a twin like crystallographic relation with its neighbor. The twins can sometimes form in two sets of colonies parallel to either (110) or $(1\bar{1}0)$ direction. The twin spacing varies somewhat is usually is of the order of 100 nm.

The twin formation can be understood as due to a need to accommodate shape change resulting from the tetragonal to orthorhombic transformation. (Fig. 11) The need to accommodate i.e. to reduce strain energy however must be balanced by the need to use as few twinning surfaces as possible. The twin spacing d can thus be obtained by minimizing the sum of strain energy (proportional to d) and surface energy (which is proportional to $1/d$). (52)

Twin boundaries are planar defects and are found in sufficient densities in many specimens. They are therefore expected to influence J_c in these materials. (53) There is some evidence that this is indeed so. (54) J_c in twinned single crystals were found to be higher than in the same crystal made twin free by the application of appropriate pressures. Also evidence of pinning by twin boundaries have been found by decoration techniques. (55) Theoretical calculation predicting a relation between twin interfaces and flux pinning are now available, (53) but need experimental verification.

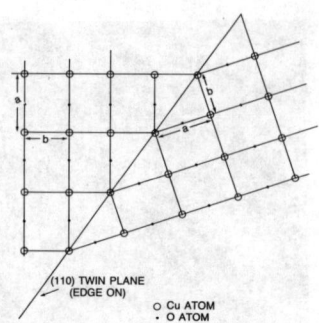

Fig. 11: A schematic diagram of a twin interface. The difference between a and b axes is exaggerated, in fact a ≅ b.

4.6.2 Cracks

Most high T_c materials show microcracks which are probably produced during processing. In case of YBaCuO, the tetragonal to orthorhombic transformation produce stresses which are reduced in ab plane by twin formation. However there can be some shrinkage along C axis, which can be relieved by microcracks. Microcracks can pin fluxiods, but they are not in sufficient number to contribute significantly to J_c. It is known that dispersion of Ag in these materials reduces cracking, but J_c is hardly affected, supporting the above conclusions.

4.6.3 Dislocations and Stacking Faults

Dislocation densities observed in high T_c materials like $YBa_2Cu_3O_7$ is usually low ($< 10^5 cm/cm^3$). Unless introduced by deformation the commonly observed (by TEM) dislocations are perfect dislocations with a [100], [010], or [110] Burgers vector. (56-58) The slip plane is always found to be (001) i.e. the a-b plane. The dislocations density can be increased somewhat by deformation processes, especially at elevated temperatures, but their role in determining both normal and superconducting properties is rather limited.

Stacking faults have been frequently observed in YBaCuO (59-60) though their density is not usually high. TEM studies have shown that they lie on (001) i.e. a-b plane and are bounded by a pair of partial dislocations. These faults are relatively wide indicating that the material has low stacking fault energy. These faults are probably chemical faults produced by the presence or (less likely) absence of a CuO plane. These defects should contribute to J_c but not much is known about this contribution. The flux pinning is expected to be maximum when the field is parallel to the stacking faults i.e. perpendicular to C axis. Typical estimates of the number of stacking faults is $\sim 10^{20}/m^3$ and hence a rough estimate is that contribution to J_c is of the order of $10^{11} A/m^3$.

4.6.4 Structure Modulation

All Bi based superconductors exhibit superstructure modulation which is not commensurate with the basic structure. The modulation has been related with the oxygen content of the specimen, and hence may have a significant role in determining superconducting properties. Fig. 12 shows a high resolution transmission electron micrograph of a 2212 Bi-Sr-Ca-Cu-O oxide superconductor. The modulations are clearly visible in this micrograph. The corresponding diffraction pattern is shown on the right hand corner.

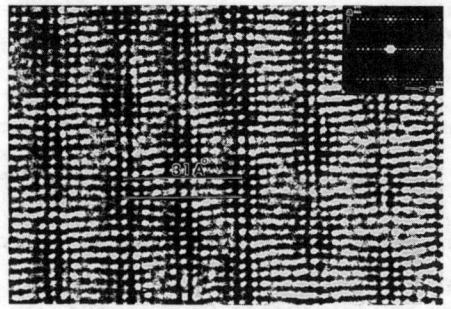

Fig. 12: Structure image of BiSCO (2212) in [010] projection. The inset shows the corresponding diffraction pattern. Modulation in structure is seen along the horizontal planes, as well as within the perovskite block shown between two lines with a spacing of 31Å. (Micrograph courtesy of Dr. Pratibha Gai.)

The structure modulation observed is due to repeated shortening and widening of Bi-Bi distances, which could be periodic. For further details see references (61-63). These modulations are also expected to alter the J_c in these materials but the flux pinning mechanism has not as yet been worked out.

4.6.5 Critical Current in Polycrystals

It is well established that the critical current of even carefully prepared polycrystalline high T_c material is extremely low (in the range of 100-1000Å/cm^2 at 77K). This value drops further in the presence of even a small magnetic field of a small fraction of a Tesla. On the other hand the J_c value for single crystals, in bulk as well as in thin film form, is at least three order of magnitude higher.

The popular view has been that this limitation in critical current is due to the behavior of the grain boundaries as a "weak link" or i.e. having a Josephson coupling between the grains (see for example Clem (64) and Jagannadham and Narayan (65). Behavior of the bicrystal transport currents under a magnetic field (usually a sharp drop is seen) is suggestive of a Josephson coupling. Much higher value of critical current J_c of single crystals in bulk as well as in thin films $YBa_2Cu_3O_7$ compared with polycrystals of the same material suggests strongly the degrading role of grain boundaries. Two papers by Dimos and co-workers (66,67) have firmly established that J_c is a very sensitive function of the grain boundary at least in thin films. However recent experimental results indicate this may not be true for all boundaries (68).

Due to the presence of grain boundaries in polycrystals, which act as weak links sintering alone cannot give high J_c values. Also weak link behavior is largely intrinsic, a better way to enhance J_c is by texturing (69).

4.7 Microscopic Theory of High T_c Superconductivity

It looks likely that most, if not all oxide superconductors may have a common microscopic mechanism operating. However what this mechanism is, is still a matter of debate. Some facts that a microscopic theory might include are the following:

(1) High T_c superconductors are highly anisotropic having quasi-two dimension CuO_2 layers.

(2) They are very sensitive to doping especially with respect to oxygen.

(3) The resistivity in a-b plane is linear with temperature and at higher temperatures the resistivity does not saturate.

There are many other observations which indicate the system cannot be treated simply as a collection of nearly free electrons.

Where does the BCS theory fail? The BCS theory is one of the most successful theories in Physics. The notion of formation of a condensate of Cooper pairs has never been challenged before, though it was modified somewhat to take into account strong coupling effects. What is being challenged now, in the case of high T_c superconductivity is the nature of the interaction which is responsible for the formation of Cooper pairs. In BCS model this interaction is due to an exchange of phonons, i.e. an attractive interaction due to the polarization of the lattice. In high T_c the Cooper pairing has been established, but the coherence length is very small (about 1 nm). Various interaction mechanisms have been proposed. These can be grouped under four categories, viz, (1) a modified phonon mechanism, (2) an interaction due to magnetic coupling, (3) electronic coupling, (4) exotic coupling. However further studies are required before any of these mechanisms could be accepted. The role of material scientists in this endeavor is to develop new materials possibly without copper, or oxygen or both, to try to raise T_c, by appropriate doping etc., to characterize the structure etc. However, their major contribution in the future is likely to be in developing microstructure to raise critical current.

4.8 Conclusions

In conclusion, we note that superconducting materials offer attractive advantages for a wide variety of applications, both small scale and large scale.

The situation has dramatically changed with the discovery of high T_c superconductors. However before these potentials could be realized, a great deal of research in the microstructure-property relationship and materials engineering needs to be carried out. In addition the search for superconductors with even higher transition temperature is proceeding briskly with the ultimate goal of a room temperature superconductor.

REFERENCES

1. H. Kamerlingh Onnes Akad. van Wetenschappen, Proceedings from the Section of Sciences (Amsterdam) **14** (1911) 113-115 and 818-821.
2. W. Meissner and R. Ochsenfeld, Naturwissensch **21** (1933) 787.
3. F. London and H. London, Physica **2** (1935) 341.
4. V. L. Ginzburg and L. Landau, Zh. Eksp. Teor. Fiz **20** (1950) 1064.
5. J. Bardeen, L. N. Cooper, and J. R. Schrieffer, Phys. Rev. **108** (1957) 1175.
6. B. D. Josephson, Phys. Lett. **1** (1962) 251.
7. J. G. Bednorz and K. A. Muller, Z. Phys. B **64** (1986) 189.
8. M. K. Wu, J. R. Ashburn, C. J. Torng, P. H. Hor, R. L. Meng, L. Gao, Z. J. Huang, Y. Q. Wang, C. W. Chu, Phys. Rev. Lett **58** (1987) 908.
9. Randy Simon and Andrew Smith, "Superconductors-Conquering Technology's Frontier," Plenum Press, New York (1988).
10. A. A. Abrikosov, Sov. Phys. JETP (English Translation) **5** (1957) 1174.
11. M. Tinkham, Introduction to Superconductivity, McGraw Hill, New York (1975).
12. A. M. Campbell and J. E. Evetts, Adv. Phys. **21** (1972) 199.
13. T. Luhman and D. Dew-Hughes (Eds.), Metallurgy of Superconducting Materials, Academic, Press New York (1979).
14. R. A. Johnson and A. N. Orlov (eds), "Physics of Irradiation Effects in Crystals," North Holland, Amsterdam (1985).
15. I. Pfeiffer and H. Hillmann, Acta Metall. **16** (1968) 1429.

16. D. C. Larbalestier and A. W. West, Acta Metall. **32** (1984) 1871.
17. M. I. Buckett and D. C. Larbalestier, IEEE Trans. Mag. **23** (1987) 1638.
18. P. J. Lee and D. C. Larbalestier, Acta Metall. **35** (1987) 2526.
19. P. J. Lee, J. C. McKinnell and D. C. Larbalestier, IEEE Trans. Mag. **25** (1989) 1918.
20. H. Gotoda, K. Osamura, M. Furusaka, M. Arai, J. Sujuki, P. J. Lee, d. C. Larbalestier and Y. Monju, Phil. Mag. B **60** (1989) 819.
21. C. Meingast and D. C. Larbalestier, J. App. Phys. **66** (1989) 5971.
22. G. F. Hardy and J. D. Hulm, Phys. Rev. **89** (1953) 884.
23. J. R. Gavaler, App. Phys. Lett. **23** (1973) 480.
24. C. S. Pande in T. Luhman and D. Dew Hughes (ed.), Metallurgy of Superconducting Materials, Academic Press, New York (1979).
25. E. Nembach and K. Tachikawa, J. Less Common Met. **19** (1969) 359.
26. R. M. Scanlan, W. A. Fietz and E. F. Koch, J. App. Phys. **46** (1975) 2244.
27. B. J. Shaw, J. App. Phys. **47** (1976) 2143.
28. A. W. West and R. D. Rawlings, J. Mater. Science **12** (1977) 1962.
29. E. J. Kramer, J. App. Phys. **44** (1973) 1360.
30. D. Dew Hughes, Phil. Mag. **55** (1987) 459.
31. C. S. Pande and M. Suenaga, App. Phys. Lett. **29** (1976) 443.
32. C. S. Pande, Unpublished work.
33. C. S. Pande and R. A. Masumura, Materials Science and Engineering (1995) (in press).
34. A. T. Santhanam, private communication.
35. A. R. Sweedler and D. E. Cox, Phys. Rev. **12** (1975) 147.
36. C. S. Pande, Solid State Commun. **24** (1977) 241.
37. S. Mahajan, s. Nakahara, J. H. Wernick and G. Y. Chin, (unpublished work).
38. L. E. Toth, Transition Metal Carbides and Nitrides, Academic Press, New York (1971).
39. W. A. Little, Phys. Rev. **134** (1964) 1416.
40. J. M. Williams, A. J. Schultz, U. Geiser, K. D. Carlson, A. M. Kini, H. H. Wang, W. K. Kwok, M. H. Whangbo and J. E. Schirber, Science **252** (1991) 1501.
41. A. E. Underhill, J. Mater. Chem. **2** (1992) 1.
42. H. W. Kroto, J. R. Heath, S. C. O'Brien, R. F. Curl and R. E. Smalley, Nature **318** (1985) 162.
43. W. Kratschmer, K. Fostiropoulos and D. R. Huffman, Chem. Phys. Lett. **170** (1990) 167.
44. R. C. Haddon, A. F. Hebard, M. J. Rosseinsky, D. W. Murphy, S. J. Duclos, K. B. Lyons, B. Miller, J. M. Rosamilia, R. M. Fleming, A. R. Kortan, S. H. Glarum, A. V. Makhija, A. J. Muller, R. H. Eick, S. M. Zahurak, R. Tycko, G. Dabbash and F. A. Thiel, Nature **350** (1991) 320.

45. A. F. Hebard, M. J. Rosseinsky, R. C. Haddon, D. W. Murphy, S. H. Glarum, T. T. M. Palstra, A. P. Ramirez and A. R. Kortan, Nature **350** (1991) 600.
46. R. Chevrel, M. Sergent and J. Prigent, J. Solid State Chem. **3** (1971) 515.
47. B. T. Matthias, M. Marezio, H. E. Barz Corenzwit and A. S. Cooper, Science **175** (1972) 1465.
48. B. Seeber, M. Decroux and O. Fischer, Physica B **155** (1989) 129.
49. T. A. Vanderah (Ed.), Chemistry of Superconducting Materials, Noyes Publications, Park Ridge, NJ (1992).
50. R. Beyers, G. Lim, E. M. Engler, R. J. Savoy, T. M. Shaw, T. R. Dinger, W. J. Gallaghar and R. L. Sandstrom, App. Phys. Lett. **50** (1987) 1918.
51. C. S. Pande, A. K. Singh, L. E. Toth, D. U. Gubser and S. A. Wolf, Phys. Rev. B **36** (1987) 5669.
52. A. G. Khachaturyan, Theory of Structure Transformations in Solids, John Wiley, New York (1983).
53. V. B. Geshkenbein, JEPT **57** (1988) 2166.
54. C. H. Chen, J. Kwo and M. Hong, App. Phys. Lett. **52** (1988) 841.
55. P. L. Gammel, D. J. Bishop, G. J. Dolan, J. R. Kwo, C. A. Murray, L. F. Schneemeyer and J. V. Waszczak, Phys. Rev. Lett. 59 (1987) 2592.
56. S. Nakahara, S. Jin, R. C. Sherwood and T. H. Tiefel, App. Phys. Lett. **54** (1989) 1926.
57. J. Rabier and M. F. Denanot, Revue Phys. Appl. **25** (1990) 55.
58. T. Yoshida, K. Kuroda and H. Saka, Phil. Mag. A **62** (1990) 573.
59. H. W. Zandbergen, R. Gronsky, K. Wang and G. Thomas, Nature **331** (1988) 596.
60. J. Tafto, M. Suenaga and R. L. Sabatini, App. Phys. Lett. **52** (1988) 667.
61. T. M. Shaw, s. A. Shivshankar, S. J. LaPlaca, J. J. Cuomo, T. R. McGuire, R. A. Ray, K. H. Kellher and D. S. Yee, Phys. Rev. B **37** (1988) 9856.
62. P. L. Gai and P. Day, Physica C **152** (1988) 335.
63. Y. Matsui, H. Maeda, S. Tanaka and S. Horiuchi, Japanese J. Appl. Phys. **27** (1988) 361.
64. J. R. Clem, Phys. Rev. B **43** (1991) 7837.
65. J. Jagannadham and J. Naryan, Phil. Mag. B **61** (1990) 129.
66. D. Dimos, P. Chandhari, J. Mannhart and F. K. LeGoues, Phys. Rev. Lett. **61** (1988) 219.
67. D. Dimos, P. Chandhari and J. Mannhart, Phys. Rev. B **41** (1990) 4038.
68. S. E. Babcock, X. Y. Cai, D. L. Kaiser and D. C. Larbalestier, Nature **347** (1990) 167.
69. S. Jin, T. H. Tiefel, R. C. Sherwood, M. E. Davis, R. B. vanDover, G. W. Kammlott, R. A. Fastnacht and H. D. Keith, App. Phys. Lett. **52** (1988) 2074.

FURTHER READING

Specific references to various topics have been provided at the appropriate places in the text for additional information. The following additional references mostly to text books are being provided for further reading.

Section 1: In addition to reference (9), a good introduction to high T_c superconductors can be found in A. M. Wolsky, R. F. Giese and E. J. Daniels, The New Superconductors: Prospects for Applications, Scientific American, Feb. 1989, p. 45.

Section 2: Two textbooks cover very well topics in this section including thermodynamics and BCS theory. P. G. Gennes, Superconductivity of Metals and Alloys, W. A. Benjamin, New York, 1966. E. M. Lifschitz and L. P. Pitaevsky, Statistical Physics Part 2, Pergamon Press, 1980.

Section 3: The following two books cover applications of superconductors in great detail. E. W. Collings, Design and Fabrication of Conventional and Unconventional Superconductors, Noyes Publications, Park Ridge, 1984. S. T. Ruggiero and D. A. Rudman (Eds.), Superconducting Devices, Academic Press, London, 1990.

Section 4: An excellent source for information in general, and especially for topics in section 4, is the handbook edited by Evetts. Jan Evetts (Ed.), Concise Encyclopedia of Magnetic and Superconducting Materials, Pergamon Press, Oxford, 1992.

Topics in high T_c superconductivity are also included in this handbook. For details on physical properties of high T_c superconductors, a good source is D. M. Ginsberg (Ed.), Physical Properties of High T_c Superconductors I, II, III, World Scientific, Singapore, 1989, 1990, 1992.

ACKNOWLEDGEMENT

The author is grateful to Dr. Peter Lee, Dr. Steven Erwin and Dr. Pratibha Gai for providing figures and to Mrs. Diana Buck for assistance in preparing this chapter.

PROBLEMS

1. Given two conductors, one metallic and the other a superconductor, how would you distinguish the two?

2. Calculate the number of vortices per cm^2 in a given superconductor at 10 T. Given $\Phi_o = 2.1 \times 10^{-15}$ Weber, $B_{c2} = 20$ T.

3. A A15 superconductor has J_c of 6×10^6 Amp/cm^2 and has a cross section of 1 mm^2. How much thicker a copper wire has to be at 4°K to carry the same current?

4. Why are critical current of high T_c polycrystalline superconductors so low? Suggest a few ways that you would try to raise it.

5. Explain why grain refining is used to increase critical current in conventional superconductors, but not in high T_c superconductors. What determines this difference in behavior?

6. Explain why for producing very high magnetic fields, some superconducting materials might be better than others.

7. Explain why superconductors find use in making high field magnets.

8. Explain why vortices need to be pinned in superconductors by grain boundaries, precipitates, etc.

9. Use London equation (4) and

$$m \frac{dv}{dt} = qE$$

to show that

$$\lambda_L^2 = \frac{mc^2}{4\pi n q^2}$$

(Hint: Differentiate London equation)

$$\frac{\partial J}{\partial t} = \frac{c^2}{4\pi \lambda_L^2} E$$

$$E = \frac{\partial A}{\partial t}$$

10. In equations (20) and (21) $\alpha = 3.2$, $S_o = 0.915$, $g = 1.86 \times 10^{-20}$ cm^2 second/neutrons. Calculate the order parameter and fluence at which T_c of Nb$_3$Al is reduced to half its value by radiation. Calculate the % of Nb sites occupied by Al atoms at that value of order parameter.

Chapter 8
Magnetic Materials

C. D. Graham, Jr.
Dept. Materials Science and Engineering
University of Pennsylvania, Philadelphia

1. Introduction

The behavior of strongly magnetic materials has been a source of fascination for centuries. The readily observable force between two permanent magnets, which can be made attractive or repulsive by simply rotating one of the magnets, is striking and memorable.

The practical applications of magnetism, except for the magnetic compass, are relatively recent but of great importance. The entire electric power industry depends on magnetic materials for the construction of generators and transformers, and a large fraction of the electric power produced is converted to mechanical power in electric (magnetic) motors, which run industrial machinery, air conditioners and ventilators in buildings of all sizes, domestic appliances, copying machines, etc. Permanent magnets hold refrigerator doors closed (and children's drawings stuck to the outside), operate small motors in windshield wipers and disk drives, and drive loudspeakers and headphones. Magnetic materials store audio, video, and digital information on tapes and disks; apply toner to create images from laser printers and copying machines; and focus electron beams in analytical instruments, television sets, and computer monitors.

2. Basic Ideas

All materials are magnetic, in the sense that they acquire a non-zero magnetization in the presence of an applied magnetic field. In all but a few materials, the magnetization is very small, measurable only with appropriate instruments, and we say loosely that such materials are *non-magnetic*. Of all the elements in the periodic table, only three, Fe, Co, and Ni, are strongly magnetic at room temperature; they are called *ferromagnetic*. Many alloys and compounds of these three elements, and a few alloys and compounds of Mn and Cr, are also ferromagnetic. There is also a class of compounds of these elements, mostly oxides, which are less strongly but still usefully magnetic; they are classified as *ferrimagnetic*. These terms will be formally defined later.

Only the ferro- and ferrimagnets are useful as engineering materials for their magnetic properties (unless, of course, an engineering material is required to be non-magnetic), so this chapter will concentrate almost entirely on strongly magnetic materials.

3. Magnetic Quantities and Units

3.1 Units

Two systems of units are in common use for the description of magnetic behavior. The *SI* (*Système Internationale*) or *mks* (meter-kilogram-second) system is used in almost all physics textbooks, and extensively in research publications and commercial specifications, especially in Europe. The older *cgs* (centimeter-gram-second) *Gaussian* system is still favored for research publications worldwide, and for materials specifications in the US and to some extent in Japan. Thus to work in magnetism it is necessary to be bilingual in the matter of units. In this chapter we will use SI units primarily, but will show cgs Gaussian units as well.

3.2 Magnetic Quantities

A *magnetic field* may be defined as a region of space in which a moving electric charge experiences a force. As discovered by Oersted in 1819, a wire carrying an electric current produces a circumferential magnetic field. A useful way to produce modest magnetic fields is to wind an insulated copper wire into a spiral or *solenoid*, and pass a current through the wire (see Fig. 1). The field at the center of the solenoid is given by

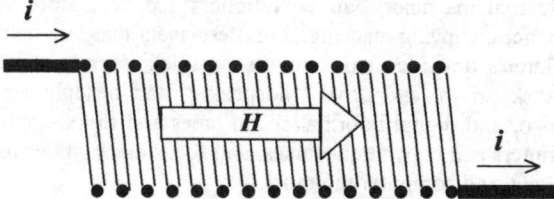

Fig. 1. Magnetic field H produced by current i in solenoid.

$$SI: H = \frac{ni}{l} \text{ (A/m)} \quad \text{or } cgs: H = \frac{4\pi}{10}\frac{ni}{l} \text{ (oersted)} \tag{1}$$

where n is the number of turns in the spiral, i is the current in ampere, and l is the length of the solenoid (meter in *SI*, cm in *cgs*). It follows that

$$\frac{(A/m)}{Oe} = \frac{10^3}{4\pi} = 79.577 \approx 80. \tag{2}$$

Without forced cooling, the maximum field produced by a solenoid is about 20 kA/m or 250 Oe. For reference, the total magnetic field of the earth (midway between the equator and a pole) is about 40 A/m [0.5 Oe], of which somewhat less than half is directed parallel to the surface to act on a magnetic compass.

The basic quantity that describes magnetic properties of matter is the *magnetic moment*. A magnetic moment m placed in a uniform magnetic field experiences a torque given by

$$L = m \times H, \qquad (3)$$

and when placed in a magnetic field gradient experiences a translational force given by

$$F = m \text{ grad } H. \qquad (4)$$

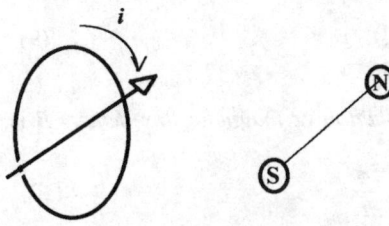

Fig. 2. Two representations of a magnetic moment m.

The magnetic moment m may be thought of as a conducting loop carrying a fixed current, or equivalently as a north and a south magnetic *pole* separated by a fixed distance, as shown in Fig. 2. The *SI* unit of moment comes directly from the first concept; it is the $A \cdot m^2$. A unit magnetic moment is represented by a current of one ampere flowing around an area of one square meter (the direction of the moment is normal to the area bounded by the current loop). A unit moment in the *cgs* system experiences a torque of one dyne-cm when placed in a uniform field of one oersted. The *cgs* unit of moment may then be taken as the erg/Oe (remembering that one dyne-cm = one erg). An unofficial but widely-used name for the erg/Oe is the *emu*, standing for *electromagnetic unit*. Strictly, *emu* only specifies that one is using the *cgs* system, but commonly it is used to mean specifically the *cgs-emu* unit of magnetic moment. Using this definition,

$$\frac{Am^2}{emu} = 10^3. \qquad (5)$$

The *magnetization* of a material is its total net magnetic moment per unit volume:

$$M = \frac{\sum m}{V}. \qquad (6)$$

In *SI*, this quantity is measured in $A \cdot m^2/m^3 = A/m$, which is the same as the unit for field; in *cgs*, it is in *emu*/cm^3. To convert,

$$\frac{(A/m)}{(emu/cm^3)} = 10^{-3}. \qquad (7)$$

When working with small samples, the sample mass is usually more accurately known than the sample volume; also the sample mass is independent of temperature, while the volume changes with temperature due to thermal expansion. Therefore it is frequently useful to specify magnetization per unit mass, which is given the symbol σ:

$$\sigma = \frac{\sum m}{W}. \tag{8}$$

where W is the sample mass. Conveniently,

$$\frac{(Am^2/kg)}{(emu/g)} = 1. \tag{9}$$

A magnetic quantity known as the *magnetic induction* or *magnetic flux density B* is defined by Faraday's Law:

$$E = -NA\frac{dB}{dt} = -N\frac{d\phi}{dt}. \tag{10}$$

Here B is the flux density, regarded as a *magnetic flux* ϕ per unit area A. This law states that a voltage E is generated in a coil of N turns when the magnetic flux in the coil changes with time. The number of turns N is regarded as dimensionless, so the SI unit of flux is the volt-second, or *weber*; and the SI unit of flux density is the Wb/m², or *tesla*. The equivalent cgs units are the *maxwell*, and the maxwell/cm² = *gauss*. The conversion is then

$$\frac{tesla}{gauss} = 10^{-4}. \tag{11}$$

Lines of magnetic flux are continuous; they must always form closed loops. Mathematically, div$B = 0$.

In empty space, and to a very good approximation in air or any other gas, the magnetic field is the only contribution to the magnetic flux density. Therefore H and B are physically the same, but are expressed in different units: H in A/m (SI) or Oe (cgs), and B in tesla (SI) or gauss (cgs). The ratio between them is called the *magnetic constant* or the *permeability of free space*, μ_0. The value of μ_0 in *SI* is defined to be

$$\mu_0 = 4\pi \cdot 10^{-7} T/(A/m) = 4\pi \cdot 10^{-7} \text{ henry/meter}, \tag{12}$$

since 1 henry=1 V·sec/A.

In cgs units, the magnetic constant is unity (this is one of the virtues or defects of the cgs system, depending on one's point of view), and in free space or air,

$$1 \text{ gauss} = 1 \text{ oersted}. \tag{13}$$

In matter, the magnetization of the material contributes to the flux density, so that in *SI*

$$B = \mu_0(H + M), \tag{14}$$

and in *cgs* (less obviously)

$$B = H + 4\pi M. \tag{15}$$

The equation relating B, H and M is a frequent source of confusion; various authors interpret it using different (although ultimately equivalent) language. The quantity B is useful and necessary because it enters Faraday's Law, and Faraday's Law is a fundamental relationship that describes the operation of electric generators and transformers. The field H is the contribution to B arising from electric currents, or from magnetized bodies at a distance (not lying in the Faraday coil). The magnetization M is the contribution to B from magnetized bodies located in the Faraday coil. One way to put it is: "H is what you pay for; B is what you get." In most situations, this is exactly true. The field H is produced by current passing through copper wires, which dissipates energy as heat, and the added contribution M is obtained free from the presence of a magnetic material. This is why strongly magnetic materials like iron are so useful.

4. Classification of Magnetic Materials

The magnetic properties of materials arise almost entirely from the behavior of the electrons; the magnetic properties of the nucleus are of interest mainly for scientific purposes such as neutron diffraction. Most atoms have no net magnetic moment in the absence of an applied field. The electrons of any atom react to an applied field by developing a very small negative moment; that is, a moment directed opposite to the applied field. The moment, and hence the magnetization of the material, increases linearly with the applied field. The proportionality constant, or the ratio of magnetization to field, is called the *magnetic susceptibility* χ, and materials with a negative susceptibility are called *diamagnetic*.

The units of susceptibility are a frequent source of trouble. In SI, if volume magnetization is in $A \cdot m^2/m^3 = A/m$ and field is in A/m, the *volume susceptibility* χ_V is dimensionless. But magnetization can also be expressed in units of tesla ($\mu_0 M$), called *magnetic polarization J*; and magnetic field can be expressed in tesla ($\mu_0 H$). So there are two units for magnetization and two for field, giving four possible ratios for susceptibility, of which two (M/H and J/B) are equal and dimensionless. In *cgs* units, volume magnetization M is in emu/cm^3 and H in Oe, giving χ_V in $emu/cm^3 \cdot Oe$. The conversion is

$$\frac{\chi_V \text{ [SI] (dimensionless)}}{\chi_V \text{ [cgs] } (emu/\text{cm}^3 \cdot \text{Oe})} = 4\pi. \tag{16}$$

There is an analogous set of units for mass susceptibility, when magnetization is measured per unit mass.

A great many materials, including most organic compounds, are diamagnetic. But in other materials, especially those containing metal atoms, a positive magnetic moment appears when a field is applied. The positive moment is often large enough to overcome the diamagnetism of the core electrons and result in a net positive susceptibility. Such a material is called *paramagnetic*. In metals, a paramagnetic contribution comes from the conduction electrons, and is known as the *Pauli paramagnetism*; it is approximately independent of temperature.

Quantum theory predicts that in the transition metal series there should be a non-zero magnetic moment permanently attached to each atom except those with filled 3d shells, arising from the orbital motion of the electron or from a kind of rotation of the electron known as the *electron spin*. Experiment shows that such moments are indeed present, and when the transition-metal atoms exist as ions in relative isolation (in inorganic compounds such as oxides, sulfates, etc.) the measured values of the atomic moments agree reasonably well with the predicted values for the electron spin moment only; the orbital moment contribution is usually small and is said to be *quenched*. It is customary to express the atomic moment in units of the theoretical moment of a single electron spin; this is called the *Bohr magneton* μ_B, and has the value

$$\mu_B = \frac{eh}{4\pi m_e} = 9.274 \times 10^{-24} \text{ A} \cdot \text{m}^2 \, [SI]; \tag{17}$$
$$= 9.274 \times 10^{-21} \text{ emu } [cgs],$$

where e is the electron charge, h is Planck's constant, and m_e is the electron mass.

The existence of these atomic moments gives rise to a relatively large paramagnetism which is strongly temperature dependent. The temperature dependence arises because thermal energy acts to keep the atomic moments randomly aligned; at high temperature a given field produces less alignment of the moments and the susceptibility is lower. When there is no interaction between the atomic moments, the susceptibility is inversely proportional to temperature:

$$\chi_V = \frac{C}{T}. \tag{18}$$

This is known as the *Curie Law*, and C is the *Curie constant*. If there is an interaction tending to align the atomic moments, the susceptibility follows the *Curie-Weiss Law*

$$\chi_V = \frac{C}{(T-\theta_p)}, \tag{19}$$

where C is again the Curie constant and θ_p is called the *paramagnetic Curie temperature*. Below a temperature approximately equal to θ_p, the atomic moments align parallel, and the material becomes *ferromagnetic*.

The interaction between moments may be negative, meaning that the moments tend to align antiparallel. In this case the value of θ_p is negative, and the structure is *antiferromagnetic*. Figure 3 illustrates these three possible magnetic structures. We will return to the behavior of such materials shortly.

When the transition elements exist as metals, either pure metals or alloys, the atomic moments in general no longer agree even approximately with the calculated free-ion values. This is attributed to the fact that the electrons in metals are not bound to individual atoms, but are shared in a conduction band. As noted above, of the transition metals, only Fe, Co, and Ni are ferromagnetic, with non-integral numbers of Bohr magnetons per atom (Fe $2.2\mu_B$, Co 1.7, Ni 0.62). Cr and Mn are antiferromagnetic, although some of their alloys and compounds are ferromagnetic, notably the *Heusler alloys* Cu_2MnSn, Cu_2MnAl, and some related compositions; and chromium dioxide CrO_2.

5. Ferromagnetic Metals and Alloys

For our purposes, a ferromagnetic metal may be regarded as an array of atoms, each bearing a fixed atomic moment, and with an interaction energy called the *exchange energy* acting to align the moments parallel (Fig. 3c). At temperatures approaching absolute zero, the saturation magnetization M (all atomic moments aligned parallel) will then be given by the moment per atom times the number of atoms per unit cell, divided by the volume of the unit cell. For the case of iron,

$$M(A/m) = 2.2\mu_B \cdot 9.27 \times 10^{-24} \frac{Am^2}{\mu_B} \cdot \frac{2 \text{ atoms/unit cell}}{(2.86 \times 10^{-9})^3 \text{ m}^3/uc} = 1.75 \times 10^6 \text{ A/m} \tag{20}$$
$$= 1750 \text{ emu/cm}^3.$$

The equivalent values for Ni and Co are 0.62×10^6 and 1.45×10^6 A/m, respectively. Table 1 gives some magnetic properties of the ferromagnetic elements Fe, Co, and Ni.

As the temperature is increased from absolute zero, thermal energy acts to randomize the directions of the atomic moments, and the saturation magnetization decreases, at first slowly and then faster, dropping to zero at the *ferromagnetic Curie temperature* T_c. Fig. 4 illustrates this behavior. Curves of this kind are measured in strong applied fields, for reasons that will shortly become clear. At temperatures approaching the Curie temperature, the magnitude of the applied field significantly affects the measured values. The ideal curve of Fig. 4 (solid line) thus becomes in practice multi-valued near T_c, (dotted lines), depending on the experimental field of the measurement. This makes the exact determination of the Curie temperature uncertain. Various methods have been proposed to avoid this problem, some based on theoretical models and other purely empirical, but no generally-accepted procedure has emerged.

The behavior of an assembly of non-interacting atomic moments in an applied field was worked out before 1900 by Langevin, who found

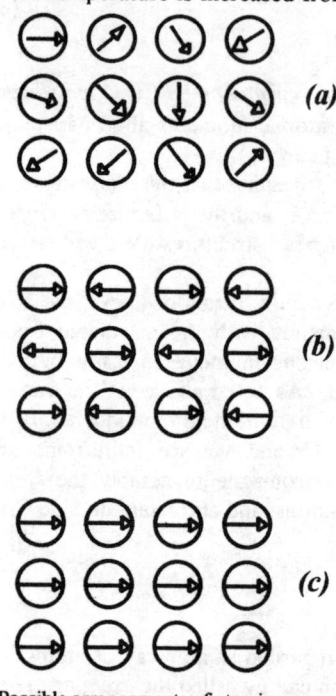

Fig. 3. Possible arrangements of atomic moments. (a) paramagnetic, (b) antiferromagnetic, (c) ferromagnetic.

$$\frac{M}{M_0} = \text{ctnh}\,\alpha - \frac{1}{\alpha}, \qquad (21)$$

Table 1. Properties of ferromagnetic elements

	Fe	hcp Co	fcc Co	Ni
M_0, A/m	1752x10³	1446x10³		520x10³
σ_0, Am²/kg	221.7	162.5		58.6
$M_{S\,20C}$, A/m	1710x10³	1431x10³		494x10³
n_B, μ_B	2.22	1.72	1.75	0.615
T_c, °C	770		1121	358

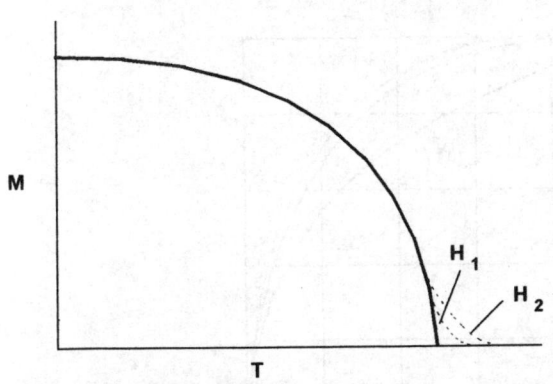

Fig. 4. Temperature dependence of magnetization. H_1 and H_2 are increasing fields.

where $a = \mu_a H/kT$; μ_a is the atomic moment, H is the applied field, k is Boltzmann's constant, and T is the absolute temperature. The function $\ctnh \alpha - \frac{1}{\alpha} = L(\alpha)$ is known as the *Langevin function*. The classical Langevin theory allows the atomic moments to point in any direction, but quantum theory says that only a finite number of orientations are allowed, each corresponding to a quantum energy state. In the simplest case, (spin quantum number J=½) only two orientations are permitted, with the moment either parallel or antiparallel to the field. In this case, the Langevin equation simplifies to

$$\frac{M}{M_0} = \tanh \alpha. \qquad (22)$$

The effect of an interaction energy acting to align the moments parallel to one another was treated by Weiss. He assumed that this effect could be treated as a fictional field, proportional to the net magnetization M: $H_M = NM$. H_M is called the *molecular field*, and N is the *molecular field constant*. The terminology is not ideal, since we are dealing with atoms rather than molecules, but is well-established. To account for the observed behavior of ferromagnetic materials, the molecular field H_M must be very much larger than ordinary laboratory applied fields. If the applied field is neglected in comparison to the molecular field, we have simply

$$\frac{M}{M_0} = \tanh\left(\frac{\mu_a NM}{kT}\right) = \tanh\left(\frac{M}{M_0}\right)\left(\frac{kT}{\theta}\right), \qquad (23)$$
$$\theta = \mu_a NM_0.$$

This equation describes the temperature dependence of the saturation magnetization in the case of spin quantum number $J = \frac{1}{2}$, and as seen in Fig. 5, it agrees reasonably well with the experimental data for iron and nickel.

Above the Curie temperature, ferromagnetic materials behave as strong paramagnets, with susceptibility χ given approximately by $C/(T-T_c)$, the Curie-Weiss Law.

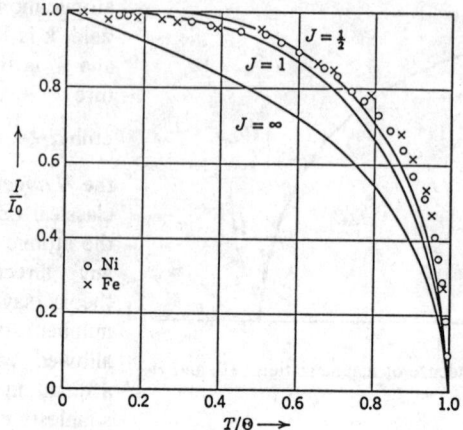

Fig. 5. Theoretical and experimental temperature dependence of magnetization. $J=\infty$ corresponds to the classical (Langevin) case. Data from R. Becker and W. Döring, *Ferromagnetismus*, Springer (1939).

5.1 Crystalline Anisotropy

It is observed that in single-crystal ferromagnets, the magnetization M has preferred crystallographic directions. In iron, the lowest-energy state occurs when M is parallel to <001>, while in nickel, the lowest-energy state is for M parallel to <111>. The energy per unit volume can be expressed in terms of the direction cosines α_1, α_2, α_3 of M relative to the crystal axes as

$$E_K = K_1(\alpha_1^2\alpha_2^2 + \alpha_2^2\alpha_3^2 + \alpha_3^2\alpha_1^2) + K_2(\alpha_1^2\alpha_2^2\alpha_3^2) \tag{24}$$

The quantity E_K is called the *magnetocrystalline anisotropy*, and K_1 and K_2 are the first and second *crystal anisotropy constants*. They are usually determined by measuring the torque exerted on a circular or spherical single crystal as it is rotated in a strong applied field; the torque L is given by

$$L = -\frac{dE_K}{d\theta} \tag{25}$$

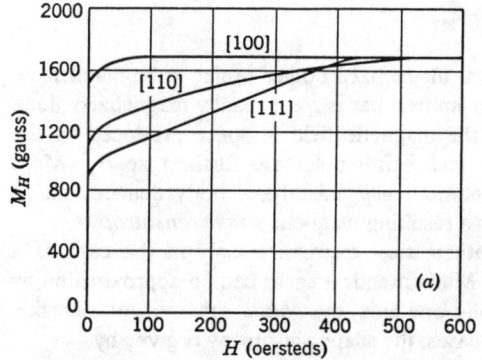

Fig. 6. Measured magnetization curves in the three primary crystallographic directions in iron. Data from K. Honda and S. Kaya, *Sci. Repts. Tohoku Imp. Univ.*, **15**, 721 (1926).

In most cases, K_2 is small compared to K_1. The effect of crystal anisotropy is most easily seen from curves of M vs. H (*magnetization curves*) measured in different crystallographic directions (see Fig. 6). The magnetization rises rapidly to its saturation value M_s when the field is applied in the *easy direction*, but approaches M_s slowly, by rotation, when the field is applied in other directions. The crystallographic direction requiring the highest field to reach saturation is called the *hard direction*.

It is sometimes useful to treat the anisotropy as an equivalent field, called the *anisotropy field* H_K, acting to keep the magnetization pointing in the low energy or *easy direction*. The anisotropy field is defined by

$$H_K = -\left(\frac{1}{M_s}\right)\left(\frac{d^2 E_K}{d\theta^2}\right), \tag{26}$$

where θ is a small angle of rotation away from the easy direction, and K_2 has been neglected. For $E_K = K \sin^2 \theta$, $H_K = 2K/M_s$. Table 2 gives values of K_1 and H_K for Fe, Ni, and Co.

In a random polycrystalline material, the macroscopic effects of crystal anisotropy average to zero, but the anisotropy within each crystal (grain) still affects the magnetic behavior.

Table 1. Magnetic anisotropy and anisotropy field at room temperature

Material	K_1 (J/m³)	H_K (A/m)	K_1 (erg/cm³)	H_K (Oe)
Fe	50x10³	46x10³	500x10³	570
Ni	-4.5x10³	16x10³	-45x10³	200
Co*	530x10³	590x10³	5.3x10⁶	7400

*Co at room temperature is hexagonal, and the anisotropy energy is given by $E_K = K_1 \sin^2\theta$, where θ is the angle between the magnetization and the c-axis.

5.2 Shape Anisotropy and Demagnetizing Field

It is common experience that nonspherical magnetized bodies prefer to be magnetized in their long direction. A compass needle or an iron bar is most readily magnetized along its long axis. This comes about because the magnetic field in space produced by the needle or the bar is least when its North and South poles are furthest apart. More formally, the *magnetostatic energy* of a geometrically anisotropic body depends on its direction of magnetization, and the body has a resulting magnetic *shape anisotropy*.

The shape anisotropy has an exact mathematical expression only in the case of a sample which is an ellipsoid of revolution. Many practical cases can be approximated by ellipsoids having circular cross-sections along one axis: the *oblate* spheroid, or pancake; and the *prolate* spheroid, or cigar. In these cases, the shape anisotropy is given by

$$E_{sh} = \frac{M_s^2}{2}\Delta D = \frac{M_s^2}{2}(D_{par} - D_{perp}), \qquad (27)$$

where D_{par} and D_{perp} are the *demagnetizing factors* for magnetization parallel and perpendicular to the long axis, which depend on the degree of eccentricity of the ellipsoid. Values of demagnetizing factors for a range of ellipsoids are given in Table 3. The table gives only the factor parallel to the long axis, but the factors along the other axes can in general be found from the relationship $D_x + D_y + D_z = 1$.

Demagnetizing factors, or more broadly demagnetizing effects, are important in magnetic measurements and in many applications of magnetic materials. Any magnetized body other than a closed ring magnetized circumferentially creates a magnetic field in space, sometimes called the *stray field*, even though it is not necessarily unwanted (see Fig. 7). The same field acts within the magnetized body, always in a direction opposite to the direction of magnetization. Here it is called the *demagnetizing field*. It is a real field, and may be large. The magnitude of the demagnetizing field in a sample of given shape and direction of magnetization is given by the product of the appropriate demagnetizing factor and the magnetization:

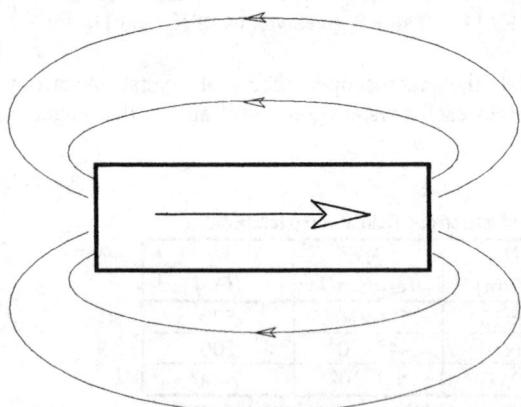

Fig. 7. Stray field produced by a magnetized body.

$$SI: \quad H_D = -DM \qquad cgs: H_D = -4\pi DM. \tag{28}$$

The negative sign indicates that the demagnetizing field is opposite to the direction of magnetization. The applied field must exceed the maximum demagnetizing field (DM_S) in order for a sample to reach magnetic saturation. See Fig. 8.

Table 3. Demagnetizing factors parallel to long axis of ellipsoids.

Axial Ratio (length/dia)	Prolate Ellipsoid	Oblate Ellipsoid
0	1.0	1.0
1	0.333	0.333
2	0.1735	0.2364
5	0.0558	0.1248
10	0.0203	0.0696
20	0.00675	0.0369
50	0.00144	0.01472
100	0.000430	0.00776
200	0.000125	0.00390
500	0.0000236	0.001567
1000	0.0000066	0.000784
2000	0.0000019	0.000392

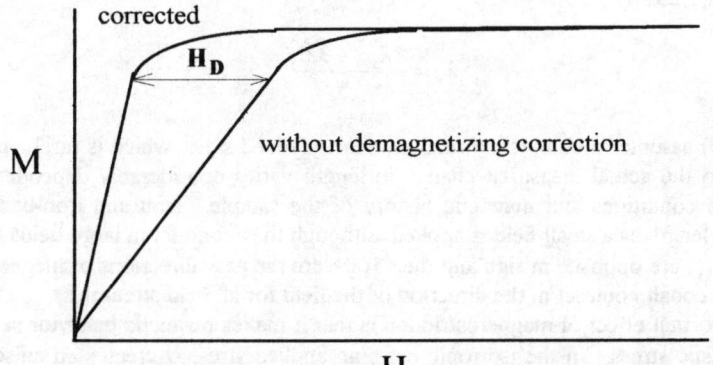

Fig. 8. Magnetization curves with and without correction for demagnetizing field.

5.3 Magnetostriction

In general, a magnetic material undergoes a small crystallographic distortion as it cools through the Curie temperature, cubic materials becoming very slightly tetragonal. The distortion is in most cases too small to be measured by conventional x-ray techniques, but it appears as a change in specimen dimensions when the magnetization is changed in magnitude or rotated. The simplest equation that describes the distortion of a cubic crystal as the magnetization is rotated with respect to the crystal axes is

$$\frac{\Delta \ell}{\ell} = \lambda_s = \tfrac{3}{2} \lambda_{100}\left(\alpha_1^2 \beta_1^2 + \alpha_2^2 \beta_2^2 + \alpha_3^2 \beta_3^2 - \tfrac{1}{3}\right) + 3\lambda_{111}\left(\alpha_1 \alpha_2 \beta_1 \beta_2 + \alpha_2 \alpha_3 \beta_2 \beta_3 + \alpha_3 \alpha_1 \beta_3 \beta_1\right) \quad (29)$$

where λ_{100} and λ_{111} are the *magnetostriction constants*, the α_n are the direction cosines of the saturation magnetization, and the β_n are the direction cosines of the measured strain. (The factors 3/2 and 1/3 result from the fact that the reference state of zero strain is taken to be the state where the sample is divided into a large number of regions magnetized in random directions.) Magnetostrictive strains are usually small, of order 10^{-5}, and difficult to measure; therefore magnetostriction constants are often not very accurately known.

For polycrystalline materials, magnetized from the demagnetized state to saturation, the strain measured in the direction of magnetization is known as the *engineering magnetostriction*. It is usually written

$$\frac{\Delta \ell}{\ell} = \frac{3}{2} \lambda_s \left(\cos^2 \theta - \frac{1}{3}\right), \quad (30)$$

where λ_s is the *saturation magnetostriction* and θ is the angle between the magnetization and the strain axis Subject to some assumptions, λ_s is related to the single crystal constants of Eq. (29) by

$$\lambda_s = \frac{2\lambda_{100} + 3\lambda_{111}}{5} \quad (31)$$

Eq. (30) assumes a completely random demagnetized state, which is rarely attained in practice, so the actual measured change in length varies considerably depending on the preparation conditions and magnetic history of the sample. Iron and iron-based alloys increase in length as a small field is applied, although they contract in large fields (because λ_{100} and λ_{111} are opposite in sign and the <100> are the easy directions of magnetization). Nickel and cobalt contract in the direction of the field for all field strengths.

An important effect of magnetostriction is that it makes magnetic behavior sensitive to applied elastic stress. In the isotropic case, an applied stress σ creates an anisotropy of magnitude

$$E_\sigma = \frac{3}{2}\lambda_s \sigma \cos^2 \theta, \tag{32}$$

where θ is the angle between the magnetization and the stress axis. So the application of a fairly modest applied stress of 150 MPa (22,000 lb/in²) to a material with $\lambda_s = 25 \times 10^{-6}$ creates an anisotropy of about 5.6×10^3 J/m³, comparable to the crystal anisotropy of nickel.

5.4 Magnetic Domains

The existence of the molecular field would seem to require that a ferromagnetic material remain fully magnetized at all temperatures below the Curie temperature. Clearly, this is not the case; strongly magnetic materials, such as soft iron, can be *demagnetized*, with net magnetization zero. The explanation for this behavior was suggested by Weiss, who proposed that a ferromagnetic material is subdivided into regions, called *domains*, each of which is magnetized to saturation, but with local domain magnetizations pointing in various directions with a net value lying anywhere between positive and negative saturation, including a value of zero. This seemed plausible, but it was about fifty years before direct observation of domains confirmed the hypothesis.

A magnetized body of any shape other than a closed ring creates a magnetic field in space, as shown in Fig. 8a. There is an energy density associated with this field (the *magnetostatic energy*), so the system can lower its energy by subdividing into two or more domains, as shown in Fig. 8b and c. The energy cost of this process is the creation of boundaries between domains, known as *domain walls*. If the

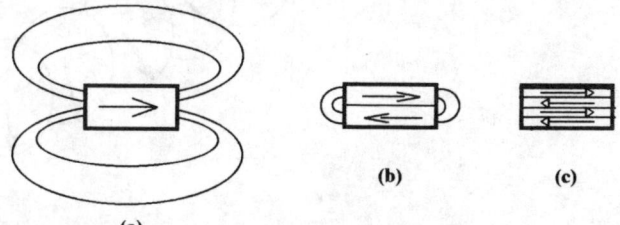

Fig. 8. Reduction in external field by the creation of domains.

system can nucleate and grow domain walls (which is usually not a problem), the number of walls will increase until the decrease in magnetostatic field energy produced by the addition of another wall is just balanced by the energy cost of the added wall. In practice, there are usually many domains in each grain or crystal of a polycrystal, and there are a large number of equivalent-energy domain configurations.

Weiss was unaware of the existence of magnetic anisotropy, and so his domain picture was highly schematic. Later it was realized that each domain should be magnetized in an easy crystallographic direction, and that the domain walls will normally be planes lying either parallel or at 45° to the easy directions. Therefore a relatively simple domain pattern is observed in the case that the sample surface is parallel to an easy crystallographic direction.

5.5 Domain Wall Energy and Thickness

The behavior of domain walls largely governs the behavior of magnetic materials, so it is important to understand the nature and energy of the wall. Consider a line of atoms lying perpendicular to a domain wall, as in Fig. 9. The atomic moments (usually referred to as *spins*) in the wall are subject to two competing torques. The exchange energy acts to keep each spin parallel to its neighbors, but does not prefer any crystallographic direction. The exchange energy can be written

$$E_{ex} = A\left(\frac{d\phi}{dx}\right)^2, \tag{33}$$

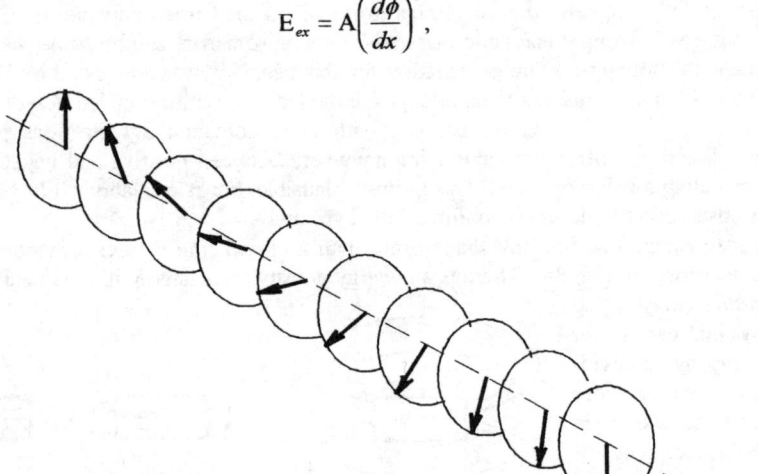

Fig. 9. Spin structure of a 180° domain wall. Each arrow represents the moment on a single atom. The line of atoms lies perpendicular to the domain wall. A domain magnetized vertically upward is at the left, a domain magnetized vertically downward at the right.

where ϕ is the angle of the spin with respect to some arbitrary axis, x is the position of the atom (spin) in the wall, and $(d\phi/dx)^2$ comes from the small-angle expansion of $\cos \phi$. The quantity A is a measure of the exchange energy, called the *exchange stiffness*. The torque acting between neighboring spins is given by

$$L_{ex} = \frac{dE_{ex}}{dx} = 2A\frac{d\phi}{dx}\frac{d\left(\frac{d\phi}{dx}\right)}{d\phi} = 2A\frac{d^2\phi}{dx^2} \tag{34}$$

If the angle between neighboring spins were constant ($d\phi/dx$=const), the torque acting on each spin would be positive due to one neighbor and equal but opposite from the other, canceling to zero. So there is an exchange torque only when the angles are unequal, that is, when the second derivative is non-zero.

The opposing torque arises from the magnetocrystalline anisotropy, which acts to turn each spin toward the nearest easy direction. In the case of a wall separating two oppositely-magnetized domains (called a *180° wall*) in a hexagonal (or uniaxial) crystal with <0001> easy directions, this anisotropy torque L_K can be written

$$L_K = \frac{dE_K}{d\phi} = K_1 \sin 2\phi \tag{35}$$

At equilibrium, the net torque acting on each spin is zero, so

$$2A \frac{d^2\phi}{dx^2} = K_1 \sin 2\phi \tag{36}$$

Noting that

$$\frac{d^2\phi}{dx^2} = \frac{1}{2} \frac{d}{d\phi}\left(\frac{d\phi}{dx}\right)^2, \text{we can write: } d\left(\frac{d\phi}{dx}\right)^2 = \frac{K_1}{A} \sin 2\phi \, d\phi \tag{37}$$

or

$$\left(\frac{d\phi}{dx}\right)^2 = \int \frac{K_1}{A} \sin 2\phi \, d\phi = \frac{K_1}{A} \sin^2 \phi$$

$$\frac{d\phi}{dx} = \sqrt{\frac{K_1}{A}} \sin \phi \tag{38}$$

which defines the variation of ϕ with position x. Rewriting,

$$dx = \sqrt{\frac{A}{K_1}} \frac{d\phi}{\sin \phi}$$

$$\int dx = x = \sqrt{\frac{A}{K_1}} \int_0^\pi \frac{d\phi}{\sin \phi} = \sqrt{\frac{A}{K_1}} \ln\left|\tan \frac{\phi}{2}\right|. \tag{39}$$

Taking x=0 at $\phi=\pi/2$, the variation of φ with x is shown in Fig.10. Note that the width of the wall is mathematically infinite, but that most of the rotation occurs over a distance of about $\pm 2\sqrt{\frac{A}{K}}$. The quantity $\sqrt{\frac{A}{K}}$ is called the *wall thickness parameter*.

The domain wall energy γ is given by the integral of the anisotropy and exchange energies over the thickness of the wall:

$$\gamma = E_K + E_{ex} = \int_{-\infty}^{+\infty} \left[A\left(\frac{d\phi}{dx}\right)^2 + K\sin^2\phi \right] dx$$

$$= \int \left[K\sin^2\phi + K\sin^2\phi \right] dx \qquad (40)$$

$$= \int 2K\sin^2\phi \, dx.$$

Recalling that $dx = \sqrt{\dfrac{A}{K}} \dfrac{d\phi}{\sin\phi}$,

$$\gamma = \int 2K\sin^2\phi \sqrt{\frac{A}{K}} \frac{1}{\sin\phi} d\phi = 2K\frac{\sqrt{A}}{\sqrt{K}} \int \sin\phi \, d\phi$$

$$= \sqrt{KA} \int_0^\pi \sin\phi \, d\phi = 2\sqrt{KA}. \qquad (39)$$

The quantity $\sqrt{KA}$ is the *wall energy parameter*. Domain wall energies in typical magnetic materials are a few mJ/m² (or erg/cm²), and wall thicknesses are several hundred atom diameters. A normal domain wall is therefore large compared to a vacancy, or to the stress field of an isolated dislocation, but small compared to most second-phase particles or grains. Both the domain wall thickness and energy depend on the values of the anisotropy K and the exchange stiffness A. The value of K can be directly measured, as noted above. There is, however, no direct measure of exchange stiffness. An approximate value can be deduced from the Curie temperature (clearly a material with a large exchange interaction should have a high Curie temperature), but this is not very satisfactory. The best experimental values of A are probably obtained by deducing a value for the domain wall energy γ and working backward from Eq. 41.

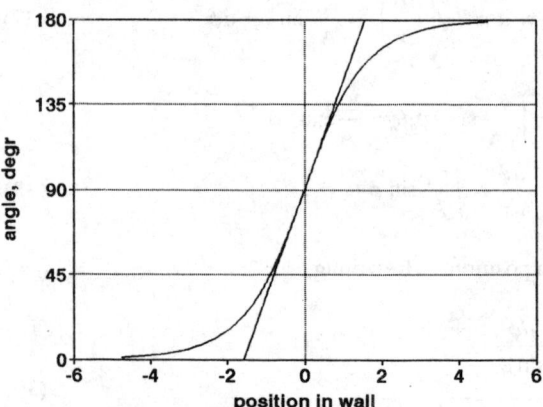

Fig. 10. Variation of angle of atomic moment φ with position in domain wall. Wall thickness in units of √(A/K).

5.6 Magnetization Processes

The value of the saturation magnetization depends only on the magnitude of the moment on the magnetic atoms, and the number of magnetic atoms per unit volume (and on the temperature). The usefulness of a magnetic material depends strongly on the way in which the material reverses the direction of its saturation magnetization. If it reverses easily (in low fields) and with little energy loss, the material may be useful in ac applications. Such a material is called a *soft magnetic material*, although it is not necessarily soft mechanically. If it reverses with difficulty (only in high fields), it is a potential *permanent magnet*, or *hard magnetic material*. It is the behavior of the domain walls that determines which category a magnetic material belongs to. In high-purity, well-annealed materials, the domain walls move easily. A small applied field moves domain walls so as to increase the volume of material magnetized with a component of magnetization parallel to the field (Fig. 11). This leads to a steep slope of a plot of M vs H or B vs H (both are used), and a correspondingly high value of *susceptibility* (M/H) or *permeability* (B/H), as illustrated in Fig. 12. Note that in contrast to paramagnetic or diamagnetic materials, the susceptibility and permeability of strongly magnetic materials are not constants, but depend on the point on the curve where the determination is made. The magnetic quality of soft magnetic materials is sometimes specified by the value of *maximum permeability*.

When the domain walls have moved as far as they can, each crystal or grain of the sample will be a *single domain* magnetized along its easy crystallographic direction which is most nearly parallel to the applied field. Further increase in net

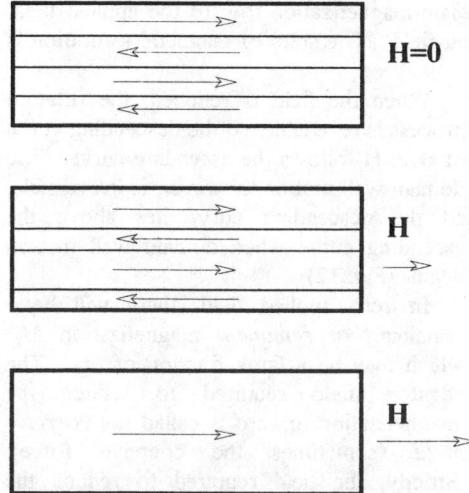

Fig. 11. Magnetization by domain wall motion.

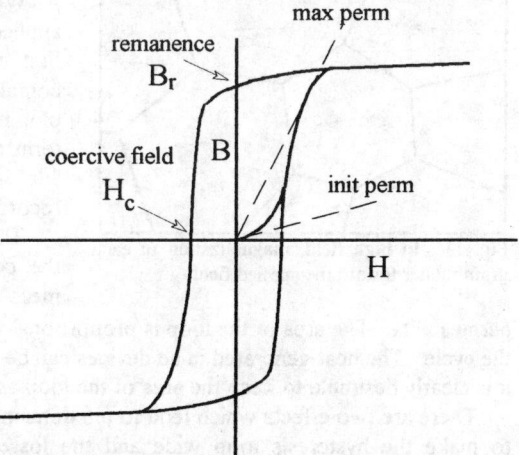

Fig. 12. Hysteresis loop, B vs H, showing initial and maximum permeability, remanence, and coercive field.

magnetization occurs by *rotation* of the domain magnetization toward the applied field, which occurs relatively slowly with increasing field. The state of magnetic saturation is approached asymptotically (Fig. 13).

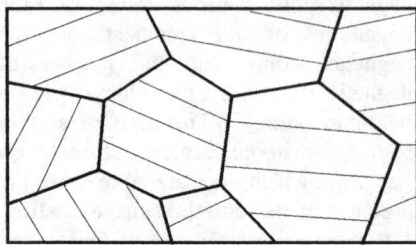

Fig. 13a. Polycrystalline material, each grain demagnetized and containing multiple domains.

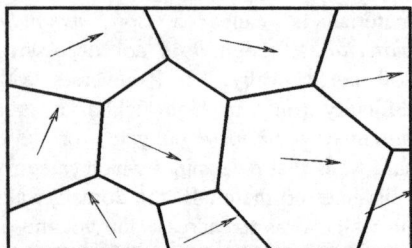

Fig. 13b. Domain walls removed, each grain magnetized to saturation in local easy direction.

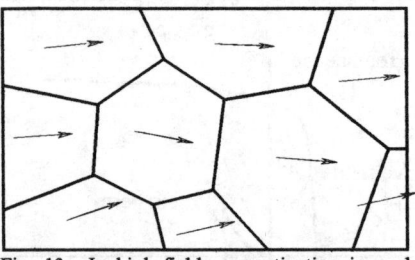

Fig. 13c. In high field, magnetization in each grain rotates toward the applied field.

When the field is reduced, the rotation process is reversible, so the descending curve of B vs H follows the ascending curve. The domain wall motion, however, is irreversible; so the descending curve lies above the ascending curve when domain wall motion begins (Fig. 12)

In zero applied field, there will be a remaining or *remanent* magnetization M_r, which may be a large fraction of M_s. The negative field required to reduce the magnetization to zero is called the *coercive field* (sometimes the coercive force). Strictly, the field required to reduce the magnetization M to zero is the *intrinsic coercive field* H_{ci} or $_iH_c$ while the ordinary or induction coercive field H_c is the field required to reduce the magnetic induction or flux density B to zero. The distinction is not significant except in the case of strong permanent magnets, where it may be substantial.

When a negative field greater than H_c is applied, the magnetization reverses direction, first by wall motion, then by rotation. A complete traverse of the M vs H (or B vs H) plot is called a *hysteresis loop*; usually this term refers to a slow cycle, loosely called a "dc" loop, but the loop may be traversed and recorded at any speed.

The width of the loop is equal to twice the coercive field, which is often a better measure of magnetic quality than the permeability. The area of the loop is proportional to the energy lost (as heat) in traversing the cycle. The heat generated in ac devices can be the limiting factor in their operation, so it is clearly desirable to keep the area of the loop as small as possible.

There are two effects which tend to pin domain walls or retard their motion, and so act to make the hysteresis loop wide and the losses large. First, domain walls may be attracted to or repelled by inhomogeneities in the material. Either interaction tends to interfere with wall motion. Second phase particles of a non-magnetic material will act to pin domain walls, since an area of wall is eliminated when the wall intersects the particle. Grain boundaries impede wall motion, since the wall must change direction at the boundary. Plastic deformation creates inhomogeneous dislocation arrays that impede domain wall motion.

Second, a moving domain wall in a conducting material generates local electrical currents (*eddy currents*), according to the Faraday equation. These currents generate fields that oppose the wall motion, and which dissipate energy by Joule heating $P = i^2 R$. In the "classical" limit of very small domain size, corresponding to uniform magnetization, the power loss due to eddy currents in a sheet material is given by

$$P = C \frac{d^2 f^2 B_{max}^2}{\rho}, \qquad (42)$$

where d is the sheet thickness, f is the ac frequency, B_{max} is the maximum flux density during the cycle, and ρ is the electrical resistivity; C is a constant depending on the units used. Because it ignores the existence of domains, this equation underestimates the actual losses, often by substantial amounts, but it illustrates four important trends: losses increase with the square of sheet thickness, frequency, and maximum induction; losses decrease with increasing resistivity.

To reduce eddy-current losses, conducting magnetic materials are made in the form of thin sheets or laminations, whose thin dimension is perpendicular to the axis of magnetization. This increases the length and hence the resistance of the eddy current paths. Alloy additions are often made to increase the electrical resistivity.

Thus, large-scale motors and transformers are commonly made from sheets about 0.01 inch (250 μm) thick of Fe+3%Si. The silicon increases the electrical resistivity of iron by almost a factor of five, at the cost of slightly lower saturation magnetization. The silicon also suppresses the formation of the fcc γ phase, which makes possible a rolling/heat treatment cycle that produces very large grains ($\approx$1 cm) with a strong crystallographic texture aligning the easy [001] axes with the rolling direction.

The best crystalline soft magnetic metals are nickel-iron alloys called *permalloys*. They can be made with very high permeability and low losses, but have saturation magnetization less than half that of iron. The highest saturation magnetization is found in Fe-Co alloys near the 50-50 composition; the value of M_s is about 10% higher than pure iron, and the Curie temperature is over 200 K higher, at about 1000° C.

A number of ferromagnetic alloys, mostly containing 15 to 25 at% B+Si, can be made *amorphous* or *glassy*; that is *non-crystalline*. This is done by very rapid cooling ($\approx 10^6$ K/sec) from the melt, producing ribbons or sheets with thicknesses less than 35 μm. In the absence of crystal structure, the anisotropy of these materials is very small, and the domain walls are consequently wide and of low energy. Furthermore, the absence of grain boundaries leads to large domains, and the high alloy content gives high electrical resistivity, although at the cost of lowered saturation magnetization. These are in many ways ideal magnetic materials for applications at power and audio frequencies. Their use is limited by cost, by mechanical properties (they are very hard, and lose ductility when annealed), and by the fact that their maximum thickness is so small.

6. Ferrimagnetic Materials

A special case of antiferromagnetism arises when the structure contains two kinds of magnetic atoms with different magnetic moments. A negative exchange interaction aligns the moments antiparallel, but the net moment may be non-zero, as shown in Fig. 14. Such materials are called *ferrimagnets*. The best-known case is *magnetite*, Fe_3O_4. The chemical formula may be rewritten as $Fe_2O_3 \cdot FeO$ to emphasize that two kinds of iron ions are present, Fe^{2+} and Fe^{3+}, which have different atomic moments (4 and 5 μ_B, respectively). In this particular case, the moments of the Fe^{3+} ions are aligned antiparallel and cancel, leaving a net moment of one Fe^{2+} ($4\mu_B$) per formula unit. There is a large family of related magnetic oxides, all having the *spinel* structure and the general formula MFe_2O_4, known collectively as *ferrites* (not to be confused with metallurgical ferrite, or bcc α-iron). The crystallographic reasoning that connects the location and interaction of the magnetic ions with the measured saturation magnetization is somewhat complicated, but quite successful. The details are given in most texts on magnetic materials.

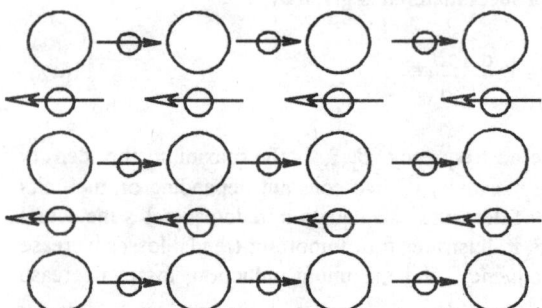

Fig. 14. Ferrimagnetic structure. Large open circles are non-magnetic atoms such as oxygen. There are two kinds of magnetic atoms, with different moments per atom, aligned antiparallel. The net magnetization is not zero.

The great advantage of the ferrites over metallic magnets is their high electrical resistivity, which greatly reduces the eddy currents and their associated energy loss. The ferrites are used for magnetic applications where the frequency exceeds a few tens of kHz. They need not be subdivided into thin laminations. The saturation magnetization, however, is greatly reduced from that of iron, both because not all the atomic moments are aligned parallel and because a substantial fraction of the volume is occupied by non-magnetic atoms. The highest room-temperature magnetization available in ferrites is about 0.6T [6kG], less than 1/3 the value of iron. The most widely used ferrites are $(NiZn)Fe_2O_4$ and $(MnZn)Fe_2O_4$, known as nickel-zinc ferrite and manganese-zinc ferrite.

Anisotropy, magnetostriction, domain walls, and domain behavior are much the same in ferrites as in metals. Because ferrites are normally made by pressing and sintering oxide powders, they typically contain several percent porosity which acts to pin domains walls. Therefore the low-frequency magnetic properties are not especially good. At very high frequencies, so-called *intrinsic losses* appear, whose physical origin is not entirely clear; these limit the maximum frequency at which the ferrites can operate successfully.

Table 4 gives approximate values of magnetic properties for some commonly-used soft magnetic materials.

Table 4. Soft magnetic materials

Material	B_{sat} (T)	H_c (A/m)
commercial purity iron	2.17	80
low carbon steel	2.17	100
3% Si steel, unoriented	2.0	40
3% Si steel, oriented	2.0	8
Mo-Permalloy (79%Ni, 17%Fe, 4%Mo)	0.8	1
MnZn ferrite	0.4	5
NiZn ferrite (high electr. resistivity)	0.3	80
amorphous $Fe_{81}B_{13.5}Si_{3.5}C_2$ (Allied-Signal Metglas 2605SC)	1.6	3

7. Permanent Magnets

In a soft magnetic material to be used under ac excitation, the aim is to make the domain walls move as easily as possible. In a permanent magnet, or hard magnetic material, the aim is just the opposite; the magnetization should remain constant even in a strong reverse field, so the domain walls should be immobile. High values of magnetocrystalline anisotropy raise the domain wall energy and decrease its thickness; both effects tend to make wall pinning sites more effective. Anisotropy generally increases as crystal symmetry decreases, so good permanent magnet materials often have hexagonal or other non-cubic symmetry.

A fundamentally different approach to permanent magnets arose from the idea that if a magnetic particle is made small enough, the appearance of even a single domain wall is energetically unfavorable. In zero field, such a *single-domain particle* is always magnetized to saturation, and has no obvious way of reversing its magnetization except by uniform rotation against crystalline and/or shape anisotropy. The calculated particle size for single domain behavior depends on the particle shape as well as on its saturation magnetization; for ellipsoidal Fe, Ni, and Co particles the critical size (length of the long axis) is in the range 250 to 2500 Å (0.025 to 0.25 μm). The single-domain theory led to a number of advances in permanent magnet materials, although it appears that ideal single-domain particles are rarely achieved in practice. The best permanent magnet materials are usually made by pressing and sintering particles that are rather larger than the calculated single-domain size limit.

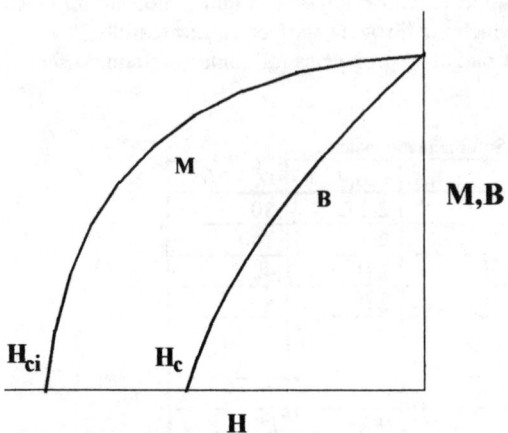

Fig. 15. Demagnetization curves of a permanent magnet material. The two curves are related by B=H+M; in the second quadrant, shown here, M>0, H<0. Note that in this case $H_{ci}>H_c$.

A good permanent magnet should, of course, have high saturation magnetization and high coercive field. It should in addition have high remanence, and should be insensitive to temperature changes. Permanent magnets are always used to create a field in space, and so they are always subject to their own demagnetizing field. In use, therefore, the magnetic material of the magnet is operating in a negative field, and the important part of the hysteresis loop is the second quadrant, with B (or M) positive and H negative (Fig. 15). This figure shows the difference between the B and M hysteresis loops, and the difference between the intrinsic coercive field H_{ci} and the ordinary coercive field H_c.

The product (BH) is known as the *energy product*; it is a measure of the volume of permanent magnet material required to produce a given field in a specified region of space (higher energy product means less material required). The value of (BH) varies from point to point along the B vs H plot in the second quadrant (Fig. 16). The magnetic material is most efficiently used if the working point coincides with the maximum value of (BH), called the *maximum energy product* $(BH)_{max}$. The numerical value of $(BH)_{max}$ is the most widely-used figure of merit for permanent magnet materials.

The first permanent magnet materials (after the naturally-occurring lodestones) were hardened steels, with varying alloy additions. They were neither very strong nor very stable magnetically, and were replaced in the 1930s by the *alnicos*, a family of complex two-phase alloys containing ALuminum, NIckel, and CObalt in addition to iron (and often Nb and Ti as well). Somewhat later a family of hexagonal ferrimagnetic oxides, the *barium* and *strontium ferrites*, were developed; they combine high coercive fields with low cost, and are the most widely-used permanent magnet materials today despite their relatively low saturation magnetization.

More recently, a number of ferromagnetic compounds of rare-earth elements with transition metals have been discovered. The rare-earth elements (atomic numbers 57-71) are mostly magnetic, but only below room temperature; however, a number of their intermetallic compounds with Fe and Co have Curie temperatures well above room temperature. They also commonly have very high magnetocrystalline anisotropy, arising mainly from the rare earth atoms and from their hexagonal or orthorhombic crystal structures. The three most commonly used compositions are $SmCo_5$, Sm_2Co_{17} (with various alloy additions), and $Nd_2Fe_{14}B$. Table 5 gives some representative magnetic properties of permanent magnet materials.

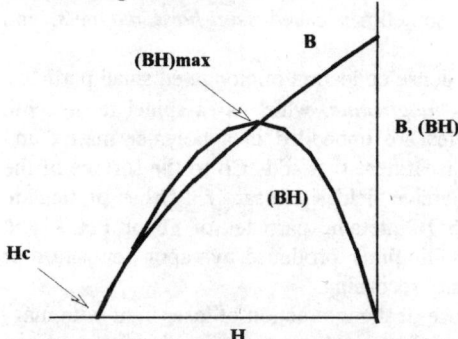

Fig. 16. The energy product (BH) is obtained by taking the product BH at each point along the B vs H curve in the second quadrant (B>0,H<0). The graph shows B and (BH) plotted vs H. For the most efficient use of material, the magnet should operate at the point where (BH) has its maximum value $(BH)_{max}$.

Since permanent magnet materials are highly anisotropic, and since in service the magnetization is always in a fixed direction, it is advantageous to make magnets with a strong crystallographic texture. This can be done invarious ways; directional solidification can be used for alnico. The rare-earth magnets are usually prepared in the form of single-crystal particles with sizes below 10 μm, and the loose powder is placed in a die and subjected to a strong dc magnetic field. The field physically rotates the powder particles so that their easy magnetic axes are parallel to the field, and while the field is applied the particles are compressed into a solid. Subsequent heating causes the particles to sinter together, but retains the crystallographic alignment.

Table 5. Typical values of magnetic properties of common permanent magnet materials

Material	B_r (T)	H_c (kA/m)	$(BH)_{max}$ (kJ/m^3)
cobalt steel	1.0	12.5	6
isotropic alnico	0.6-0.8	60-100	40-60
anisotropic alnico	1.1-1.4	60-120	50-70
Ba or Sr. ferrite	0.4	180-240	35
$SmCo_5$	0.85	650	150
Sm_2Co_{17} $(Sm_2(Co_{.68}Cu_{.10}Fe_{.21}Zr_{.01})_{14.8})$	1.1	525	250
$Nd_2Fe_{14}B$	1.25	900	300

8. Recording Materials

Magnetic materials are widely used for recording of analog and digital information: products include audio, video, and digital tapes; floppy disks; hard disks; and magneto-optical disks. The information is stored as magnetization varying in magnitude and/or direction in a band or track on the magnetic material, usually written by a very localized field from the gap of a *recording head*. The recording medium must retain the signal in the presence of various external and internal fields, but at the same time must be erased and rewritten with the relatively modest fields available from a recording head. The required material is neither a good soft magnetic material nor a good permanent magnet, but something in between. Such materials are sometimes called *semi-hard magnets*, and sometimes simply called *recording materials*.

The most common recording material is a dense collection of elongated small particles, in the single-domain size region, of γ-Fe_2O_3, *maghemite*, which is a spinel ferrite with some unoccupied oxygen sites. The particles are imbedded in a polymer matrix and supported on a flexible or rigid substrate. A treatment that adds Co to the surface of the particles can increase the anisotropy and coercive field significantly. Other particulate recording materials include ferromagnetic CrO_2, metallic particles of Fe or FeCo, and ferrimagnetic barium ferrite. Continuous metallic films, produced by vapor deposition or sputtering, are increasingly used for high-density recording.

Magneto-optic recording materials make use of the interaction of laser light with magnetized matter. A major advantage of this technique is that no physical contact with the recording medium is required; only a tiny beam of laser light hits the recording surface. This technology is developing rapidly, with new materials being investigated.

References

The following are brief introductions:

J. Crangle, *The Magnetic Properties of Solids*, Edward Arnold (1977).
J. Jacubovics, *Magnetism and Magnetic Materials*, Institute of Metals, London (1992)

General texts:

B. J. Cullity, *Introduction to Magnetic Materials*, Addison Wesley (1972). Well-written, materials science (as opposed to physics) orientation.
S. Chikazumi, *Physics of Magnetism*, Wiley, (1964). Introductory; more emphasis on physics.
David Jiles, *Introduction to Magnetism and Magnetic Materials*, Chapman and Hall (1991).

Journals and Conferences:

There are two major annual conferences: *Magnetism and Magnetic Materials*, with papers published in the *Journal of Applied Physics*, and *Intermag*, papers published in the *IEEE Transactions on Magnetism*.
The *Journal of Magnetism and Magnetic Materials* (North Holland) also publishes papers from some conferences.

Magnetic Materials 441

Problems

1. Using the values of saturation magnetization in Table 1, confirm the atomic moments of Fe, Co, and Ni (in Bohr magnetons).

2. A cylindrical sample of Ni measures 2.5 mm in diameter and 100 mm long. Calculate the demagnetizing field when the rod is magnetized to saturation a) parallel, and b) perpendicular to its long axis.

3. If a coil of 100 turns of insulated copper is wrapped around the rod of Problem 2, and the magnetization is switched from $+M_S$ to $-M_S$ in a time of 0.1 sec, find the average voltage induced in the coil.

4. Consider a single crystal disk of iron, with [001] normal to the disk surface. A strong field H is applied parallel to the plane of the disk, and parallel to the [010]. The disk is then rotated about the [001] axis, with θ measuring the angle between the field and the [010]. Apply Eq. 25 to Eq. 24, and calculate the torque acting on the sample as a function of the angle θ. [Note: you may assume that the field is sufficiently strong so that $M=M_S$, and M is parallel to H.]

5. In rare-earth permanent magnet materials, the crystal anisotropy K may be 40 times higher than in iron, although the exchange constant A is about 1/3 lower. Compare the relative domain wall energy and thickness of iron and of a rare-earth permanent magnet.

6. An ordinary mild steel may have a value of magnetostriction λ_S (Eq. 30) of 20×10^{-6}. Compare the dimensional changes (strain) produced by a) applying a saturating magnetic field, b) heating the sample by 20°C, and c) applying a stress of 70 MPa (10,000 lb/in^2).

Subject Index

123 or YBCO superconductor, 402, 403
2122 superconductor, 403
2223 superconductor, 403

A

A15 superconductors, 391
   critical temperatures, 392
   microstructure, 393, 394
      dislocations, 398
      grain boundaries, 393
      precipitates, 397
      radiation damage, 397
   structure, 392
Activation energy
   for creep in ceramics and metals, 339
   for creep in superalloys, 82, 83
   for crystallization in glass, 368
   for diffusion in Al, 6
   for nucleation of particle in Al, 14
Alkali silicates, 358
Alkaline earth silicates, 358
Alumina, 256, 259, 260
Aluminum, 1
   fcc lattice, 2
   grain structure, 2
   oxide layer, 1
   production, 1
   properties, 1
Aluminum alloys, 1
   2016-T6, 26
   2024, 33
   2024-T3, 40
   2024-T8, 39
   2024-T851, 26
   2090, 36-37
   2090-T81, 39
   2090-T8X, 36
   2124-T351, 36
   2XXX series, 37, 40
   5XXX series, 40
   7050, 25
   7075. 34, 43
   7075-T6, 40
   7079, 43
   7079-T6, 40
   7079-T651, 26
   7079-T7351, 26
   7150-T651, 36
   7178, 23
   7XXX series, 40
   8090-DA, 39
   8090-T851, 33
   8090-T877, 39

Aluminum alloys (cont'd)
   8090-T8X alloy, 36
   AA2024T8, 38
   AA2090T81, 38
   AA8090 DA, 38
   AA8090T8771, 38
   age hardenable alloys, 14
   Al-6Zn-2Mg-XCu alloys, 42
   Al-Cu alloys, 18
   Al-Cu-Mg alloys, 18
   Al-Cu-Mg-Li alloy, 2
   Al-Cu-Li-Mg-Ag alloy, 18
   Al-Li alloys, 19
   Al-Mg alloys, 18
   Al-Mg-Si alloys, 18
   Al-Zn-Mg alloys, 18, 19
   alloy designation systems, 6, 7, 12
   amorphous alloys, 3-5
   binary phase diagrams, 3, 4
   casting alloys, 7
   ceramic phases, 3
   corrosion fatigue, 35
   crack closure, 35
   cross slip of dislocations, 2
   damage tolerent design, 35
   density, 1, 22
   diffusion data, 6
   environmental effects, 35
   fatigue crack growth, 35, 36
   fatigue damage, 33
   fatigue resistance, 32
   fracture toughness, 22-32
   heat treatable alloys, 12
   heterogeneous precipitation,
   intercrystalline SCC, 1
   intermetallic compounds, 3, 11
   metallic glasses, 3
   microstructure, 1
   non-equilibrium structures, 5
   non-heat treatable alloys, 11
   nucleation of precipitates, 14-16
   nucleation on dislocations, 16
   Paris law, 35
   particle cracking, 26
   precipitate-free zone (PFZ), 16, 28
   precipitation kinetics, 16
   precipitation and plastic deformation, 17
   property microstructure relations, 21
   quasicrystals, 3, 5
   slip plane softening, 29
   solubility data, 6
   stacking fault energy, 2
   strain localization, 29
   strengthening mechanisms, 7-10
   stress corrosion cracking, 40
   types of solid phases, 1, 4
   void necleating particles, 29

444   Subject Index

Aluminum alloys (cont'd)
  wrought alloys, 7
  X-2095, 37
  X-2095-T8, 39
  X-2095-T8, 20, 38
  X-7075, 34
  Young's modulus, 22
Aluminum nitride, 259
Amorphous or glassy metals
  aluminum alloys, 3-5
  magnetic alloys, 435
Annealing under stress, 84
Antiphase boundaries
  in Ni-base superalloys, 71-73
Aramid fibers, 187
  fabrillated, 189
  Kevlar fiber, 187
  manufacturing, 187
  properties, 184
  structure, 188
  Technora fiber, 187
  Twaron fiber, 187
ASGlass, Al_2O_3,MgO,CaO,SiO_2, 253

B

Barium and strontium ferrites, 438
Bauschinger effect
  in Al alloys, 33
  in metal matrix composites, 161, 166
BCS theory, 384
  failure of, 408
Binary phase diagrams
  of Al alloys, 3, 4
Bismaleimides resins, 191
BMAS glass ceramic, BaO, MgO, Al_2O_3, SiO_2, 253
Boron oxide, 260
Borosilicate glass (BSG), 247
BSG, see Borosilicate glass

C

C-glass
  composition, 183
C/BSG composite
  matrix cracking, 264
Carbon fibers, 185
  3-D structure, 185
  dry oxidation, 198
  manufacturing from PAN, 186
  pitch-based, 186
  polar surface energy, 199
  production of, 186
  properties of, 184
  Rayon-based, 186
  wet oxidation, 198

Carbon nitride superconductors, 398
CAS glass ceramic, CaO, Al_2O_3, SiO_2, 253
Celion C fiber, 252
Ceramic matrix composites(CMC), 247
  applications, 250, 251
  ASGlass, Al_2O_3,MgO,CaO,SiO_2, 253
  BMAS glass ceramic, 253
  C-BSG, matrix cracking, 264
  CAS glass ceramic, 253
  Celion C fiber, 252
  ceramic matrix properties, 254
  chemical vapor infiltration, 256
  compression, 277
  cracking of Nicalon-CAS, 271
  cracking sequence, 271
  creep, 338
  critical stress for matrix cracking
    cross-ply laminates, 320
    unidirectional composites, 299
    woven composites, 325
  critical volume fraction, 249
  cross-ply laminates
    critical stress for cracking, 320, 322
    damage modes, 319
  damage evolution, 276
  debounding at interfaces, 249
    crack front debounding, 249
    crack wake debounding, 249
    criterion, 248
  directed metal oxidation, 258
  fatigue damage, 336
    fatigue crack growth, 337
    fatigue limit, 337
    gradual modulus degradation, 337
    modulus recovery, 337
    Nicalon-aluminosilicate glass, 337
    Nicalon-CAS composite, 337
  fiber/matrix interfaces, 285
    interlaminar shear properties, 286-287
    measurements, 288-292
    properties, 293-296
    TEM of SiC-CAS interface, 287
  fiber bridged crack, 300
  fiber properties, 252
  fibers and matrices, 251
  flexural deformation, 282
  FP (alpha Al_2O_3 fiber), 252
  glass ceramic properties, 282
  glass properties, 253
  high temperature capability, 249, 255
  HMU(C fiber)-BSG composite
    stress-strain curve, 265, 267
    ultimate tensile strength, 265, 269
    Young's modulus, 265, 266, 268
  HPZ fiber (Si + N + O + C), 252
  HSGlass, SiO_2 with some B_2O_3, 253

Subject Index 445

Ceramics matrix composites (cont'd)
  impact and fracture properties, 284
  LAS glass ceramic, 253
  Magnamite C fiber, 252
  MAS glass ceramic, 253
  modulus recovery
    in fatigue, 337
  modulus reduction
    cross-ply laminates, 322
    due to fatigue, 337
    in SiC/SiC composite, 332
    Nicalon-CAS, 265, 315
    unidirectional composites, 311
    woven composites, 330
  MPDZ fiber (β SiC+C+O+N), 252
  MPS fiber (β/α SiC + O), 252
  Nextel fiber (Mullite + B_2O_3), 252
  Nicalon fiber (β SiC + C + O), 252
  Nicalon-aluminosilicate glass composite
    thermal fatigue, 337
  Nicalon-BSG composite, 247
    matrix cracking, 264
    ultimate tensile strength, 316
  Nicalon-CAS composite
    debonding, 264
    degradation of Young's modulus, 315
    fatigue crack propagation, 337
    flexural strength, 333
    matrix cracking, 264, 269
    micrograph of cracking, 263
    modulus reduction, 315
    Poisson ratio, 264, 266
    tensile stress strain curve, 262, 270
    thermomechanical fatigue, 338
    ultimate tensile strength, 317
    Young's modulus, 264, 265, 270, 273
  Nicalon-CAS II composite
    thermomechanical fatigue, 338
  Nicalon-LAS composite
    flexural strength, 333
    stress-strain behavior, 334, 335
    ultimate tensile strength, 314, 316, 317
  Nicalon-SiC(0°/90°) composite
    steady state creep rate, 339
  perhydropolysilazone, 261
  Poisson ratio
    for Nicalon-CAS composites, 266, 274
  polymer pyrolysis, 260, 261
  polysilazane structure, 261
  porosity reduction, 262
  processing of CMC, 255
    chemical vapor infiltration, 256
    directed metal oxidation, 257, 258
    polymer pyrolysis, 260
    reaction bonding, 262

Ceramics matrix composites (CMC)(cont'd)
  processing of CMC (cont'd)
    slurry infiltration and hot pressing, 258, 259
    sol-gel, 259
  properties of fibers, 252-253
  properties of glass and glass ceramic, 253
  properties of ceramic matrices, 254
  reaction bonding, 262
  Saffil fiber (αAl_2O_3), 252
  SCS(β/α SiC on Si)fiber, 252
  shear deformation, 283
  SiC-RBSN composite, UTS, 316
  slurry infiltration and hot pressing, 259
  sol-gel, 259
  static mechanical behavior
    compressive behavior, 277
    creep, 338
    flexural behavior, 282
    high temperatures, 332
    impact and fracture, 284
    room temperature, 262
    shear behavior, 283
    steady state creep, 339
    tensile behavior, 264
  steady state creep rates, 339
  stress strain behavior, 311
    SiC/SiC composite, 275, 283
    aged SiC/SiC composite, 336
    HMU-BSG, 269
    Nicalon-CAS, 262, 270, 282
    Nicalon-LAS composite, 335
  tensile deformation, 264
  tensile properties, 278-281
  tensile strength, 314
    Coleman model, 315
    cumulative weakening model, 316
    Curtin model, 318
    fiber break propagation model, 316
    Prewo model, 314
    Sutcu model, 318
    TSSE model, 316
  thermal expansion effect, 250
  thermal fatigue, 336
  thermal mechanical fatigue, 336
  thermal residual stresses, 250
  Thornel C fiber, 252
  Tyranno (β SiC +Ti+C+O) fiber, 252
  ultimate tensile strength
    of HMU-BSG, 269
  Young's modulus
    of HMU-BSG, 268
    of Nicalon-CAS, 273
    rule of mixtures, 264
Ceramic phases
  in Al alloys, 3

Chemical Vapor Infiltration, 256
Chevrel phase superconductors, 401
CMC, see ceramic matrix composites
Coherence length
  in superconductors, 385
Compression testing,
  in ceramic matrix composites, 277
Continuous fiber composites, 217
  fiber orientation effects, 221
  fracture strength, 222
  longitudinal compressive strength, 220
  longitudinal tensile strength, 217
  properties, 218
  transverse tensile strength, 219
Corning 9606, 9607, 9608, 9658, 372
Corrosion fatigue
  in Al alloys, 35
Corrosion resistance
  of superalloys, 88
Crack closure
  in Al alloys, 35
Crack tip opening displacement
  in Al alloys, 30
Creep deformation
  grain size effect in superalloy, 85
  of ceramic-matrix composites, 339
  of crystalline, single phase solid, 82
Creep resistance
  in nickel-base Superalloys, 81, 82
Critical magnetic field
  for superconductivity, 382, 383
Critical volume fraction, 249
Critical strain to fracture, 30
Critical temperature for superconductors, 381
Cross-ply ceramic matrix composites
  critical stress for cracking, 320
    effect of fiber volume fraction, 323
    effect of fiber Young's modulus, 324
    effect of ply thickness, 322
  damage modes, 319
Crystallization
  in glass, 360
    rates measured, 366
Crystalline anisotropy, 424
Curie temperature, 438
Cyanate resins, 191
CVI process, see Chemical Vapor Infiltration
Cyclic stress-strain behavior
  of Al alloys, 33
Cyclic hardening, 33
  effect of GP zones, 33
  of 8090-T851, 33
Cyclic softening, 33

D

Damage tolerent design

  of Al alloys, 35
Debye temperature, 384
Debounding at interfaces, 249
  crack front debounding, 249
  crack wake debounding, 249
Density
  of Al alloys, 1, 22
  of ceramic matrices, 254
  of E glass fibers, 184
  of epoxy resin, 192
  of fiber composites, 218
  of fibers in ceramic composites, 252
  of glasses, 253
  of glass-ceramic matrices, 253
  of Kevlar 49 fiber, 184
  of liquid crystal polymer, 194
  of polycarbonate, 194
  of polyester resin, 192
  of polyimides (PMR-15), 192
  of polysulfone, 194
  of S glass fiber, 184
  of thermoplastic resins, 194
  of Type I carbon fiber, 184
  of Type III carbon fiber, 184
  of vinyl ester resin, 192
Differential scanning calorimetry(DSC)
  for glasses, 368
  for fluoride glasses, 369-371
Diffusion data
  in Al alloys.6
Digital data processing
  high speed, 381
Digital signal processing
  high speed, 381
Dimpled rupture, 31
Directional solidification
  for permanent magnets, 439
  for superalloys, 87
Dislocation density
  in metal matrix composites, 112-123
  162
Dislocation generation
  from particles, 112-123
Dislocations in superconductors, 406
  dislocation loops, 387
  dislocation pinning, 398
Ductility
  of metal matrix composites, 165

E

E-glass
  composition, 183
  properties, 184
Elastic modulus
  in nickel-base Superalloys, 66
Electrical power storage

lossless, 381
Electrical power transmission
  lossless, 381
Electromagnetic radiation detectors, 381
Environmental effects
  in Al alloys, 35
Epoxy resins, 190
  compressive strength, 192
  curing agents, 191
  density, 192
  elongation to failure, 192
  epoxide group, 191
  glycidyl ether and amines, 191
  Poisson ratio, 192
  tensile strength, 192
  thermal expansion, 192
  Young's modulus, 192

F

Failure analysis
  of polymer matrix composites, 226
Fatigue crack growth
  in 2090-T8X alloy, 36
  in 2124-T351, 36
  in 7150-T651, 36
  in 8090-T8X alloy, 36
  in AA2024T8, 38, 40
  in AA2090T81, 38, 40
  in AA8090 DA, 38, 40
  in AA8090T8771, 38, 40
  in Al alloys, 35
  in Al-Li alloys, 36
  in Al-Zn-Mg 7150 alloy, 36
  in ceramic-matrix composites, 337
  in X2095-T8, 38, 40
Fatigue damage
  in Al alloys, 33
  in ceramic-matrix composites, 336
Fatigue failure of polymer matrix composites, 229
  angle-plied laminates, 231
  continuous fiber composites, 229, 230
Fatigue life
  and grain size of Al alloys, 34
  and slip length in Al alloys, 35
  in polymer matrix composites, 230
Fatigue resistance
  of Al alloys, 32
Ferrimagnetic materials, 436
Ferromagnetic metals and alloys, 421
Fiber strength and modulus, 182
  alumina, 182
  boron, 182
  E-130 (pitch based carbon), 182
  E-glass, 182
  M60J (PAN based carbon), 182

P-100 (pitch based carbon), 182
PBO (polyphenylene benzobisoxazole), 182
Kevlar 149, 182
S-glass, 182
SiC, 182
spectra 1000 (polyethylene), 182
steel, 182
T-300 (PAN based carbon), 182
T-1000 (PAN based carbon), 182
Finite element modeling
  of metal matrix composites, 135, 137
Fluoride glasses, 369
  composition, 370
  crystallization, 371
  DSC measurements, 369-371
Flux line pinning in superconductors, 386
Forced chemical vapor infiltration, 256,257
FP (alpha Al_2O_3 fiber), 252
Fracture
  cracking of Nicalon-CAS, 271
  cracking sequence, 271
  critical stress for ceramic matrix cracking
    cross-ply laminates, 320
    unidirectional composites, 299
    woven composites, 325
  cross-ply laminates
    critical stress for cracking, 320, 322
    damage modes, 319
  damage evolution, 276
  fiber bridged crack in ceramics, 300
  of ceramic-matrix composites, 284
Fracture Surfaces
  of 2024-T8 Al alloy, 39
  of 2090-T81 Al alloy, 39
  of 8090-T877 Al alloy, 39
  of 8090-DA Al alloy, 39
  of X2095-T8 Al alloy, 39
Fracture toughness
  effect of composition in Al alloys, 25
  influence of second phase particles, 27
  of 7050 Al alloy, 25
  of 7075 Al alloy, 28
  of Al alloys, 22-32
  of metal matrix composites, 168
  of polymer matrix composites, 228
  relation to slip band width and spacing, 29-30
Free edge delamination, 237
Fullerene, 399, 400

G

General Eectric Re-X glass, 372
GLAG theory for type II superconductors, 385
Glass ceramics, 249, 253, 371
  properties and uses, 372

Subject Index

Glass fibers
  compositions, 183
  manufacturing, 184
  properties, 184
Glasses, see inorganic glasses
Grain boundary precipitates, 32
Grain boundary sliding, 81
Graphite
  moduli in different directions, 185
Graphite lattice structure, 185
Guinier-Preston (GP) zones, 15, 16, 18, 19

H

Hall-Petch relation, 81
Hastelloy, 67
High T_C superconductors, 401
  cracks, 406
  critical current, 407
  dislocations, 406
  microscopic theory, 408
  stacking faults, 406
  structure modulation, 407
  twins, 403
HMU(C fiber)-BSG composite
  stress-strain curve, 265, 267
  ultimate tensile strength, 265, 269
  Young's modulus, 265, 266, 268
HPZ fiber (Si + N + O + C), 252
HSGlass, SiO$_2$ with some B$_2$O$_3$, 253

I

Impact energy
  of liquid crystal polymer (Vectra), 194
  of thermoplastic resins, 194
  of polycarbonate, 194
  of polysulfone, 194
Inconel alloy, 67
Inorganic Glasses, 351
  3-D Ising model, 358
  absorption edge, 375
  Ag coated CdS particles in glass, 375
  alkali silicates, 358
  alkaline earth silicates, 358
  applications, 372
  coated particles, 375
  color chages with particle size, 374
  combined nucleation and growth, 367
  copper coated CuCl particles in glass, 376
  Corning 9606, 9607, 9608, 9658, 372
  crystallization, 360
    interface control, 364
    maximum velocities, 364
    rate equation, 365
    rates measured, 366

Inorgaanic glasses (cont'd)
  electrical conductivity of glass, 360
  experimental proof of nucleation theory, 359
  fiber optical networks, 373
  fluoride glasses, 369
    composition, 370
    crystallization, 371
    DSC measurements, 369-371
  General Eectric Re-X glass, 372
  glass ceramics, 371
  gold ruby glass, 373
  growth of particles, 355
  growth of crystal, 363
    effect of entropy of fusion, 364
    interface control, 364
    maximum growth velocities, 364
  high speed optical switching, 373
  interfacial energy, 362
    temperature dependence, 362
  kinetics of nucleation, 358
  lithium silicates, 357
  molecular glasses, 367
  network glasses, 367
  nonlinear optical properties, 373, 374, 375
    effect of coated particles, 375
  nucleating particles and agents, 363
  nucleation of particles, 351
  nucleation rates, 361
  optical computing, 373
  optical properties, 360
  Ostwald ripening, 355
  Owens-Illinois Cer-vit glass, 372
  phase diagrams, 356
    lithium silicates, 357
    sodium silicates, 357
  phase separation, 356
  plasma resonance band, 373
  pyrex borosilicates, 356
  quantum confinement, 375
  renormalization group methods, 358
  Schott Zerodur glass, 372
  semiconductor particles in glass, 374
  silicate glasses, 358
  Slagsitall glass, 372
  small particles in glass, 373
    coated particles, 375
    metallic particles, 373
    semiconductors, 374
  sodium borosilicates, 356, 357
  sodium silicates, 357
  soda-lime glass, 358
  soda-lime silicates, 356, 357
  soda-baria-silica, 357
  soda-lithia-silica, 357
  spinodal decomposition, 354
  subregular solution model, 358
  surface crystallization, 351, 360

Inorganic glasses (cont'd)
  theories of miscibility, 358
  translucent glasses, 360
  viscosity of glass, 360, 363
  Vycor process, 351, 356
  ZB glass, 370, 371
  ZBL glass, 369-371
  ZBLA glass, 370, 371
  ZBLALi glass, 370, 371
  ZBLALiPb glass, 370, 371
  ZBLAN glass, 370
  ZBLLi glass, 370
  ZBLN glass, 370
  ZBLPb, glass, 370, 371
Intergranular fracture, 31
  and localized deformation, 34
Intermetallic compounds
  in Al alloys, 3, 11
Investment casting, 51
Ising model, 358

K

Kevlar 49
  hydrogen bonding, 188
  properties, 184
  structure, 188
Kevlar fiber, 187
  PPTA, poly(phenylene terephthalamide), 187
Kramer's model, 394

L

LAS glass ceramic, 253
Lawson-Miller plot
  in nickel-base Superalloys, 93, 95
Lexane (polycarbonate, PC), 194, 195
Liquid crystal polymer(Vectra),194, 195
  density, 194
  elongation to fracture, 194
  heat distortion temperature, 194
  impact energy, 194
  melting temperature, 194
  properties, 194
  tensile strength, 194
  thermal expansion, 194
  Young's modulus, 194
Liquid metal infiltration, 109
Liquid state fabrication, 110
Lithium aluminum silicates, 363
London penetration length, 383
Lorentz force, 386

M

Maghemite, 440

Magnamite C fiber, 252
Magnesium
  strengthening Al, 7
Magnetic domains, 429
Magnetic materials, 415
  amorphous or glassy metals, 435
  barium and strontium ferrites, 438
  basic ideas, 415
  classification, 419
  crystalline anisotropy, 424
  Curie temperature, 438
  demagnetizing field, 426
  domain wall energy and thickness, 430
  energy product, 438, 439
  ferrimagnetic materials, 436
  ferromagnetic metals and alloys, 421
  maghemite, 440
  magnetic domains, 429
  magnetic quantities and units, 416
  magnetite, 436
  magnetization process, 433
  magneto-optic recording material, 440
  magnetostriction, 428
  permalloys, 435
  permanent magnets, 437
  recording materials, 439
  semihard magnets, 440
  shape anisotropy, 426
  single domain particle, 437
Magnetic sensors, 381
Magnetite, 436
Magnetization process, 433
Magneto-optic recording material, 440
Magnetostriction, 428
MAS glass ceramic, 253
McMillan equation, 384
Meissner effect, 381, 382
Metal matrix composites, 107
  Arsenault-Shi model, 115, 117
  back stress factor(BSF), 163, 164
  Bauschinger effect, 161, 166
  Bauschinger stress factor, 163, 166
  clustered composites, 147, 149
  compocasting, 110
  continuous metal matrix composites (CMMC), 108
  discontinuous metal matrix composites (DMMC), 108
  dislocation density, 112-123, 162
    effect of particle size, 116
    near the failure surface, 169
  dislocation energy, 122
  ductility, 165
    effect of aging, 167
    effect of deformation localization, 169
  Eshelby model, 128, 132
    yield stress of composite, 134

Metal matrix composites (cont'd)
  Eshelby model (cont'd)
    work hardening of composite, 134-135
  Eshelby tensor, 129
  fiber clustering, 147, 149
    interaction with homogeneous region, 170, 171
  finite element modeling, 135, 137
    predicted stress strain relation, 144
  fracture toughness, 168
    and cold work, 168
  foundry techniques, 110
  high strenth fibers in, 107
  liquid metal infiltration, 109
  liquid state fabrication, 110
  localization of plastic deformation, 169
    effect of heat treatment, 172
    in particle clusters, 172
  matrix microstructure, 112
  maximum local strain, 169
  mean field approximation, 171
  micromechanics, 135
  particle clustering, 171, 172
  periodic arrangement of fibers, 145
  periodic array of clusters, 146
  periodic clustering model, 169, 170
  powder cloth technique, 109
  powder metallurgy procedures, 109, 111
  plastic zone around fiber, 140
  processing of, 108
  production of, 111
  reinforcement aspect ratio, 117
    effect on dislocation density, 117
  reinforcement clustering, 173
  residual stresses in, 150
    deformation induced changes, 150
  shear lag model, 123
  squeeze casting, 112
  strengthening mechanisms in, 123
    due to work hardening of matrix, 158
    particle size effect, 161
  stress strain relations
    effect of fiber clustering, 149
    from Eshelby model, 136
    from finite element calculation, 144
    tension vs. compression, 153
  stresses in the fiber, 125-126
  subgrain size, 115, 116, 162
    effect of particle size, 116
  surface activation techniques, 109
  tensile ductility
    effect of matrix strain hardening, 165
    overaged composites, 167
    underaged composites, 167
  thermal residual stresses, 115, 137, 148, 163
  vortex method, 110

Metal matrix composites (cont'd)
  whiskers in, 107
  work hardening, 132
  yield stress, 152
    effect of volume fraction of SiC, 159
  yielding, 132
  Young's modulus of composite, 128, 132
    tension vs. compression, 152
Microstructure
  in A15 superconductors, 393
  in Al alloys,1
  in ceramic matrix composites, 271, 287, 319
  in inorganic glasses, 351
  in metal matrix composites, 112
  in magnetic materials, 430, 434
  in nickel-base Superalloys, 54, 55, 60-64, 66, 68, 74, 75, 85, 91
  in Nb-Ti superconductors, 389
  in polymer matrix composites, 221, 223, 235
  in superconductors, 389
Mixed mode fracture, 31
Mo_3Re supersonductor, 392
Mode I and Mode II fracture testing, 232, 236
Modulus recovery in ceramic-matrix composites
  in fatigue, 337
Modulus reduction in ceramic-matrix composites
  cross-ply laminates, 322
  due to fatigue, 337
  in SiC/SiC composite, 332
  Nicalon-CAS, 265, 315
  unidirectional composites, 311
  woven composites, 330
MPDZ fiber (β SiC+C+O+N), 252
MPS fiber (β/α SiC + O), 252

N

Nabarro-Herring diffusion, 87
Nb_3Al superconductor, 392
Nb_3Ga superconductor, 392
Nb_3Ge superconductor, 392
Nb_3Si superconductor, 392
Nb_3Sn superconductor, 392, 393
Nextel fiber (Mullite + B_2O_3), 252
Nicalon fiber (β SiC + C + O), 252
Nicalon-aluminosilicate glass composite
  thermal fatigue, 337
Nicalon-BSG composite, 247
  matrix cracking, 264
  ultimate tensile strength, 316
Nicalon-CAS composite
  debonding, 264
  degradation of Young's modulus, 315
  fatigue crack propagation, 337

Subject Index 451

Nicalon-CAS composite (cont'd)
   flexural strength, 333
   matrix cracking, 264, 269
   micrograph of cracking, 263
   modulus reduction, 315
   Poisson ratio, 264, 266
   tensile stress strain curve, 262, 270
   thermomechanical fatigue, 338
   SiC-RBSN composite
   ultimate tensile strength, 317
   Young's modulus, 264, 265, 270, 273
Nicalon-CAS II composite
   thermomechanical fatigue, 338
Nicalon-LAS composite
   flexural strength, 333
   stress-strain behavior, 334, 335
   ultimate tensile strength, 314, 316, 317
Nicalon-SiC($0°/90°$) composite
   steady state creep rate, 339
Nickel-base intermetallics, 94
Nickel-base Superalloys, 51
   alloying for creep resistance, 81
   alloying for surface stability, 95
   alloys with lattice mismatch, 78
   alloys with no lattice mismatch, 76
   aluminum for surface protection, 96
   alumina filament strengthening, 95
   antiphase boundaries in, 71-73
   antiphase boundary energy, 73
   borides in, 55, 64, 68
   carbides in, 55, 60-63, 68
   chromium addition, 95
      effect on corrosion rate, 96
   CMSX4, 52, 89
   coherency strains in, 68, 74
   composite strengthening in, 93
   compositions
      of DS nickel-base eutectic alloys, 92
      of selected nickel-base alloys, 52
   corrosion resistance
      comparison of polycrystals, single crystals and DS alloys, 88
      Hastelloy-C, 95
      Inconel-600, 95
   creep rate, 80
      component thickness/grain size, 86
      effect of volume fraction of γ', 85
      grain size effect, 85
   creep rupture life, 82, 85
      comparison of polycrystals, single crystals and DS alloys, 88
      effect of B and Zr, 87
      effect of volume fraction of γ', 86
   creep rupture strength, 85
      effect of B and Zr, 87

Nickel-base Superalloys (cont'd)
   creep rupture strength (con'd)
      effect of grain aspect ratio, 91
      of DS Mar M200 + Hf, 92
      of ODS alloy MA 6000E, 92
      of single crystal PWA 454, 92
      of TD nickel, 92
   critical resolved shear stress (CRSS), 74
   directional solidification process, 87
   directionally solidified (DS) alloys, 88
      thermal fatigue of, 89
   directionally solidified eutectics, 92, 93
   dislocation bypass models, 75
   dispersion strengthened alloys, 90
   effects of alloying elements, 53
   elastic modulus, 66
   flow stress, temperature effects, 77
   forgeability in wrought superalloys, 87
      effect of Mg addition, 87
   gamma double prime in, 55, 59-60
   gamma matrix of, 54, 55
   gamma prime in, 55, 56-59, 68
      effect on creep, 84
      flow stress and temperature, 59
   gas turbine alloys, 52
   grain aspect ratio, 91
   grain boundary γ', 55
   grain boundary chemistry on creep, 86
   grain boundary diffusion
      affected by misfitting atoms, 87
   grain boundary sliding, 81, 88
   grain size effect on yield, 81
   grain size effect on creep, 81, 85
      component thickness effect, 85
   hafnium in DS alloys, 88
   Hall-Petch relation, 81
   hardening mechanisms, 69
   Hastelloy-C, 95
   Hastelloy-X, 67
      oxidation and hot corrosion resistance, 96
   hot corrosion resistance
      CMSX4, 52, 89
   Inconel alloy-600, 95
   Inconel alloy-625, 67
   intermetallic compounds, 94
   lattice parameter, 64
   Lawson-Miller plots
      of γ/γ'-δ (p.92) and others, 93
      of fiber reinforced NiCrAlY, 93
      of Rene 80 and others, 95
   low temperature corrosion of, 95
   major constituent phases in, 54, 55
   Mar-M200 single crystals
      Cr content and corrosion of, 96
      high temperature strength of, 83
   mechanically alloyed materials, 91
   misfit hardening in, 74-75

Nickel-base Superalloys (cont'd)
   molybdenum addition, 95, 97
   Nabarro-Herring diffusion, 87
   new superalloys, 90
   Ni-base intermetallics, 94
   Ni_3Al, properties, 94
      flow stress and temperature, 59
   NiAl, properties, 94
      Lawson-Miller plots, 95
   Nimonic series of alloys, 67
   ODS alloys and composites, 81
   order in particles, 68
   order strengthening of, 71-73
   Orowan bowing in, 75
   oxidation and hot corrosion of, 96
   polycrystals of, 81
   precipitation hardening in, 68
      hardening mechanisms, 69
      particle cutting, 68
   rhenium addition, 89
      CMSX4, 52, 89
   rupture life, 82, 85
      effect of B and Zr, 87
      effect of volume fraction of γ', 86
   rupture strength, 85
      effect of B and Zr, 87
      effect of grain aspect ratio, 91
      of DS Mar M200 + Hf, 92
      of ODS alloy MA 6000E, 92
      of single crystal PWA 454, 92
      of TD nickel, 92
   service life
      comparison of polycrystals, single crystals and DS alloys, 88
   short range order in, 66, 67
   single crystals of, 79
   solid solution hardening of, 64, 65, 67
   stacking fault energy in, 68, 81, 82
   steady state creep of, 82
   tantalum addition, 90
   TD (ThO_2 dispersed) nickel, 92
   thermal fatigue, 80
      comparison of polycrystals, single crystals and DS alloys, 88, 89
   topologically close packed phases in, 64
   tungsten addition, 97
   Waspaloy, 94
   wire reinforced nickel-base alloys, 93
   yield stress
      and stacking fault, 64
      effect of stress annealing, 84
      influence of γ' morphology, 84
      model for single crystal, 80
   Young's modulus of superalloy PWA 1480
      effect of orientation, 79
Nimonic alloys, 67

Niobium-titanium superconductor, 389
Nitride and carbonitride superconductors, 398
Nucleation
   in Al alloys, 14-16
   in glasses, 359, 361

O

Order hardening
   in Al alloys, 29
   in Ni-base superalloys, 71
Organic fibers, 186
   chemical structure, 187
   high strength-high modulus, 186
Organic superconductors, 399
Orthotropic laminate, 215
Ostwald ripening, 355
Overaging, 30
Owens-Illinois Cer-vit glass, 372

P

PAI(polyamideimide, Torlon), 194, 195
PAN, polyacrylonitrile, 186
Pande and Suenaga model, 395, 396
Paris law
   in Al alloys, 35
Particle cracking
   in 2016-T6 Al alloy, 26
   in 2024-T851 Al alloy, 26
   in 2124-T851 Al alloy, 26
   in 7079-T651 Al alloy, 26
   in 7079-T7351 Al alloy, 26
   in 4340 steel, 26
   of cementite in iron, 26
   with deformation in Al alloys, 26
   with deformation in 4340 steel, 26
PC (polycarbonate, Lexane), 194, 195
PEEK(polyetheretherketone, Victrex), 194, 195
PEI(Polyether imide, Ultem), 194-196
Peierls stress, 29
PEK(polyetherketone, victrex) 194, 195
Permalloys, 435
Permanent magnets, 437
PES(polyethersulfone, Victrex), 194, 195
Phase diagrams
   of Al alloys, 3, 4
   of lithium silicates, 357
   of sodium silicates, 357
Phenol-formaldehyde, 190
Phosphorous pentoxide, 363
Plastic zone
   around fiber in metal, 140
Poisson ratio
   for Nicalon-CAS composites, 274
Polyamideimide(PAI, Torlon), 194, 195
   compressive strength, 194

Polyamideimide(PAI, Torlon) (cont'd)
   elongation to fracture, 194
   glass temperature, 194
   properties, 194
   tensile strength, 194
   thermal expansion, 194
   Young's modulus, 194
Polycarbonate(PC, Lexane), 194, 195
   compressive strength, 194
   elongation to fracture, 194
   glass temperature, 194
   properties, 194
   tensile strength, 194
   thermal expansion, 194
   Young's modulus, 194
Polyetheretherketone(PEEK,Victrex),194,195
   compressive strength, 194
   elongation to fracture, 194
   glass temperature, 194
   properties, 194
   tensile strength, 194
   thermal expansion, 194
   Young's modulus, 194
Polyether imide(PEI,Ultem), 194-196
   compressive strength, 194
   elongation to fracture, 194
   glass temperature, 194
   properties, 194
   tensile strength, 194
   thermal expansion, 194
   Young's modulus, 194
Polyetherketone (PEK, victrex) 194, 195
   compressive strength, 194
   elongation to fracture, 194
   glass temperature, 194
   properties, 194
   tensile strength, 194
   thermal expansion, 194
   Young's modulus, 194
Polyethersulfone (PES, Victrex), 194, 195
   compressive strength, 194
   elongation to fracture, 194
   glass temperature, 194
   properties, 194
   tensile strength, 194
   thermal expansion, 194
   Young's modulus, 194
Polyethylent Spectra 900 fiber, 184
   properties, 184
Polyester resin
   compressive strength, 192
   density, 192
   elongation to failure, 192
   Poisson ratio, 192
   tensile strength, 192
   thermal expansion, 192
   Young's modulus, 192

Polyimides (PMR-15)
   compressive strength, 192
   density, 192
   elongation to failure, 192
   Poisson ratio, 192
   tensile strength, 192
   thermal expansion, 192
   Young's modulus, 192
Polymer matrix composites, 181
   Aramid fibers, 187
      fabrillated, 189
      structure, 188
      properties, 184
   carbon fibers, 185
      3-D structure, 185
      dry oxidation, 198
      elongation to fracture, 184
      manufacturing from PAN, 186
      pitch-based, 186
      polar surface energy, 199
      production of, 186
      properties of, 184
      Rayon-based, 186
      tensile strength, 184
      thermal expansion, 184
      wet oxidation, 198
      Young's modulus, 184
   composite constituents, 182
   continuous fiber composites, 217
      fiber orientation effects, 221
      fracture strength, 222
      longitudinal compressive strength, 220
      longitudinal tensile strength, 217
      properties, 218
      transverse tensile strength, 219
   delamination, 231
      free edge delamination, 237
      interlamella fracture modes, 231
      interlamella fracture testing, 232
   double cantilever beam specimens, 232
   epoxy resins, 190, 191
      properties, 192
   failure analysis, 226
   fatigue failure, 229
      angle-plied laminates, 231
      continuous fiber composites, 229, 230
   fiber bridging, 235
   fiber properties, 184
   fracture of, 226
      fiber pullout, 228
      fracture toughness, 228
      interface debonding, 226
      post debonding friction, 227
      stress redistribution, 227
   free edge delamination, 237
   glass fibers, 183
      composition, 183

Polymer matrix composites (cont'd)
  glass fibers (cont'd)
    elongation to fracture, 184
    manufacturing, 184
    properties of, 184
    sizing materials, 185
    tensile strength, 184
    thermal expansion, 184
    Young's modulus, 184
  gradphite structure, 185
  heat distortion temperature (HDT), 195
  high modulus PE fibers, 188
  interfaces, 196
    characterization, 199
    engineered interfaces, 201
    single fiber tests, 200
    surface treatment of fibers, 196
      carbon fibers, 198, 199
      glass fibers, 197
  interlaminar fracture testing, 232
    area method, 235
    compliance method, 233
    double cantilever beam test, 233
    Mode I, 232
    Mode II, 236
    strain energy release rate, 233
  interlaminar shear strength, 198
  losipescu shear test, 201
  Kevlar 149, 188
  Kevlar 29, 188
  Kevlar 49. 188
    hydrogen bonding, 188
    properties, 184
    structure, 188
  Kevlar fiber, 187
    elongation to fracture, 184
    PPTA,poly(phenylene terephthalamide), 187
    tensile strength, 184
    thermal expansion, 184
    Young's modulus, 184
  Lexane (polycarbonate, PC), 194, 195
  liquid crystal polymer(Vectra),194-196
  matrix systems, 189
    epoxy resins, 191
    polyester resins, 192
    repeat units of thermoplastics, 195
    thermoplastics, 190, 193-194
    thermosets, 190, 192
  mechanics of composites, 214
    laminate theory, 215
    shear lag model, 214
    stress transfer, 214
    Young's modulus, 216
  micromechanics of stress transfer, 214
  molding processes, 205
    autoclave molding, 205, 206

Polymer matrix composites (cont'd)
  molding processes (cont'd)
    bulk molding compound, 207, 224
    contact molding, 205, 206, 224
    compressive molding, 207, 208
    injection molding, 208, 209
    open molding, 205, 206
    resin transfer molding, 208, 209
    sheet moding compound, 207, 224
  organic fibers, 186
    chemical structure of, 187
    properties of, 184
  orthotropic laminate, 215
  PAI(polyamide imide, Torlon), 194, 195
  PC (polycarbonate, Lexane), 194, 195
  PEEK(polyetheretherketone,Victrex),194 195
  PEI(Polyether imide,Ultem), 194-196
  PEK(polyetherketone, victrex) 194, 195
  PES(polyethersulfone, Victrex), 194, 195
  phenol-formaldehyde, 190
  polyamideimide(PAI, Torlon), 194, 195
  polycarbonate(PC, Lexane), 194, 195
  polyester resins, 192
    properties, 192
  polyetheretherketone(PEEK,Victrex),194, 195
  polyether imide(PEI,Ultem), 194-196
  polyether ketone (PEK, victrex) 194, 195
  polyethersulfone (PES, Victrex), 194, 195
  polyethylene fibers, 188
  polyethylene Spectra 900, 184
  polyimides (PMR-15)
    properties, 192
  polyphenylene sulfide(PPS, Raton), 194, 195, 196
  polysulfone (PS, Udel), 194, 195
  PPS(polyphenylene sulfide,Raton),194-196
  processing and fabrication of, 202
    compression molding, 208
    diaphragm forming, 213
    direct roving, 202, 203
    filament winding, 204
    injection molding, 208
    molding, 205
    pultrusion, 204
    resin transfer molding, 208
    spray up, 203
    stamping, 212
    thermoplastic matrix, 209
  processing of thermoplastic composites, 209
    co-woven sheet, 210, 211
    commingled yarn, 210, 211
    continuous consolidation, 212
    continuous forming, 212
    diaphragm forming, 210, 213
    film stacking, 211

Subject Index 455

Polymer matrix composites (cont'd)
   processing of thermoplastic composites (cont'd)
      hydro-rubber forming, 210, 213
      powder coating, 212
      roll forming, 210, 211
      stamping, 212
      step pressing, 210
   properties, 218
   PS (polysylfone, Udel), 194, 195
   Raton(PPS,polyphenylene sulfide),194-196
   reinforcements, 182
   single fiber tests, 200
   short beam shear test, 201
   short fiber composites, 223
      aligned fibers, 223
      critical transfer length, 223
      fiber efficiency, 224, 225
      random fibers, 224
   silane, 197
   specific tensile strength, 182
   static fracture of composites, 226
      fiber-matrix interface debounding, 226
      fiber pull-out, 228
      fracture toughness, 228
      post debounding friction, 227
      stress redistribution, 227
   strain energy release rate, 232-237
   strength of continuous fiber composites,217
      fiber buckling, 221
      fiber orientation effects, 221
      fracture strength, 222
      longitudinal compression, 220
      longitudinal tensile strength, 217
      transverse tensile strength, 219
   surface treatment of fibers, 196
      carbon fibers, 198
      glass fibers, 197
   thermoplastics, 190, 193-194
      advantages, 193
      disadvantages, 195
      mechanical properties, 194
      processing, 209
   thermosetting resins, 190, 192
      bismaleimides resins, 191
      cyanate resins, 191
      epoxy resins, 190, 191
      phenol- formaldehyde, 190
      polyimide resins, 191
      properties, 192
      unsaturated polyester resins, 190, 191
      vinylester resins, 190, 191
   Torlon(PAI,polyamideimide), 194, 195
   Udel (PS, polysulfone), 194, 195
   Ultem(PEI, Polyether imide), 194-196
   unsaturated polyester resins, 190, 191
   vectra (liquid crystal polymer), 194-196

Polymer matrix composites (cont'd)
   Victrex(polyetheretherketone,PEEK),194, 195
   Victrex(polyetherketone,PEK)194,195
   Victrex(polyethersulfone,PES)194, 195
   vinylester resins, 190, 191
      properties, 192
   Young's moduli of fibers, 184
   Young's moduli of resins, 192
Polyphenylene sulfide(PPS, Raton), 194-196
   compressive strength, 194
   elongation to fracture, 194
   glass temperature, 194
   properties, 194
   tensile strength, 194
   thermal expansion, 194
   Young's modulus, 194
Polysulfone (PS, Udel), 194, 195
   compressive strength, 194
   elongation to fracture, 194
   glass temperature, 194
   properties, 194
   tensile strength, 194
   thermal expansion, 194
   Young's modulus, 194
Porous glasses, 351
Portland cement reinforced by steel
   stress strain curve, 314
Powder cloth technique, 109
PPS(polyphenylene sulfide, Raton),194-196
Precipitate-free zone
   next to grain boundaries, 16, 19
Precipitation hardening
   of Ni-base superalloys, 68
Precipitation kinetics, 16
Property microstructure relations
   in Al alloys, 21
PS (polysylfone, Udel), 194, 195
Pyrex borosilicates, 356

Q

Quasicrystals
   Al-Mn alloy, 4, 5
   Al-Cu-Fe alloy, 4, 5
   electron diffraction, 5
   of Al alloys, 3-5

R

R-curve behavior, 247
Raton(PPS,polyphenylene sulfide),194-196
Reaction bonding, 262
Reciprocating engines, 51
Recording materials, 439
Residual stresses
   deformation induced changes, 150

Residual stresses (cont'd)
  in metal-matrix composites, 150
  thermal residual stresses in metal-matrix composites, 115, 137, 148. 163
Retrogression and reaging
  of Al alloys, 43

S

S-glass
  composition, 183
  properties, 184
Saffil fiber (αAl_2O_3), 252
Schott Zerodur glass, 372
SCS(β/α SiC on Si)fiber, 252
SCS-6 SiC/Si_3N_4 ($0°$) composite
  steady state creep, 339
Semihard magnets, 440
Shear deformation, 283
Shear lag model
  in metal-matrix composites, 123
Short fiber composites, 223
  aligned fibers, 223
  critical transfer length, 223
  fiber efficiency, 224, 225
  random fibers, 224
$Si_3N_4+4\%Y_2O_3$
  steady state creep, 339
SiC-RBSN composite
  ultimate tensile strength, 316
SiC/SiC composite
  stress strain after ageing, 336
Silane, 197
Silicate glasses, 358
Silicon carbide, 249, 259, 261
  Alumina composite, 258, 259
  glass composite, 259
Silicon dioxide, 260
Silicon nitride, 249, 256, 259, 261, 262
Sizing materials, 185
Slagsitall glass, 372
Slip plane softening in Al alloys, 29
Slip plane strength in Al alloys, 29
Smooth bar life
  of Al alloys, 34
S-N curves
  of 7075 and X-7075 Al alloys, 34
Sodium borosilicates, 356, 357
Sodium silicates, 357
Soda-lime glass, 358
Soda-lime silicates, 356, 357
Soda-baria-silica, 357
Soda-lithia-silica, 357
Solid Solution hardening
  in nickel-base Superalloys, 64-67
Solubility data

  in Al alloys, 6
Spectra 900 (polyethylene) fiber
  properties, 184
Spinodal decomposition, 354
Squeeze casting
  of metal-matrix composites, 112
Stacking fault energy
  in Al alloys, 2
  in Ni-base superalloys, 68, 81, 82
  and stress exponent in creep, 82
Stacking faults in high T_c superconductors, 406
Static fracture of composites, 226
  fiber-matrix interface debounding, 226
  fiber pull-out, 228
  fracture toughness, 228
  post debounding friction, 227
  stress redistribution, 227
Steam generators, 51
Strain energy release rate, 232-237
Strain localization
  in Al alloys, 29. 32
  in the PFZ of Al alloys, 30, 31
Strain to failure, 247
Strengthening mechanisms
  in Al alloys, 7-10
  in ceramic-matrix composites, 262-340
  in metal-matrix composites, 123-161
  in Ni-base superalloys, 64-94
  in polymer-matrix composites, 214-237
Stress corrosion cracking
  of 2024-T3 Al alloy, 40
  of 2XXX series Al alloys, 40
  of 5XXX series Al alloys, 40, 44
  of 7075-T6 Al alloy, 40
  of 7079-T6 Al alloy, 40
  of 7XXX series Al alloys, 40, 41
  of Al alloys, 40
  of Al-Cu-Mg-Li alloy, 2
  of Al-6Zn-2Mg-XCu alloys, 42
  of pure Al, 41
  retrogression and reaging, 43
Stress differential effect
  in Metal-matrix composite, 163, 166
Stress intensity factor, 30
Stress strain behavior of:
  aged SiC/SiC composite, 336
  borosilicate glass, 247
  ceramic-matrix composites, 311
  effect of fiber clustering, 149
  from Eshelby model, 136
  from finite element calculation, 144
  HMU-BSG, 269
  Nicalon-BSG, 247
  Nicalon-CAS, 262, 270, 282
  Nicalon-LAS composite, 335
  SiC/SiC composite, 275, 283

Subject Index 457

tension vs. compression, 153
Stress strain curves
  see tensile stress strain curves
Superconducting materials, 381
  123 or YBCO superconductor, 402, 403
  2122 superconductor, 403
  2223 superconductor, 403
  A15 superconductors, 391
    critical temperatures, 392
    microstructure, 393, 394
    structure, 392
  applications, 381, 388
  BCS theory, 384
    failure of, 408
  brief historical introduction, 381
  carbon nitride superconductors, 398
  cascades, 387
  chevrel phase superconductors, 401
  coherence length, 384, 385
  Cooper pairs, 384
  critical current in polycrystals, 407
    grain boundary effect, 408
    Josephson coupling, 408
    weak links, 408
    texture effect, 408
  critical magnetic field, 382, 383
  Debye temperature, 384
  dislocations, 406
  dislocation loops, 387
  dislocation pinning, 398
  doped fullerene, 399, 400
  electron-phonon interaction parameter, 385
  flux tubes or flux lines, 385
  fluxiods or vortices, 385
  GLAG theory, 385
  grain boundary pinning, 393
  high T_C superconductors, 401
    cracks, 406
    critical current, 407
    dislocations, 406
    microscopic theory, 408
    stacking faults, 406
    structure modulation, 407
    twins, 403
  historical introduction, 381
  Josephson coupling, 408
  Kramer's model, 394
  London penetration length, 383
  long range order effect, 397
  McMillan equation, 384
  Meissner effect, 381, 382
  microscopic theory, 408
  microstructure, 389
    A15 superconductors, 391
    chevrel phase superconductors, 401
    high T_C superconductors, 401

Superconducting materials (cont'd)
  Niobium-Titanium alloys, 389
    nitride and carbonitrides, 398
    organic superconductors, 399
  Nb_3Al superconductor, 392
  Nb_3Ga superconductor, 392
  Nb_3Ge superconductor, 392
  Nb_3Si superconductor, 392
  Nb_3Sn superconductor, 392, 393
  niobium-titanium alloys, 389
  nitride and carbonitride superconductors, 398
  organic superconductors, 399
  Pande and Suenaga model, 395, 396
  penetration depth, 384
  pinning force, 386
  pinning interactions, 386
  point defects, 387
  precipitate pinning, 397
  radiation effects, 387, 397
  role of microstructure, 385
  stacking faults, 406
  structure modulation, 407
  texture effect of critical current, 408
  theoretical background, 381
  twins, 403
    tweed like structure, 405
    twin boundaries, 405, 406
    twin misorientation, 404
  Type I superconductors, 382, 385
  Type II superconductors, 382, 385
  V_3Ga superconductor, 392, 394
  V_3Si superconductor, 392
  Vortex state, 384
  weak link effect, 408
Superconducting transition temperature
  or critical temperature, 381
Surface crystallization, 351

T

Tensile deformation
  of ceramic-matrix composites, 264, 278-281
Tensile strength
  of ceramic matrices, 254
  of fibers in ceramic composites, 252
  of unidirectional ceramic composites, 314
Tensile stress strain curves
  borosilicate glass, 247
  from finite element calculation, 144
  HMU C/7740 glass, 265,267
  Nicalon-CAS composite, 262, 270
  Nicalon fiber reinforced BSG, 247
  Portland cement/steel, 314

Thermal fatigue
  in Nickel-base Superalloys, 89
Thermal residual stresses
  in metal-matrix composites, 115, 137, 148. 163
Thermoplastics, 190, 193-194
  advantages, 193
  mechanical properties, 194
Thermosetting resins, 190, 192
  bismaleimides resins, 191
  cyanate resins, 191
  epoxy resins, 190, 191, 192
    additives, 192
  phenol- formaldehyde, 190
  polyester resins, 192
  polyimide resins, 191, 192
  properties, 192
  unsaturated polyester resins, 190, 191
  vinylester resins, 190, 191, 192
Thornel C fiber, 252
Titanium boride, 256
Titaium carbide, 256, 259
Titanium nitride, 259
Titanium dioxide, 363
Titanium oxide, 260
Torlon(PAI,polyamideimide), 194, 195
Toughness of 7075 Al alloy, 28
Toughness tree, 24
Transgranular fracture, 29
Tsai-Hill criterion, 266, 267
Turbine materials, 51
Twins in superconductors, 403-405
Type I carbon fiber
  properties, 184
Type I superconductors, 382
Type II carbon fiber
  properties, 184
Type II superconductors, 382
Type III carbon fibers, 186
Tyranno (β SiC +Ti+C+O) fiber, 252

U

Udel (PS, polysulfone), 194, 195
Ultem(PEI, Polyether imide), 194-196
Ultimate tensile strength (UTS)
  HMU-BSG with angle, 265, 269
Unsaturated polyester resins, 190, 191

V

V_3Ga superconductor, 392, 394
V_3Si superconductor, 392
Vectra (liquid crystal polymer),194, 195
Vacuum melting, 51

Victrex(polyetheretherketone,PEEK),194,195
Victrex(polyetherketone,PEK) 194, 195
Victrex(polyethersulfone,PES)194, 195
Vinylester resins, 190, 191
  compressive strength, 192
  density, 192
  elongation to failure, 192
  Poisson ratio, 192
  tensile strength, 192
  thermal expansion, 192
  Young's modulus, 192
Void nucleating particles, 29
Vortex method for metal-matrix
  composites, 110

W

Weibull distribution
  of fiber strength, 315
Weibull shape parameter, 315
Work hardening
  in metal-matrix composites, 132

Y

Yield stress
  difference in tension and compression, 80
  of metal-matrix composites, 152
Yielding
  in metal-matrix composites, 132
  model for single crystal superalloy, 80
Young's modulus
  degradation in Nicalon-CAS, 315
  difference in tension and compression, 143, 157, 158
  effect of alloying on Al,22
  effect of orientation in superalloy, 79
  of Al alloys, 22
  of ceramic matrices, 254
  of E-glass, 184
  of epoxy, 192
  of fibers in ceramic composites, 252
  of glasses, 253
  of glass-ceramic matrices, 253
  of HMU-BSG, 268
  of Kevlar 49, 184
  of liquid crystal polymer (vectra), 194
  of metal-matrix composite, 128
  of Nicalon-CAS, 273
  of polyamide imide, 194
  of polycarbonate, 194
  of polyester, 192
  of polyether ketone, 194
  of polyether imide, 194
  of polyether sulfone, 194
  of polyetherether ketone, 194
  of Polyethylene Spectra 900, 184

Young's modulus (cont'd)
    of polyimides (PMR-15), 192
    of polymer fiber composites, 216, 218, 264, 266
    of polyphenylene sulfide, 194
    of polysulfone, 194
    of S-glass, 184
    of thermoplastic resins, 194
    of thermosetting resins, 192
    of Type I carbon fiber, 184
    of Type III carbon fiber, 184
    of vinyl ester, 192
    of woven composite, 330
        Iso-phase model, 330
        random phase model, 331
    rule of mixtures, 264

Z

ZB glass, 370, 371
ZBL glass, 369-371
ZBLA glass, 370, 371
ZBLALi glass, 370, 371
ZBLALiPb glass, 370, 371
ZBLAN glass, 370
ZBLLi glass, 370
ZBLN glass, 370
ZBLPb, glass, 370, 371
Zirconia, 256
Zirconium carbide, 259
Zirconium nitride, 259

Author Index

A

Abbe, F., 339, 346
Abel, A., 33, 35, 47
Abrikosov, A. A., 384, 409
Aden, A. L., 376, 380
Advani, S. H., 292, 344
Ahearn, J. S., 44, 49
Allen, S., 82, 83, 102
Altenpohl, D., 1, 46
Anderson, W. E., 24, 29, 47
Antolovich, S. D., 55, 59, 60, 61, 62, 99
Anton, D. L., 95, 104
Arai, M., 391, 409
Ardell, A. J., 78, 79, 101
Arderson, S. I., 177
Argon, A. S., 11, 46, 169, 178
Arnberg, L., 9, 46
Arridge, R. G. C., 292, 297, 344
Arsenault, R. J., 82, 107, 108, 115, 117, 129, 131, 132, 135, 137, 143, 148, 158, 160, 161, 163, 164, 168, 169, 170, 172, 174, 176, 177, 178
Ashley, S., 250, 251, 257, 258, 341, 342
Ashton. R. F., 23, 29, 47
Ashburn, J. R., 382, 402, 409
Atkins, A. G., 228, 240
Augis, J. A., 367, 379
Aveston, J., 248, 288, 299, 301, 302, 311, 313, 314, 341
Avrami, M., 367, 378

B

Baaklini, G. Y., 264, 342
Babcock, S. E., 408, 411
Backmann, V., 36, 48
Bader, M. G., 202, 206, 239
Bailey, J. E., 269, 343
Baillie, C., 192, 198, 199, 201, 238
Baker, J. F., 55, 59, 60, 99
Ballard, D., 16, 46
Bansal, N. P., 368, 369, 371, 379
Barber, J. R., 288, 306, 307, 337, 344
Bardeen, J., 382, 384, 409
Barsoum, M. W., 264, 301, 342, 344
Baruch, T. R., 110, 175
Barz Corenzwit, H. E., 401, 411
Batt, J., 333, 345
Beall, G. H., 359, 362, 372, 378, 379
Beardmore, P., 77, 101
Beaumont, P. W. R., 227, 240
Bednorz, J. G., 382, 402, 409

Beeston, B. E. P., 65, 100
Beevers, C. J., 35, 36, 48
Bell, G. H., 365, 379
Benjamin, J. S., 91, 103
Bennett, J. E.,368, 379
Bennett, S. C., 185, 238
Benson, R. A., 237, 240
Bentzen, J. J., 333, 334, 335, 336, 345
Benzeggagh, M. L., 235, 240
Bergeron, C. G., 378
Berry, J. P., 235, 240
Bever, M. B., 372, 379
Beyerley, D. S., 268, 343
Beyers, R., 403, 411
Beyrens, D., 358, 378
Bhagat, R. B., 299, 303, 312, 313, 344
Bhatt, R. T., 262, 264, 283, 288, 290, 316, 333, 334, 335, 342, 343, 344, 345, 346
Biancaniello, F., 16, 46
Binger, W. W., 43,48
Bird, J. E., 82, 102
Birnboim, M. H., 375, 376, 380
Bischoff, E., 264, 317, 342
Bishop, D. J., 405, 411
Blackburn, D. H., 357, 377
Blankenship, C. P. Jr.,1, 9, 18, 31, 38, 39, 40, 46, 47, 48
Bleay, S. M., 339, 342, 346
Block, S., 358, 377
Boesch, W. J., 59, 60, 64, 99
Boettinger, W., 16, 46
Bohlen, J., W., 285, 288, 343
Boisvert, R. P., 285, 343
Bomford, M. J., 91, 103
Bonney, L. A., 285, 343
Borrelli, N. F., 374, 375, 380
Bousuge, M., 33, 334, 345
Bowden, C. M., 375, 380
Bowen, D. H., 248, 258, 259, 341, 342
Bradley, W. L., 237, 240
Brandes, E. A., 49
Brennan, J. J., 248, 253, 255, 258, 259, 260, 264, 265, 284, 285, 317, 341, 342, 343
Brenner, S. S., 108, 174
Bretz, P. E., 36, 48
Bridge, J. E., 79, 83, 102
Briggs, A., 258, 259, 316, 342, 345
Bright, J. D., 290, 292, 344
Brindley, P. K., 110, 175
Brook, G. B., 49
Brown, L. M., 69, 70, 71, 73, 79, 101
Brown. R. H., 43, 48
Bruce, A. J., 368, 369, 379
Brun, M. K., 299, 344
Brus, L. E., 375, 380

Bryant, P. E. C., 52, 99
Buckett, M. I., 389, 409
Budiansky, B., 299, 303, 344
Buljan, S-T, 250, 251, 341
Bunk, W. G. B., 256, 257, 259, 342
Burke, J. E., 363, 378
Burke, P. M., 82, 102
Burnett, D. G., 357, 358, 359, 377
Busby, J., 43, 49
Butkus, L. M., 337, 338, 346
Byrnes, R., 47

C

Cahn, J. W., 359, 364, 378, 379
Cai, X. Y., 408, 411
Calabrese, C., 33, 47
Campbell, A. M., 387, 409
Camus. G., 333, 334, 335, 336, 345
Canada, H. B., 59, 60, 64, 99
Cannon, W. R., 339, 346
Cao, H. C., 264, 316, 317, 342, 345
Caputo, A., J., 256, 257, 342
Carlson, K. D., 399, 410
Carlsson, L. A., 236, 240
Caron, S., 110, 175
Carpenter, H. W., 285, 288, 343
Carsson, L. A., 202, 206, 239
Carter, R. D., 35, 36, 48
Castillo, R., 85, 86, 103
Castino, F., 201, 239
Cerny, G. A., 254, 255, 276, 342
Chamin, R. A., 237, 240
Chamis, C.C., 222, 239
Champier, G., 36, 48
Chandhari, P., 408, 411
Chang, K. M., 95, 103
Charles, R. J., 357, 358, 377
Chatterjee, A., 250, 264, 341
Chawla, K. K., 183, 238
Chen, C. H., 405, 411
Chen, O., 285, 343
Chen, P. E., 225, 239
Chen, T-K., 169, 172, 178
Cheremisinoff, N. P., 199, 200, 201, 239
Chermant, J. L., 339, 346
Cheruva, N. S., 52, 85, 86, 99, 103
Chesney, P., 110, 175
Chevrel, R., 401, 410
Chiang, Y-C., 260, 300, 308, 310, 342, 344
Chikazumi, S., 440
Chin, G. Y., 398, 410
Cho, C. D., 288, 306, 307, 337, 344, 346
Chou, T. W., 131, 177, 196, 224, 239, 247, 248, 250, 252, 258, 260, 261,

262, 263, 264, 265, 266, 268, 269, 270, 271, 272, 273, 274, 275, 277, 283, 284, 285, 299, 300, 303, 308, 310, 312, 313, 314, 315, 316, 317, 320, 321, 322, 323, 324, 325, 326, 328, 330, 332, 333, 337, 339, 341, 342, 343, 344, 345, 346
Christin, F., 277, 343
Christman, T., 169, 172, 178
Christodoulou, L., 37, 40, 41, 43, 44, 48
Chu, C. W., 382, 402, 409
Chyung, K., 285, 333, 343
Clauer, A. H., 83, 91, 92, 102
Cleave, J. F., 43, 49
Clem, J. R., 408, 411
Cogswell, F N., 210, 211, 239
Cojean, D., 285, 287, 343
Coleman, B. D., 314, 315, 345
Collins, J. M., 260, 342
Comninou, M., 337, 346
Cook, J. L., 110, 175
Cooke, D. C., 44, 49
Cooke, R. G., 247, 264, 268, 277, 282, 284, 285, 337, 342, 343
Coole, D. C., 49
Cooper, A. S., 401, 411
Cooper, G. A., 248, 288, 299, 301, 302, 311, 313, 314, 341
Cooper, L. N., 382, 384, 409
Cooper, R. F., 285, 333, 343
Copley, S. M., 69, 72, 73, 77, 78, 83, 87, 101
Corleto, C. R., 237, 240
Cormia, R. L., 366, 378
Corne, P., 288, 291, 344
Cornie, J. A., 112, 175, 260, 342
Cotterell, B., 200, 239
Cottrell, A. H., 228, 240
Cox, B., 299, 308, 337, 344
Cox, D. E., 397, 410
Cox, H. L., 123, 126, 176, 214, 239
Coyle, T. W. 288, 290, 292, 344
Crangle, J., 440
Cranmer, D. C., 288, 292, 344
Crowe, C. R., 168, 178
Cudd, R. L.,43, 49
Cullity, B. J., 440
Cuomo, J. J., 407, 411
Curl, R. F., 399, 410
Curtin, W. A., 316, 317, 345
Curtis, P. T., 269, 343
Curwick, L. R., 58, 99
Cybulsky, M., 52, 99

D

Dabbash, G., 400, 410

Author Index

Daeubler, M. A., 28, 30, 42, 47
Daniel, A. M., 346
Daniel, I. M., 314, 320, 345
Darolia, R., 89, 92, 95, 104
Davidge, R. W., 316, 345
Davidson, D. L., 169, 170, 172, 178
Davies, P., 235, 240
Davies, R. G., 57, 58, 59, 76, 77, 78, 94, 99
Davis, M. E., 408, 411
Day, P., 407, 411
Dayananda, M. A., 47
De Charentenay, F. X., 235, 240
Decker, R. F., 56, 57, 61, 73, 79, 85, 87, 99, 101
Decroux, M., 401, 411
Delannay, F., 112, 175
Deluca, J. P., 378
Denanot, M. F., 406, 411
Derby, B., 346
Deruyttere, A., 112, 175
Deshmukh, U. V., 288, 290, 292, 344
Despres, J. F., 285, 287, 343
Dew-Hughes, D., 387, 391, 393, 395, 409, 410
Dhignra, A. K., 108, 109, 112, 168, 174, 175, 178
DiCarlo, J. A., 288, 290, 344
Diefendorf, R. J., 285, 343
Dillamore, I. L., 65, 100
Dimos, D., 408, 411
Dinger, T. R., 403, 411
Divecha, A. P., 108, 109, 110, 174
Dix, E. H., 43, 48
Dix, E. M., 40, 48
Dobb, M. G., 187, 188, 238
Dobbs, J. R., 95, 103
Doherty, R. D., 32, 37, 40, 41, 43, 44, 47, 48
Dolan, G. J., 405, 411
Donachie, M. J., 85, 103
Donald, I. W., 253, 255, 341
Doremus, R. H., 351, 352, 353, 355, 356, 358, 359, 363, 364, 365, 367, 368, 369, 371, 373, 374, 375, 376, 377, 379
Dorn, J. E., 82, 102
Douglas, R. W., 357, 358, 359, 377
Drake, 342
Drexhage, M. G., 369, 370, 379
DuBost, B., 36, 48
Duclos, S. J., 400, 410
Dudzinski, N., 21, 22, 46
Duhl, D. N., 64, 79, 80, 81, 82, 83, 87, 88, 89, 92, 100, 102, 103
Duquette, D. J., 64, 100
Duva, J. M., 28, 30, 42, 47

E

Eagan, R. J., 378
Eick, R. H., 400, 410
Einstein, A., 352, 377
Eiselstein, H. L., 59, 60, 99
Eisenmann, J. R., 237, 240
Eldridge, J. I., 262, 342, 346
Elliott, J. F., 76, 101
Ellis, J. R., 338, 346
Engler, E. M., 403, 411
English, G. C., 43, 48
Englnd, R. O., 44, 49
Erdogen, F., 35, 36, 48
Erickson, G. L., 51, 64, 99, 100
Erturk, T., 338, 346
Eshelby, J. D., 128, 131, 176
Evans, A. G., 248, 264, 268, 299, 303, 308, 316, 317, 333, 337, 341, 342, 343, 344, 345, 346
Evans, U. R., 44, 49
Evetts, J. E., 387, 409
Ezaki, H., 64, 100
Ezz, S. S., 80, 102

F

Faber, K. T., 292, 344
Fang, N., 283, 284, 343
Faraday, M., 369, 373, 379
Fastnacht, R. A., 408, 411
Fehrenbacher, L. L., 254, 255, 276, 342
Felbeck, D. K., 228, 240
Feng, C. R., 135, 148, 168, 169, 172, 177, 178
Field, R. D., 55, 59, 60, 89, 92, 99, 103
Fietz, W. A., 393, 410
Filipuzzi, L., 333, 334, 335, 336, 345
Fine, M. E., 33, 47, 48
Fischer, O., 401, 411
Fisher, J. C., 67, 76, 100
Fisher, R. M., 107, 108, 160, 174
Fishman, S. G., 108, 109, 110, 112, 174, 175
Fistiropoulos, K., 410
Fitz-Randolph, J., 227, 240
Fleischer, R. L., 65, 66, 74, 100
Fleming, R. M., 400, 410
Flemmings, M. C., 110, 112, 175
Fostiropoulos, K., 399, 410
Foulds, W. T., 337, 346
Frenkel, D., 364, 378
Frety, N., 333, 334, 345
Freiman, S. W., 288, 292, 344
Frischat, G. H., 369, 379
Froyen, L., 112, 175

Fukuda, H., 316, 345
Fukunaga, H., 314, 322, 324, 345
Funayama, O., 260, 261, 342
Furusaka, M., 391, 409

G

Gai, P. L., 407, 411
Gallaghar, W. J., 403, 411
Gamble, R. P., 84, 102
Gammel, P. L.,405, 411
Gangloff, R. P., 36, 38, 39, 40, 48
Gannon, J. A., 191, 238
Gao, L., 382, 402, 409
Gao, Y. C., 200, 239
Garcia, R., 371, 373, 379
Garg, A. C., 238
Gavaler, J. R., 391, 410
Gayle, F. W., 18, 23, 29, 46, 47
Geiser, U., 399, 410
Gell, M., 73, 85, 87, 88, 89, 101, 103
Genzel, L., 373, 379
Gerold, V., 69, 79, 101
Gerosshen, T. J., 103
Geshkenbein, V. B., 405, 411
Gessinger, G. H., 96, 103
Gest, R. J., 41,48
Giamei, A. F., 64, 87, 88, 100, 103
Gibbons, T. B., 85, 103
Gibbs, J. W., 352, 377
Gill, M., 82, 83, 102
Gillespie, J. W., 202, 206, 236, 239, 240
Gilmer, G. H., 364, 378
Ginty, C. A., 250, 251, 341
Ginzburg, V. L., 381, 384, 409
Glarum, S. H., 400, 410
Gleiter, H., 69, 71, 72, 73, 75, 101
Goan, J. C., 198, 239
Goetteler, R. W., 292, 344
Goldman, E. H., 95, 103
Gorden, J. R., 49
Gotoda, H., 391, 409
Graham, C. D. Jr., 415
Grande, D. H., 288, 292, 297, 343
Grant, N. J., 61, 62, 65, 67, 99, 100
Gray, H. R., 250, 251, 341
Gray, R. A., 168, 178
Green. J. A. S., 44, 49
Greenwood, G. W., 355, 377
Greer, A. L., 362, 367, 378
Griffin, C. W., 290, 292, 344
Grimes, G. C., 314, 320, 345
Gronsky, R., 406, 411
Grosskreutz, J. C., 35, 48
Grossman, J. E., 103
Grossmann, D. G., 372, 379

Gruhl, W., 40, 48
Gubser, D. U., 403, 404, 411
Gungor, M. N., 110, 175
Gurland, J., 25, 47
Gurson, A. L., 169, 178

H

Haberkorn, H., 69, 79, 101
Habib, F. A., 247, 264, 268, 277, 282, 284, 285, 337, 342, 343
Haddon, R. C., 400, 410
Hagashi, T.,110, 175
Hagel, W. C., 56, 57, 61, 62, 63, 64, 68, 69, 73, 79, 80, 86, 89, 99, 100, 102
Hahn, G. T., 25, 26, 27, 30, 47
Hall, C. M., 1
Hall, D. W., 374, 375, 380
Hall, E. O., 81
Haller, W., 357, 358, 377
Ham, R. K., 33, 35, 47, 69, 70, 71, 73, 79, 101
Hamel, F. G., 110, 175
Hammel, J. J., 358, 359, 378
Hancox, N. L., 193, 194, 195, 199, 238
Hannache, H., 277, 343
Hardy, G. F., 391, 410
Harrigan, W. C., 168, 178
Harris, B., 226, 227, 239, 247, 264, 268, 277, 282, 284, 285, 337, 342, 343
Hartman, G. A., 337, 346
Hashemi, S., 234, 240
Hashin, Z., 277, 343
Hasson, D. F., 168, 178
Hatch, J. E., 49
Haus, J. W., 375, 380
Hawkins, W. L., 54, 99
Hayashi, 235, 240
Haymet, A. D., 364, 378
Hazlett, T. H., 65, 100
He, M. Y., 248, 341
Headinger, M., 250, 251, 254, 255, 341, 342
Heath, J. R., 399, 410
Hebard, A. F., 400, 410
Hecht, N., 254, 255, 341
Hergenrother, P. M., 190, 238
Herman, H., 378
Heroult, P., 1
Herricks, R. J., 85, 103
Herring, C., 87
Heubaum, F. H., 18, 46
Hildeman. G. J., 44, 49
Hillig, W. B., 365, 379
Hillman, H., 389, 409
Hindman, D. L., 250, 251, 341

Holland, H. J., 374, 375, 380
Hollingsworth, E. H., 43, 48
Holroyd, N. J. H., 37, 40, 41, 43, 44, 48
Homeny, J., 285, 343
Holmes, J. W., 288, 306, 307, 337, 338, 339, 344, 346
Honda, K., 288, 344
Hong, K. C. C., 288, 292, 297, 343
Hong, M., 405, 411
Hoover, W. R., 108, 110, 174, 175
Hopkins, B. E., 85, 103
Hopkins, S. W., 73, 101
Hor, P. H., 382, 402, 409
Horiuchi, S., 407, 411
Hornbogen, E., 1, 9, 30, 46, 47, 69, 71, 72, 73, 101
Howe, J. M., 167, 178
Howson, T. E., 92, 103
Hoyt, J. T., 256, 342
Hsueh, C-H., 292, 344
Hu, C. L., 260, 342
Hu, X. Z., 236, 240
Huang, X. G., 301, 344
Huang, Z. J., 382, 402, 409
Huffman, D. R., 399, 410
Hull, D., 107, 174, 183, 197, 217, 220, 221, 238
Hulm, J. D., 391, 410
Hunt, J. D., 365, 379
Hunt, W. H. Jr., 109, 174
Hutchinson, J. W., 248, 299, 303, 341, 344
Hutter, R. K., 285, 343, 344
Hyatt, M. V., 24, 29, 47
Hyde, J. M., 356, 377

I

Ichihara, M., 76, 101
Im, J., 11, 46, 169, 178
Inguva, R., 375, 380
Inoue, S., 64, 100
Ishai, O., 236, 240
Ishida, H., 193, 197, 238, 239
Isoda, T., 260, 261, 333, 342, 345
Itoh, T., 260, 261, 342
Ives, L., 16, 46
Iwata, M., 333, 345

J

Jablonski, D. A., 333, 345
Jackson, J. J., 85, 103
Jackson, K. A., 364, 365, 379
Jacubovics, J., 440
Jaffe, M., 186, 187, 238
Jagannadham, J., 408, 411

Jain, R. K., 374, 380
James, P. F., 359, 362, 377, 378
Jata, K. Y., 29, 47
Jensen, R. R., 76, 101
Jero, P. D., 292, 344
Jiles, D., 440
Jin, S., 406, 408, 411
Jinn, T-J., 292, 344
Johnson, B., 333, 335, 345
Johnson, D. D., 250, 252, 341
Johnson, D. J., 185, 187, 188, 238
Johnson, R. A., 387, 409
Johnson, W. S., 236, 240
Johnston, N. J., 190, 238
Johnston, T. L., 57, 58, 59, 77, 78, 94, 99
Jonas, J. J., 161, 177
Jones, J. W., 337, 346
Jones, R. M., 215, 239
Josephsen. B. D., 382, 409

K

Kagawa, Y., 288, 344
Kaiser, D. L., 408, 411
Kaiser, J. D., 262, 342, 344, 346
Kalyaniwalla, N., 375, 380
Kamerlingh Onnes, H., 381, 409
Kamiya, A., 260, 261, 342
Kammlott, G. W., 408, 411
Kancheev, O. D., 79, 102
Kanei, A., 288, 292, 344
Kangutkar, P., 264, 342
Kannedy, J. M., 344
Kao, S. C., 373, 375, 376, 379, 380
Karandikar, P. G., 247, 248, 250, 262, 263, 264, 265, 266, 268, 269, 270, 271, 272, 273, 274, 283, 284, 285, 312, 313, 316, 317, 321, 322, 324, 325, 333, 337, 341, 342, 343, 345, 346
Karmarker, S. D., 108, 110, 174
Katz, A. P., 333, 334, 345
Kawamoto, Y., 356, 377
Kawata, K., 110, 175
Kawata, K., 316, 345
Kear, B. H., 69, 72, 73, 77, 78, 83, 84, 101, 102
Keer, J. G., 288, 311, 344
Keith, H. D., 408, 411
Kellher, K. H., 407, 411
Kelly, A., 64, 75, 101, 123, 126, 176, 185, 186, 187, 193, 194, 195, 199, 223, 227, 238, 239, 240, 248, 288, 299, 301, 302, 311, 313, 314, 341
Kelly, M., 70, 71, 73, 79, 101
Kelly, M. A., 285, 343

Kelton, K. F., 362, 367, 378
Kennedy, J. M., 288, 292, 337, 344, 346
Kerans, R. J., 250, 264, 288, 291, 292, 297, 298, 341, 344
Kerker, M. L., 376, 380
Ketterer, M. E., 212, 239
Khachaturyan, A. G., 405, 411
Kim, C. T., 118, 121, 176
Kim, H. J., 250, 251, 341
Kim, J., 339, 341, 346
Kim, J. K., 181, 187, 189, 192, 196, 198, 199, 200, 201, 205, 226, 227, 228, 238, 239, 240
Kim, R. Y., 237, 240, 264, 265, 268, 342, 343
Kimber, A. C., 288, 311, 344
Kini, A. M., 399, 410
Kinloch, A. J., 234, 240
Kinzig, B., 200, 239
Kishi, T., 285, 343, 345
Kissinger, H. E., 368, 379
Knowles, K. M., 316, 317, 345
Ko, F., 250, 252, 341
Ko, S. H., 371, 379
Koch, E. F., 393, 410
Koczak, M. J., 44, 49
Koenig, J. L., 193, 238
Konitzer, D. G., 95, 103
Konur, O., 229, 240
Kortan, A. R., 400, 410
Kotil, T., 337, 346
Koul, A. K., 85, 86, 103
Kramer, E. J., 394, 410
Kratschmer, W., 399, 410
Krawitz, A. D., 137, 148, 177
Kreibig, U., 373, 379
Kroschwitz, J. I., 191, 238
Kroto, H. W., 399, 410
Kuhlmann-Wilsdorf, D., 108, 174
Kumar, P., 299, 303, 312, 313, 344
Kumar, P. M., 110, 174
Kung, C. Y., 48
Kuo, W. S., 299, 303, 312, 313,315, 320, 322, 323, 324, 325, 326, 328, 330, 332, 344, 345
Kuroda, K., 406,411
Kwo, J. R., 405, 411
Kwok, W. K., 399, 410

L

Labusch, R., 66, 100
Lackey, W. L., 254, 255, 276, 342
Lahrman, D. F., 89, 92, 95, 103
Laird, C., 33, 35, 47, 277, 343
Laird, J. A., 378
Lamb, S., 54, 99

Lamouroux, F.,333, 345
Lamon, J., 288, 291, 344
Landau, L., 381, 384, 409
Langdon, T. G., 339, 346
Lankford, J., 277, 343
Lannoo, M., 380
Laplaca, S. J., 407, 411
Larbalestier, D. C., 389, 390, 391, 408, 409, 411
Larsen, D. C., 253, 255, 341
Laughlin, D. E., 30, 31, 47
Laughner, J. W., 288, 290, 344
Lavernia, E. J., 110, 175
Lawless, B. H., 55, 59, 60, 99
Lawrence, C. W., 346
Lawrence, P., 297, 344
Layden, G. K., 248, 253, 255, 258, 259, 265, 341, 342
Leatham, A., 110, 175
Lee J. W., 314, 320, 345
Lee, E. W., 35, 36, 48
Lee, J. K., 118, 121, 176, 292, 344
Lee, P. J., 390, 391, 409
Lee, S., 201, 239
Lee, S. M., 205, 206, 239
Lees, L. H., 225, 239
Legouse, F. K., 408, 411
Le Guillou, J-C., 358, 378
Lehman, R. H., 248, 250, 262, 341
Leming,R. M., 410
Lemkey, F. D., 84, 102
Leontewa, A., 378
Leverant, G. R., 73, 83, 101, 102
Levin, E. M., 358, 377
Levitt, S. R., 248, 284,341
Levy, M., 358, 378
Levy, S. A., 23, 29, 47
Lewandowski, J. J., 165, 167, 169, 178
Lewendowski, J. J., 110, 174
Lewis, M. H., 285, 343, 346
Liang, F. L., 112, 176
Liang, W. W., 250, 251, 341
Lifshitz, E. M., 355, 377
Liholt, H., 131, 135, 176
Lim, G., 403, 411
Limaye, S. Y., 290, 292, 344
Lind, R. C., 374, 380
Lindley, T. C., 35, 36, 48
Linger, K. R., 337, 345
Lippens, P. E., 380
Lissart, N., 288, 291, 344
Little, W. A., 410
Liu, C., 110, 174
Liu, C. T., 112, 176
Liu, H. K., 260, 342
Liu, L. C., 380

Llorca, J., 161, 169, 170, 172, 177, 178
Lloyd, D., 161, 163, 164, 177
Lloyd, D. J., 167, 169, 170, 172, 178
Lohne, O., 9, 46
London, F., 381, 383, 384, 409
London, H., 381, 383, 384, 409
Low, J. R. Jr., 25, 47
Lowden, R. A., 256, 257, 342
Lu, C-C., 260, 261, 342
Ludtka, G. M., 30, 31, 47
Luh, E. Y., 333, 345
Luhman, T., 387, 391, 393, 409, 410
Lulay, K. E., 161, 163, 164, 177
Lütjering, G., 28, 30, 33, 34, 42, 47, 48
Lyman, T., 43, 48
Lynch, S. P., 47
Lyons, K. B., 400, 410

M

MaCartney, L. N., 344
MacCrone, R. K., 378
MacDowell, J. F., 359, 378
Macedo, P. B., 357, 358, 377
Mackay, D. B., 201, 239
Mackenzie, J. D., 366, 378
Mackin, T. J., 288, 292, 344
Maeda, H., 407, 411
Mah, T., 333, 334, 345
Mahajan, S., 398, 410
Mahon, G. J., 167, 178
Mai, Y. W., 181, 187, 189, 192, 196, 198, 199, 200, 201, 205, 226, 227, 228, 236, 238, 239, 240
Makel, D. D., 174
Makhija, A. V., 400, 410
Malakondaiah, G., 33, 48
Mall, S., 250, 264, 268, 337, 341, 343, 346
Mallet, C., 277, 343
Mallon, P. J., 213, 239
Maloney, L. D., 250, 251, 341
Mandell, J. F., 288, 292, 297, 343
Manders, P. W., 314, 345
Maniar, G. N., 79, 83, 102
Mannhart, J., 408, 411
Manoharan, M., 165, 167, 178
Marek, M., 41, 43, 48
Marezio, M., 401, 411
Margolin, H., 112, 176
Margolis, W. S., 237, 240
Margraf, T., 371, 379
Maribo, D. W., 174

Marshall, D. B., 248, 249, 264, 288, 289, 299, 308, 316, 317, 337, 341, 342, 344
Marston, T. U., 228, 240
Martens, V., 72, 101
Martin, J. W., 33, 47
Martin, T. W., 198, 239
Masounave, J., 110, 175
Masumura, R. A., 395, 410
Masur, L., 112,175
Matsui, Y., 407, 411
Matthews, F. L., 229, 240
Matthias, B. T., 401, 411
Matusita, K., 368, 378, 379
Maurer, R. D., 363, 378
Mazdiyasni, K. S., 333, 334, 345
Mazurin, O. V.,377
Mehrebian, R., 110, 175
Mendelson, A., 121, 176
Mendiratta, M. G., 333, 334, 345
Merchant, R. H., 25, 47
Mervyn, D. A., 92, 103
Meshii, M., 33, 34, 47
McAdams, L. V., 191, 238
McCarthy, G. P., 103
McCartney, L. N., 299, 308, 344
McCullum, D., 254, 255, 341
McDanels, D. L., 109, 170, 174, 178
McEvily, A. J., 43,49
McGahay, V., 358, 377
Mcguire, T. R., 407, 411
Mckinnell, J. C., 390, 409
McMeeking, R. M., 337, 346
McMillan, P. W., 253, 255, 341, 359, 363, 372, 377, 378
McQueen, H. J., 161, 177
McTague, J. P., 364, 378
Mehrabian, R., 16, 46
Mehrebian, R., 175
Meingast, C.,391, 410
Meissner, W., 381, 384, 409
Mendelson, A., 176
Mendiratta, M. G., 345
Meng, R. L., 382, 402, 409
Merchant, R. H., 47
Mervyn, D. A., 103
Meshii, M., 47
Miannay, D., 36, 48
Michaud, V. J., 112, 175
Mie, G., 373, 379
Mihalisin, J. R., 100, 102
Miller, A. C., 36, 48
Miller, B., 400, 410
Miller, K, J., 109, 174
Minford, E., 337, 346
Mirkin, I. L., 79, 102
Mishima, H., 65, 66, 67, 100

Mitchell, W. J., 78, 101
Monju, Y., 391, 410
Monthioux, M., 285, 287, 343
Mori, T., 117, 176
Morimoto, T., 131, 135, 176
Morinaga, M., 64, 100
Moriya, Y., 357, 359, 377
Morral, J. E., 43, 49
Mortensen, A., 112, 175, 260, 342
Moschler, J. W., 250, 264, 341
Mostovoy, S., 232, 240
Moynihan, C. T., 368, 369, 370,379
Mukherjee, A. K., 82, 102
Muller, A. J., 400, 410
Muller, K. A.,382, 402, 409
Munjal, V., 78, 79, 101
Munro, M., 201, 239
Mura, T., 117, 129, 131, 176
Murata, Y., 64, 100
Murphy, D. W., 400, 410
Murphy, M. C., 226, 240
Murray, C. A., 405, 411
Murthy, V. S. R., 285, 343

N

Nabarro, F. R. N., 87
Nakahara, S., 398, 406, 410, 411
Nakajima, H., 359, 378
Nakano, K., 260, 261, 342
Napolitano, A., 358, 377
Nardone, V. C., 93, 103, 264, 265, 267, 268, 269, 342
Naryan, J., 408, 411
Naslain, R., 256, 257, 259, 277, 333, 334, 335, 336, 342, 343, 345
Naudin, F., 359, 377, 378
Needleman, A., 161, 169, 170, 172, 177, 178
Neevs, A. E., 375, 376, 380
Neilson, G. F., 362, 369, 370, 378
Neite, G., 100
Nell, J. M., 61, 62, 99
Nembach, E., 68, 72, 75, 76, 100, 101, 393, 410
Nes, E., 9, 46
Nethercott, R. B., 47
Nicholson, R. B., 64, 70, 71, 73, 75, 79, 101
Noguchi, O., 58, 99
Nordheim, R., 67, 100
Nordone, V. C., 128, 176
Nourbakhsh, S., 112, 176

O

O'Bradaigh, C. M., 213, 239

O'Brien, S. C., 399, 410
O'Brien, T. K., 237, 240
Ochiai, S., 57, 58, 99
Ochsenfeld, R., 381, 384, 409
Ogilvy, A., 110, 175
Oliver, W. C., 288, 289, 344
Orlov, A. N., 387, 409
Orowan, E., 68, 69, 75, 90, 100
Osamura, K., 391, 409
Otter, W., 374, 380
Outwater, J. D., 226, 240
Oxtoby, D. W., 364, 378
Oya, Y., 58, 99,

P

Pachalis, J. R., 339, 346
Pagano, N. J., 237, 240, 250, 264, 265, 288, 291, 341, 342, 344
Paidar, V., 80, 102
Palstra, T. T. M., 400, 410
Pande, C. S., 381, 393, 395, 398, 403, 404, 410, 411
Pantano, C. G., 343
Paresh, J., 250, 252, 341
Paris, P. C., 35, 36, 48
Parker, E. R., 65, 100
Parthasarathi, T. A., 288, 291, 292, 297, 298, 344
Parvizi-Majidi, A., 247, 248, 250, 260, 261, 262, 250, 252, 264, 265, 268, 269, 274, 275, 277, 283, 284, 285, 316, 317, 333, 341, 342, 343, 344, 345
Pasto, A. E., 250, 251, 341
Patano, C. G., 285, 337, 343, 346
Patrick, R. L., 232, 240
Pearson, D. D., 84, 103
Pedersen, O. B., 161, 163, 164, 177
Pelloux, R. M., 33, 47
Pelloux, R. M. N., 65, 100
Penty, R. A., 109, 174
Petch, N. J., 81
Peters, M., 18, 36, 46, 47, 48
Peters, P. W. M., 322, 324, 345
Petrasek, D. W., 109, 174
Petter, R. J., 109, 174
Pfeiffer, I., 387, 409
Phillip, D. C., 227, 240
Phillips, D. C., 248, 258, 259, 341, 342
Phillips, R. E., 283, 316, 343
Phillips, V. A., 78, 101, 102
Piascik, R. S., 35, 36, 48
Pickens, J. R., 18, 44, 46, 49
Piearcey, B. J., 63, 100
Piggott, M. R., 200, 239
Pillai, U. T. S., 163, 178

Pincus, A. G., 358, 377
Pipers, R. B., 236, 240
Plichta, M. R., 118, 121, 176
Plueddemann, E. P., 197, 239
Pluvinage, P., 275, 277, 283, 333, 343, 345
Poh, J., 192, 198, 199, 201, 238
Polmear, I. J., 13, 46
Pope, D. P., 80, 102
Porai-Koshits, E. A., 377
Poulose, P. K., 43, 49
Powell, P. C., 210, 211, 239
Prasad, N. E., 33, 48
Pratt, P. L., 35, 48
Prel, Y. J., 235, 240
Prescott, R., 198, 239
Prewo, K. M., 128, 176, 248, 253, 255, 258, 259, 260, 264, 265, 267, 268, 269, 282, 283, 284, 285, 314, 316, 317, 332, 333, 335, 337, 341, 342, 343, 345, 346
Prigent, J., 401, 411
Pryce, A. W., 264, 265, 268, 269, 342
Psioda, J. A.,27, 47

Q

Qi, G., 285, 343
Quenisset, J. M., 277, 343
Quist, W. E., 24, 29, 47

R

Rabier, J., 406, 411
Rack, H. J.,110, 175
Raju, K. N., 33, 48
Ramakrishnan, V., 337, 346
Ramirez, A. P., 400, 410
Rao, K. T. V., 36, 37, 48
Rao, P. G. N., 374, 375, 376, 379, 380
Rao, P. R., 33, 48
Rawlings, R. D., 393, 410
Ray, R. A., 407, 411
Raynor, D., 78, 102
Rechiniac, C., 288, 291, 344
Reifsnider, K. L., 237, 240
Reno, R., 16, 46
Reuss, A., 152, 177
Rice, R. W., 248, 341
Richards, C. E., 35, 36, 48
Richards, E. G., 86, 103
Ridder, S., 16, 46
Riek, R. G., 110, 175
Ripling, E. J., 232, 240
Risbud, S. H., 380
Ritchie, R. O., 35, 36, 48

Ritter, A. M., 110, 174, 299, 303, 312, 313, 344
Roach, D., 337, 346
Roberts, B. W., 364, 379
Roberts, S., 374, 376, 380
Robertson. W. D., 44, 49
Robinson, M. 369, 370, 379
Rohatgi, P. K., 110, 175
Rosamilia, J. M., 400, 410
Rosen, B. W., 220, 239, 277, 316, 343, 345
Rosen, M., 16, 46
Rosenfield, A. R., 25, 26, 27, 30, 47
Ross, E. W., 56, 57, 62, 63, 64, 86, 99, 100
Rosseinsky, M. J.,400, 410
Rossignol, J. Y., 277, 343
Rothschilas, R., 236, 240
Rousseau, C. Q., 337, 346
Ruh, R., 333, 334, 345
Ruhle, M., 264, 317, 342
Russell, A. J., 235, 236, 240
Ryum, N., 9, 15, 46

S

Sabatini, R. L., 406, 411
Sabetay, L., 36, 48
Safoglu, R., 11, 46
Sahm, P. R., 37, 48
Saka, H., 406, 411
Sakka, S., 368, 379
Salama, K., 33, 48
Sambell, R. A. J., 248, 341, 342
Sanders, T. H. Jr.,15, 19, 23, 29, 47
Sandor, B. I., 33, 47
Sandstrom, R. L., 403, 411
Santhanam, A. T., 397, 410
Santner, J, S., 33, 47
Sarkar, A., 357, 377
Sarkar, B., 41, 43, 48
Sato, K., 260, 261, 342
Saville, B. P., 187, 188, 238
Savoy, R. J., 403, 411
Sbaizero, O., 264, 268, 316, 317, 342, 343, 345
Scamans, G. M., 41, 48
Scanlan, R. M., 410
Schaffhauser, A. C., 254, 255, 276, 342
Schirber, J. E., 399, 410
Schiroky, G. H., 257, 258, 342
Schneemeyer, L. F., 405, 411
Schneibel, J. H., 87, 103
Schreiner, M. E., 250, 251, 341
Schrieffer, J. R., 382, 384, 409
Schulte, K.,322, 324, 345
Schultz, A. J., 399, 410

Schuster, D. M., 110, 175
Schwartz, H. S., 266, 267, 343
Schwartz, R. T., 266, 267, 343
Schweitert, H. R., 316, 345
Scott, V. D., 339, 346
Sears, G. W., 365, 379
Seeber, B., 401, 411
Sela, A., 236, 240
Sengers, J. V., 378
Sergent, M., 401, 411
Seward, T. P., 359, 378
Shah, D. M., 80, 81, 95, 102, 104
Shah, S. P., 292, 344
Shandar, K. R., 33, 48
Shaw, B. J., 410
Shaw, G. C., 35, 48
Shaw, N. J., 288, 290, 344
Shaw, T. M., 403, 407, 411
Sherby, O. D., 82, 102
Sherwood, R. C., 406, 408, 411
Shetty, D. K., 290, 292, 297, 344
Shi, N., 115, 117,135, 137, 143, 148, 158, 161, 163, 169, 170, 172, 176, 177
Shibata, S., 117, 176
Shivshankar, S. A., 407, 411
Showman, H. G., 250, 252, 341
Shuler, S. F., 337, 346
Sigl, L. S., 316, 345
Signorelli, R. A., 109, 174
Silcock, J. M., 78, 102
Simmons, J. H., 358, 362, 365, 377, 379
Simon, R., 382, 409
Sims, C. T., 56, 57, 61, 62, 63, 64, 68, 69, 73, 79, 80, 86, 89, 99, 100, 102
Sinclair, J. H., 222, 239
Singer, L. S., 185, 238
Singh, A. K., 403, 404, 411
Singh, P. M., 169, 178
Singh, R. N., 288, 290, 339, 344, 346
Skibo, M. D., 110, 175
Slavik, D. C., 38, 39, 40, 48
Slyozov, V. V., 355, 377
Smalley, R. E., 399, 410
Smallman, R. E., 65, 100
Smashey, R. W., 63, 100
Smiley, A., 236, 240
Smith, A., 382, 409
Smith, D. W., 374, 375, 380
Smith, G. C., 32, 33, 47
Smith, G. L., 369, 370, 379
Smith, L. F., 137, 148, 177
Smith, P. A., 264, 265, 268, 269, 342
Smith, R. F., 109, 174
Snyder, J. E., 250, 251, 341
Sondergaard, N. A., 174
Sorenson, B. W., 257, 258, 342

Spear, K. E., 285, 343
Spearing, S. M., 268, 342, 343
Speidel,M. O., 37, 41, 48
Springer, G. S., 110, 175, 264, 265, 268, 269, 342
Srinivasan, G., 357, 377
St. Hilaire, G. M., 338, 342, 346
Staley, J. T., 22, 23, 24, 25, 28, 29, 46, 47
Stang, H., 292, 344
Starke, E. A. Jr.,1, 9, 15, 18, 23, 28, 29, 30, 31, 33, 34, 35, 36, 38, 39, 40, 41, 42, 43, 46, 47, 48
Starrett, S., 333, 335, 345
Steif. P. S., 316, 345
Stephens, J. R., 109, 174
Stephenson, G. B., 378
Stinton, D. P., 154, 255, 256, 257, 276, 342
Stobbs, W. M., 161, 163, 164, 177
Stoloff, N., 175
Stoloff, N. S., 51, 56, 57, 58, 62, 63, 64, 68, 69, 73, 76, 79, 80, 86, 89, 99, 100, 102, 103, 110, 112, 176
Stolz, R. E., 33, 47
Stookey, S. D., 363, 373, 378, 379
Street, K. N., 235, 236, 240
Streltsina, M. W., 377
Strife, J., 168, 178
Strife, J. R., 260, 342
Stringer, J., 88, 103
Strong, A. B., 205, 206, 239
Stull, K., 285, 343
Suenaga, M., 395, 406, 410, 411
Sugamata, M., 31, 47
Sujuki, J., 391, 409
Sukumar, V., 375, 376, 380
Surappa, M. K., 110, 175
Suresh, S., 36, 48, 161, 169, 170, 172, 177, 178
Sutcu, M., 316, 317, 345
Suzuki, T., 58, 99, 260, 261, 342
Suzuki, K., 76, 101
Swaminathan, V. P., 52,85, 86, 99, 103
Swarlyengruber, L., 16, 46
Sweedler, A. R., 397, 410

T

Tachikawa, K., 393, 410
Tack, W. T., 18, 46
Tafto, J., 406, 411
Tai, N-H., 258, 342
Takaku, A., 292, 297, 344
Takamori, T., 378
Takao, Y., 131, 177
Takeda, N., 285, 343
Takeuchi, S., 101

Talreja, R., 229, 240
Tanaka, S., 407, 411
Taplin, D. M. R., 33, 48
Tashiro, M., 362, 378
Taub, A. I., 112, 176
Taya, M., 117, 129, 131, 132, 135, 148, 161, 163, 164, 176, 177
Teleman, A. S., 240
Tennery, V. J., 254, 255, 341
Thebault, J., 333, 334, 335, 336, 345
Thiel, F. A., 400, 410
Thomas, G., 406, 411
Thompson, D. S., 23, 29, 47
Thornton, P. H., 57, 58, 59, 94, 99
Thouless, M. D., 316, 317, 345
Tiefel, T. H., 406, 408, 411
Tien, J. K., 76, 83, 84, 87, 88, 92, 93, 101, 102, 103
Tien, J. T., 103
Tillman, T. D., 103
Tinkham, M., 385, 409
Tolman, R. C., 352, 377
Tomozawa, M., 356, 358, 360, 377, 378
Torng, C. J., 382, 402, 409
Totesh, A. S., 377
Toth, L. E., 399, 403, 404, 410, 411
Tracy, G. D., 337, 346
Tredway, W. K., 285, 333, 343, 345
Tressler, R., 285, 343
Tretheweyand, B., 236, 240
Troiano, A. R., 41, 48
Tsai, S. W., 110, 175, 222, 239, 264, 265, 266, 267, 268, 269, 342, 343
Turkalo, A. M., 356, 374, 377, 379
Turnbull, D., 359, 364, 366, 378, 379
Turner, C. W., 259, 342
Tweer, I., 357, 377
Tycko, R., 400, 410
Tyson, W. R., 223, 239

U

Uhlmann, D. R., 260, 342, 359, 362, 365, 378, 379
Umekawa, S., 110, 175
Underhill, A. E., 399, 410
Underwood, S. M., 379
Unwin, P. N. T., 32, 47
Upadhya, K., 250, 251, 341
Urquhart, A. W., 257, 258, 342

V

Van Stone, R. H., 25, 27, 47
Vanderah, T. A., 403, 411
Vandoverm R. B., 408, 411
VanMegan, W., 367, 379

VanValzah, J. R., 285, 343
Vasudevan, A. K., 32, 36, 37, 40, 41, 43, 44, 47, 48, 167, 178
Vedula, K., 110, 174
Vergano, P. J., 365, 378
Vigo, T., 239
Vincens, J., 339, 346
Vincenzini, P., 341
Viswanathan, R., 88, 103
Voigt, W., 152, 177
Volmer, M., 351, 377
Voyiadjis, G. Z., 168, 172, 178

W

Wagner, C., 355, 377
Wagstaff, F. E., 357, 366, 377, 378
Wakashima, K., 161, 163, 164, 177
Wang, A. S. D., 264, 300, 301, 308, 310, 342, 344
Wang, H. H., 399, 410
Wang, K., 406, 411
Wang, L., 135, 148, 168, 169, 172, 177, 178
Wang, S. S., 233, 240
Wang, S. W., 264, 265, 268, 269, 285, 287, 342
Wang, Y. Q., 382, 402, 409
Wang, Y. R., 339, 346
Wang, Z., 169, 172, 178
Wang, Z. G., 277, 343
Warren, B. E., 358, 377
Warrington, D., 357, 359, 377
Waszczak, J. V., 405, 411
Weber, A., 351, 377
Wee, D. M., 58, 99
Wei, R. P., 48
Weinberg, M. C., 369, 370, 379
Weinburg, M. C., 362, 378
Welpmann, K., 36, 48
Wernick, J. H., 398, 410
West, A. W., 389, 393, 409, 410
Westbrook, J. H., 71, 101
Westfall, L. J., 109, 174
Wetherhold, R. C., 337, 346
Weyl, W. A., 373, 379
Whangbo, M. H., 399, 410
White, C. L., 87, 103
Whitney, J. M., 237, 240
Wilcox, B. A., 83, 91, 92, 102
Wilkins, D. J., 231, 237, 240
Williams, J. C., 48
Williams, J. G., 234, 240
Williams, J. M., 399, 410
Williams, K. R., 83, 102
Wilm, Alfred, 18
Wilm, Alfred,18

Author Index

Wilner, B., 135, 143, 158, 170, 177
Wilshire, B., 83, 102
Wilson, H. A., 365, 379
Winkler, P-J., 18, 46, 47
Withers, P. J., 161, 163, 164, 177
Wlodek, S. T., 55, 59, 60, 99
Wolf, S. A., 403, 404, 411
Wood, J. V., 110, 175
Worthem, D. W., 338, 346
Wu, M. K., 382, 402, 409
Wu, S. B., 161, 163, 177

Y

Yamane, M. Y., 369, 379
Yamaoka, T., 131, 135, 176
Yang, J. M., 256, 342
Yang, X. F., 316, 317, 345
Yee, D. S., 407, 411
Yeh, J. R., 292, 344
Yen, C-F., 277, 343
Yokobori, T., 285, 343
Yokota, R., 359, 378
Yoo, M. H., 103
Yoshida, T., 406, 411
Yu, W., 36, 48
Yukawa, N., 64, 100

Z

Zahurak, S. M., 400, 410
Zandbergen, H. W., 406, 411
Zarzycki, J., 359, 377, 378
Zawada, L. P., 337, 346
Zhang, J., 110, 175
Zhou, L. M., 199, 200, 201, 205, 239
Zinn-Justin, J., 358, 378
Zok, F. W., 288, 292, 344
Zwebwn, C., 316, 345

OHIO UNIVERSITY LIBRARY
Please return this book as soon as you have finished with it. In order to avoid a fine it must be returned by the latest date stamped be-